PROTEIN ENGINEERING

TUTORIALS IN MOLECULAR AND CELL BIOLOGY

SERIES EDITOR
C. Fred Fox

Protein Engineering
Dale L. Oxender and C. Fred Fox, *Editors*

PROTEIN ENGINEERING

Editors

Dale L. Oxender
Center for Molecular Genetics
University of Michigan Medical School
Ann Arbor, Michigan

C. Fred Fox
Molecular Biology Institute
Department of Microbiology
University of California
Los Angeles, California

ALAN R. LISS, INC., NEW YORK

Address all Inquiries to the Publisher
Alan R. Liss, Inc., 41 East 11th Street, New York, NY 10003

Printed in the United States of America

Library of Congress Cataloging-in-Publication Data

Protein engineering.

(Tutorials in molecular and cell biology)
Includes bibliographies and index.
1. Protein engineering. 2. Genetic engineering
I. Oxender, Dale L. II. Fox, C. Fred. III. Series.
[DNLM: 1. Genetic Intervention. 2. Proteins—analysis.
3. Proteins—genetics. QU 55 P96624]
TP248.P77P76 1986 547.7'5 86-27741
ISBN 0-8451-4300-X

Contents

Contributors

Robin M. Adams, Research Department, Genencor, Inc., South San Francisco, CA 94080 **[279]**

Tim J. Ahern, Department of Applied Biological Sciences, Massachusetts Institute of Technology, Cambridge, MA 02139; present address: Genetics Institute, Cambridge, MA 02140 **[213]**

Tom Alber, Institute of Molecular Biology, University of Oregon, Eugene, OR 97403 **[289]**

R.J. Almassy, Molecular Biology Institute, University of California, Los Angeles, CA 90024 **[354]**

Edward Arnold, Department of Biological Sciences, Purdue University, West Lafayette, IN 47907 **[350]**

Robert L. Baldwin, Department of Biochemistry, Stanford University Medical Center, Stanford, CA 94305 **[125, 127]**

Anne M. Beasty, Department of Chemistry, Pennsylvania State University, University Park, PA 16802; present address: Department of Chemistry, Lehigh University, Allentown, PA 18104 **[91]**

Stephen J. Benkovic, Department of Chemistry, Pennsylvania State University, University Park, PA 16902 **[251]**

J. Michael Bishop, G.W. Hooper Foundation and Department of Microbiology and Immunology, University of California, San Franscisco, CA 94143 **[315]**

Richard R. Bott, Biocatalysis Department, Genentech, Inc., South San Francisco, CA 94080 **[279]**

Richard R. Burgess, Biotechnology Center and McArdle Laboratory for Cancer Research, University of Wisconsin, Madison, WI 53706 **[71]**

Marvin H. Caruthers, Department of Chemistry and Biochemistry, University of Colorado, Boulder, CO 80309 **[65]**

Joseph E. Coleman, Department of Molecular Biophysics and Biochemistry, Yale University, New Haven, CT 06510 **[323]**

R. John Collier, Department of Microbiology and Molecular Genetics, Harvard Medical School and Shipley Institute of Medicine, Boston, MA 02115 **[299]**

Charles S. Craik, Departments of Pharmaceutical Chemistry and Biochemistry and Biophysics, University of California, San Francisco, CA 94143-0448 **[257]**

Thomas E. Creighton, MRC Laboratory of Molecular Biology, Cambridge CB2 2QH, United Kingdom **[83]**

Brian C. Cunningham, Biocatalysis Department, Genentech, Inc., South San Francisco, CA 94080 **[279]**

Deborah DeFeo-Jones, Department of Virus and Cell Biology, Merck Sharp and Dohme Research Laboratories, West Point, PA 19486 **[307]**

William F. DeGrado, Central Research and Development, E.I. du Pont de Nemours and Co., Inc., Wilmington, DE 19898 **[201]**

Ken A. Dill, Departments of Pharmaceutical Chemistry and Pharmacy, University of California, San Francisco, CA 94143 **[187]**

The number in brackets is the opening page number of the contributor's article.

David Eisenberg, Department of Chemistry and Molecular Biology Institute, University of California, Los Angeles, CA 90024 **[125, 127, 181, 354]**

Susan Erickson-Viitanen, Central Research and Development, E.I. du Pont de Nemours and Co., Inc., Wilmington, DE 19898 **[201]**

David A. Estell, Research Department, Genencor, Inc., South San Francisco, CA 94080 **[279]**

Richard J. Feldmann, Division of Computer Research and Technology, National Institutes of Health, Bethesda, MD 20892 **[348]**

Alan R. Fersht, Department of Chemistry, Imperial College of Science and Technology, London SW7 2AY, United Kingdom **[221, 269]**

Barry C. Finzel, Protein Engineering Division, Genex Corporation, Gaithersburg, MD 20877; present address: Central Research and Development, E.I. du Pont de Nemours and Co., Inc., Wilmington, DE 19898 **[165]**

Robert J. Fletterick, Hormone Research Institute, Department of Biochemistry and Biophysics, University of California, San Francisco, CA 94143 **[257]**

C. Fred Fox, Molecular Biology Institute, Department of Microbiology, University of California, Los Angeles, CA 90024 **[xi]**

Christin A. Frederick, Department of Biology, Massachusetts Institute of Technology, Cambridge, MA 02139 **[237]**

Stephen J. Gardell, Hormone Research Institute and Department of Biochemistry and Biophysics, University of California, San Francisco, CA 94143 **[257]**

John Grable, Department of Biological Sciences, University of Pittsburgh, Pittsburgh, PA 15260 **[237]**

Thomas P. Graycar, Research Department, Genencor, Inc., South San Francisco, CA 94080 **[279]**

Patricia Greene, Department of Biochemistry and Biophysics, University of California, San Francisco, CA 94143 **[237]**

Cameron Haase, Department of Biology, Massachusetts Institute of Technology, Cambridge, MA 02139 **[109]**

Wayne A. Hendrickson, Department of Biochemistry and Molecular Biophysics, College of Physicians and Surgeons, Columbia University, New York, NY 10032 **[3, 5]**

Herb H. Heyneker, Research Department, Genencor, Inc., South San Francisco, CA 94080 **[279]**

Donald Hilvert, Laboratory of Bio-organic Chemistry and Biochemistry, Rockefeller University, New York, NY 10021 **[257]**

Elizabeth E. Howell, Agouron Institute, La Jolla, CA 92037 **[251]**

Mark Hurle, Department of Chemistry, Pennsylvania State University, University Park, PA 16802 **[91]**

C.A. Janson, Molecular Biology Institute, University of California, Los Angeles, CA 90024 **[354]**

Catherine M. Joyce, Department of Molecular Biophysics and Biochemistry, Yale University, New Haven, CT 06511 **[227]**

Emil Thomas Kaiser, The Rockefeller University, New York, NY 10021 **[193]**

Martin Karplus, Department of Chemistry, Harvard University, Cambridge, MA 02138 **[35]**

Brad A. Katz, Research Department, Genencor, Inc., South San Francisco, CA 94080 **[279]**

Garry C. King, Department of Molecular Biophysics and Biochemistry, Yale University, New Haven, CT 06510 **[323]**

Jonathan King, Department of Biology, Massachusetts Institute of Technology, Cambridge, MA 02139 **[109]**

Alexander M. Klibanov, Department of Applied Biological Sciences, Massachusetts Institute of Technology, Cambridge, MA 02139 **[213]**

William H. Konigsberg, Department of Biophysics and Biochemistry, Yale University, New Haven, CT 06510 **[323]**

Anthony A. Kossiakoff, Biocatalysis Department, Genentech, Inc., South San Francisco, CA 94080 **[279]**

Joseph Kraut, Department of Chemistry, University of California/San Diego, La Jolla, CA 92093 **[251]**

Robin J. Leatherbarrow, Department of Chemistry, Imperial College of Science and Technology, London SW7 2AY, United Kingdom **[269]**

Richard H. Lee, Department of Biological Chemistry, Hershey Medical Center, Pennsylvania State University, Hershey, PA 17033; present address: Glenview, IL 60025 **[175]**

Glenn J. Lesser, Department of Biological Chemistry, Hershey Medical Center, Pennsylvania State University, Hershey, PA 17033 **[175]**

Joanna T. Manz, Department of Chemistry, Pennsylvania State University, University Park, PA 16802; present address: Department of Immunology, University of Washington, Seattle, WA 98195 **[91]**

John L. Markley, Department of Biochemistry, College of Agricultural and Life Sciences, University of Wisconsin, Madison, WI 53706 **[15]**

Brian W. Matthews, Institute of Biology and Department of Physics, University of Oregon, Eugene, OR 97403 **[225, 289]**

C. Robert Matthews, Department of Chemistry, Pennsylvania State University, University Park, PA 16802 **[91]**

Ruth J. Mayer, Department of Chemistry, Pennsylvania State University, University Park, PA 16902 **[251]**

Judith McClarin, Department of Biological Sciences, University of Pittsburgh, Pittsburgh, PA 15260 **[237]**

Andrew D. McLachlan, MRC Laboratory of Molecular Biology, Cambridge CB2 2QH, United Kingdom **[181]**

Jeff V. Miller, Research Department, Genencor, Inc., South San Francisco, CA 94080 **[279]**

Douglas H. Ohlendorf, Protein Engineering Division, Genex Corporation, Gaithersburg, MD 20877; present address: Central Research and Development, E.I. du Pont de Nemours and Co., Inc., Wilmington, DE 19898 **[165]**

Karyn T. O'Neil, Central Research and Development, E.I. du Pont de Nemours and Co., Inc., Wilmington, DE 19898 **[201]**

Dale L. Oxender, Center for Molecular Genetics, University of Michigan Medical School, Ann Arbor, MI 48190 **[xiii]**

Carl O. Pabo, Department of Biophysics, The Johns Hopkins School of Medicine, Baltimore, MD 21205 **[xv]**

Scott D. Power, Research Department, Genencor, Inc., South San Francisco, CA 94080 **[279]**

David B. Powers, Biocatalysis Department, Genentech, Inc., South San Francisco, CA 94080 **[279]**

Richard V. Prigodich, Department of Chemistry, Trinity College, Hartford, CT 06106 **[323]**

Norbert Reich, Department of Biochemistry and Biophysics, University of California, San Francisco, San Francisco, CA 94143 **[237]**

David C. Richardson, Department of Biochemistry, Duke University Medical Center, Durham, NC 27710 **[149]**

Jane S. Richardson, Department of Biochemistry, Duke University Medical Center, Durham, NC 27710 **[149]**

Steven Roczniak, Hormone Research Institute and Department of Biochemistry and Biophysics, University of California, San Francisco, CA 94143 **[257]**

George D. Rose, Department of Biological Chemistry, Hershey Medical Center, Pennsylvania State University, Hershey, PA 17033 **[175]**

John M. Rosenberg, Department of Biological Sciences, University of Pittsburgh, Pittsburgh, PA 15260 **[237]**

John J. Rossi, Department of Molecular Genetics, Beckman Research Institute of the City of Hope, Duarte, CA 91010 **[51]**

Michael Rossmann, Department of Biological Sciences, Purdue University, West Lafayette, IN 47907 **[350]**

William J. Rutter, Hormone Research Institute and Department of Biochemistry and Biophysics, University of California, San Francisco, CA 94143 **[257]**

F. Raymond Salemme, Protein Engineering Division, Genex Corporation, Gaithersburg, MD 20877; present address: Central Research and Development, E.I. du Pont de Nemours and Co., Inc., Wilmington, DE 19898 **[165]**

Edward M. Scolnick, Department of Virus and Cell Biology, Merck Sharp and Dohme Research Laboratories, West Point, PA 19486 **[307]**

Yousif Shamoo, Department of Biophysics and Biochemistry, Yale University, New Haven, CT 06510 **[323]**

David Shortle, Department of Biological Chemistry, The Johns Hopkins University School of Medicine, Baltimore, MD 21205 **[103]**

Paul Sigler, Department of Biochemistry and Molecular Biology, University of Chicago, Chicago, IL 60637 **[352]**

Stephen Sprang, Hormone Research Institute and Department of Biochemistry and Biophysics, University of California, San Francisco, CA 94143 **[257]**

Thomas Stackhouse, Department of Chemistry, Pennsylvania State University, University Park, PA 16802; present address: Department of Biochemistry, University of California, Davis, CA 95616 **[91]**

Thomas A. Steitz, Department of Molecular Biophysics and Biochemistry, Yale University, New Haven, CT 06511 **[227]**

Mark A. Synder, California Biotechnology, Inc., Mountain View, CA 94043 **[315]**

Kelly Tatchell, Department of Biology, University of Pennsylvania, Philadelphia, PA 19104 **[307]**

Mark H. Ultsch, Biocatalysis Department, Genentech, Inc., South San Francisco, CA 94080 **[279]**

Jesus E. Villafranca, Agouron Institute, La Jolla, CA 92037 **[251]**

Gerrit Vriend, Department of Biological Sciences, Purdue University, West Lafayette, IN 47907 **[352]**

Christopher T. Walsh, Departments of Chemistry and Biology, Massachusetts Institute of Technology, Cambridge, MA 02139 **[47]**

Bi-Cheng Wang, Biocrystallography Laboratory, V.A. Medical Center, Pittsburgh, PA 15260 **[237]**

Mark S. Warren, Department of Chemistry, University of California/San Diego, La Jolla, CA 92093 **[251]**

Patricia C. Weber, Protein Engineering Division, Genex Corporation, Gaithersburg, MD 20877; present address: Central Research and Development, E.I. du Pont de Nemours and Co., Inc., Wilmington, DE 19898 **[165]**

James A. Wells, Biocatalysis Department, Genentech, Inc., South San Francisco, CA 94080 **[279]**

William Wilcox, Molecular Biology Institute and Department of Chemistry and Biochemistry, University of California, Los Angeles, CA 90024 **[181]**

Kenneth R. Williams, Department of Molecular Biophysics and Biochemistry, Yale University, New Haven, CT 06510 **[323]**

Myeong-hee Yu, Department of Biology, Massachusetts Institute of Technology, Cambridge, MA 02139 **[109]**

Michael H. Zehfus, Department of Biological Chemistry, Hershey Medical Center, Pennsylvania State University, Hershey, PA 17033 **[175]**

Mark Zoller, Cold Spring Harbor Laboratory, Cold Spring Harbor, NY 11724 **[51]**

Preface

Major scientific movements are sometimes born when areas that have developed independently undergo fusion. Molecular genetics and structural biophysics are two such areas which currently are merging to form a hybrid field: protein engineering. This volume represents an attempt to capture the intense interest this new field has generated, and the promise it holds.

Our understanding of protein structure-function relationships is based firmly on x-ray crystallographic analysis of protein structure. The number of high resolution crystallographic solutions of protein structure has increased substantially as the result of detector technology that has enhanced the speed of data collection. This recent accumulation of data has facilitated parallel advances in capabilities for predicting protein structures from primary amino acid sequence information. Such theoretical approximations are crude when applied to sizable amino acid sequences in cases where no crystallographic data exist, but are relatively effective in predicting structural alterations resulting from changes in one or a few amino acid residues when applied to proteins of known structure.

Advances in molecular genetics, and more precisely, site specific mutagenesis, are the other main element in this unique and powerful fusion of technologies. Predicting changes that will arise as a consequence of amino acid substitutions is one thing. Creating that mutant protein, obtaining an empirical solution of its structure, and performing tests for altered functional qualities are quite another. Now, there are relatively routine procedures for introducing nonrandom single site changes in structural genes, and these mutant genes can be coupled to promoters that direct the synthesis of a purposely altered protein at high yield. Advanced procedures can be utilized to obtain that protein in a correctly folded form, leading to analysis of both its structure and its functional properties.

This book presents the reader with a systematic approach to protein engineering from both fundamental and operational points of view. The first two sections introduce methods for determining protein structure, and for performing site specific mutagenesis and other procedures to obtain purposely modified proteins and analyze their properties. The third section presents fundamental principles for determining where and how proteins should be altered, and for predicting, with the help of advanced computer graphic technology, the changes in structure which are likely to result. *Protein Engineering* concludes with examples of proteins which have been altered purposely, and provides previews of the beneficial information that can be derived from rational application of these principles and technologies. In one example, the altered protein has been expressed not only in cultured cells, permitting an assessment of its regulatory properties, but also in animals. This yielded results that were not anticipated from studies of the altered protein in vitro or of its properties in cultured cells. If this is a harbinger of things to come, protein engineering will have many biological surprises in store for us.

The concept for this book was developed at the Genex-UCLA Symposium on "Protein Structure, Folding and Design," held at Keystone, Colorado, in April of 1985. Even before the opening session, it was apparent that a unique synthesis of technologies had been achieved. Structural biophysicists and molecular biologists had found common ground, and the practitioners of each discipline were seized by strong desires to create a medium for sharing their knowledge. Most authors conferred in two sessions called for the purpose of determining how this could best be accomplished.

We first established the goal of creating a book that would be useful as a supplement to introductory biochemistry textbooks, and as a means for independent study by students and others with formal biochemical training. We decided to model this book after *Biophysical Science—A Study Program* [1]. That classic volume of short, readable position papers was created 25 years ago to stimulate the development of biophysics, but helped give birth to molecular biology instead. The authors also concluded that *Protein Engineering* should contain a didactic section on protein structure principles, written specifically for the nonexpert. We tried to maintain that tone throughout so that the contents of the book would be comprehensible to both specialists and nonspecialists.

We are especially grateful to Carl Pabo for his introductory chapter, to Wayne Hendrickson, David Eisenberg, Buzz Baldwin, Alan Fersht and Brian Matthews for their overviews which summarize the content of each section, to 30 authors who managed to produce their articles mainly on time, to Bill Konigsberg who read the book in its entirety and gave us valuable input, to Irving Geis for the cover illustration, and to Paulette Cohen, our editor at Alan R. Liss, Inc., who solved the hoary and exacting production problems required by the essential use of color and stereo illustrations.

Dale L. Oxender
C. Fred Fox

1. Oncley JL, Schmitt FO, Williams RC, Rosenberg MD, Bolt RH (eds) (1959): "Biophysical Science—A Study Program." New York: John Wiley and Sons.

Introduction

Proteins play central roles in all life processes, catalyzing biochemical reactions with remarkable specificity and serving as key structural elements in all cells and tissues. As enzymes, proteins catalyze both the digestion of foodstuffs and the construction of new macromolecules. As collagen, actin, myosin, and intermediate filaments, they control the structure and motion of cells and organisms. Antibodies protect us against disease; membrane proteins regulate ion transport and intercellular recognition. Repressors regulate gene expression, polymerases replicate genes, and histones help package DNA into chromosomes. Proteins are involved in every aspect of life: catalysis, structure, motion, recognition, regulation.

In recent decades, tremendous progress has been made in understanding protein structure and function. The amino acid sequence of insulin was reported in the early 1950s; today many thousands of protein sequences are known. The three-dimensional structure of myoglobin was reported in 1960; today hundreds of protein structures are known. In the past several decades, methods were also developed to help determine the role that particular amino acids played in a protein's function. Each method yielded a wealth of information, but each method had serious shortcomings: 1) Isolation and biochemical characterization of mutant proteins was helpful, but was limited by the random nature of mutagenesis. 2) Chemical modification could help reveal the role of particular residues, but useful reactions could only be designed for certain amino acids. Side reactions and the difficulties involved in separating modified species were a constant problem. 3) Solid-phase peptide synthesis allowed scientists to directly test the effects of sequence changes upon peptide structure and function, but some sequences were difficult to synthesize, and the overall yields were too low to allow routine synthesis of modified proteins.

Recent advances in gene synthesis and genetic engineering have made it possible, in principle, to construct any desired amino acid sequence. If the desired sequence is similar to an existing protein, oligonucleotides can be used to introduce the desired mutations and thus change the appropriate residues. If the desired sequence is not similar to an existing protein or if the original gene cannot be isolated, it is possible to synthesize a gene encoding the entire protein. These methods have overcome many of the limitations inherent in studies relying on random mutagenesis, chemical modification, or peptide synthesis. Progress in genetic engineering and gene synthesis has been so rapid that the era of "protein engineering" has arrived. In many cases, oligonucleotide synthesis and the required genetic manipulations are routine enough that we can focus on the more fundamental and challenging question: What amino acid sequence will give a protein with the desired properties?

Protein engineering should have tremendous *theoretical* and *practical* implications. It can be used both to explore fundamental questions about protein folding, structure, and function and to design useful proteins for medical and industrial applications. When used as an adjunct to other biochemical and structural investigations, directed mutagenesis gives the biochemist or crystallographer a chance to address the recurring question: "I wonder what would happen if that residue were a. . . ?"

Countless practical applications of protein engineering should be possible. It should be possible to alter the substrate specificity and pH optimum of enzymes, to increase the thermal stability of proteins, and to design proteins that can be used in non-aqueous solvents. In addition to modifying existing proteins, it should be possible to design entirely novel peptides and proteins with useful properties. Molecular design is the natural limit of microtechnology: The possibilities seem endless since the amino acid sequences found in existing proteins represent an infinitesimal fraction of all possible amino acid sequences.

In spite of the tremendous promise of this field, predicting appropriate sequence changes is very difficult, and this is the real challenge of protein engineering. There are fundamental problems: 1) We cannot, in general, predict how a protein will fold. Although the amino acid sequence of a protein determines its three-dimensional structure, knowledge of the sequence does not yet allow us to predict the structure. 2) We have only a partial understanding of the relationship between protein structure and function.

Given these problems, how does one proceed? What approaches are most likely to be successful? What experiments will teach us the most about protein modification and design? To take advantage of the real power of directed mutagenesis, one must know the three-dimensional structure of a protein in atomic detail, and X-ray crystallography can be used to determine the structure. This level of detail, where one knows the positions of residues in space, gives one the opportunity to relate structure, function, and the effects of possible mutations. At a minimum, structure determination will let one see which residues are near the active site. Even if it is not possible to predict which residues will be most favorable at a given position, one can at least predict which residues will be the most critical and where substitutions should be tested. Given the complexity of protein structure, computer graphics provides the most convenient method for visualizing a protein structure and examining proposed changes. Empirical energy calculations can be used to simulate the effects of a mutation, but can only serve as a rough guide because we have only a partial understanding of the forces that control protein structure.

What are the prospects for the long-range future of protein engineering? Attempting to predict scientific progress is a risky undertaking at best, but several areas seem ripe for progress over the next several decades: 1) Continued analysis of protein structures will give us a deeper understanding of folding motifs, of forces that stabilize folded proteins, and of the "rules of protein structure". 2) Data bases with information about known structures and advanced programming methods—such as those used in artificial intelligence research—may play critical roles in protein design. These approaches may help us plan modifications and design new proteins more effectively, since a program can test many more sequence changes than a person could test at a computer graphics system. 3) De novo design of proteins may become practical. Current attempts at de novo design have focused on designing stable, folded structures, but it should eventually be possible to design entirely novel proteins with desired binding properties or catalytic activities. 4) Combining strategies for protein modification and design with classical organic chemistry—for example, attaching peptides and proteins to other molecules or to synthetic polymers—will open new vistas for molecular design.

In summary, developments in gene synthesis and genetic engineering have now opened the fields of protein engineering and protein design. This volume is intended to capture the excitement of these developing fields: to outline our current understanding of protein folding, structure and function; to describe methods used for planning sequence changes and constructing mutants; and to summarize the results of some of the first investigations of modified peptides and proteins. We must recognize that progress in this field will be difficult at times because we do not yet really understand protein folding or the relationship between protein structure and function. However, protein engineering and protein design will eventually have enormous impacts on medicine, on industry, and on biological research. Enormous challenges and rewards await.

Carl O. Pabo

I
METHODS FOR DETERMINING PROTEIN STRUCTURE

Protein Engineering, pages 3–4

Overview: Section I

Wayne A. Hendrickson

Molecular biology has recorded explosive scientific and technological development in recent years. This progress is now also having a major impact on physical studies of macromolecules. For one thing, by introducing recombinant DNA into appropriate expression vectors, it is now possible to produce plentiful supplies of interesting but intrinsically scarce proteins. This markedly enhances the prospects for structural analysis by physical methods. Moreover, it is practical to incorporate specific isotopes or certain heavy-atom labels into proteins overproduced in efficient bacterial expression systems, and this can facilitate the experimental analysis. Just as important as these advantages in supplying material for study is the benefit of gene manipulation to structure–function studies. The possibility to design specific mutants of proteins essentially at will opens up unique opportunities to examine fundamental thermodynamic principles, to explore the structural basis of macromolecular action, and to test specific mechanistic hypotheses.

This section of the book focuses on physical methods that have an especially important role in protein modification and design and in the analysis and exploitation of these products. The armory of biophysical techniques for the study of macromolecules is quite extensive. Measurement of hydrodynamic properties—e.g. diffusion, viscosity, ultracentrifugal sedimentation, chromatographic migration—give important data on overall size and shape, but resolution on the atomic scale is needed here. Electron microscopy and solution scattering of light, X-rays, or neutrons, although powerful in their own right, also tend to suffer a disadvantage in resolution. Various spectroscopies—e.g., electronic absorption, fluorescence, circular dichroism, infrared, Raman, electron spin resonance, X-ray absorption—are exquisitely sensitive but tend either to give averages over many residues or to see only specialized sites. While all of these techniques can have a role in protein engineering, greater power residues in the high-resolution techniques that give comprehensive information. These include X-ray crystallography and nuclear magnetic reasonance on the experimental side, and the theoretical simulation and molecular graphics on the calculation side.

The three chapters in this section speak well for themselves on the basic characteristics of the methods and on the nature of results from applications to protein moleculars—especially as regards intentionally designed (factitious) proteins. It should be obvious that these techniques complement one another. X-ray diffraction analysis gives the detailed and comprehensive structures that form the essential basis of most design work. However, crystals are required for detailed X-ray work, whereas NMR studies have the decided advantage of examining the structure of molecules in solution. Both techniques can be used to describe dynamic properties, but NMR is distinctly more powerful in this regard. These experimental methods are essential, since the definition of the three-dimensional structure of a protein is beyond the scope of present-day theory. However, it is the essence of protein modification and design to build on the patterns found in natural products in order to obtain molecules with desired alternative properties. While biological or chemical selection from among a plethora of mutants can be effective, predictions from basic principles and theoretical calculations provide a most elegant and increasingly competitive solution to the question of which polypeptides to synthesize. Molecular graphics has an important place in all of these activities; in fact, it is a discipline in its own right. We have no chapter here on the use of such tools, but the fruit of these instruments is amply evident in the book. The strength and weakness of molecular graphics is, at once, that its most important computer is the human brain. This operator brings a potential for reactivity and insight not found in computer programs, but it is also prone to succumb to the foible of subjectivity: "What I've built is so pretty that it must be right."

Recent advances in these techniques of molecular biophysics have been no less spectacular than those seen in molecular biology. Here, it is high-technology instruments—synchrotrons, interactive displays, high-field NMR spectrometers, supercomputers, X-ray area detectors, etc.—that have led the way. These have been accompanied by major advances in methodology, such as two-dimensional NMR, molecular dynamics, and novel phasing methods for diffraction. The coalescence of physical biochemistry of such sophistication with molecular cloning in all its ramifications offers marvelous opportunities for structural biology generally and for protein engineering in particular.

Protein Engineering, pages 5–13

1

X-Ray Diffraction

Wayne A. Hendrickson

Department of Biochemistry and Molecular Biophysics, Columbia University, New York, New York 10032

INTRODUCTION

It is advantageous, if not essential, to know how the atoms of major molecules of life—proteins and nucleic acids—are disposed in space if the way they perform their vital functions is to be understood in detail. Knowledge of three-dimensional structure will usually also be indispensible in the intelligent design of proteins with modified properties. At present, such structural information is most effectively attained by analysis of diffraction from crystals of the macromolecules.

The successes of X-ray diffraction studies in biology have been spectacular. Holdings of the Protein Data Bank [1] now exceed 260 coordinate sets from over 150 distinctive macromolecules, and the structures of at least another 70 proteins, oligonucleotides, and viruses are known at, or near to, atomic resolution. These structures have had a profound impact on biochemistry and related sciences. The results have not been won easily, however; typical studies have absorbed several research years. Fortunately, the pace is quickening due to two major factors: First, advances in instrumentation and methods are beginning to be felt; second, plentiful supplies of inherently scarce but intrinsically interesting proteins often can be made by way of recombinant DNA.

This chapter introduces the theoretical basis for diffraction methods, describes how this elegant theory is implemented for macromolecules, and discusses how results from such crystallographic studies can be evaluated.

THEORETICAL FOUNDATION

Elementary Concepts

X-rays are a form of electromagnetic radiation, like visible light; but this light is distinguished by its relatively short wavelength and by its ability to propagate through matter with barely perceptible deviation from a straight line. X-ray wavelengths (on the order of 1 Å) are commensurate with atomic dimensions (1.5 Å for carbon–carbon single bonds); thus, X-ray diffraction—the interference between waves scattered from individual atoms in a substance—is sensitive to the atomic structure within molecules. However, images of molecules cannot be formed directly with X-rays. Whereas visible light that is scattered from an object can be collected, as in the human eye, to form an image of the subject; there are no lenses to bend and focus the scattered X-rays. X-ray images must be reconstructed computationally by the theory of diffraction, as described below, from the intensities of the diffracted waves, which is what can be recorded experimentally.

Crystals, or other periodic arrays, are needed for X-ray diffraction analysis of macromolecules, because the scattering from individual molecules is far too weak to be measurable. Diffraction from crystals concentrates the scattering exclusively into discrete directions (reflections) and further enhances the signal by the multiplicity of the lattice (more than 10^{14} aligned molecules in a typical protein crystal). As a result of this averaging, diffraction intensities from protein crystals can be measured very precisely.

Beams of electrons or neutrons can also be used for diffraction studies and have advantages for certain purposes. However, the scattering cross-section for neutrons is very low, and radiation damage from electrons is severe. The periodicity within fibrous macromolecules, in two-dimensional arrays of molecules, or in stacks of membranes can also be exploited—sometimes even at atomic resolution. Still, X-ray crystallography remains the most suitable means for determining the atomic structures of proteins and other biological macromolecules.

Dependence of Diffraction on Structure

X-rays scatter from the electrons clouds of atoms and, apart from resonance effects known as anomalous scattering, this scattering is directly proportional to that expected from a classical free electron. Thus, a volume element, **dv**, located at a position **r**, scatters in proportion to the electron density, $\rho(\mathbf{r})$, at that point. The phase of a wave scattered from this volume element, relative to that from a point at the origin, depends critically on the position **r** as well as on the wave-vector direction, $\hat{\mathbf{k}}$, in relation to the incident wave vector direction, $\hat{\mathbf{k}}_o$. Thus the dependence of diffraction on structure.

This position-dependent phase shift is the essence of diffraction, and it is given by $2\pi\ \mathbf{r}\cdot\mathbf{S}$, where $\mathbf{S} = (\hat{\mathbf{k}} - \hat{\mathbf{k}}_o)/\lambda$ ($|\mathbf{S}| = 2\,s = 2 \sin \Theta/\lambda$) is the scattering vector in an internal coordinate frame, λ being the X-ray wavelength and Θ the scattering angle. When the scattering contributions of all elements of the object are added up, the total scattered wave is given by the integral

$$F(\mathbf{S}) = \int_V\!\!\int \rho(\mathbf{r}) \exp(2\pi i\, \mathbf{S}\cdot\mathbf{r})\, dv, \quad (1)$$

where F is normalized to the scattering from a free electron.

The integral of Eq. (1) quite naturally expresses the physics of diffraction in the elegant, and very useful, mathematical form of a Fourier transform. This equation is general and, if the object is nonperiodic, it is a continuous function of the scattering vector. Since the electron density function is composed of the electron clouds of discrete atoms, it is quite equivalent to sum over the individual atomic integrals. These terms are characterized by atomic scattering factors, f_j, that can be computed once and for all for each kind of atom; by the atomic position, $\mathbf{r}(x_j,y_j,z_j)$, for each atom j; and by a factor that takes account of atomic mobility.

In the case of a crystal, the periodicity introduces a discrete character to the diffraction. The locations of the diffraction maxima are on a lattice (h, k, l) dictated by the crystal parameters in a manner reciprocal to the unit cell dimensions; that is, spots are closer together as cell lengths get longer. Now, if the volume of integration in Eq. (1) corresponds precisely to a unit cell of the crystal, the values obtained at integral values of $\mathbf{S}(h, k, l)$ are directly proportional to those for the crystal as a whole. These values are known as the structure factors of the crystal and are given by

$$F_{hkl} = \sum_{\substack{\text{atoms in}\\ \text{unit cell}}} f_j(s_{hkl}) \exp(-B_j s_{hkl}^2) \cdot \exp\{2\pi i(hx_j + ky_j + kz_j\} \quad (2)$$

Here B_j is the isotropic atomic mobility (or temperature) parameter and it is related by $B = 8\pi^2\, \overline{u^2}$ to the atomic displacements, u.

The actual integrated intensities of diffraction maxima from real crystals are related to the square of the structure factor magnitudes,

$$I_{hkl} = C\,|F_{hkl}|^2, \quad (3)$$

with a proportionality factor, C, that depends on a number of experimental factors that are either constant or knowable. These include such things as incident beam intensity, crystal volume, and geometry of diffraction. All dependence upon the structural parameters is embodied in the F's. However, in the observational interaction, a crucial part of this information is lost. The structure factors are in general complex variables, $F = |F| \exp(i\phi)$, and the magnitude, but not the phase (ϕ), is preserved in the measurement.

Recovery of Structure From Diffraction

It is the excellence of the theory represented by Eqs. (2) and (3) that makes crystallography an essentially definitive science. There can in general be many intensity observations in comparison to the number of atomic parameters. This overdetermination is somewhat diminished in the case of proteins because of the exponential fall-off with scattering angle that depends on atomic mobility (which is often quite large in marcomolecular crystals). Nonetheless, typical protein crystals provide at least as many diffraction observations as there are x, y, z and B variables, and oftentimes there are four or more times that number.

The problem is to extract the atomic parameters from the diffraction measurements. In contrast to many spectroscopic techniques where each observation can be assigned to a particular structural feature, in the diffraction experiment all atoms contribute to every measurement. This distributed character of the structural information, coupled with the nonlinearity of the equations, makes it impossible to recover the structure directly by solution of the system of equations. However, once an approximate model is known, these equations do provide an excellent basis for refinement.

A key to the recovery of atomic parameters from diffraction data lies in the structure factor as expressed in Eq. (1). It is a property of the Fourier transform that the function can be readily inverted mathematically by way of the inverse transform. Then if $F(S)$ is known, completely, $\rho(\mathbf{r})$ can be calculated everywhere. In the case of a crystal, where the diffraction values are discrete rather than continuous, this inverse Fourier integral reduces to the Fourier series:

$$\rho(xyz) = \frac{1}{V} \sum_h \sum_k \sum_l F_{hkl} \cdot \exp\{-2\pi i(hx + ky + lz)\}, \quad (4)$$

whereby the electron density at a particular point is related to all of the structure factors. Although the summations in principle extend to infinity, in practice useful images can be produced with those terms within a sphere of observation $s < s_{max}$ that corresponds to a resolution limit of $d_{min} = 0.5/s_{max}$.

The electron density equation (Eq. 4) is extremely useful but it by itself is not sufficient. The complete structure factor is required, including phase (ϕ), whereas only the magnitude ($|F|$) can be measured directly. recovery of structure from diffraction is thus reduced to the phase problem—the problem of evaluating the phases lost in the observational interaction. Various methods have been used to overcome this central conceptual difficulty in crystal structure determination and some of these are described below. More information on the theoretical foundations of X-ray diffraction can be found in texts and reviews [2–4].

PRACTICAL IMPLEMENTATION

Crystallographic analyses of macromolecular structures normally proceed through a sequence of experimental or analytical operations that generate various tangible objects of the kind shown in Figure 1. Crystals are grown, diffraction data are measured, phases for the structure factors are evaluated, an electron density map is synthesized, an atomic model is fitted to this map, and then this model is refined. The pathway often includes cycles that repeat some of these operations, as in refinement, or shunts as alternative attempts at solution are tried. The process is seldom routine, but there are several rather standard procedures. Some of these are described briefly below. More detail is given in a recent

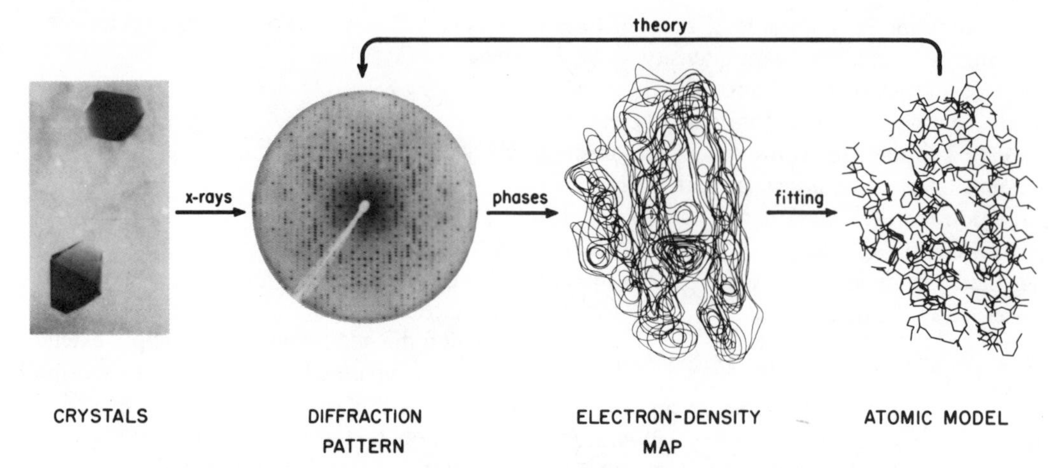

Fig. 1. *The pathway of structure analysis in protein crystallography. The tangible objects used here to illustrate this process are taken from the study of myohemerythrin. (From Hendrickson [3]. Reproduced with permission.)*

review [3], in *Protein Crystallography* by Blundell and Johnson [4] and in two new volumes of *Methods in Enzymology*, edited by Wyckoff, Hirs, and Timasheff [5].

Crystallization

This first, and perhaps most crucial, of the steps in a protein structure analysis is also the least well understood. The physics of crystal growth for simple systems is somewhat in hand; some of this transfers, but the process is clearly more complex for macromolecules. The current effort to grow protein and nucleic-acid crystals on the space shuttle is bringing new attention to crystal growth theory for proteins. However, earth-bound empirical methods are likely to continue to predominate. Over one thousand well-characterized kinds of protein crystals have been grown. It is widely felt that, with sufficient perseverance by a knowledgeable investigator, any reasonably homogenous globular protein can be crystallized.

Purity and protein concentration are two of the most critical factors in crystallization. Both of these are favorably affected when otherwise scarce proteins are produced in abundance by recombinant DNA techniques. Nevertheless, microtechniques are very important in the crystallization of macromolecules. Microdialysis and vapor diffusion, usually in hanging drops, are most commonly used. The protein mass of a single X-ray-quality crystal (0.2–0.4 mm on an edge) is actually quite small—perhaps 10–20 μg. However, many trials and several crystals are required, so that most successful studies start with at least 10 mg of pure material. Many different reagents have been used to grow crystals from protein solutions, but widest success is found with ammonium sulfate, phosphate salts, polyethylene glycol (PEG), and 2-methyl-2,4-pentanediol (MPD).

Data Measurement

Prior to the systematic measurement of diffraction intensities, it is important to characterize the unit cell dimensions, internal symmetry, and extent of diffraction from a crystal. This is most readily done on precession cameras—instruments specially designed to produce an undistorted picture of the diffraction lattice. However, other equipment is better suited for collection of the three-dimensional data, which often comprise hundreds of thousands of reflections. Three main techniques are used: automated diffractometry, rotation photography, or electronic area-detector diffractometry.

Recent advances in instrumentation are dramatically changing the rate and character of data acquisition in protein crystallography.

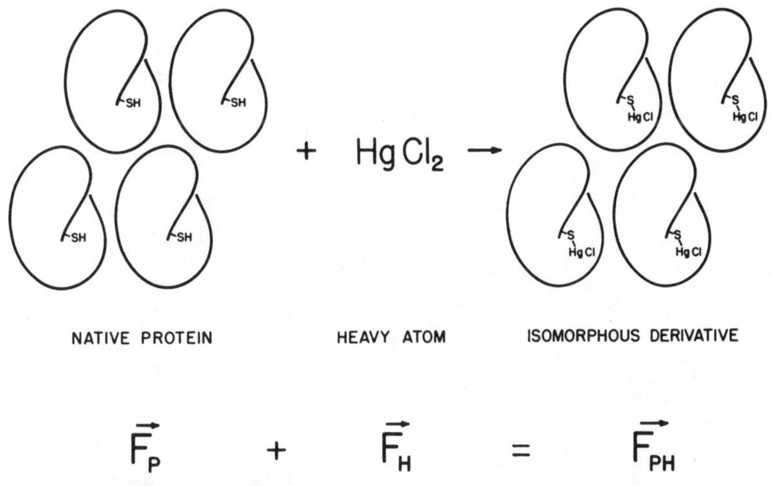

Fig. 2. *Schematic illustration of the isomorphous replacement reaction of heavy atoms with a protein crystal. Crystallographic structure factors for the corresponding components are related ideally as shown below. (From Hendrickson [3]. Reproduced with permission.)*

Area detectors at national resources and in some laboratories are now producing all of the data needed for a complete structure determination in two weeks' time at conventional sources. Synchrotron radiation is also having a dramatic impact. X-ray beams from these sources are already some thousand times more intense than conventional sources, and incredibly brighter beams are in prospect. The continuously tunable nature of synchrotron radiation has special advantages for phase determination and for time-resolved studies.

Phase Determination

Various methods have been used to evaluate, or approximate, the phases needed to reconstruct a structural image from diffraction data. All of these are somewhat indirect (i.e., phases do not come straightforwardly from diffraction measurements on a single crystalline species) and lack generality. However, some of the approaches are extremely powerful, and rapid progress is being made toward enhanced effectiveness. Phase determination is a critical step, since errors in phases usually contribute much more to the error in electron density maps than do those from the diffraction measurements.

Three categories of problems arise, and all are pertinent to the design of modified proteins:

1) If the crystal of a modified protein is nearly isomorphous with a parent crystal structure, then the difference Fourier method can be used with phases from the parent structure to show up the changes in structure.

2) If the crystal is novel but is composed of molecules that are similar to one of known structure, then the molecular replacement method can often be used to build a model of properly positioned components from which phases can be computed by the structure factor equation.

3) If the molecule is the first of its class to be analyzed, then the crystallographic phase problem must be solved ab initio.

The method of isomorphous replacement has played a central role in the diffraction analysis of nearly all truly new protein structures. The essence of the method is shown in Figure 2. It depends upon the preparation of heavy atom derivatives; this is usually possible by rational design or with reference to prior experience. Collection of the additional data presents no problem, but lack of isomorphism or difficulties in finding heavy atom positions do frequently present stumbling blocks. Molecular averaging when noncrystallographic symmetry is present can be an extremely powerful way to refine approximate phases. The uniformity of interstitial solvent and positivity of electron density provide important additional constraints for phase refinement. A new class of procedures for essentially direct determination of phases exploits resonance effects in scattering. These anomalous scattering methods, when based on multiple wavelength measurements, offer the unique advantage of an algebraically exact solution of the phase problem from a single crystalline species. Synthrotron radiation is especially important here. Direct methods based on maximum entropy principles are also under development.

Density Map Interpretation

The most exciting part of a crystal structure determination comes when a new electron density map is synthesized. All of the work that precedes this point is justified by the anticipation of marvelous discoveries about what a molecule looks like and how it works. The goal at this stage is to build an initial atomic model that fits the experimental density. Methods for interpreting density maps have evolved greatly in recent years. Minimaps (transparent stacks of contoured sheets at a scale of 0.5 cm/Å or less) are still often used to trace the polypeptide chain, but computerized graphic displays are used increasingly for chain tracing and almost universally for detailed model building.

The ease of interpretation depends most critically on two factors—the accuracy of phasing and the resolution limit. If phases are accurate, a polypeptide chain can readily be traced at about 3.5 Å resoution or, if highly helical, even at lower resolution. At about 1.5 Å resolution, individual atoms can be located. However, since the detailed structures of polypeptide components are well known, the actual resolution of individual atoms is not needed for interpretation. It does help greatly, though, to know the amino acid sequence. Full atomic models can usually be fitted to reasonably accurate maps at about 3 Å resolution. Electron density maps at various resolution limits are compared in Figure 3.

Model Refinement

The ultimate test of a crystallographic model is its agreement with the observed diffraction pattern. The most commonly used measure of this match is the *R*-value,

$$R = \frac{\Sigma\left|\ |F_{obs}(\underset{\sim}{h})| - |F_{calc}(\underset{\sim}{h})|\ \right|}{\Sigma |F_{obs}(\underset{\sim}{h})|}, \tag{5}$$

where $|F_{obs}|$ related to the diffraction observations (Eq. 3) and $|F_{calc}|$ is from the calculated structure factors (Eq. 2). The theory that relates an atomic model to its diffraction pattern also provides the basis for refinement of the model to optimize the agreement. Parameters such as *x, y, z,* and *B* are adjusted to minimize residuals in a least squares sense and thus to reduce the *R*-value.

It is now standard practice to carry out rather thorough least-squares refinements of macromolecular structures. Adequate starting models can be built at intermediate resolution—e.g., 2.8 Å—and can then be extended to higher resolution—e.g., 1.8 Å—by refinement. The effectiveness of current refinement procedures depends heavily on the incorporation of stereochemical restraints to supplement the diffraction data. Interactive molecular graphics is also a crucial ingredient of the process. Initial models usually contain errors that are outside the radius of convergence of nonlinear least squares, and periodic manual intervention with graphics rebuilding is essential. The final important component is high-performance computing machinery. The coupling of restrained refinement methods with rapidly improving computer and graphics instrumentation has an extremely positive impact on the accuracy of crystallographic models.

JUDGING THE RESULTS

Atomic models obtained from diffraction data are a gold mine of information. Even the smallest of proteins are amazingly complex

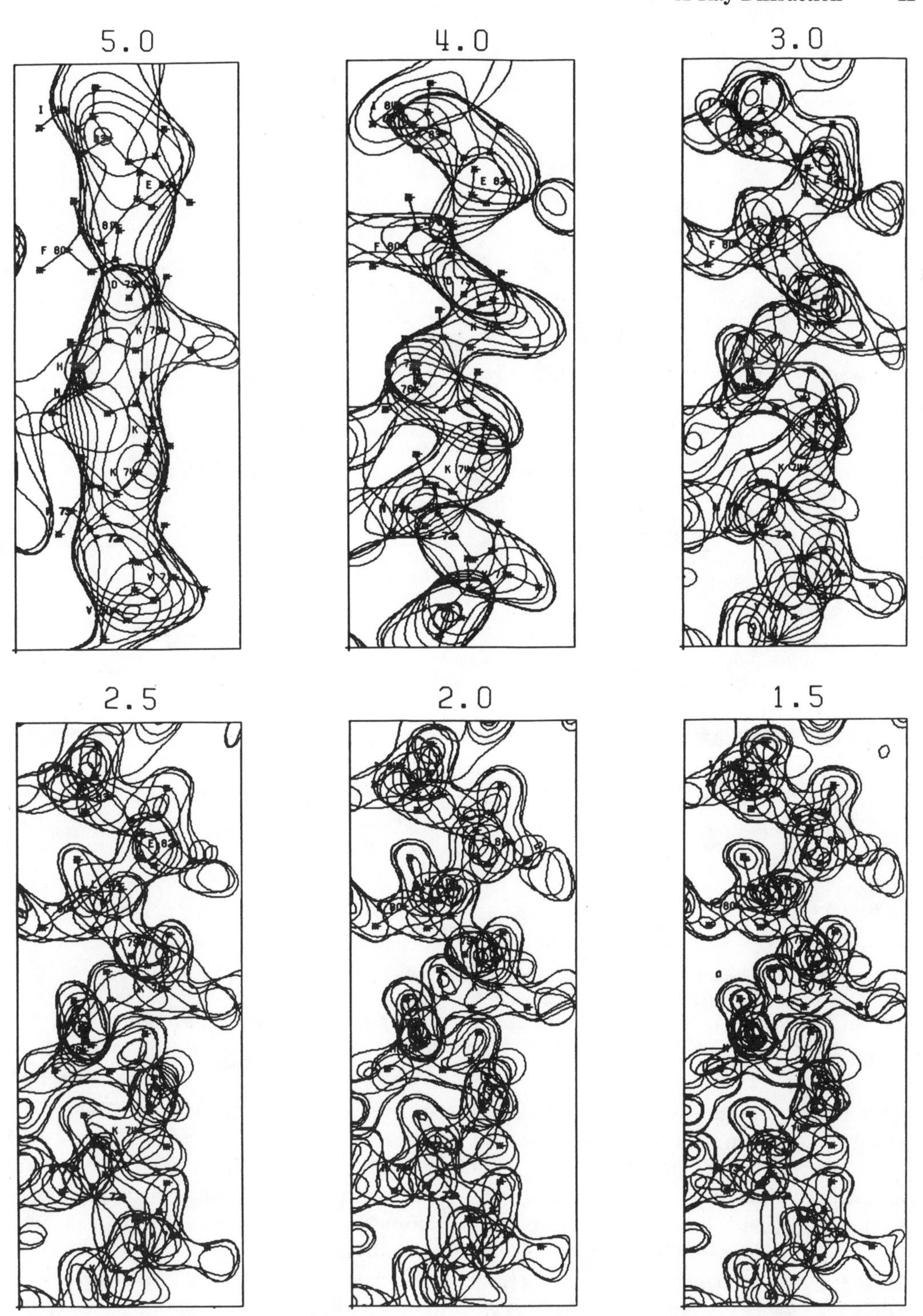

Fig. 3. *Electron density distributions for the C helix of myohemerythrin at various resolutions. Each frame shows the projected electron density distribution from a Fourier synthesis truncated at the indicated nominal resolution. The number of terms included approximately doubles with each successive step in resolution: 638 unique reflections contribute at 5.0 Å resolution, 1208 at 4.0 Å, 2699 at 3.0 Å, 4531 at 2.5 Å, 8618 at 2.0 Å and 16747 at 1.5 Å. In each case the final refined backbone model, including C_β atoms from the side chains, is shown superimposed on the density map. (Drawings by Steven Sheriff.)*

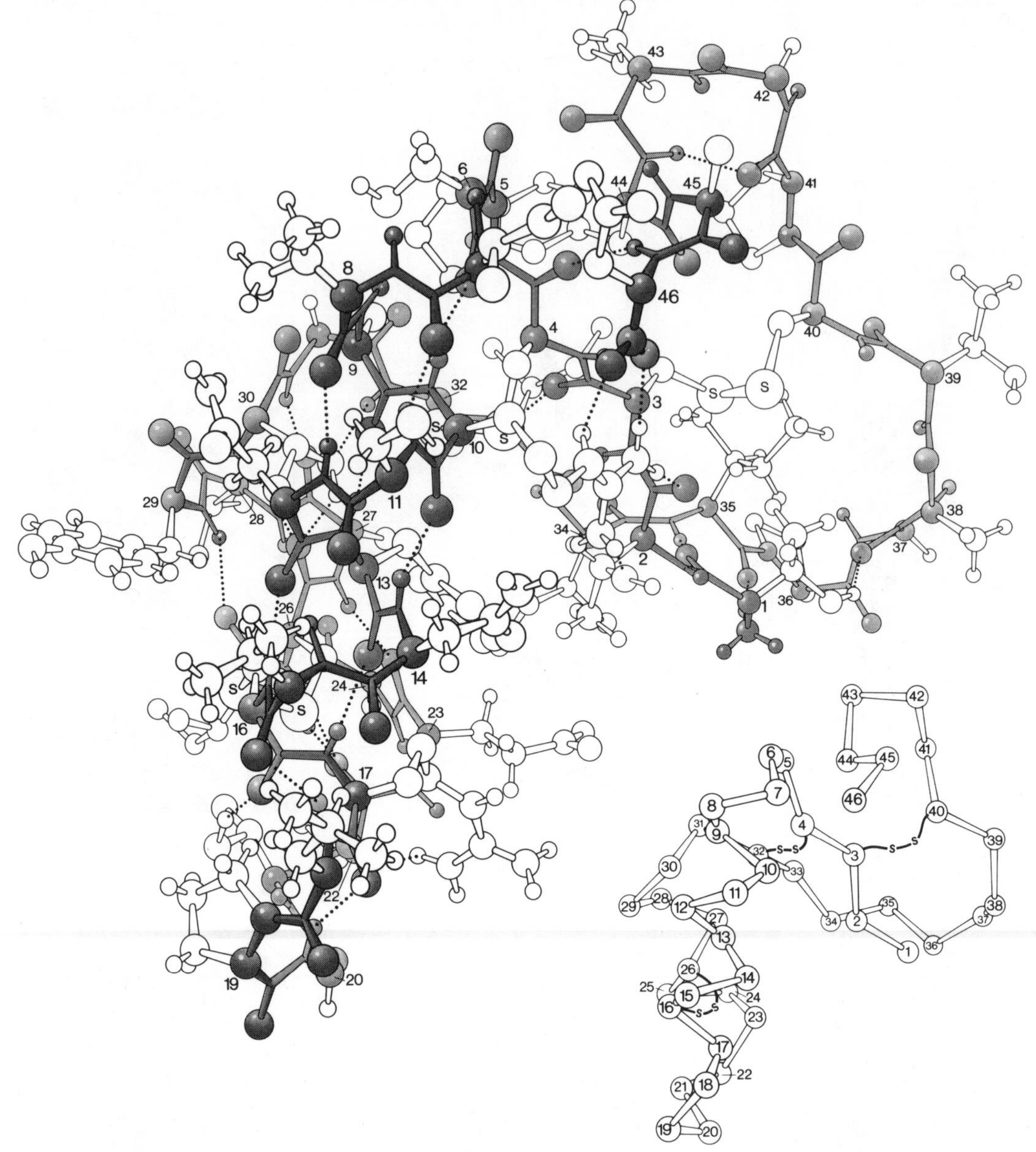

Fig. 4. *The atomic structure of crambin. Atomic positions derive from a crystallographic analysis at 0.945 Å resolution and are those for the predominant conformational state. Selected hydrogen atoms are included. (This drawing is copyright 1986 by Geis and Hendrickson and is reproduced here with permission.).*

when seen in full glory, as in the drawing of crambin shown in Figure 4. Such pictures bear a distinct air of certainty. But one must also wonder about the genuine validity of this seemingly definitive information before making detailed chemical interpretations. Indeed, drawings of a model with substantial errors might also look very convincing. How accurate are the atomic coordinates? What is the possibility that the chain tracing is wrong? Is the structure different in solution? How does the molecule change with time? These questions are not easy to answer and might especially trouble the noncrystallographer who uses diffraction results.

Perhaps the most important criteria by which to judge a structural analysis are its resolution and its state of refinement. The nominal resolution of a diffraction study is set by minimal Bragg spacing ($d_{min} = ½s_{max}$) of the data set. Atomic mobility and disorder in protein crystal structures naturally curtail the extent of measurable diffraction, but the data might also be truncated prematurely for some expedience. In either case, limited resolution in turn limits the accuracy of an atomic model. However, since the centroids of density features can be located much more precisely than the limit of point resolution, positional accuracy can still be quite good. For example, at 2 Å resolution, standard deviations on the order of 0.2 Å can be achieved, but only after refinement.

The quality of initial density maps, at whatever resolution, is quite variable, and it is usually rather difficult to assess quality from the published data. Stereodrawings of maps are probably the best guide. Unless the phasing has been exceptionally good, atomic models fitted to experimental maps usually have some errors of conformational detail and may even have incorrect connections of polypeptide segments. However, these errors are eliminated in well-refined structures, and it is now standard practice to carry out such refinements.

The *R*-value is a good index of the quality of a crystal structure. The agreement factor for initial models at 3 Å resolution is typically greater than 0.40, whereas this is reduced to under 0.20 for well-refined structures. Geometric ideality (or potential energy) is also an important criterion, and to some extent diffraction agreement can be gained at the expense of good geometry. In a properly weighted refinement, the root–mean–square deviation in bond lengths is a good measure of ideality, and this should be 0.20 Å or under for good structures. When the model is right, both the *R*-value and stereochemical ideality will be excellent simultaneously.

The atomic mobility and structural disorder that limit the extent of diffraction are also the most difficult aspects to model. The dynamic character of proteins is such that atoms typically execute large excursions in the crystal. These movements can be highly anisotropic, so that "*B*-values" must be generalized to multiparameter forms to match reality. Sophisticated treatment of atomic mobility and disorder is still under development. However, even isotropic B values have been shown to have functional meaning.

Solvent structure is another important component of a refined structure. Protein crystals are highly hydrated—usually about half of the volume is solvent of crystallization— and for this reason molecules in crystals are, for the most part, relevant to the situation in solution. Even most of the lattice interactions are mediated by water molecules. Water is also frequently involved in active sites. Obviously, careful attention must be paid to solvent structure if a protein crystal is to be realistic and relevant.

In summary, diffraction studies have the potential (now commonly realized) to give a comprehensive and definitive picture of protein molecules. Important advances in methodology and instrumentation are in progress. An increased flow of exciting structural stories from protein crystallography can be expected, and this should be an important component of protein modification and design.

REFERENCES

1. Bernstein FC, Koetzle TF, Williams GJB, Meyer EF Jr, Brice MD, Rodgers JR, Kennard O, Shimanouchi T, Tasumi M (1977): The Protein Data Bank: a computer-based archival file for macromolecular structures. J Mol Biol 112:535–542.
2. Guinier A (1963): "X-ray Diffraction in Crystals, Imperfect Crystals and Amorphous Bodies." San Francisco: Freeman.
3. Hendrickson WA (1986): X-ray diffraction methods for the analysis of metalloproteins. In Wright JR, Hendrickson WA, Osaki S and James GT (eds): "Physical Methods for Inorganic Chemistry." New York: Plenum Publishing Corp, pp 215–259.
4. Blundell TL, Johnson LN (1976) "Protein Crystallography." London: Academic Press.
5. Wyckoff HW, Hirs CHW, Timasheff SN (eds) (1985): "Diffraction Methods for Biological Macromolecules." Methods in Enzymology, Vols. 114–115. Orlando: Academic Press.

Protein Engineering, pages 15–33

2

One- and Two-Dimensional NMR Spectroscopic Investigations of the Consequences of Amino Acid Replacements in Proteins

John L. Markley

Department of Biochemistry, College of Agricultural and Life Sciences, University of Wisconsin–Madison, Madison, WI 53706

INTRODUCTION

An important goal of protein chemistry is the rational design of polypeptide sequences that have desired structural, chemical, or catalytic properties. An essential component of this effort is the ability to determine the chemical and structural properties of designed proteins. Of the spectroscopic techniques available, nuclear magnetic resonance (NMR) spectroscopy offers perhaps the most versatile approach to studies of biomolecules on a molecular level. Several recent developments in NMR methodology can be coupled with advances in protein engineering to provide a powerful means of elucidating structure–function relationships in solution. NMR spectroscopy of small proteins has advanced to the point where one can routinely obtain site-specific information such as conformational changes, apparent pK_a values, hydrogen-bonding patterns, hydrogen-exchange rates, ligand-binding geometry, and side-chain mobility. Traditionally, NMR spectroscopy has been an effective tool for studying protein folding, protein–ligand interactions, and enzyme mechanisms. The methods of protein engineering and design make it vastly easier to address sequence-dependent questions. Genetic engineering techniques also provide a general solution to the serious problem of obtaining enough material for NMR analysis. In cases where proteins of interest can be cloned and overproduced in cell culture, the introduction into proteins of stable isotopes as NMR-active probes becomes simplified and more economical.

PROGRESS IN NMR SPECTROSCOPY OF PROTEINS

The development of NMR as a physical technique and its application to proteins have experienced a parallel evolution. Each major advance in NMR instrumentation or methodology has opened up a new category of applications to protein chemistry and has relaxed previous practical limitations of molecular size or structural complexity. A characteristic feature of NMR spectroscopy has been the regularity of such breakthroughs. The most important of these have been: signal averaging, which led to the increased sensitivity required for studies of single nuclei (or sets of equivalent nuclei) in proteins; higher magnetic fields achieved by superconducting solenoids with associated increases in resolution and sensitivity; and one- and two-dimensional pulse-Fourier transform methods, which have broadened the scope of NMR experiments and increased the efficiency of data acquisition and analysis. Key steps in NMR methodology and its application to proteins in solution are listed in Table I. Complete discussions of these and related developments may be found in textbooks and review articles [1,2, and references therein]. Many of these improvements have involved more sophisticated computerization of the experiment, from control of the radiofrequency signals exciting the sample to the acquisition, processing, and interpretation of NMR spectral data. This trend undoubtedly will accelerate in the future. Complementary advances in protein chemistry have had their impact on NMR investigations—most notably, methods for selective labeling of proteins with NMR active isotopes, improvements in protein isolation, purification, and, most recently, the exploitation of recombinant DNA technology.

TWO-DIMENSIONAL AND MULTIPLE-QUANTUM NMR METHODS

In the conventional one-dimensional pulse-Fourier transform NMR experiment [3], the sample, which is placed in a uniform magnetic field, is subjected to a short pulse of electromagnetic energy (transmitter signal) whose frequency (radio band) is centered within the spectral region of interest. This pulse is designed to excite all the nuclei of interest (e.g., all ^{1}H or all ^{13}C). The transitory magnetization of the sample, produced at right angles to the static magnetic field, generates a signal in the receiver coil which is detected in the audio frequency range as an offset from the radio frequency of the transmitter. This signal is digitized by the analog-to-digital converter and stored in the computer memory. Fourier transformation of this signal (amplitude as a function of the time t_2 from the beginning of signal acquisition) yields the NMR spectrum. NMR signal intensity is plotted as a function of the offset frequency from a reference line; this frequency normally is divided by the spectrometer carrier frequency and presented as the "chemical shift" in dimensionless units of parts per million. The time axis of the experiment is divided into two periods (Fig. 1a): a preparation period during which the system equilibrates, and a data acquisition period following the excitation pulse. The typical one-pulse 1DFT NMR protocol is shown in Figure 1b. In the past ten years, new pulse sequences have been devised that make use of a more complicated experimental time axis (Fig. 1c). The signal intensity is expressed as a function of two time delays (t_1 and t_2) rather than just one, and the NMR data are collected in a two-dimensional array in computer memory. This array is processed by double Fourier transformation (in practice, two sequential Fourier transformations: the first with respect to t_2 and the second with resepct to t_1) to obtain the spectral intensity as a function of two frequencies. The general technique is referred to as two-dimensional Fourier transform (2DFT) NMR [4,5 for reviews]. Depending on the particular experiment, the two 2DFT NMR frequencies may provide chemical shifts vs. chemical shifts, chemical shifts vs. coupling constants, or spectral intensity vs. distance along some dimension of the sample (as in a 2DFT NMR imaging experiment). A few 2DFT NMR protocols that are important in protein studies (Figure 1d–g) are discussed below. For simplicity, we shall omit here mention of numerous experimental details required for the optimal acquisition and processing of 2DFT NMR data such a phase cycling and digital filter functions [4,5].

Since the acquisition of a 2D NMR spectrum is equivalent to obtaining a series of one-dimensional spectra, the number of which de-

TABLE I. Key Developments in NMR Methodology and Its Application to Proteins in Solution

Year	Development or application
1958	First ^{1}H NMR spectrum of a protein
1963	Signal averaging
1964	Resolution of signals from individual groups in proteins
1966	Pulse Fourier transform NMR; pH titration studies as monitored by NMR
1967	NMR spectrometer with superconducting solenoid
1968	Simplification of NMR spectra by isotopic labeling
1970	^{13}C NMR spectrum of a protein
1972	Saturation transfer NMR; NMR relaxation studies; wide-bore NMR spectrometers
1975	Computer control of many spectrometer functions; multiple-pulse NMR methods
1976	Two-dimensional Fourier transform NMR
1979	Concept of specification of protein structure by use of NOE data and a distance geometry algorithm
1982	Multiple-quantum techniques; sequence-specific assignment of NMR data; heteronuclear two-dimensional Fourier transform techniques

termines the resolution in the second dimension, generation of 2DFT spectral data typically may require hours rather than minutes (for 1DFT). Analysis and plotting of 2DFT spectra is more time consuming than for 1DFT NMR, but both can be facilitated by the use of a fast minicomputer equipped with an array processor and digital plotter. Higher resolution can be achieved in 2DFT plots than in a 1D spectrum through the dispersal of peaks over a plane rather than along a line. Where line widths are independent of magnetic field, as in ^{1}H NMR to a first approximation, the resolution of a 2DFT plot increases as the square of the magnetic field strength rather than linearly as in a 1DFT spectrum. The resulting spectral simplification of overlapping regions frequently permits complete analysis of spin systems. The major classes of two-dimensional NMR experiments provide efficient means of: 1) *separation* of NMR spectral parameters so that they are more easily measured; 2) *correlation* of two sets of NMR data so as to identify networks of magnetic nuclei (homonuclei or heteronuclei) that are mutually spin–spin coupled; and 3) *identification* of pairs of magnetic nuclei related by spatial proximity or chemical exchange.

The basic experiment of first type is two-dimensional J spectroscopy, which provides a separation of chemical shift and coupling information. Although this approach has been applied to proteins in homonuclear [6] and heteronuclear versions [7], it has not proved as versatile as other 2DFT NMR approaches, largely because it is possible to extract coupling constants and identify spin systems by using the correlation experiments discussed below.

The second type of two-dimensional NMR experiment includes Spin–Echo Correlated Spectroscopy (SECSY) and COrrelated SpectroscopY (COSY) [8]. COSY utilizes the pulse sequence shown in Figure 1d. The absolute value mode COSY plot obtained for a small protein is shown in Figure 2a. The normal 1DFT spectrum lies along the diagonal, and the off-diagonal peaks (symmetrical above and below the diagonal) represent pairs of protons that are spin–spin coupled to one another. The same information might be obtained from a series 1DFT NMR spectra employing single-frequency decoupling at a

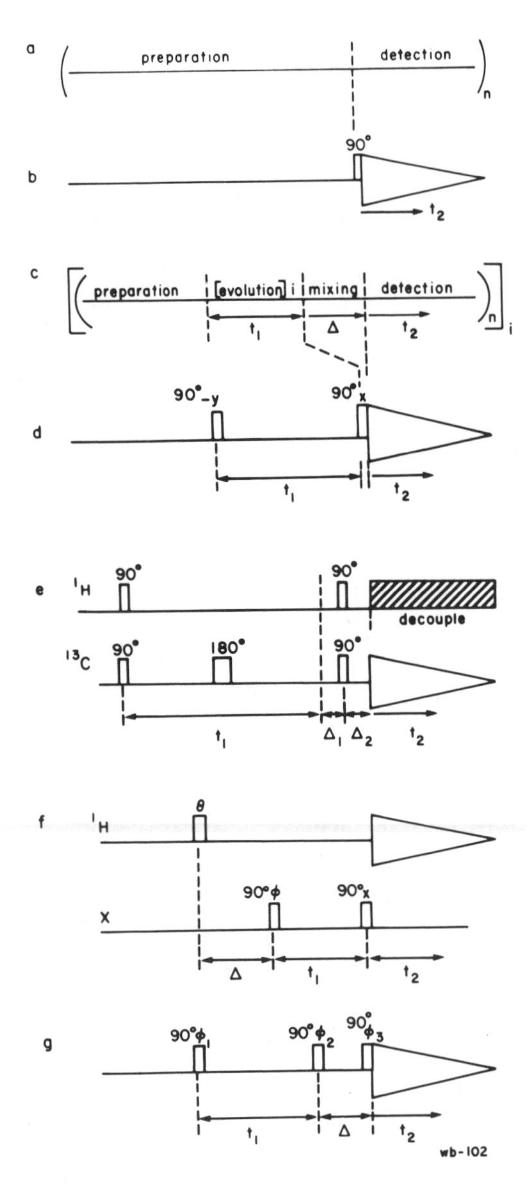

Fig. 1. *Representative pulse sequence employed in one- and two-dimensional Fourier transform NMR spectroscopy of proteins. (a) General subdivision of the time axis in a one-dimensional pulse Fourier transform NMR experiment; the $_n$ refers to the number of times the sequence is repeated to time-average the data in order to increase the signal-to-noise ratio. (b) Pulse scheme for the one-dimensional NMR single pulse experiment. (c) General subdivision of the time axis in a two-dimensional pulse Fourier transform NMR experiment. The $_n$ refers to the number of times the sequence is repeated for signal averaging; the index i indicates that the time t_1 is incremented from one subexperiment to the next. (d) Pulse sequence used for the homonuclear correlated spectroscopy (COSY) experiment. (e) Pulse scheme used for two-dimensional heteronuclear (^{13}C,^{1}H) correlated spectroscopy (HETCOR). (f) Pulse sequence used for two-dimensional heteronuclear (^{1}H,X) multiple quantum spectroscopy (HMQ). (g) Pulse scheme used for the two-dimensional NMR experiment used to detect cross-relaxation and chemical exchange (NOESY).*

series of closely spaced points in the spectrum. The 2DFT experiment provides the results more efficiently and more effectively because the decoupler in the 1DFT experiment may excite neighboring resonances in crowded spectral regions so as to produce ambiguous correlations. Modifications of the basic 2DFT NMR correlation experiment as applied to proteins include heteronuclear versions [9–11] (Figure 1e), multiple quantum versions [12,13], relay experiments [14,15], and versions incorporating multiple-quantum filters [16]. The two-dimensional Fourier transform heteronuclear multiple quantum (HMQ) experiment [17] (Figure 1f) promises a powerful means of simplifying the proton spectrum of a protein in order to extract specific kinds of information. Since proton detection is employed, this can be achieved without a great sacrifice in sensitivity. Figure 3 illustrates a specific example in which an NMR active metal ion ($^{113}Cd^{2+}$) was substituted for the copper ion in plastocyanin. 2DFT (^{1}H, ^{113}Cd) HMQ contour plots are shown in Figure 3a–c. Whereas the full ^{1}H NMR spectrum of the protein is complex (Figure 3d), the HMQ data yield subspectra (Fig. 3e,f) which contain peaks only from the metal ligands which are spin–spin coupled to ^{113}Cd. The sensitivity of the method is such that the correlations can be detected even when spin–spin coupling is not resolved in the 1DFT spectrum [12].

The fundamental 2DFT experiment of the third type employs a sequence of three pulses (Figure 1g). In addition to the delay times (t_2

and t_1), a mixing time ($\Delta = t_m$) is employed which is essentially constant during the experiment. After Fourier transformation in two dimensions, the data yield a plot that can be interpreted in terms of cross-relaxation (NOESY or Nuclear Overhauser Effect SpectroscopY), chemical exchange, or a combination of both [18]. As in the COSY plot, the normal spectrum lies along the diagonal. The cross-peaks, however, arise from relaxation effects arising from dipole–dipole interactions between pairs of magnetic nuclei that are very close to one another in space or through a chemical exchange process affecting the same spin. If a chemical exchange process takes an NMR active nucleus from one magnetic environment to another in a time comparable to or somewhat shorter than the longitudinal relaxation time, a cross-peak will appear that correlates the chemical shifts of that nucleus in each state. With proteins, the chemical exchange information can be used to correlate the chemical shifts of protons in two oxidation states [19,20] or two conformational forms—for example, native and denatured states [21]. In the absence of exchange, cross-peak intensities are determined by the nuclear Overhauser effect which falls off as $1/r^6$, where r is the distance between the pair of protons. The maximum distance sampled depends on the mixing time (eg., a mixing time of 100 ms for a small protein yields cross-peaks from pairs of protons closer than about 3 Å); however, identification of pairs of nuclei more distant than about 5 Å becomes unreliable owing to a mechanism of relayed dipole–dipole interaction known as "spin diffusion" [22]. NOESY data frequently are used to deduce conformations of macromolecules in solution. The NOESY plot of a small protein is shown in Figure 2b.

STRATEGIES FOR ASSIGNING RESONANCES IN PROTEIN SPECTRA

A useful methodological distinction is made between "first-order" and "second-order" assignments of NMR peaks in a protein. A "first-order" assignment, which identifies the type of residue to which the NMR peak corresponds (in general one of the 20 normal varieties of amino acid) but not its position, typically is based on unique chemical shifts and coupling patterns (spin systems) observed for each type of amino acid residue. Certain 2DFT experiments lend themselves particularly to first-order assignments: relayed COSY and multiple-quantum methods can be used to extend the spin correlation one or more steps beyond that obtained by COSY; TOtal Correlated SpectroscopY (TOCSY) [23] or 2D Hartmann-Hahn spectroscopy [24] provide a means of identifying or selecting for all the resonances from a particular residue in the peptide sequence. Multiple-quantum filtering methods [25], by which one can generate subspectra corresponding to spins that are coupled in a particular geometrical pattern or "spin topology" [26], offers a powerful new approach to first-order assignments.

Implicit in a "second-order" assignment is the position of the assigned nucleus in the amino acid sequence of the protein. Such assignments may be based on knowledge that the protein contains only one residue of a given type or only one with characteristic chemical features. Second-order assignments also may be based on spin correlations that bridge the peptide bond and thus provide a link from an assignment on one residue to one on its neighbor. Sequential backbone assignments need to be put into register with the amino acid sequence in order to qualify as second-order assignments. The additional information needed for this can come from a single second-order backbone assignment or from two or more first-order backbone assignments that uniquely determine the residue numbers of the amino acids [27,28].

At present, the most effective sequential assignment method relies on comparisons of COSY and NOESY spectra of the protein [28]. Both through-space and through-bond information are needed for sequential assignments based on ^1H NMR (see the connectivities illustrated in Figure 4). Since ^1H–^1H spin–spin

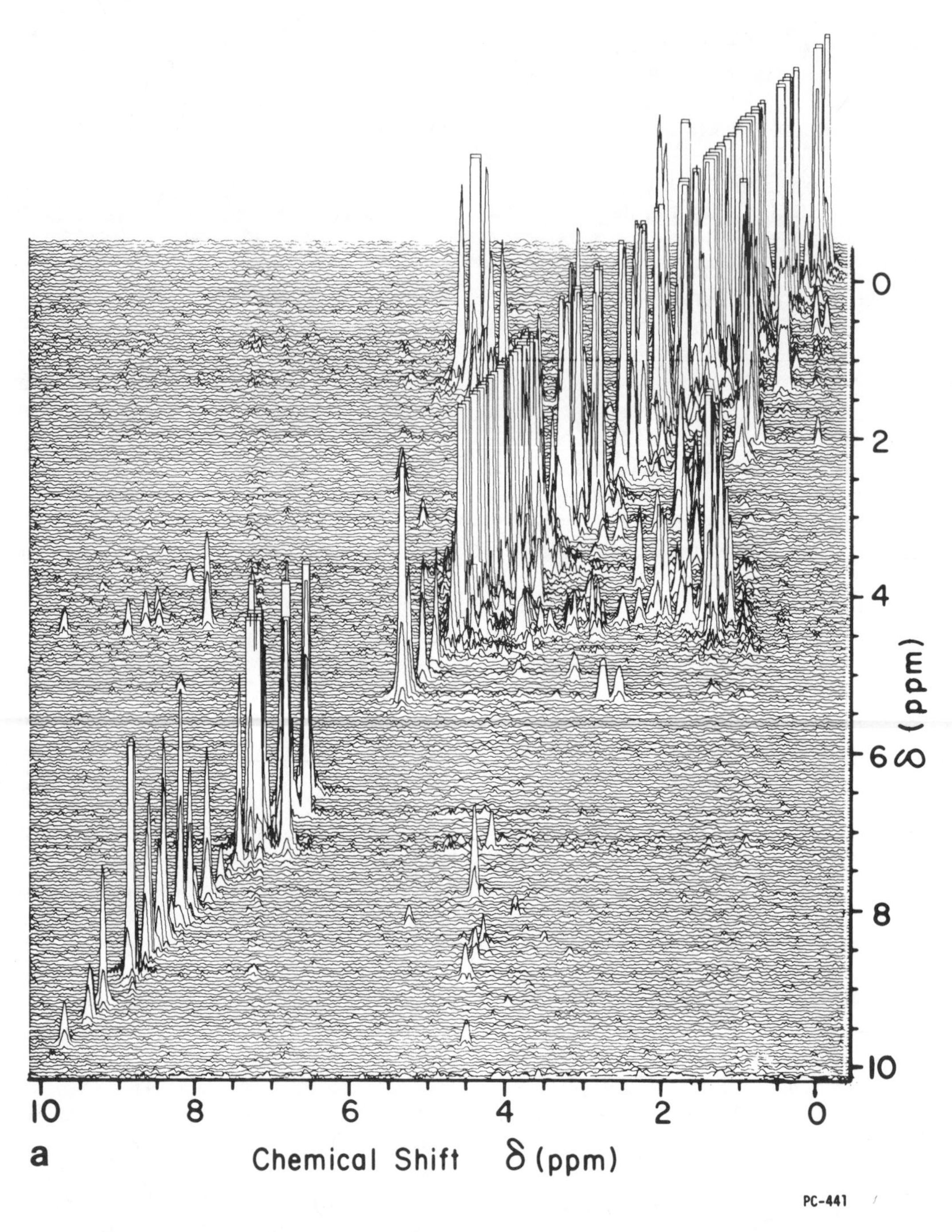

Fig. 2. *(a) Two-dimensional correlated spectroscopy (COSY) plot of a small protein. The ^1H NMR spectral data were obtained in the absolute value mode at 470 MHz. The sample was 12 mM turkey ovomucoid third domain (M_r 6,000) dissolved in 0.2 M KCl in ^2H_2O; 25°C, pH 8.0. (b) Two-dimensional nuclear Overhauser spectroscopy (NOESY) plot of the same sample. (From Markley et al. [34].)*

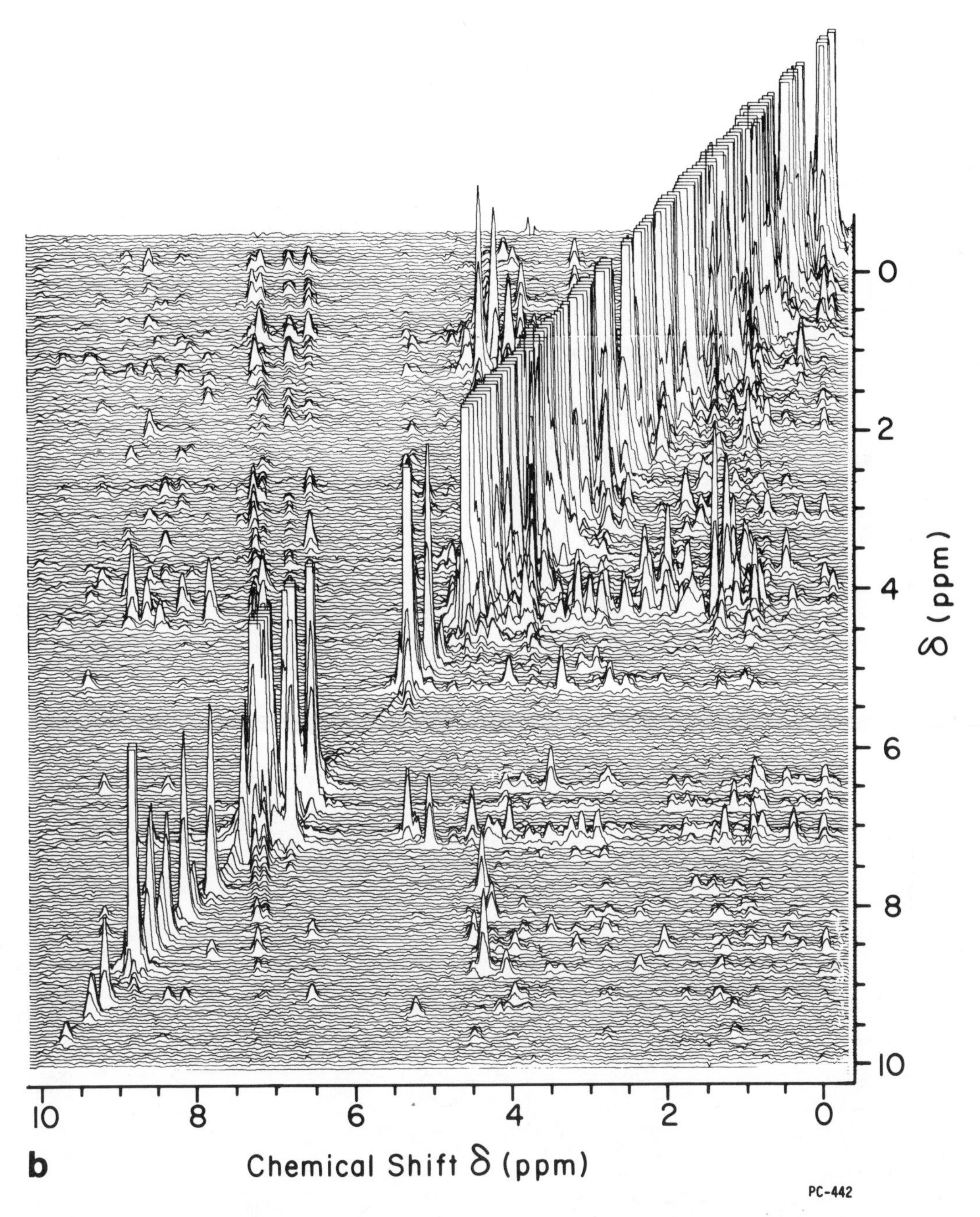

coupling across the peptide bond does not yield reliable correlation peaks, the link between one residue and its neighbor is provided by NOESY cross-peaks resulting from the spatial proximity of the peptide amide N-^{1}H of one residue to protons on the adjacent residue. The COSY plot provides the information needed to bridge the intraresidue connectivity between the N-^{1}H and the C_α-^{1}H. Both the NOESY and COSY data sets must be obtained with the protein dissolved in ^1H_2O (rather than ^2H_2O) because signals from each N-^{1}H are

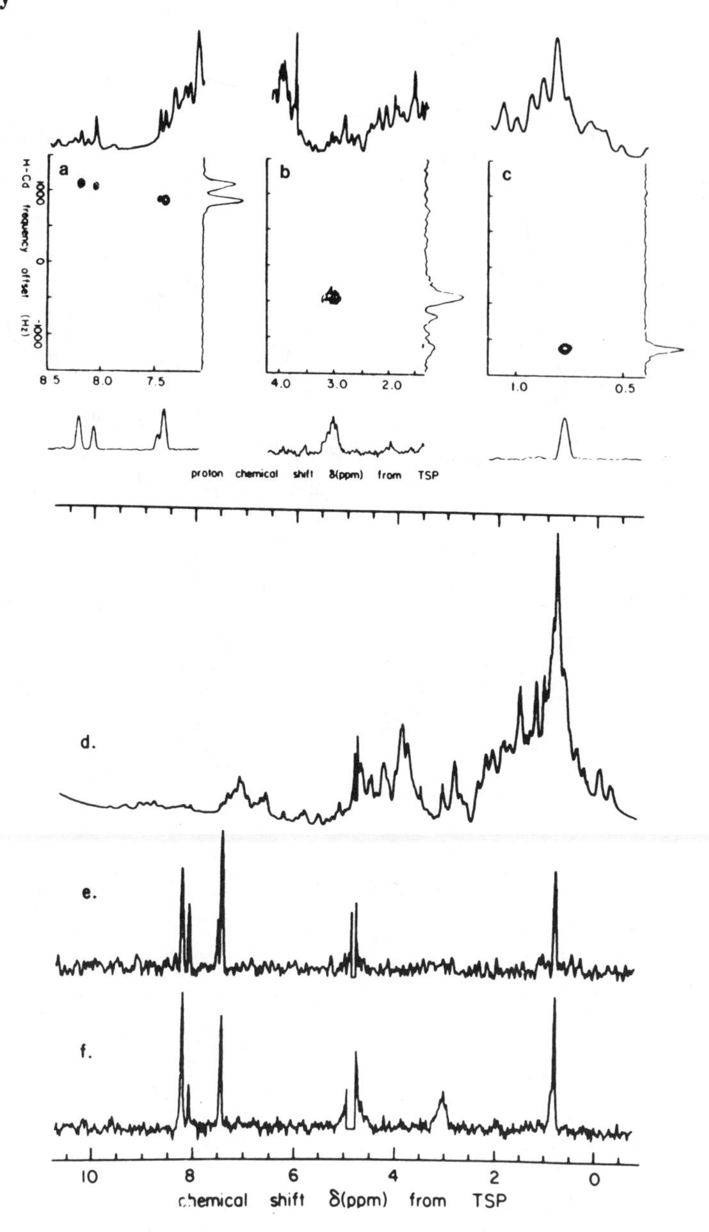

Fig. 3. (Top) *Proton-detected (^{1}H–^{113}Cd) double-quantum coherence in cadmium-substituted plastocyanin from spinach. The contour plots are of two-dimensional Fourier transform NMR spectra. Projections of the two-dimensional surface are shown at the bottom and side of each spectral region displayed. Corresponding one-dimensional ^{1}H NMR spectra obtained at 470 MHz are shown for reference at the top of each region. The sample contained 10 mM protein with 27 mM phosphate buffer at pH 6.6 in $^{2}H_2O$. See reference [12] for other experimental details. Assignments of ^{1}H–^{113}Cd coherences to metal ligands: (a) histidine ring protons, (b) cysteine methylene protons, (c) methionine methyl protons.* **(Bottom)** *Comparison of (d) the one-dimensional ^{1}H NMR spectrum at 300 MHz of ^{113}Cd plastocyanin with the full ^{1}H projection of the two-dimensional spectra (e) and (f). Projection (e) was derived from the two-dimensional spectrum shown above in contour plots (a) and (c); projection (f) was derived from the data that yielded contour plot (b). The sharp intensities near 5 ppm are artifacts from the zeroed water signal. (From Live et al. [12].)*

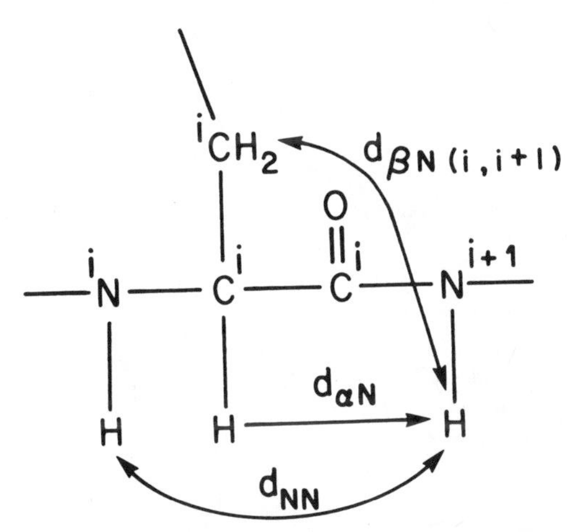

Fig. 4. *Through space connectivities used to establish inter-residue connectivities for sequential assignments in a protein [28]. These can be obtained from two-dimensional Fourier transform nuclear Overhauser (NOESY) analysis of the protein in* ^1H_2O *solution.*

essential to the assignment algorithm. In favorable cases, this method can provide complete assignment of the backbone proton resonances. These assignments then can be extended to the side-chain protons by the methods outlined above (e.g., COSY, relayed COSY, 2D Hartmann-Hahn).

In practice, the COSY–NOESY sequential assignment technique suffers from several limitations. Larger proteins contain more amino acids and, hence, potentially larger numbers of ambiguous interresidue connectivities resulting from overlapping chemical shifts. NMR peaks from larger proteins are broader because the molecules rotate more slowly; hence, the cross-peaks are characterized by lower intensity and lower resolution. The largest proteins for which this sequential assignment method has been successful have been cytochrome c (monomer M_r of 10,000) [29] and cro protein (dimer M_r of 14,000) [30]. Information about the secondary structure of the protein can make sequential assignments of proteins of this size more tractable.

Even small proteins may present sequential assignment difficulties. Prolines, which have no peptide N–H, have to be bridged by some other connectivity or be approached sequentially from both sides. The intensity of NOESY and COSY cross-peaks depends on the molecular geometry and may be missing in some instances. Unfavorably overlapping resonances may lead to insoluble ambiguities in the sequential tracing of NMR peaks.

Stable isotope-labeling strategies provide a proven way of extending the molecular weight barrier and facilitating assignments [1]. Since deuteron (^2H) and proton (^1H) resonances occur in widely separated spectral regions, deuteration (which may be accomplished by chemical reaction or exchange by biological synthesis from labeled metabolites) can be used to simplify ^1H NMR spectrum or to make assignments. A convenient approach is to feed a microorganism that produces the desired protein a mixture of perdeuterated amino acids to which one or more protonated amino acids are added in excess [31]. If further simplification is desirable, the added amino acid itself can be deuterated selectively. Selective deuteration has been used to study proteins as large as immunoglobulins (M_r 150,000) [32]. Developments in the field of biotechnology which permit the overproduction of interesting proteins by microorganisms that tolerate high levels of isotopes make labeling of proteins by stable isotopes much more feasible. ^1H NMR linewidths from C-*H* groups in proteins can be lowered by deuterium labeling neighboring hydrogen sites that relax the proton(s) of interest. One should be able to make more selective NOE measurements (and hence more reliable distance estimates) by observing cross-relaxation of pairs of protons surrounded only by deuterons.

Analogous labeling procedures employing carbon-13 and nitrogen-15 labeling also are useful for larger proteins. Even with total enrichment, the NMR sensitivity of ^{13}C and that of ^{15}N are considerably lower than that of ^1H (by factors of 63 and 962, respectively, at the same magnetic field strength). Heteronuclear 2DFT NMR methods have made ^{13}C and ^{15}N labeling of proteins still more attractive. If the labeled carbon or nitrogen (X) has an attached proton, it can be advantageous to use ^1H NMR

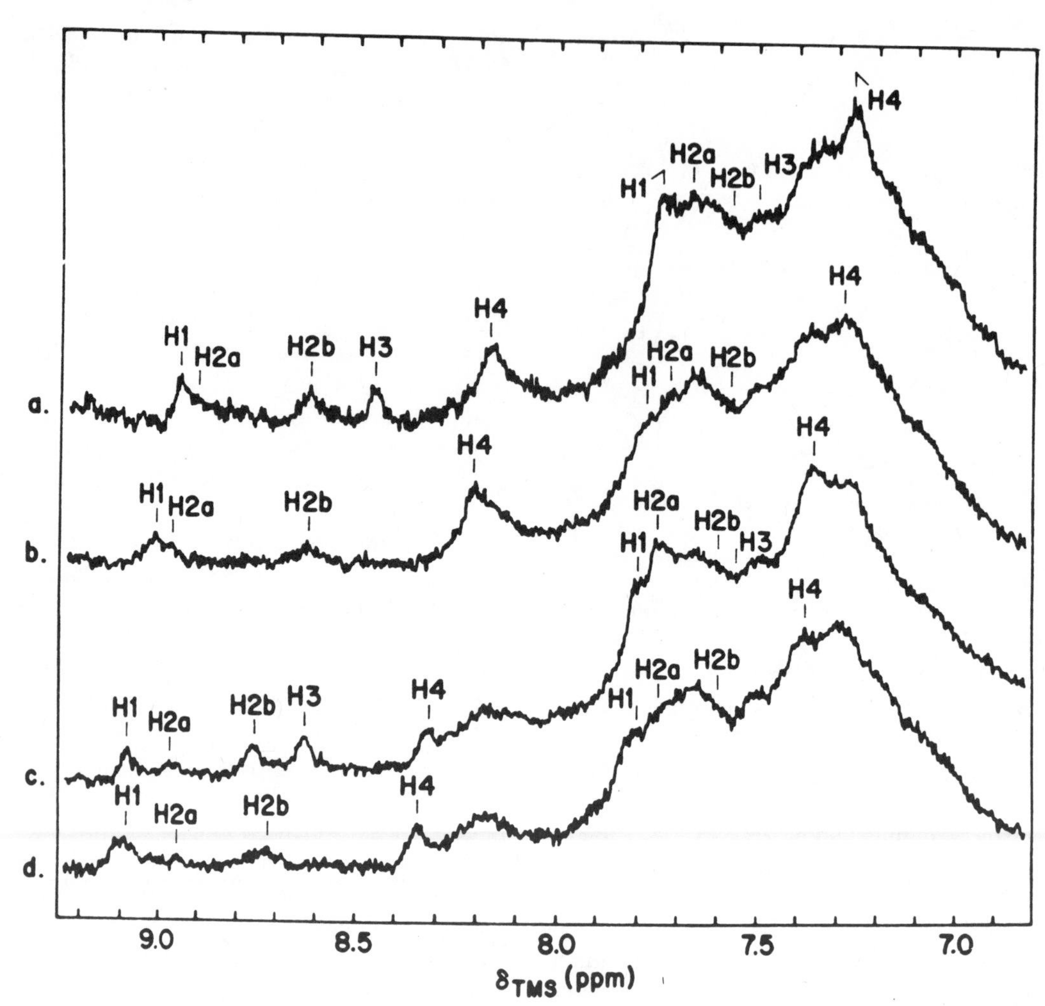

Fig. 5. *The aromatic region of 100 MHz ^1H NMR spectra of two variants of straphylococcal nuclease that differ at position 124. Peak H3 from a histidine C_ϵ-H is present in spectra of nuclease from the Foggi strain which has His-124 but is missing in spectra of the nuclease from the V-8 strain which has Leu-124. Nuclease from the Foggi strain, pH 6.00; (b) nuclease from the V-8 strain, pH 6.00; (c) nuclease from the Foggi strain, pH 7.25; nuclease from the V-8 strain, pH 7.25.*

detection [17]. Observation of the ^{1}H–X coupling may lead to a specific assignment, or heteronuclear COSY or HMQ spectroscopy may be used to obtain a ^{1}H NMR/X subspectrum containing resonances of the ^{1}H and X nuclei.

Sequence-specific assignments have been achieved by feeding microorganisms two labeled amino acids, one with a ^{15}N in the α-amino position and another with a ^{13}C in the backbone carbonyl position. Detection of ^{15}N–^{13}C coupling identifies dipeptides with the appropriate –^{13}C–^{15}N– peptide linkage. Unique dipeptide sequences are assigned unambiguously [27,33]. The method has been demonstrated with a protein as large as M_r 24,000 [27]. Sequential NMR assignment strategies based on double labeling with ^{15}N and ^{13}C have been proposed [4,34] which may provide a means of assigning larger proteins. The high cost and lack of availability of labeled amino acids has presented a barrier to such experiments.

Amino acid replacement provides a widely used approach to the assignment of NMR peaks in proteins. The first application of this strategy involved the study of two naturally occurring variants differing by substitution of a single amino acid residue: the enzyme staphylococcal nuclease with the replacement histidine-124/leucine-124 (Fig. 5) [35]. In favorable cases, comparison of spectra from variants with multiple substitutions may yield several assignments. It must be remembered, however, that this assignment strategy, unlike those discussed above, involves a chemical modification. The method can be effective only if this modification leads to a minor spectral (structural) perturbation so that the change observed can be assigned to the site of substitution itself. A genetically engineered mutant of staphylococcal nuclease provides a demonstration of this cautionary point. Two variant proteins produced by mutagenesis, which are identical except at one residue (phenylalanine-76/valine-76), exhibit numerous changes in the aromatic and methyl regions of the ^1H NMR spectrum that preclude assignments to residue 76 (Fig. 6a,b). The spectral differences appear to arise from a major sequence-dependent difference in the conformations of the two proteins at room temperature; the spectra of the two proteins above the thermal transition temperature (thermally denatured state) differ only in regions that can be assigned to the side chain of residue 76 (Fig. 6c,d) [36]. It will be interesting to learn just how the structures of these two variants differ.

INFORMATION CONTENT OF PROTEIN SPECTRA

An illustration of the kind of information that can be derived from NMR analysis of a small protein is provided by our studies of ovomucoid third domain [37]. Sequences of several variants of this protein that have been studied by NMR are shown in Figure 7. The third domain is a fragment of a larger protein (ovomucoid) which is the second most abundant protein in avian egg white. The third domain (like the isolated first and second domains of ovomucoid) functions as a reversible inhibitor of serine proteinases. The set of avian ovomucoid third domains is the subject of intensive structure–function investigations [38]. Ovomucoid third domains have been sequenced from over 100 species of birds, and 41 pairs of these sequences differ from one another by a single amino acid substitution [39]. An attempt is being made to devise a sequence-to-reactivity algorithm for the association constant between a serine proteinase and members of this set of protein proteinase inhibitors. As a first-order approximation, it is assumed that the contributions of single substitutions are independent of others so that they can be treated in a linear fashion. The results to date indicate that this approximation works remarkably well, at least for the naturally occurring variants that have been studied, and suggests that a given amino acid replacement generally causes only minor perturbations at remote sites [38]. These ideas are supported by NMR results with several ovomucoid third domains [37] and by a recent comparison [40] of the X-ray structures of ovomucoid third domains from Japanese quail (OMJPQ3), silver pheasant (OMSVP3), and turkey (OMTKY3) (as determined in complex with the proteinase B from *Streptomyces griseus*) [41]. Both the NMR and X-ray data indicate that the structures of those ovomucoids that have been studied are strikingly similar.

As stressed above, extraction of information from an NMR spectrum requires, as a first step, the resolution and assignment of the relevant spectral features. Assigned NMR peaks can provide a variety of information about chemical and geometrical properties of a protein. NMR spectroscopy is well suited for determining: 1) bond type and order [43], 2) hydrogen-exchange rates at individual sites [44], 3) apparent pK_a values of individual protein groups [45], and 4) ligand binding kinet-

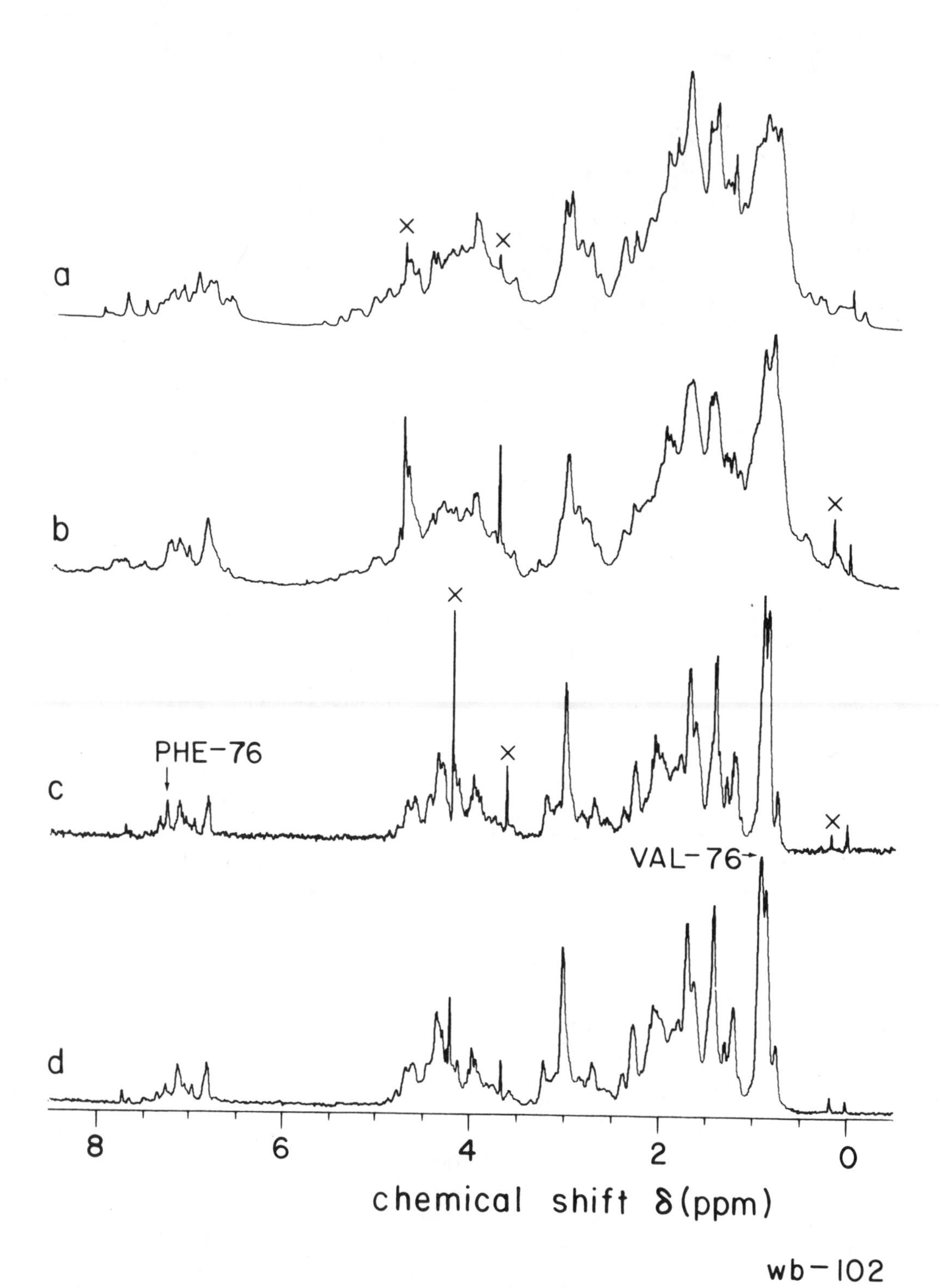
a
b
c
d
PHE-76
VAL-76
8
6
4
2
0
chemical shift δ(ppm)
wb-102

```
        1        10        20        30        40        50

OMTKY3  LAAVSVDCSEYPKPACTLEYRPLCGSDNKTYGNKCNFCNAVVESNGTLTLSHFGKC
OMIPF3  LAAVSVDCSEYPKPACTLEHRPLCGSDNKTYGNKCNFCNAVVESNGTLTLSHFGKC
OMSVP3  LAAVSVDCSEYPKPACTMEYRPLCGSDNKTYGNKCNFCNAVVESNGTLTLSHFGKC
OMGMQ3  FAAVSVDCSEYPKPACTLEYRPLCGSDNKTYANKCNFCNAVVESNGTLTLSHFGKC
OMJPQ3  LAAVSVDCSEYPKPACPKDYRPVCGSDNKTYSNKCNFCNAVVESNGTLTLNHFGKC
OMCHI3  LAAVSVDCSEYPKPDCTAEDRPLCGSDNKTYGNKCNFCNAVVESNGTLTLSHFGKC
OMFTD3  VAT--VDCSDYPKPACTLEYMPLCGSDNKTYGNKCNFCNAVVDSNGTLTLSHFGKC
```

Fig. 7. *Sequences of seven ovomucoid third domains that have been studied by NMR spectroscopy: OMTKY3, turkey; OMIPF3, peafowl; OMSVP3, silver pheasant, OMGMQ3, Gambel's quail; OMJPQ3S, Japanese quail; OMCHI3, chicken; and OMFTD3, fulvous tree duck [37]. Residues that differ from the corresponding ones in OMTKY3 are underlined.*

ics and thermodynamics [2]. We have made primary assignments of backbone ^{1}H NMR peaks by comparing COSY plots of spectra of ovomucoid third domain variants differing by one or a few amino acid residue replacements (Figure 8). These have served as starting points for sequential NMR assignments based on comparison of COSY and NOESY plots (Figure 9) [37]. This approach has led to assignments of signals from over 70% of the residues of OMTKY3. The ^{1}H NMR assignments have been extended to ^{13}C NMR peaks by means of 2D heteronuclear correlation spectroscopy (Figure 10) [10] and to ^{15}N NMR peaks by means of 2D heteronuclear multiple-quantum spectroscopy [13].

Figure 11 illustrates a pH titration study as monitored by ^{13}C NMR spectroscopy that reveals properties of the tyrosine residues of two ovomucoid third domain variants [46]. With a protein of this size it should be possible to determine the pK_a values of all groups that ionize in the pH range in which the protein is stable.

Geometric information is obtained from three-bond coupling constants which define peptide dihedral angles according to extensions of the Karplus relationship [47] or from nuclear Overhauser enhancements [48]. Coupling constants in spectra of small proteins can be extracted from two-dimensional NMR data in favorable cases [49]. The NOESY spectrum of a protein contains enough information about close contacts between residues to define the structure in general terms [50]. It has been found that α-helices and β-sheets give rise to characteristic patterns of NOESY cross peaks [28]. Figure 12 shows the NOESY peaks that arise from a stretch of α-helix in OMTKY3. An area of intense activity at present is concerned with determinations of protein topology from sets of NOE or other NMR parameters [51]. Two approaches have been

Fig. 6. *Comparison of 470 MHz $_1$H NMR spectra [36] of two genetically engineered variants of staphylococcal nuclease. Both proteins were studied at two temperatures in ^2H_2O solution containing 0.3 M NaCl. The pH of the solutions was 7.6 at room temperature. The parent protein as expressed in Escherichia coli (Figure 6a and c) has the sequence of the nuclease from the Foggi strain of* Staphylococcus aureus. *The mutant protein (Figure 6b and d), which has lowered thermal stability, has a (phenylalanine → valine) replacement at residue 76. At 80°C, where both proteins are thermally denatured, the two proteins yield similar spectra whose differences can be interpreted in terms of the substitution at position 76: additional intensity in the aromatic region attributed to a phenyl group is present in the spectrum of the Phe-76 variant (c) compared to that of the Val-76 variant (d), whereas additional intensity in the methyl region is present in the spectrum of the Val-76 variant (d) comparted to that of the Phe-76 variant (c). At 30°C, which is below the thermal denaturation temperature for both proteins, the spectra of the two proteins are very different; the extent of these differences indicates that the resonances from a number of amino acid residues are affected by the single amino acid replacement.*

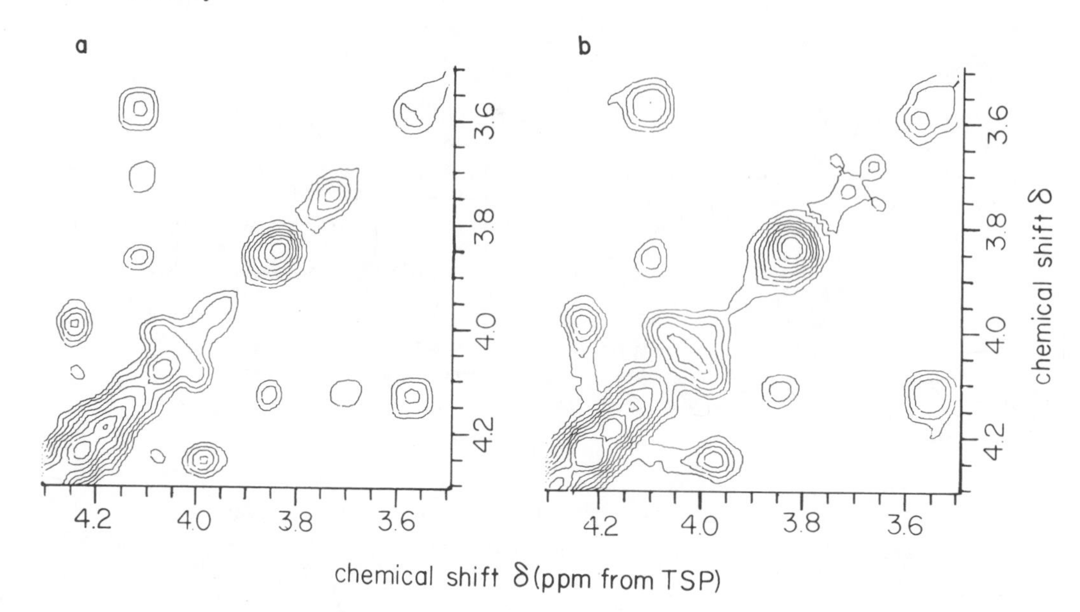

Fig. 8. *The glycine region of the two-dimensional Fourier transform correlated spectrum (COSY) of (a) turkey ovomucoid third domain (OMTKY3) and (b) Gambel's quail ovomucoid third domain (OMGAM3). The two proteins differ by the substitution at position 32 of glycine (present in OMTKY3) for alanine (present in OMGAM3). The cross peak present at (4.12 ppm, 3.70 ppm) in the spectrum of OMTKY3, but absent in the corresponding region of the spectrum of OMGAM3, is assigned to glycine-32. The chemical shifts of this cross peak are those of the two methylene protons of the glycine which are spin-spin coupled to one another. (From Markley et al. [34].)*

used: 1) a modification of the distance-geometry algorithm of Havel, Crippen, and Kuntz [52], and 2) a method that combines molecular dynamics and energy minimization [53].

In addition, NMR spectra provide information about the dynamic properties of protein groups [44]. NMR observables span a broad time scale for the study of rate processes: correlation times derived from NMR data (ns), line shapes (ms to seconds), saturation transfer rates (ms to seconds), and comparison of spectra obtained at different times (minutes to years).

METHODOLOGICAL REQUIREMENTS AND LIMITATIONS

For ^{1}H NMR detection, a one-dimensional spectrum requires on the order of 0.05 μmole protein at a concentration of 10^{-4} M, and a two-dimensional spectrum requires 1 μmoles of protein at 2×10^{-3} M. The one-dimensional ^{1}H NMR spectrum can be acquired in 5 minutes or less, whereas the 2DFT NMR spectrum may take 8 hours. One can trade reduced concentration or samples size for longer acquisition times up to a point, but sample solubility and stability requirements must be met for a successful NMR experiment. Detection of ^{13}C or ^{15}N resonances at natural abundance requires substantial amounts of protein and high solubility: typical values are 30 μmoles of protein at a concentration of 10^{-2} M.

Another tradeoff is between spectral detail and molecular weight. Since detailed conformational analysis of a molecule requires numerous coupling constant and/or nuclear Overhauser effect measurements, this will of necessity be limited to smaller proteins.

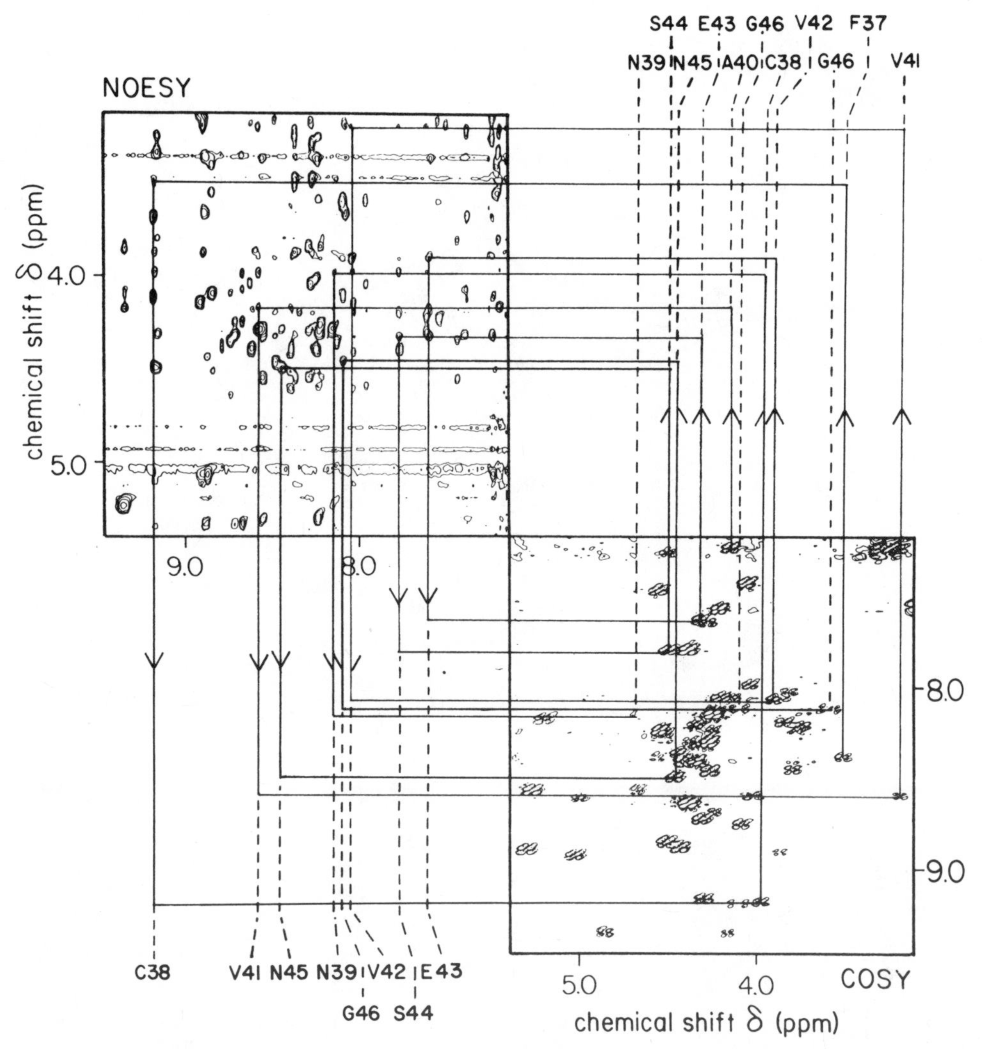

Fig. 9. *Sequential assignment spiral for residues 40 to 46 in turkey ovomucoid third domain, pH 4.1, obtained by the method of Reference [28]. Both COSY and NOESY data were collected in pure phase mode at 500 MHz. The sample contained 15 mM protein and 0.2 M KCl; the temperature was maintained at 25°C in both experiments. The water solvent contained 10% D_2O for the lock. The water peak was suppressed by pre-irradiation. 512 blocks of 2048 data points were collected; each free induction decay was the average of 64 acquisitions. The total data collection time was approximately 20 hours for each experiment. The NOESY mixing time was 100 msec. Spectral analysis of this kind at 500 MHz has led to assignments of all backbone peaks. (From Robertson [42].)*

Whereas one may now anticipate achieving a complete analysis of the ^{1}H NMR spectrum of a protein of M_r 10,000, with a protein twice the size or larger, one may need to be content at present with resolving and assigning a few key resonances, unless one resorts to the introduction of stable isotopes. Spectra achieved at natural abundance may nonetheless contribute important information, depending on the design of the experiment.

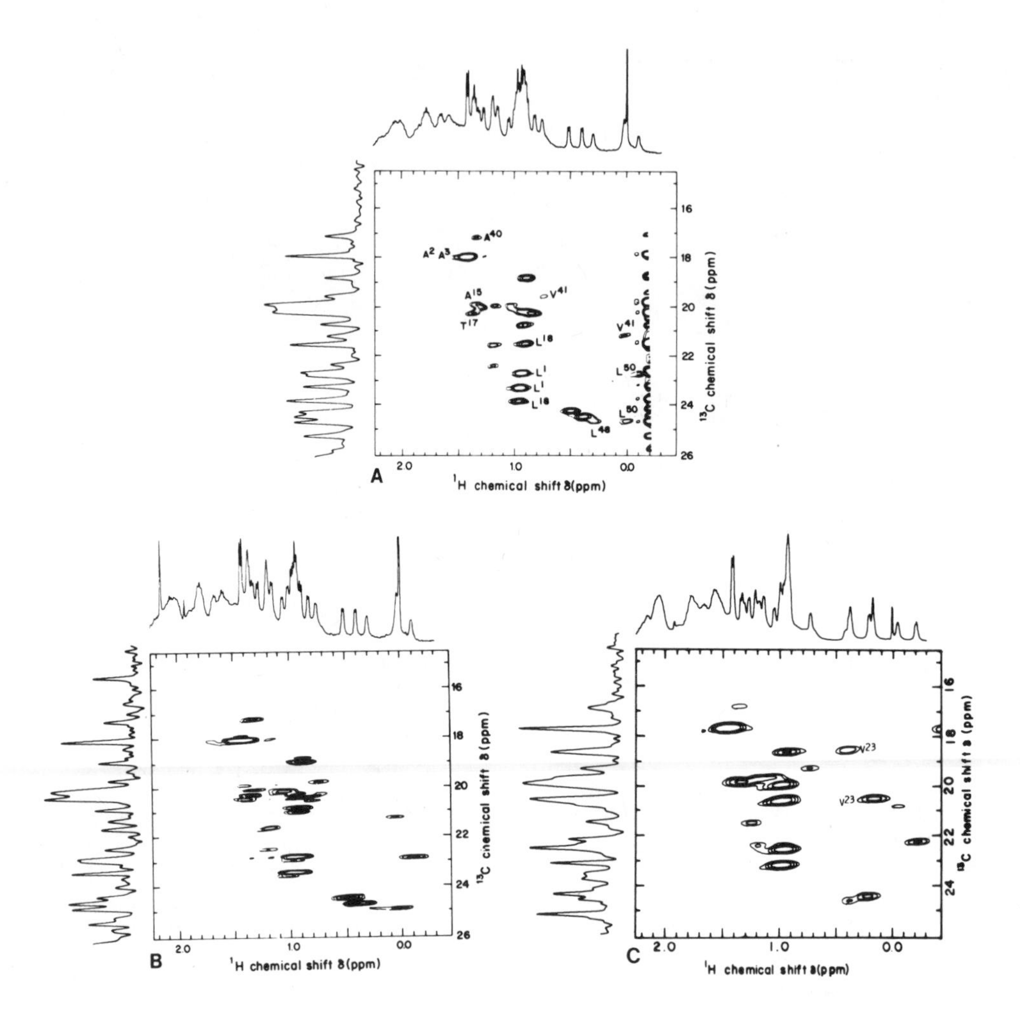

Fig. 10. *Natural abundance two-dimensional Fourier transform* $^{13}C\{^1H\}$ *chemical shift correlated spectra, obtained at 50 MHz for* ^{13}C *and 200 MHz for* ^1H*, of the methyl region of ovomucoid third domain from (a) turkey, (b) silver pheasant, and (c) Japanese quail. Along the left side of each contour plot is the corresponding 50.3 MHz* ^{13}C *NMR spectrum, and along the top is the corresponding 470 MHz* ^1H *NMR spectrum. The samples were dissolved in 2.5 ml of 0.2 M KCl in* ^2H_2O *to give final protein concentrations of 12 mM, OMTKY3; 15 mM, OMSVP3; 14 mM, OMJPQ3. The pH* of each sample was 8.0, and the probe temperature was approximately 25 °C. Contour peaks are labeled in (a) to show the* ^1H-^{13}C *cross assignments for the methyl groups of several amino acid residues in the proteins. The expected cross peak in (b) from the S-*CH_3 *of* Met^{18} *of OMSVP3 is not observed in the figure, but it shows up if a longer experimental delay time is used to prevent saturation. Comparison of contour plots (a) and (b) confirms the assignment of cross peaks to the methyls of* Leu^{18} *in OMTKY3 (OMSVP3 has* Met^{18}*). Comparison of contour plots (a) and (c) reveals the presence of additional cross peaks in the spectrum of OMJPQ3 that are assigned to the methyls of* Val^{23} *(OMTKY has* Leu^{23}*) and confirms the assignment of the methyl of* Thr^{17} *in spectrum (a) of OMTKY3 (OMJPQ has* Pro^{17}*). (From Markley et al. [37].)*

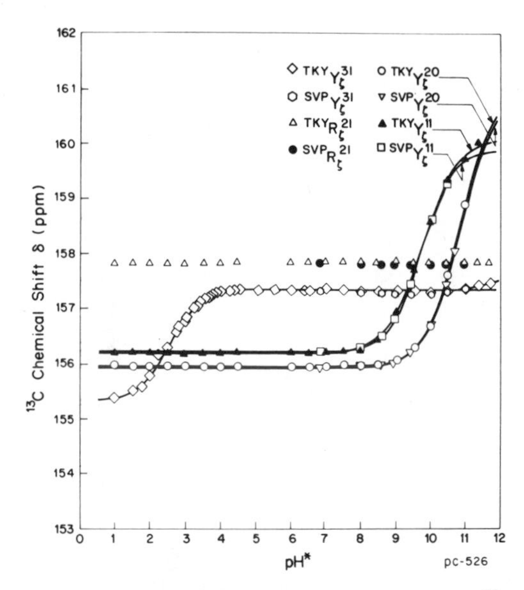

Fig. 11. *The pH titration curves obtained from ^{13}C NMR data at 50 MHz for the tyrosine residues of two ovomucoid third domains. The two proteins differ only at the reactive site residue: that from turkey (OMTKY3) has leucine-18; and that from silver pheasant (OMSVP3) has methionine-18. The chemical shifts of the side chain carbons of tyrosines 11, 20, and 31 are essentially unaffected by the replacement at residue 18. The data indicate that the pK_a' values of tyrosines 11 and 20 are the same within experimental error. The pK_a' value of tyrosine 31 is abnormally high (above pH 12.5) in both proteins. (From Ortiz-Polo [54].)*

An important difference exists between the effects of disorder or internal motions on NMR data and single crystal diffraction data. It frequently is difficult to distinguish between static disorder and dynamic effects when reduced electron density is observed in X-ray diffraction studies. Where diffraction peaks are resolved, one can assume that they represent the average positions of the atoms, and their temperature factors can be interpreted in terms of molecular dynamics. On the other hand, the sharpest peaks in the NMR spectrum of a protein arise from the regions of highest mobility, but NMR parameters such as coupling constants and the NOE, which give reliable structural information for a rigid molecule, lose their precision for flexible peptides because the observed values do not necessarily represent the average structure [2]. Possible errors inherent in an NMR structural analysis can be modeled so that a family of structures consistent with the NMR data is displayed. Although particular NMR experiments may yield quite precise structural information about interatomic distances, a high-resolution X-ray analysis is expected to give a more accurate picture of the atomic coordinates. NMR holds the advantage for certain ligand-binding studies and investigations of conformational equilibria in proteins. Ideally, the advantages of both methods should be exploited to achieve a more complete understanding of protein structure and function.

OUTLOOK

Further advances in NMR instrumentation and methodology are on the horizon. The improved sensitivity of newer NMR spectrometer designs will be a critical factor for practical applications of two-dimensional and multiple-quantum NMR approaches. One may predict confidently that the analysis of NMR data will become more computerized and more standardized. The trend is toward transfer of raw data to a powerful general-purpose computer, and computer programs written in standard languages such as FORTRAN and PASCAL are becoming available for 1D and 2DFT NMR. Initial results suggest that pattern-recognition algorithms may make the extraction of NMR parameters from 2DFT spectra faster and more reliable. NMR spectroscopy of nuclei other than the proton has had an explosive growth in recent years. This trend is likely to continue as it is fueled by more sensitive NMR instrumentation, improved methods for detecting less sensitive nuclei such as ^{13}C and ^{15}N, and more creative use of the methods of molecular biology for producing large quantities of protein and for introducing NMR active nuclei in specific locations.

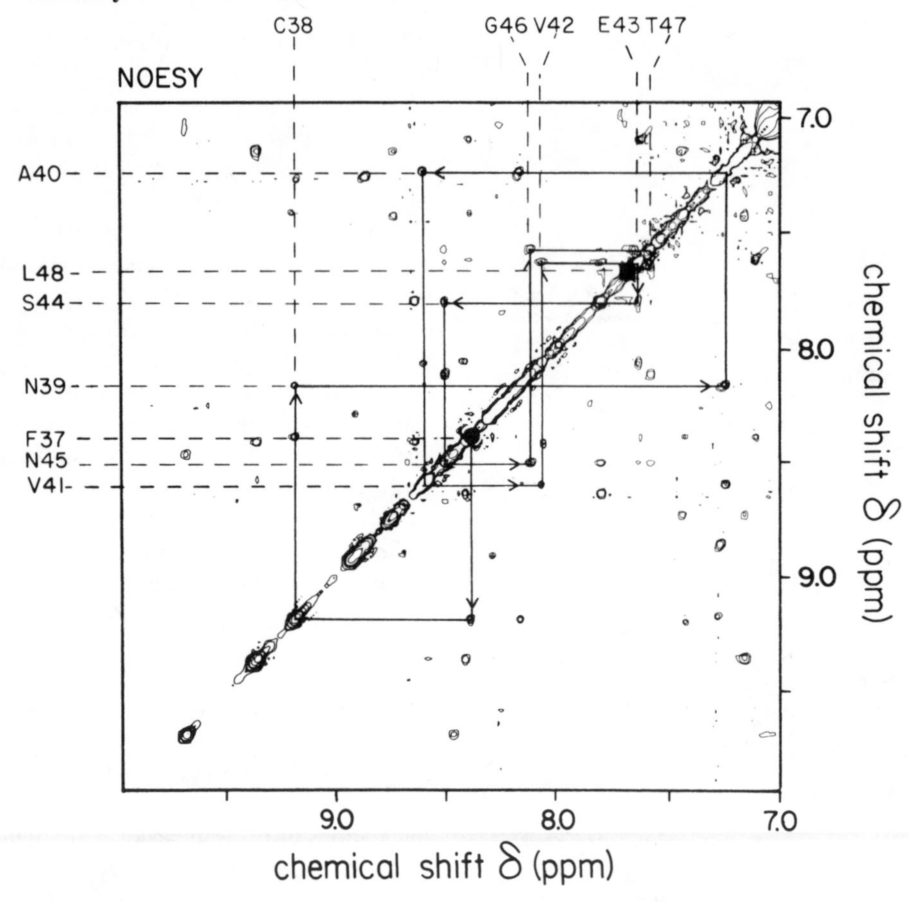

Fig. 12. *A portion of the phase-sensitive NOESY of 15 mM turkey ovomucoid third domain obtained under the conditions described in Fig. 9. This region contains cross peaks from pairs of peptide N-*H *that are close to one another in space. A sequence of cross peaks from N-*H *in adjacent residues is expected for an α-helix. This map shows a sequence of N-*H*-N-*H *connectivities spanning residues 37–48. This is in agreement with the crystal structure of turkey ovomucoid third domain which shows that residues 33–45 are helical [41]. (From Robertson [42].)*

REFERENCES

1. Markley JL, Ulrich EL (1984): Ann Rev Biophys Bioeng 13:493.
2. Jardetzky O, Roberts, GCK (1981): "NMR in Molecular Biology," New York: Academic Press.
3. Shaw D (1976): "Fourier Transform NMR Spectroscopy." Amsterdam: Elsevier.
4. Bax A (1982): "Two Dimensional Nuclear Magnetic Resonance in Liquids." Amsterdam: Reidel.
5. Bax A, Lerner L (1986): Science 232:960.
6. Nagayama K, Wüthrich K, Bachmann P, Ernst RR (1977): Biochem Biophys Res Commun 78:99.
7. Chan T-M, Westler WM, Santini RE, Markley JL (1982): J Am Chem Soc 104:4008.
8. Nagayama K, Kumar A, Wüthrich, Ernst RR (1980): J Magn Reson 40:321.
9. Chan T-M, Markley JL (1983): Biochemistry 22:5996.
10. Kojiro CL, Markley JL (1983): FEBS Lett 162:52.
11. Westler WM, Ortiz-Polo G, Markley JL (1984): J Magn Reson 58:354.
12. Live DH, Kojiro CL, Cowburn D, Markley JL (1985): J Am Chem Soc 107:3043.
13. Ortiz-Polo G, Krishnamoorthi R, Markley JL, Live DH, Davis DG, Cowburn D: (1986) J Magn Reson 68:303.
14. Wagner G, Zuiderweg, ERP (1983): Biochim Biophys Acta 113:854.
15. King G, Wright PE (1983): Biochem Biophys Res Commun 106:559.
16. Wagner G (1983): J Magn Reson 55:151.
17. Bax A, Griffey RH, Hawkins BL (1983): J Magn Reson 55:301.

18. Kumar A, Ernst RR, Wüthrich K (1980): Biochem Biophys Res Commun 95:1.
19. Body, J, Dobson CM, Redfield C (1983): J Magn Reson 55:170.
20. Santos H, Turner DL, Xavier AV (1984): J Magn Reson 58:344.
21. Dobson CM, personal communication; see Fox RO, Evans PA, Dobson CM (1986): Nature 320:192.
22. Kalk A, Berendsen HJC (1976): J Magn Reson 24:343.
23. Braunschwieler L, Ernst RR (1983): J Mag Reson 53:521.
24. Davis DG, Bax A (1985): J Am Chem Soc 107:2820.
25. Shaka AJ, Freeman R (1983): J Magn Reson 51:169.
26. Levitt MH, Ernst RR (1985): J Chem Phys 83:3297.
27. Kainosho M, Tsuji T (1982): Biochemistry 21:6273.
28. Wüthrich K, Wider F, Wagner G, Braun W (1982) J Mol Biol 155:311.
29. Wand AJ, Englander SW (1985): Biochemistry 24:5290.
30. Weber PL, Wemmer DE, Reid BR (1985): Biochemistry 24:4553.
31. Markley JL (1972): Methods Enzymol 26:605.
32. Anglister J, Frey T, McConnell HM (1984): Biochemistry 23:5372.
33. LeMaster DM, Richards FM (1985): Biochemistry 24:7263.
34. Markley JL, Westler WM, Chan T-M, Kojiro CL, Ulrich EL (1984): Fed Proc 43:2648.
35. Jardetzy O, Markley JL (1970): Il Farmaco, Ed Sci 25:894.
36. Alexandrescu AT, Ulrich EL, Grissom CB, Mills D, Shortle D, Markley JL (1986): Fed Proc 45:1921.
37. Markley JL, Croll DH, Krishnamoorthi R, Ortiz-Polo G, Westler WM, Bogard WC, Laskowski M Jr. (1986): J Cellular Biochem 30:291.
38. Laskowski M Jr, Empie MW, Kato I, Kohr WJ, Ardelt W, Bogard WC Jr, Weber E, Papamokos E, Bode W, Huber R (1981): In Eggerer H, Huber R (eds): "Structural and Functional Aspects of Enzyme Catalysis," vol 32, Berlin, Heidelberg: Springer Verlag, pp 136–152.
39. Laskowski M Jr, Kato I, Ardelt W, Cook J, Denton A, Empie MW, Kohr WJ, Park SJ, Parks K, Schatzley DL, Schoenberger OL, Tashiro M, Vichot G, Whatley HE, Wieczorek A, Wieczorek M (1987): Biochemistry 26:202.
40. Bode W, Epp O, Huber R, Laskowski M Jr, Ardelt W (1985): Eur J Biochem 147:387.
41. Fujinaga M, Read RJ, Sielecki A, Ardelt W, Laskowski M Jr, James MNG (1982): Proc Natl Acad Sci USA 79:4868.
42. Robertson AD, Markley JL: unpublished data.
43. Baillargeon MW, Laskowski M Jr, Neves DE, Porubcan MA, Santini RE, Markley JL (1980): Biochemistry 19:5703.
44. Wagner G (1983): Quart Rev Biophys 16:1.
45. Markley JL (1975): Accts Chem Res 8:70.
46. Ortiz-Polo G, Markley JL: unpublished data.
47. Bystrov, VF (1976): Prog NMR Spectr 10:41.
48. Noggle JH, Shirmer RE (1971): "The Nuclear Overhauser Effect." New York: Academic Press.
49. Pardi A, Billeter M, Wüthrich K (1984): J Mol Biol 180:741.
50. Wüthrich K, Billeter M, Braun W (1983): J Mol Biol 169:949.
51. Braun W, Bösch C, Brown LR, Gō N, Wüthrich K (1981): Biochim Biophys Acta 667:377.
52. Havel TR, Crippen GM, Kuntz ID (1979): Biopolymers 18:73.
53. Kaptein R, Zuiderweg ERP, Scheek RM, Boelens R, van Gunsteren WF (1985): J Mol Biol 182:179.
54. Ortiz-Polo G (1985): MS Thesis, Purdue University.

Protein Engineering, pages 35–44

3

The Prediction and Analysis of Mutant Structures

Martin Karplus

Department of Chemistry, Harvard University, Cambridge, Massachusetts 02138

INTRODUCTION

It is now possible to make mutant and modified proteins almost at will. The essential question is which ones to make. To choose one would like to be able to predict the properties of the modified protein before trying to prepare it by synthetic or molecular biological techniques. Unfortunately, predicting the properties of a modified protein is a very difficult task, even if much is known about the unmodified or "parent" proteins. In the short history of site-directed mutagenesis, there are already cases where the mutant behaved very differently from the expectations of the person who made it. It is important, therefore, to develop predictive strategies and to use all the available information—structural, thermodynamic, biochemical—in any attempt at a prediction.

Experience with native and naturally occurring mutant proteins makes clear that a prediction of the structure of the modified protein is a prerequisite to predicting properties. This must be based on knowledge of the structure of the parent, since it is not possible to predict the secondary and tertiary structure of a protein from the sequence alone, particularly when one realizes that to predict the function or the changes in function requires structural results of high accuracy. In determining experimental structures, high-resolution X-ray crystallography has played and continues to play a paramount role (see Chapter 1 by W. Hendrickson on X-ray Diffraction, this volume). Once a modified protein has been made, X-ray analysis is the best approach to its structure, though nuclear magnetic resonance and other spectroscopic techniques can be useful, particularly for finding out where the structure of the mutant differs from that of the parent. However, neither X-ray diffraction nor nuclear magnetic resonance can be employed until after the mutant has been made. It is the objective of this chapter to examine theoretical approaches to predicting the structure and properties of a mutant protein from a knowledge of the parent structure and of the specific amino acid changes in the mutant. The first part introduces the methodology, based on empirical energy functions and their utilization for minimization and molecular dynamics simulations of macromolecules. Applications of this methodology to mutant structure prediction are then described. Finally, some limitations of the present methodology and a prognosis for its utility are outlined.

THEORETICAL METHODOLOGY

To study theoretically the energetics, dynamics, and thermodynamics of a protein, one must have a knowledge of the protein potential energy surface, which corresponds to the energy of the molecule as a function of the atomic coordinates. The potential energy can be used directly to determine the relative stabilities of different possible structures of the protein. The forces acting on the atoms are obtained from the first derivatives of the potential energy with respect to the atomic positions. These forces can be used to calculate dynamical properties of the system, e.g., by solving Newton's equations of motion to determine how the atomic positions change with time. From the second derivatives of the potential surface, the force constants for small displacements can be evaluated and used to find the normal modes; this serves as the basis for an alternative approach to the dynamics in the harmonic limit. The first and second derivatives of the energy are also used for minimizing the energy of the protein or for finding its response to a perturbation.

Although quantum mechanical calculations can provide potential surfaces for small molecules, empirical energy functions of the molecular mechanics type are the only possible source of such information for proteins and their solvent surroundings. Most of the motions that occur at ordinary temperatures leave the bond lengths and bond angles of the polypeptide chains near the equilibrium values, which appear to vary little throughout a protein (e.g., the standard dimensions of the peptide group first proposed by Pauling et al., in 1951 [1]). Therefore, the energy-function representation of the bonding can be hoped to have an accuracy on the order of that achieved in vibrational analyses of small molecules. Where globular proteins differ from small molecules is that the contacts among nonbonded atoms play an essential role in the potential energy of the folded or native structure. From the success of the pioneering conformational studies of Ramachandran et al., in 1963 [2], which used hard-sphere nonbonded radii, it is likely that relatively simple functions (Lennard-Jones nonbonded potentials supplemented by electrostatic interactions) can adequately describe the interactions involved.

The energy functions used for proteins generally are composed of terms representing bonds, bond angles, torsional angles, van der Waals interactions, electrostatic interactions and hydrogen bonds. The resulting expression has the simple form shown in Eq. (1)

$$\begin{aligned} E(\vec{R}) = {} & \frac{1}{2}\sum_{bonds} K_b(b-b_0)^2 + \frac{1}{2}\sum_{\substack{bond\\angles}} K_\theta(\theta-\theta_0)^2 \\ & + \frac{1}{2}\sum_{\substack{torsional\\angles}} K_\phi[1+\cos(n\phi-\delta)] \\ & + \sum_{\substack{nb\ pairs\\ r<8\text{Å}}} \frac{A}{r^{12}} - \frac{C}{r^6} + \frac{q_1q_2}{Dr} \\ & + \sum_{\substack{H\\bonds}} \left(\frac{A'}{r^m} - \frac{C'}{r^p}\right) f(\theta',\theta'') \end{aligned} \tag{1}$$

The energy is a function of the Cartesian coordinate set $\vec{R}$, specifying the positions of all the atoms involved, but the calculation is carried out by first determining the internal coordinates for bonds (b), bond angles (θ), dihedral angles (ϕ) and the interparticle distances (r) for any given geometry, $\vec{R}$, and using them to evaluate the contributions to Eq. (1). They depend on the bonding energy parameters K_b, K_θ, K_ϕ, Lennard-Jones parameters A and C, atomic charges q_i, dielectric "constant" D, hydrogen-bond parameters A' and C', and geometric reference values b_0, θ_0, n, and δ. In the H-bond term, f(θ', θ'') represents a function of the donor (θ') and acceptor (θ'') angles and exponents m = 12 or 6 and p = 10 or 4 have been used. The earliest protein calculations made use of the extended atom representation in which one extended atom replaces a nonhydrogen atom and any hydrogen bonded to it, to reduce the size of the problem by reducing the number of atoms considered (i.e., approximately half of the protein atoms are hydrogens). Most present calculations treat hydrogen-bonding hydrogens explicitly, and in some cases all hydrogen atoms are included. Also, the hydrogen bonding term in Eq. (1) is sometimes omitted and its effect is introduced by use of the appropriate charges on the atoms involved. Certain energy functions, particularly those applied to small molecules, introduce terms in addition to those listed in Eq. (1); for example, coupling between bond length and bond angle changes may be included.

Energy Minimization

Given a potential energy function, it is often desirable to find low-energy (relatively stable) configurations of a system. In some cases, minimization is performed to relieve strain in conformations obtained experimentally or by the averaging of several structures. In other cases, finding a local or global energy minimum may be of primary interest, e.g., for determining the configurations of a peptide. For macromolecular systems, the number of local minima and the cost of the computations prevent exhaustive search of the energy surface, so it is frequently impossible to determine the global energy minimum; generally, a local minimum in the neighborhood of the X-ray structure, if available, is examined.

The simplest minimization algorithm is steepest descents (SD). In this procedure, a displacement opposite to the potential energy gradient is added to the coordinates at each step. The step size is increased if a lower energy results; otherwise, it is decreased. Although this method suffers from poor convergence, the positional shifts are gentle; it is therefore useful for introducing small changes such as removing bad contacts. Other minimization algorithms with better convergence properties are often used. Some of these are the conjugate gradient technique, the Newton-Raphson method with its many variants, and quenched dynamics. A brief discussion of these approaches and a description of some simple applications are given in reference [3].

Molecular Dynamics

Molecular dynamics is a theoretical method for studying the internal motions of proteins [4]. Although the classic view of proteins has been static in character, it is now recognized that the atoms of which a protein is composed are in a state of constant motion at ordinary temperatures. The presence of such motional freedom implies that a native protein at room temperature samples a range of conformations. Most are in the general neighborhood of the average structure, but at any given moment an individual protein molecule is likely to differ significantly from the average. This in no way implies that the X-ray structure, which corresponds to the average in the crystal, is not important. Rather, it suggests that fluctuations about the average are likely to play a role in protein function.

The dynamic picture of proteins incorporates a variety of phenomena known to be involved in their biological activity, but whose detailed description was not possible under the static view. An example is provided by the transient packing defects due to atomic motions which play an essential role in the penetration of oxygen to the heme-binding site in myoglobin and hemoglobin. Further, the structural changes in proteins regulate the activity of many of these molecules through induced fit and allosteric effects. The chemical transformations of substrates by enzymes also typically involve significant atomic displacements in the enzyme substrate complexes. Electron transfer processes may depend strongly on vibronic coupling and fluctuations that alter the distance between the donor and acceptor.

In a molecular dynamics simulation, the energy function given in Eq. (1) is used to obtain the forces on the protein atoms by taking the appropriate derivatives; i.e., the components of the force on atom i (F_x^i, F_y^i, F_z^i) are given by

$$F_x^i = -\frac{\partial E}{\partial x_i};\ F_y^i = -\frac{\partial E}{\partial y_i}; \qquad F_z^i = -\frac{\partial E}{\partial z_i}\ (i = 1, 2, \ldots, N) \tag{2}$$

where N is the number of atoms included in the simulation. By Newton's second law of motion, the force on an atom produces an acceleration (a_x^i, a_y^i, a_z^i),

$$F_x^i = m_i a_x^i = m_i \frac{d^2x_i}{dt^2};$$
$$F_y^i = m_i a_y^i = m_i \frac{d^2y_i}{dt^2}; \qquad (3)$$
$$F_z^i = m_i a_z^i = m_i \frac{d^2z_i}{dt^2}$$
$$(i = 1, 2, \ldots, N)$$

where m_i is the mass of atom i and (x_i, y_i, z_i) is its position at time t. In a molecular dynamics simulation, Eq. (3), with the forces given by Eq. (2), is integrated simultaneously for all of the atoms in the system. After a small time step Δt, the new positions are found to be $(x_i+\Delta x_i, y_i+\Delta y_i, z_i+\Delta z_i)$. By repeating this process, a trajectory is generated that gives the atomic positions as a function of time. Normally, a simulation is performed at a specified average temperature, T, defined by

$$T = \frac{1}{N_f \, k_B} \sum_{i=1}^{N_f} \frac{1}{2} m_i \langle v_i^2 \rangle$$

where $\langle v_i^2 \rangle$ is the average velocity squared of atom i, N_f is the number of degrees of freedom of the system, and k_B is Boltzmann's constant.

APPLICATIONS

Energy minimization and dynamics have been applied to a wide range of problems that arise in the prediction and analysis of the structure and function of proteins. In what follows, we describe a few applications that are concerned particularly with questions raised by mutant or modified proteins.

Prediction of the Structure of a Single Mutant

For single amino acid substitutions, homology modeling of proteins is reduced to the simplest question: given two proteins that differ by a single amino acid, can one predict the structure of the mutant from the known (X-rays) structure of the parent? A procedure that is commonly used in homology modeling would introduce the new side chain in some arbitrary fashion, usually following the parent dihedral angles as far as possible, and then energy-minimize the entire structure with Eq. (1) to obtain a prediction for the mutant protein. An alternative approach also makes use of the empirical energy function but is closer in spirit to the model-building procedure that is used in structure determinations by X-ray crystallography. A search of the local conformational space of the substituted amino acid is made to find all structures that are of relatively low energy; each of these structures is then refined by constraint energy minimization, and the results are evaluated to determine the most likely structure(s). Such an approach is based on one assumption and one requirement. The assumption is that the conformation of a stable protein resulting from a single amino acid substitution closely resembles that of the parent protein, with any tertiary structural changes localized in the neighborhood of the substitution. Support of this assumption is provided, for example, by the X-ray structures for mutant hemoglobins and T4 lysozymes. The requirement is that the approximate empirical energy functions now available be sufficiently accurate to determine the conformation of a side chain in the presence of the interactions with the known structure of the rest of the protein. This question has been examined in a study of the bovine pancreatic trypsin inhibitor [5]. A rigid search of the full dihedral angle space for each individual side chain resulted in the correct positioning of the interior side chains (most χ angles within $\pm$ 10°); exposed residues tended to be less well defined and, in some cases, could only be localized by including the crystal environment in the calculation.

The perturbation approach has as its first step the verification that the original amino acid, which has been altered in the mutant, is in a minimum energy position in the parent structure. This is done by the same dihedral search procedure as used in the bovine pancreatic trypsin inhibitor. Assuming that the parent minimum is verified, the mutant amino acid is introduced into the parent sequence, and a grid search of its side-chain dihedral-angle space is made in the presence of the potential energy of interaction with the unaltered part of the parent structure. This initial search is restricted to varying the side-chain dihedral angles of the mutant amino acid, keeping all of its bond lengths and angles fixed and not moving any other protein atoms. Since only a small number of nonbonded interactions have to be calculated (i.e., those between the moving atoms of the amino acid and the rest of the protein), this search is very fast.

The multidimensional potential surface generated in this way is examined for low-energy regions. If one or more regions have energies that suggest a satisfactory structure (i.e., energies similar to those of the parent protein), further analysis is restricted to them. If the search has yielded only high-energy structures, most likely due to unfavorable van der Waals contacts, more extended explorations of the energy surface, including variations in the surrounding side-chain dihedral angles and the positions of local backbone atoms, may be required to obtain satisfactory starting points for the next step. This may be accomplished by constrained energy minimization, the constraints being introduced to keep the portions of the protein that are assumed to be unaltered by the perturbation close to the known structure. The search and minimization procedure may yield several reasonable structures. At best there will be one clearly defined lowest energy structure. Instead there may be several minima that are difficult to distinguish by the available methods; their existence may suggest that more than one side-chain conformation can occur.

An example of the minimum perturbation approach is provided by a blind study of an antigenic variant, V3a, of the HA glycoprotein of the influenza virus [6]. In this variant an aspartic acid has replaced a surface glycine at position 146 of the parental (X31) sequence, whose X-ray structure had been determined at 3.0-Å resolution [7]; residue 146 is part of an exposed loop region (positions 140–146) that has been implicated as an antigenic site. Since the parent sequence has a glycine in the position of interest, no preliminary search was made to verify its structure. The aspartic acid was introduced into a partly refined model of the parent structure, and the two dihedral angles (χ_1,χ_2) of the mutant side chain were varied systematically at 10° intervals. For each side-chain position, the energy was calculated. The resulting potential energy map (see Fig. 1) has a well-defined low-energy region for $-120° \leqslant \chi_1 \leqslant -150°$ and $50° \leqslant \chi_2 \leqslant 150°$. The rest of the potential surface is much higher in energy.

The low-energy region in Figure 1 has its minimum at $\chi_1 = -145°$ and $\chi_2 = 130°$. Starting with this structure energy minimization yielded a final structure with $\chi_1 = -162°$ and $\chi_2 = 153°$. In Figure 2 the minimized structure and the X-ray structure ($\chi_1 = -129°$, $\chi_2 = 128°$) [8] are shown as cross-eyed stereo diagrams. The minimization suggested no significant changes in other residues positions, except for a small rotation of Arg 141 that improves the hydrogen bond with Asp 146. Such a displacement of Arg 141 was in fact found in the X-ray structure when it was reexamined to determine whether the predicted change was visible. No other features are apparent in the experimental difference map; this is in accord with the minimum perturbation hypothesis.

Similar calculations have been made for other mutants (e.g., trypsin Asp 102 → Asn 102), and it appears that the technique described here is of general applicability, although it is approximate and therefore does not necessarily yield correct results.

Prediction of Loop Conformations

Some of the simplest projects in the engineering of new proteins consist of taking certain secondary structure elements (e.g., α-

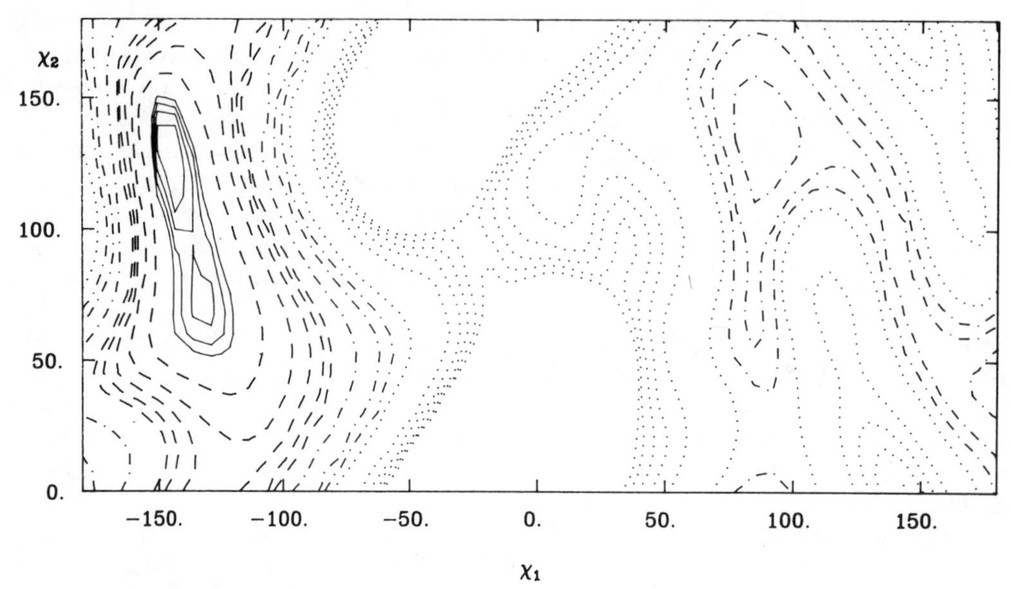

Fig. 1. *Potential energy map for variation of χ_1 and χ_2 of Asp-146; the remainder of the protein was kept rigid at the parent X-ray structure. The figure represents the dihedral-angle energy of the Asp-146 side chain plus its van der Waals, electrostatic, and H-bonding energy and interaction with the rest of the molecule. A 10-Å distance cut-off from C $^{\alpha}$ of Asp-146 was used for the calculation. The energy contours are relative to the minimum energy geometry as zero. Contours: —, at 0.5, 1.5, 2.5, and 3.5 kcal (1 cal = 4.184J); --, at 10, 20, 30, 40, and 50 kcal; -·-·-, at 100, 200, and 300 kcal; and ···, at 500, 2,300, 4,100, ... 9,500 kcal.*

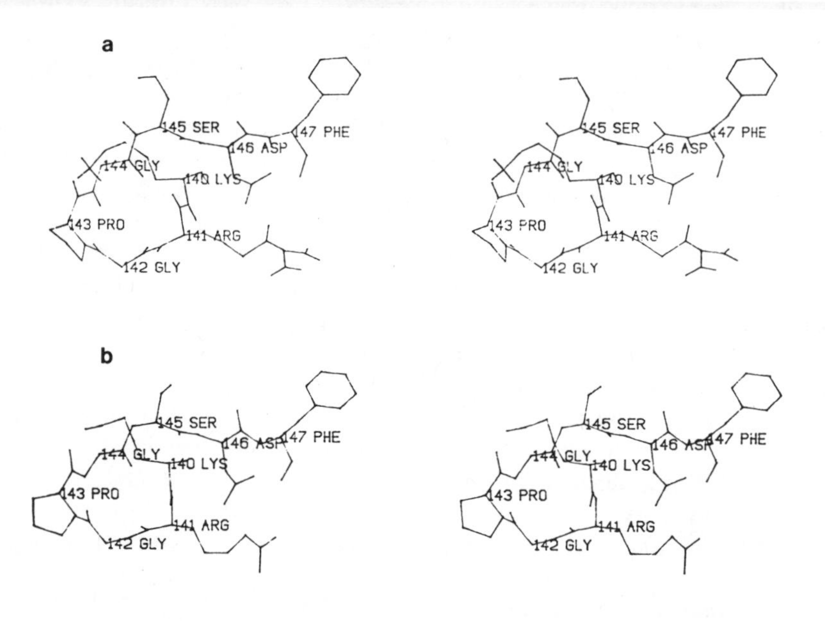

Fig. 2. *Comparison minimized and X-ray structure of Asp-146 mutant in cross-eyed stereo. a) Predicted structure [6] with polar hydrogens included. b) X-ray structure [8].*

helices) and connecting them by loop regions to construct a stable structure. One such idea involves "synthesis" of a four-helix bundle analogous to some naturally occurring proteins like the hemerythrins [9]. One aspect of the construction is the determination of what residues should be chosen for the loop regions. For this purpose one possible approach is to look at known structures and simply pick out a sequence that has been observed to occur with the desired structural attributes. An alternative to this approach is to examine a number of amino acid sequences and attempt to predict their most likely conformations, i.e., to determine whether the desired conformation is one that is likely to be stable. More generally, in trying to predict the structure of homologous proteins, one of the difficult questions concerns the loop regions that are likely to involve nonhomologous amino acids (as well as additions and deletions), even in highly homologous structures. A case where this difference is extreme and essential to the function is in the hypervariable regions of immunoglobulins. Here the specificity for different antigens is determined by the variation in amino acid sequence and the diversity of the resulting stable structures.

An approach to the prediction of the structures of such loop regions consisting of three to ten or so amino acids employs a search procedure analogous to, but considerably more complicated than, that employed for the side-chain prediction described above. The essential idea is to assume that a framework exists which specifies the positions of the two ends of the loop region and to attempt to predict for a given amino acid sequence the most likely (most stable) conformation to bridge the gap. This requires that one determine the lowest energy main-chain dihedral angle set for the residues (i.e., the values, ϕ_1, ψ_1; ϕ_2, ψ_2, . . . ϕ_n, ψ_n for the n amino acids), as well as the dihedral angles for their side chains; deviations of the peptide bond from planarity (the angles ω_i) could be included as well. This is a much more difficulty problem than that for a single side chain; e.g., a loop of six amino acids has 12 main-chain angles and, on the average, 10 side-chain dihedral angles. A complete search of such a 22-dimensional space is clearly prohibitive. With the 10° interval used for the side-chain predictions (see above) it would involve $(36)^{22} \cong 10^{18}$ energy evaluations, an impossibly large number. To reduce the problem to manageable proportions and yet perform a search that is adequate to sample the available conformational space, it is necessary to introduce ways of eliminating the portions of space that need not be searched. This can be achieved by using a tree search, in which angles are searched successively, the energies of the structures are calculated as they are constructed, and the branches that give too high energies or lead too far away from the end to permit loop closure are not followed [10]. In addition, the search time can be reduced by using a relatively coarse grid (30° instead of 10°). Tests are needed to justify such an approach (i.e., to show that the cut-off criteria and grid size choices are valid). These have been carried on known structures with an automated algorithm with no human intervention; only in this way can a known structure be used to test a predictive scheme. Table I shows results obtained for a number of five residue segments of several proteins; also one seven residue segment is included to illustrate how rapidly the number of possible structures increases with chain length. What was done was to use the known structure of the protein to specify the beginning and end of the loop position and to use the search algorithm to find possible structures to bridge the gap. Table I lists the number of structures found that satisfied the cut-off criteria. These structures were compared with the X-ray structure. The root mean square (rms) difference between each loop that was generated and the X-ray structure of the loop was calculated. Examination of the table shows that the results are encouraging. The structure with the smallest rms is always within 1 Å of the X-ray structure, indicating that the search method is satisfactory. As to the energy criterion for selecting among the possible structures, it is

TABLE I. Segment Prediction Algorithm Results[a]

Protein	Second structure	No. of structures	Rm[b]		Min energy rms[c]	
			Min	Max	Lowest	5 Lowest
Flavodoxin	α-Helix	12	0.841	1.428	0.944	0.841
Plastocyanin	β-Sheet	7	0.938	2.925	0.938	0.938
Fab Kol	Turn	18	0.766	3.497	2.240	1.054
	Turn (7 residues)	1,138	0.847	5.551	0.951	0.847

[a]The search was made with a 30° grid, a van der Waals energy cut-off of 100 kcal, and 2 kcal cut-off on the ϕ, ψ map for the particular residue type.
[b]Rms values (in Å) are given here for the structures in the generated set that were closest to (min) and furthest from (max) the X-ray structure.
[c]Listed are the rms deviation of the structure with the lowest energy (Lowest) and the smallest rms deviation found among the five lowest energy structures (5 Lowest).

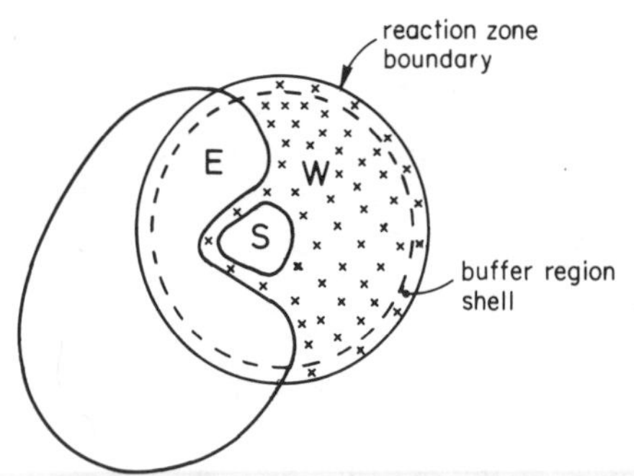

Fig. 3. *Schematic partitioning of an enzyme (E)—substrate (S)—water (W) system into a spherical reaction zone and the surroundings. A subdivision of the reaction zone into a large inner volume (reaction region) and a surrounding shell (buffer region) is also indicated.*

approximately valid in that the five lowest energy structures usually include the best possible one (the five-residue segment of Fab Kol is an exception), although it is not generally true that the lowest energy structure is closest to the X-ray result.

Molecular Dynamics of Solvated Regions

For a variety of studies of protein structure and function, it is essential to include the solvent in the simulation. This is particularly true for the active sites of enzymes that are exposed to solvent, as well as for free energy studies of drug binding and mutant stabilities. Although it is possible to do a molecular dynamics simulation of a protein in aqueous solution, including the solvent water molecules explicitly, such a calculation is very time consuming because of the large number of atoms (protein atoms and water atoms) involved. As an alternative, a stochastic boundary molecular dynamics method has been developed [11,12]. The stochastic boundary method is based on a partitioning of the system into several regions. This partitioning is illustrated schematically for an enzyme-substrate system in Figure 3. There are two major regions: a reaction zone where all atoms are treated explicitly, and the surrounding region composed of atoms distant from the volume of primary dynamical interest. Atoms in this region are eliminated from the calculation, and their effect on atoms in the reaction zone is replaced by explicit boundary forces and stochastic terms. The reaction zone contains the site of major dynamical interest; e.g., for an enzyme the reaction zone is chosen large enough to include the substrate, the residues that interact with the substrate, and some surrounding residues. It is subdivided into a large inner reaction region and a buffer region (typically a 2-Å shell) near the boundary. The reaction atoms are treated by standard molecular dynamics. The buffer region atoms represent a heat bath and are treated as Langevin particles subject to the same empirical energy function as the reaction region particles; for protein buffer atoms, a constraining potential is added to obtain fluctuations in accord with independent estimates. The buffer region allows energy fluctuations to occur in the reaction region and thus should approximate the many-body dissipative effects that would be produced by the surroundings in the complete system. The partitioning of the protein atoms into different regions is kept fixed throughout the simulation. The water molecules are confined to the reaction zone by an appropriate boundary force, but they are allowed to diffuse freely in it; e.g., they can enter into or exit from the buffer region.

The stochastic boundary approach greatly reduces the number of atoms that have to be included in the simulation relative to that required for a full protein in solution. The gain in computing time is expected to be a factor somewhere between 10 and 100, depending on the specific system. The simulation method has been applied to a series of ribonuclease-substrate-complex structures which correspond to the important steps of the reaction catalyzed by ribonuclease A [13]. The atomic fluctuations of the protein atoms relative to their average positions were found to be in good qualitative agreement with temperature factors determined by X-ray crystallography. Correlated fluctuations of protein atoms confirmed some of the interactions postulated in the catalytic mechanism of ribonuclease A. The stabilization of a set of positively charged groups in the active site by a network of water molecules was observed.

The application of the stochastic boundary method to the active site of ribonuclease A and its substrate bound complexes represents only one example of this approach to the dynamics of macromolecules. It is likely that the method will be a useful complement to the minimum perturbation approach for the determination of mutant structures. Of particular interest is the possibility of using this type of simulation for evaluating the free energy difference between the parent and the mutant protein.

CAUTIONS AND PROGNOSIS

A number of methods for the theoretical analysis of the structure and function of proteins have been outlined and some illustrative applications have been presented. It is important to remember that the methods are approximate in that the energy functions are empirical and that the minimizations, searches, and dynamics are limited by the large amounts of computer time required. This means that any results obtained by such calculations have to be regarded with a certain skepticism. Nevertheless, theoretical studies can be extremely useful if employed in the right way. Because proteins are very complex systems, it is difficult to look at them effectively without aids of various types. One of these aids is provided by molecular graphics. Another aid is the information obtained from empirical energy function calculations. There are many worthwhile things that one can do with molecular graphics and empirical energy function techniques (some of these have been described in this chapter), but it is necessary to remember that they are only aids to our thinking; if one does not think, nothing useful will come out of the methods.

As to the future, it is apparent that a wide range of biological problems involving proteins, not to mention nucleic acids and membrane lipids, are ready for study, and exciting new results can be expected as the empirical energy function methods are applied to them. In the coming years, we shall learn how to calculate meaningful rate constants for enzymatic reactions, ligand binding, and many other biologically important processes. The effects of changes in amino acid sequences on the structure, dynamics, and thermodynamics of protein will be determined by a combination of experimental and theoretical methods. As the predictive powers of the theoretical approaches increase, applications will be made to practical problems arising in areas such as genetic engineering and industrial enzyme technology.

REFERENCES

1. Pauling L, Corey RB, Bronson HB (1951): Proc Natl Acad Sci USA 37:205–213.
2. Ramachandran GN, Ramakrishnan C, Sasisekharan V (1963): J Mol Biol 7:95–99.
3. Brooks B, Bruccoleri RE, Olafson BD, States DJ, Swaminathan S, Karplus M (1983): J Comp Chem 4:87–217.
4. Karplus M, McCammon JA (1983): Ann Rev Biochem 53:263–300.
5. Gelin B, Karplus M (1979): Biochemistry 18:1256–1268.
6. Shih HH-L, Brady J, Karplus M (1985): Proc Natl Acad Sci 82:1697–1700.
7. Wilson IA, Skekel JJ, Wiley DC (1981): Nature (London) 289:368–373.
8. Knossow M, Daniels RS, Douglas AR, Skekel JJ, Wiley DC (1984): Nature (London) 311:678–680.
9. Sheriff S, Hendrickson WA, Stenkamp RE, Sieker LC, Jensen Lh (1985):
10. Bruccoleri RE, Karplus M "Biopolymers" (to be published).
11. Brooks CB III, Karplus M (1983): J Chem Phys 79:6312–6325.
12. Brünger A, Brooks CL III, Karplus M (1984): Chem Phys Lett 105:495–500.
13. Brünger A, Brooks CL III, Karplus M (1985): Proc Natl Acad Sci USA (in press).

II
BIOLOGICAL AND BIOCHEMICAL METHODS IN PROTEIN MODIFICATION

Protein Engineering, pages 47–49

Overview: Section II

Christopher T. Walsh

Chemistry and Biology Departments, Massachusetts Institute of Technology, Cambridge, Massachusetts 02139

This volume deals with the revolution under way in the study of protein structure–function relationships and the approaches available for design or redesign of primary sequences of proteins and the analyses of the effects of those purposeful changes. Other sections of the book deal with representative specific proteins that have been and are being reengineered for changes in catalytic, regulatory, or stability properties. The chapters in this section summarize current knowledge on construction of altered protein encoding DNA sequences, on the protein-folding problem, and on strategies for protein purification.

DESIGN AND REDESIGN FOR GENES ENCODING PROTEINS

Three chapters in this section address the three modern complementary methodologies available for generation of site-specific or regionally modified proteins. Rossi and Zoller provide an overview of approaches for high-yield site-directed mutagenesis by oligonucleotide methods, by far the most widely used approach over the past five years for alteration of a specific residue in a protein coding primary sequence to any of several other amino acids. (With "cassette" oligonucleotide mutagenesis one can generate a set of mutant proteins each with any of the 20 amino acids used in protein biosynthesis at a specific site in a target protein. This chapter also examines methods for deletion or insertion of a desired stretch of amino acids in a given protein region. Such deletion studies are very useful in multifunctional proteins to assess the roles of domains of the protein. For example, alanyl-t1RNA synthetase has recently been shown to have distinct catalytic activities linearly arranged in functional arrays along the 800-plus amino acid length of its primary sequence [1]. Rossi and Zoller note the development under way of nonsense-suppressing t-RNAs to put any of the 20 amino acids at a given codon in a protein. This will be an alternative to synthetic DNA-mediated site-specific mutagenesis.

The chapter by David Shortle is a complementary one focusing not on one at a time directed replacements at a given amino acid residue, but on genetic strategies to find sets of phenotypic mutations; that is, subsets of mutations with effects on some specific protein function being assayed. Shortle poses the oft-asked question: How does one analyze the way a one-dimensional linear protein sequence specifies a precise three-dimensional sequence and folding arrangement? The efforts to develop quantitative rules for those relationships remain a central goal of protein chemistry.

In contrast to the localized operations of site-directed mutageneisis, the phenotypic mutational approach is in effect a systems analysis via introduction of stable perturbations in structure by single-base substitutions in the encoding DNA at (up to) every nucleotide position. The power of this approach to the protein structure–function equation is proved by work on the lac repressor and on staphylococcal nuclease where dozens to hundreds of mutations have been isolated, sequenced, and various functional properties categorized. Of course, what you find depends on what functional change you assay in the phenotypic selection and this can be altered at any stage. Shortle also notes the utility of second site suppression or reversion analysis, long a staple of the geneticist, to factor out allele-specific vs. global suppressor sites. For example, two global suppressors are now known in lambda repressor that lead to higher affinity of the mutant repressors for operator DNA compared to wild-type protein [2]. Shortle has identified a global suppressor site in *Staphylococcus* nuclease which imparts greater stability against denaturation of native structure. The double-mutant approach is a powerful one to assess roles of residues in protein functions, as elegantly demonstrated by the studies of Fersht and Leatherbarrow, noted elsewhere in this volume.

In the third chapter Caruthers focuses on DNA alterations and the explosive developments in DNA synthesis whereby one can not only readily make oligonucleotides of 15–30 length for site-directed mutagenesis, as noted above, but can also achieve total synthesis of genes (400–1,000) nucleotides) as an operational reality. If the first activity is protein redesign, the second activity is protein design. Caruthers quotes the reaction of a commentator to the Khorana group's initial chemical synthesis of a gene, reported about a decade ago: "Like NASA with its Apollo programme, Khorana's group has shown it can be done and both feats may never be repeated." Progress a decade later has outstripped the imagination of many, and dozens of genes have now been made. The α-consensus interferon example is an instructive one where the synthetic consensus gene shows 10–20-fold higher antiviral activity than that coded by native α-interferon genes. The synthetic gene was assembled in three sections with restriction endonuclease sites built in at convenient and strategic loci so that a given section of the gene can be pulled out and replaced at will. The gene total synthesis approach allows complete control on what information the given DNA sequence contains and allows amalgamation of design and redesign strategy with a flexibility for in-depth analysis of protein structure–function relationships.

THE PROTEIN-FOLDING PROBLEM

The protein-folding problem has been one of protein chemistry's central mysteries, essentially unsolved despite persistent efforts over the past generation. Not only is this of great fundamental consequence for prediction both of native structure and eventually designable specific functions but also is the practical stumbling block when proteins overproduced for commerical purposes are generated in vivo in unfolded and/or denatured condition. As King et al. note in their chapter, more than 100 proteins have had their structure elucidated by X-ray analysis to atomic resolution, but the nature of the folding instructions remains an undeciphered aspect of the genetic code.

The chapters by Creighton and by Matthews et al. summarize principles and methods to analyze both thermodynamics and kinetics of the unfolded-folded states and their interconversions. Among the difficulties are the enormous conformer multiplicity of unfolded and partially folded states and the few methods for detection of transiently formed intermediates. The thermodymanics of folded/unfolded states may be finely balanced with as little as 5–15 Kcal/mole favoring native conformation. Site-directed mutagenesis studies on protein folding are at the stage where analyses of folding/unfolding patterns of mutants are underway to try to develop a sufficient data base to lead to predictive rules, not yet available from primary sequence or in terms of a standard protocol for refolding.

The chapter by King et al. focuses on the isolation and use of temperature-sensitive folding mutants in assembly of phage P_{22} tail spike protein. The goal is to analyze the nature and importance of intermediates, and pathways to them, which are transient structural elements which nucleate a folding pattern but where the structures may not exist in the final native folded structure. It was Creighton's work on bovine pancreatic trypsin inhibitor folding which so clearly established that a specific S–S bond was a nucleating influence in folding but a transient one because this S–S bond was rearranged to a different one later in the kinetic process of folding. The initial S–S bond is a crucial kinetic intermediate that is not present in the final structure. King's approach is again a genetic search for a specific phenotype, temperature sensitivity in folding, but no temperature sensitivity in the final protein structure, to assess which amino acid residues carry the folding instructions and where they are placed in the primary sequence. This approach is complementary to the study of mutants that affect the end point stability of native mature folded proteins.

By analysis of thermolabile folding mutants in the assembly of the 666-residue spike protein monomers into a mature thermal stable trimer (with endorhamnosidase crucial for bacteriophage penetration into the activity host cell), King et al. have begun to define "foldons" as elements, transient or permanent, of structure in protein maturation. This approach will surely have general relevance to deconvolution of the protein folding problem.

PROTEIN PURIFICATION

Burgess covers new and developing trends in protein purification, whether at the microgram level to generate enough for N-terminal sequencing for gene cloning, or at the kilogram level for protein production. He notes criteria for purity depending on usage and deals with the issue that protein overproduction in microbial cells often leads to inclusion bodies where active protein is obtainable only by denaturation/renaturation protocols (back to the folding problem!). The end point in protein design and redesign efforts is the obtention of ponderable quantities of pure protein for characterization to allow optimization of design criteria. These needs are driving a renaissance in protein purification and bioseparation methodologies.

In sum, the chapters in this section provide the rationale and guidelines that underpin the explosive growth of activity in recombinant DNA and protein engineering. They summarize current approaches, methodologies, and goals to relate protein primary sequence to three-dimensional structure and thereby specific function. They are basic to understanding the particular studies described in the next two sections of this book.

REFERENCES

1. Josin M, Regan L, Schimmel P (1983): Nature 306:441.
2. Nelson H, Sauer R (1985): Cell 42:549.

Protein Engineering, pages 51–63

4

Site-Specific and Regionally Directed Mutagenesis of Protein-Encoding Sequences

John Rossi and Mark Zoller

Department of Molecular Genetics, Beckman Research Institute of the City of Hope, Duarte, California 91010 (J.R.); Cold Spring Harbor Laboratory, Cold Spring Harbor, New York 11724 (M.Z.)

INTRODUCTION

The study of protein structure/function has been approached over the past 30 years using tools such as group-specific modification of amino acids, affinity labels, substrate analogs, limited proteolysis, synthetic proteins, spin-labeled substrates, NMR, and X-ray crystallography. In addition, the study of proteins with similar primary sequences has yielded important information about the determinants of protein structure and catalysis.

The advent of recombinant DNA technology has revolutionized the study of protein structure–function. It is now possible to clone the gene or c-DNA encoding of virtually any protein for which a partial amino acid sequence or antibody is available [11,14,26]. In many cases, even this information is not necessary, since a number of ingenious schemes such as the plus–minus approach have been developed for cloning c-DNAs encoding proteins of low abundance [9,23]. Once the gene or c-DNA is cloned, large amounts of the protein often can be manufactured by using sophisticated expression vectors which direct abundant transcription of the cloned sequence. Proteins with primary sequences differing by only a single amino acid at a predetermined position can be constructed easily. Extremely sophisticated questions now can be asked about the role of any amino acid in a protein. In addition, these powerful techniques have spawned a renaissance in the field of crystallography and structure determination. It is the purpose of this chapter to communicate some of the technological advances which now make it feasible to alter and engineer protein sequence and structural information virtually at will. The basic strategy for mutagenesis of a protein is shown in Figure 1.

GENES FOR PROTEINS

The techniques described in this chapter for altering protein sequences operate on the DNA level. The DNA encoding the protein is modified, the protein is expressed either in vitro or in vivo (the protein can be purified), and then the properties of the altered protein are studied. If the primary sequence of the protein of interest is known, a synthetic gene can be constructed by "reverse translating" the amino acid sequence. The gene is constructed using segments of oligonucleotides that are ligated together. Recognition sites for restriction enzymes can be engineered deliberately into the gene. This facilitates mutagenesis by replacement of only small regions of the gene with mutant "oligonucleotide cartridges." The synthetic gene is cloned into an appropriate expression vector for functional analysis and structural characterization. Methods for DNA and gene synthesis are covered in the chapter by M. Caruthers, this volume. For additional information see the reference by M. Gait [6].

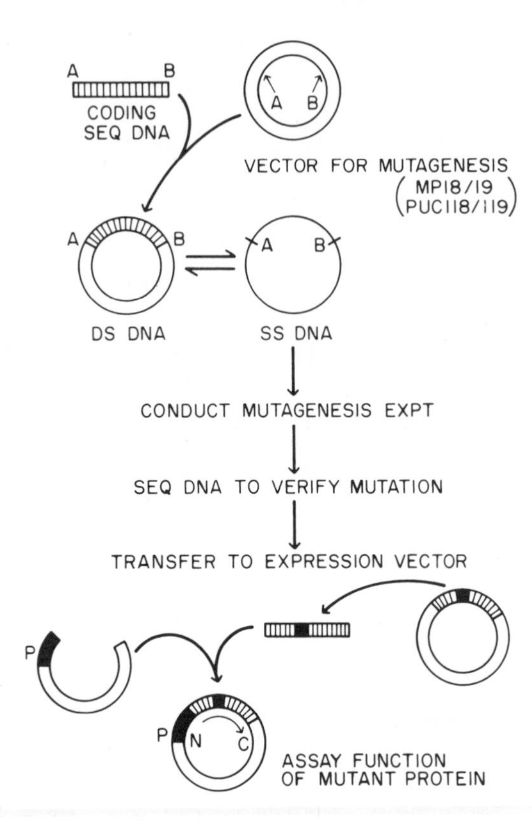

Fig. 1. *Generalized scheme for protein mutagenesis and design. The coding sequence is cloned into an appropiate vector system (see text), and single-stranded DNA is prepared either by isolation of filamentous phage plus strand DNAs or by methods which render duplex DNA single-stranded (see Fig. 5). The mutagenesis can be carried out using either synthetic DNA or chemical approaches.*

SYNTHETIC DNA-MEDIATED SITE-DIRECTED MUTAGENESIS

Prior to recombinant DNA technology, the study of altered proteins was confined to the isolation of phenotypically selectable mutants. The classical techniques of inducing mutations in genes included chemical mutagens, ultraviolet light, and ionizing radiation. However, since the entire organism was subjected to the mutagenesis regime, multiple mutations often arose. The desired mutation could then be identified only by using complex genetic manipulations. With the use of gene cloning and other molecular biological techniques, mutants can be characterized at the molecular level in lieu of genetic mapping. This is especially important for those biological systems that are not readily amenable to genetic analyses.

The construction of a protein bearing a predetermined sequence is best accomplished by either gene synthesis or oligonucleotide-directed mutagenesis. The alternative goal is construction of many different proteins containing random amino acid changes. This is best done by other techniques (see below).

Gene synthesis was described above as a way to obtain the gene from the protein sequence. Mutagenesis via gene synthesis is very simple (Fig. 2) and can be accomplished by either total gene synthesis or replacement of a limited segment of the gene encoding the region containing the desired mutation. The availability of reliable strategies for manual oligonucleotide synthesis and an increasing

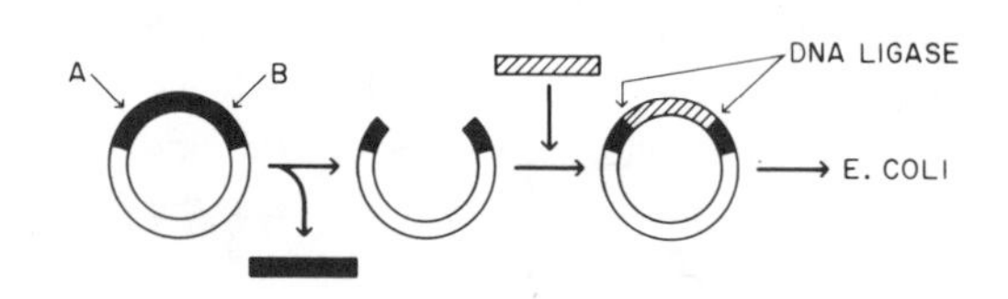

Fig. 2. *Mutagenesis by chemical synthesis of a gene. The chemically synthesized gene or gene segment, with appropriate flanking restriction endonuclease sites, is ligated into a restricted, double-stranded vector and transformed into a bacterial host.*

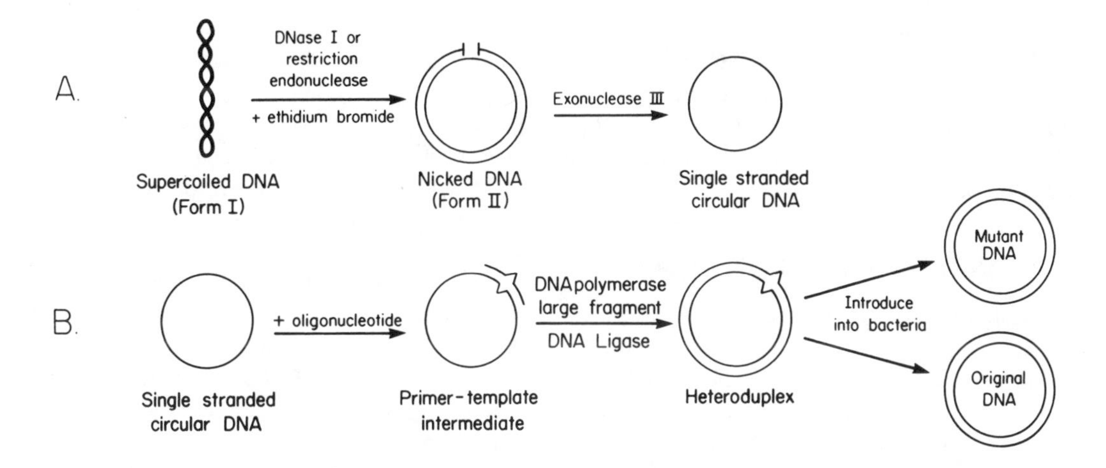

Fig. 3. *Scheme for single primer mutagenic approach: A, preparation of single-stranded template from double-stranded covalently closed circular plasmid DNAs; B, synthetic oligonucleotide-primed repair synthesis of single-stranded templates prepared from either filamentous phage or the procedure outlined in Figure 5A.*

access to automated DNA synthesizers have made gene synthesis a reasonable method to produce genes with specific amino acid changes.

The basic strategy of oligonucleotide-directed mutagenesis was developed using the single-stranded phage phiX-174. The technique stemmed from the combination of a number of observations about nucleic acids: 1) marker rescue of mutations in phiX-174 by restriction fragments annealed to single stranded DNA; 2) the stability of DNA duplexes containing mismatches; and 3) the ability of *Escherichia coli* DNA polymerase I to extend oligonucleotide primers hybridized to single stranded templates. In these pioneering studies, amber mutations were either corrected or created in vitro by using synthetically produced oligonucleotides (7–17 nucleotides in length) as primers for DNA polymerase I mediated repair synthesis on the single stranded templates. The success rate of the mutagenesis was monitored by phenotypic scoring of the resultant phage plaques [7,8,10,19]. Wallace and co-workers [24,25] applied the basic principle to create mutant genes cloned into the bacterial plasmid pBR322. In this case, mutant clones were identified using the mutagenic oligonucleotide as a hybridization probe. This method of screening does not need to rely on phenotypic changes, an important feature in the production of mutations in cloned DNA fragments.

Oligonucleotide-directed mutagenesis is similar in principle to mutagenesis by gene synthesis in that an oligonucleotide that contains the sequence of the desired change is inserted into a cloned fragment of DNA (Fig. 3). In this strategy, a short oligonucleotide consisting of the mutant sequence is hybridized to its complementary sequence in a clone of the wild-type DNA, thereby forming a mutant–wild-type heteroduplex. The oligonucleotide serves as a primer for in vitro enzymatic DNA synthesis of the rest of the template DNA. A double stranded heteroduplex is formed, which is subsequently segregated in vivo into separate mutant and wild-type clones. The two molecules can be distinguished readily by a number of screening techniques discussed below.

Current strategies for oligonucleotide directed mutagenesis can be divided into two categories based on the form of the DNA template: single-stranded templates derived from filamentous phage vectors such as the M13mp series [15,16] and bacteriophage fd and f1 [28], and double stranded templates from plasmid vectors. In both systems the oligonucleotide is annealed to a single-stranded region of the template DNA.

SINGLE-STRANDED TEMPLATES

A number of cloning vectors are available that have been derived from filamentous phages such as M13 and fd. The DNA fragment of interest is inserted into the replicative form of the vector by ligation in vitro. Upon transfection of *E. coli* with this DNA, phage particles containing single-stranded DNA can be isolated from the media in which the cells are growing. This serves as the template DNA for mutagenesis.

The oligonucleotide is designed to be complementary to the region of interest, except for the mismatch that encodes the desired change. Table I shows some general rules for the sizes of oligonucleotides used and the types of changes possible. Two procedures are generally used in conjunction with single-stranded templates: 1) the one-primer method and 2) the two-primer method (Figs. 3 and 4).

In the one-primer method, a single oligonucleotide is annealed to the single-stranded stranded DNA template. Since the oligonucleotide is short, hybridization occurs rapidly. Next, the mutagenic oligonucleotide is extended by *E. coli* DNA polymerase I (large fragment, lacking the 5′-3′ exonuclease activity) in the presence of all four deoxyribonucleoside triphosphates, and the newly synthesized strand is covalently ligated to the 5′ end of the oligonucleotide by T4 DNA ligase. The formation of covalently closed circular DNA (ccc) molecules is often incomplete, leaving some gapped molecules that generally yield wild-type clones. Therefore, to increase the efficiency of mutagenesis, ccc-DNA molecules are purified by centrifuging the in vitro extension/ligation reaction through an alkaline sucrose gradient. This effectively separates ccc-DNA from gapped molecules [29]. *E. coli* is transformed with the ccc-DNA pool and the resulting transformants are screened for mutant DNA.

The two-primer method [30] is a variation of the one-primer procedure. Two primers are annealed to the single-stranded template DNA, the mutagenic primer, and a universal primer which is completely complementary to template sequences upstream of the mutagenic oligonucleotide (Fig. 4). Priming initiates si-

TABLE I. Types of Mutations Made by Oligonucleotide-Directed Mutagenesis

1. Single-base changes, deletions, and insertions (16–18 mer)
2. Multiple-base changes within a short region (>17 mer; degenerate)
3. Large site-specific deletions (25–40 mers)
4. Large insertions (25–70 mers)
5. Multiple changes at same site (16–20 mer; degenerate)

mer: number of nucleotides in length

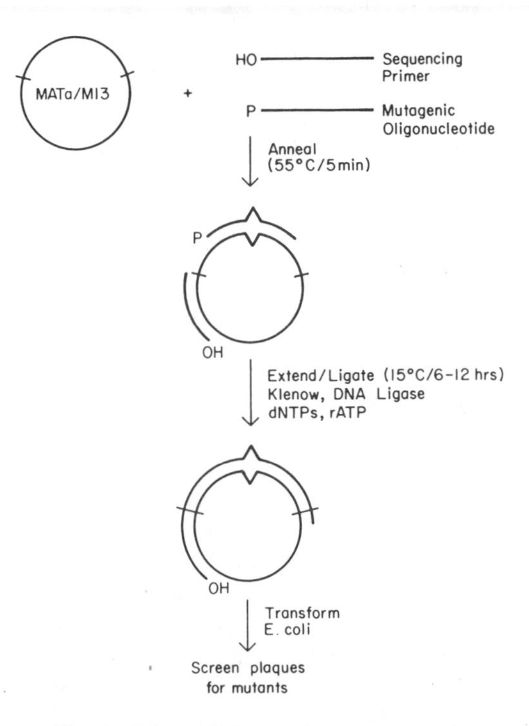

Fig. 4. *Scheme for two-primer mutagenesis.*

multaneously from both oligonucleotides. When priming from the upstream universal primer reaches the mutagenic oligonucleotide, DNA ligase will covalently join the fragments, thus forming long stretches of complementary base-paired sequences on either side of the mismatched bases at the mutagenic site. In essence, these flanking base-paired regions protect the mismatched site from exonucleolytic degradation and repair following transformation into the host. The two-primer method can also be used to excise a fragment containing the heteroduplex using restriction sites flanking the cloned target sequence. This excised fragment can then be subcloned into a plasmid vector for subsequent screening. Using this approach, Norris and co-workers [18] have achieved a 42% mutagenesis efficiency.

DOUBLE-STRANDED VECTORS

The use of vectors derived from filamentous phage arose for a number of reasons:

1) The single-stranded template DNA closely resembles the form of the template DNA in the early studies using phiX-174.

2) Pure single-stranded template DNA is easily obtained.

3) DNA sequencing from single-stranded template DNA is relatively free of artifacts.

4) Screening for the mutation by hybridization techniques is also simple.

5) The efficiency of mutagenesis with methods that use single-stranded vectors is generally higher than with methods that use double-stranded vectors.

However, subsequent experiments using the mutagenized gene are usually conducted in a plasmid-based expression vector. This requires moving the mutant DNA from the phage vector into a second vector. Isolation of double-stranded M13 DNA (RF DNA) is somewhat difficult and involves another step in order to construct the expression vector. Procedures are available that attempt to eliminate this last step by conducting the mutagenesis on the expression vector to be used in biological experiments. These procedures consist of many of the same steps used with single-stranded templates. The initial steps involve exposing a single-stranded region of the double-stranded plasmid DNA to which the

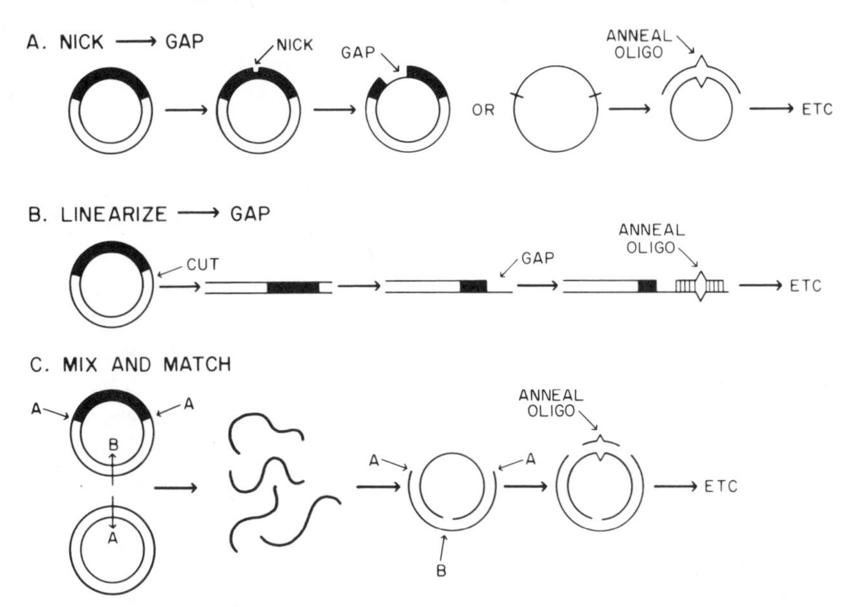

Fig. 5. *Summary of methods for generating single-stranded gaps in duplex molecules (see text for details).*

mutagenic oligonucleotide can anneal. There are several variations of this basic idea.

The various methods for creating single stranded gapped regions in duplex DNAs are depicted in Figure 5. In Figure 5A, a randomly placed single-stranded break is introduced by treating the duplex with either a restriction endonuclease or DNAse I in the presence of ethidium bromide, forcing single-stranded scissions. The nick serves as an entry point for limited exonuclease III digestion, which exposes a single-stranded region of DNA. Alternatively, complete exonuclease treatment can be done which results in single-stranded circular DNA to be used as template for mutagenesis. Only one of the two strands will be complementary to the mutagenic oligonucleotide; the other strand will yield a background of wild-type molecules.

A variation of the above method involves linearizing the vector at a restriction site near the gene fragment (Figure 5B). Limited exonuclease III digestion exposes a single-stranded region to which the mutagenic oligonucleotide can hybridize. The procedure continues, as in the two-primer M13 method.

A third method utilizes two vectors (Fig. 5C): one contains the gene fragment of interest, and the other consists of the parental cloning vector. First, the two plasmids are linearized at different sites. The parental vector is cleaved at the site at which the gene fragment was inserted, and the other plasmid is cleaved at some other site outside of the gene. The two plasmids are mixed together, the plasmids are denatured, then the strands are renatured. This yields four possible duplex molecules: the original forms of the two vectors, and two heteroduplex molecules containing a single-stranded gap spanning the gene sequence. One of these two gapped forms will be complementary to the mutagenic oligonucleotide. The oligomer is hybridized to the DNA; primer extension and ligation are followed as described above.

In addition to the single-stranded versus double-stranded vectors, a new class of bifunctional vectors has been developed recently[4,27]. In these vectors, the replication origins from filamentous phages have been cloned in bacterial plasmid vectors using antibiotic selection. When the host cells are superinfected by the wild-type parental virus, which provides replication and packaging functions in *trans*, the plasmid is replicated as a rolling circle, and the plus strand (with respect to the ss-DNA virus replication origin) is packaged along with the parental virus. Each of the ss-phage vectors can then be utilized as a template for synthetic DNA-mediated mutagenesis, as depicted in Figure 3B. The double-stranded form of these vectors can be used directly for expression.

SCREENING FOR THE PRODUCTION OF MUTANTS

Upon transformation or transfection of *E. coli* with the in vitro synthesized heteroduplex DNA, the individual strands are replicated, thereby segregating the mutant and wild-type DNAs. In the case of filamentous phage vectors, colonies harboring M13 DNA grow slower than uninfected colonies. On a plate covered with a lawn of uninfected cells, the colonies of infected cells appear clear or "plaque like" compared to the opaque lawn. In the double-strand methods, colonies harboring plasmid vectors can be selected by resistance to antibiotics in the plate media.

The simplest method to screen for mutant molecules amid a large background of nonmutant DNAs is by hybridization of the radiolabeled oligonucleotide to the phage or plasmid DNA bound to nitrocellulose or cellulose filter papers. The mutagenic oligonucleotide has one or more bases not complementary to the nonmutant target sequence; therefore, under the appropriate temperature and salt concentration conditions, the oligonucleotide becomes an exquisitely sensitive hybridization probe and will form a stable duplex with only a perfectly complementary sequence, i.e., the mutated vector. A single-base mismatch between an oligonucleotide 14mer and its template can result in a lowering of the T_d (temperature at which 50% of probe becomes dissociated from its complementary target) by as much as 10°C [25]. Such specificity in hybridized template stability enables one to choose a temperature for hybridization or filter washing that will discriminate against nonmutant sequences.

In practice, the hybridization is done under conditions of low stringency so that the oligonucleotide binds to both mutant and wild-type sequences. Hybridization is visualized by autoradiography of the filters using X-ray film. The hybridization signals show up as dark spots on the film. Upon washing the filter at a higher temperature, the mutagenic oligonucleotide remains hybridized to mutant DNA molecules and dissociates from wild-type molecules. At a certain temperature, mutant DNA can readily be distinguished from wild-type DNA. The temperature at which this occurs depends on the particular mutation and the sequence of the mutagenic oligonucleotide.

Wallace and co-workers have devised a simple formula for oligonucleotides in the range of 12–23 bases in length which roughly predicts a T_d for a completely base paired duplex. This formula uses 4 × the G+C content plus 2 × the A+T content to give the predicted T_d. In practice, this formula usually works quite well for oligonucleotides in the size range described above. Thus, for the oligonucleotide 5′ GCCATAATCGGCAT-3′, the theoretical T_d by this formula would be 42°C. If the mismatch occurs within the middle of the oligonucleotide, hybrids to nonmutant clones will wash off at or very close to this temperature. An example of oligonucleotide mutant screening is depicted in Figure 6.

Once the screening has identified putative mutant clones, these are purified and the DNAs are subjected to DNA sequencing to verify that the correct mutation has been constructed.

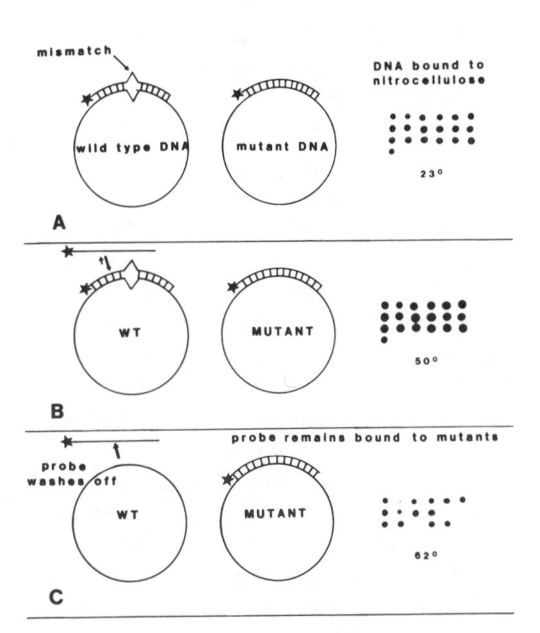

Fig. 6. *Example of synthetic DNA-mediated screening of in vitro constructed mutants: A, the synthetic oligonucleotide which was used to prime the mutagenesis experiment is radioactively labeled and used to probe plaques under low-stringency conditions (23C), both mutant and nonmutant sequences hybridize; B and C, increasing the temperature of the filter washes results in elimination of probe from wild-type sequences because of instability introduced by mismatched bases. The hybridizing spots represent an autoradiographic picture of the nitrocellulose filters following hybridization and washes.*

IMPROVEMENTS

The theoretical yield for mutants using synthetic DNA-mediated site-directed mutagenesis is 50% of the primary transformants. In practice, this efficiency is almost never realized, unless special host strains are utilized (see below). A number of improvements and modifications in the mutagenesis procedure which can result in dramatically increased mutagenesis are described below.

Perhaps the most significant improvement for increasing the efficiencies of synthetic DNA-mediated site-directed mutagenesis is based on a strong biological selection against the original, unaltered genotype. The procedure developed by Kunkel [12] takes advantage of a mutant *E. coli* host which is deficient in the enzyme dUTPase. This mutant accumulates dUTP intracellularly, which is sometimes misincorporated into DNA in place of dTTP. In addition, this host has a second mutational alteration which results in a deficiency in the enzyme uracil glycosylase, and is thus unable to excise the incorporated dUTP. The template DNA is prepared in this host under conditions where approximately 20–30 dUTP residues are incorporated per genome. The template is then used for synthetic DNA-mediated mutagensis, with repair synthesis taking place in the presence of all four dNTPs. When the heteroduplex is transfected into a host which contains a functional uracil glycosylase, the template strand is preferentially destroyed, leaving intact the in vitro–made strand (which contains dTTP), resulting in a substantial enrichment for the mutant phenotype. In some cases, greater than 50% of the transfected progeny harbor the in vitro–introduced mutation. The use of these mutants in concert with one or more of the above-mentioned mutagenesis techniques should enable the rapid and efficient production of virtually any desired mutation.

Winter and co-workers have developed a set of M13–derived vectors and primers that allow for selection of the mutant strand [2]. In one system, the vector contains an amber mutation in gene IV of M13. A two-primer mutagenesis procedure is followed: One primer is the oligonucleotide for the desired mutation, and the other primer changes the amber in gene IV to wild type. Following extension and ligation, the in vitro-synthesized DNA is transformed into a nonsuppressing strain of *E. coli*. Only progeny phage that contain a wild-type gene IV will propagate. For this procedure to yield a high efficiency, the *E. coli* strain also contains mutations in the mismatch repair system.

A second vector/primer system has also been developed by Winter and co-workers. In this system, the primer changes an Eco K restriction site to an Eco B site [2]. Selection is accomplished by transforming the DNA into an Eco K strain of *E. coli*. Progeny containing the Eco B site will survive. A second primer is available that changes the new Eco B site back to Eco K. Thus, multiple mutations can be constructed and the selection for the mutant strand can still be used simply by transforming the appropriate strain of *E. coli*. This feature was not possible with the amber to wild-type primer described above.

DELETION, INSERTION, AND REGIONALLY DIRECTED MUTAGENESIS

In many instances it is neither desirable nor instructive to create specific point alterations until more information is obtained concerning the function of a segment of DNA coding for a polypeptide.

The approaches to be described below involve either deletion, insertion, or random chemical mutagenesis within a defined target area. A generalized scheme for deletion mutagenesis involves cleaving the target area DNA with a restriction endonuclease and then excising either a defined or random segment of DNA (cloned in a circular plasmid or viral vector) by use of a second restriction endonuclease site or treatment of the DNA with an exonuclease such as Bal31, lambda-exonuclease, or a combination of exonuclease III and nuclease S1 [for review of these methods, see reference 21]. In each of the above methods, a segment of the gene of interest is deleted in vitro. The covalently closed form of the vector can be regenerated by treatment of the DNA with DNA ligase. In the case of deletion by successive restriction endonuclease cleavage, the termini generated by the restriction enzymes must be either cohesive and complementary, or rendered flush by subsequent treatment with either DNA polymerase I to fill in cohesive termini, or by nuclease S1 treatment to cleave off cohesive termini. In introducing random deletions into the coding segment for a protein, it should be remembered that two out of three random deletions will alter the translational reading frame and a certain percentage of these will create premature stop codons within the coding segment. Such a random approach has very limited use in protein structure–function studies.

A different approach to insertional mutagenesis has been developed by Barany [1]. Hexameric, self-complementary single-stranded DNA oligonucleotides are inserted into preexisting restriction sites to create a new six-base restriction site. The insertion of the linkers also creates a two-codon insertion in the protein reading frame. Barany has also constructed a Kanamycin resistance cassette which is inserted at the newly created restriction site to allow biological selection of the mutation. Once the mutants have been selected, the cassette can be readily excised and the circular plasmid containing the two codon insertion can be recircularized. An important feature of this approach is that the hexameric oligonucleotides can be ligated efficiently to restriction sites with only two protruding bases, and the procedure can be used with either 5′ or 3′ protruding termini.

Directed deletions using restriction endonuclease sites can be designed to generate in-phase, nontruncated peptides, provided that the proper sites or combinations of sites are chosen. For instance, certain restriction endonucleases generate three-base protruding 5′ termini, such as HinfI (G′ANTC). When the two HinfI termini generated by cleavage with the enzyme are treated with nuclease S1 and joined together by blunt ended ligation, a three-base pair deletion can be effected. Dependent upon whether the nucleotides deleted are entirely from within one codon or overlap with adjacent codons, a precise one-codon deletion, or a one-codon deletion and an adjacent codon alteration, will be made. For restriction enzymes which generate four-base protruding termini (such as HindIII, A′AGCTT), a number of approaches can be used to generate non-frameshift deletions. One example would involve adding a single dA to a protruding HindIII–generated terminus using DNA polymerase I (Klenow fragment). This will leave three bases in the single-stranded protruding end which can be removed with nuclease S1. In a separate experiment, all four protruding bases can be removed from the other side of the HindIII site with nuclease S1. By cleaving the two vectors at another unique restriction site and joining the two half molecules together with DNA ligase, a three-base pair deletion will result. Other ways in which this combination of restriction cleavage, partial fill-in, nuclease S-1 treatment and blunt-ended ligation can be used to create small site-directed deletions within a defined target area can be conceived quickly with a pencil and paper.

The above methods are limited by the availability of convenient restriction endonucleases sites within the cloned segment. A somewhat different approach that doesn't necessitate a conveniently placed restriction endonuclease site requires formation of a D loop by annealing a short single-stranded piece of DNA to a partially denatured duplex circle. A single-stranded looped-out region is formed which serves as a substrate for nuclease S1 nick introduction. The resultant intermediate is unstable, and the single-stranded oligonucleotide is displaced by the vector strand resulting in a nicked duplex. The nicked dulpex is then treated with an exonuclease to create a single-stranded gap. This gapped region can then be excised with nuclease S-1 followed by religation to generate a circle with an internal deletion [21].

CHEMICAL MUTAGENESIS

The mutagen sodium bisulfite catalyzes the deamination of cytosine to form uracil under mild temperature and pH conditions. Cytosine residues in a duplex molecule are essentially inert to bisulfite attack. Therefore, gapped single-stranded regions can be targeted for mutagenesis by this compound [21]. To create such single-stranded target regions, several methods have been employed. Within a duplex molecule, single-stranded nicks can be introduced by treatment of the DNA with a restriction endonuclease in the presence of ethidium bromide (Figure 5A). The unique nick, which can occur on either strand, is then converted into a short single-stranded region by the 5′ exonuclease activity of *M. luteus*

DNA polymerase. The gapped, single-stranded region is subsequently treated with sodium bisulfate. Conversion to a double-stranded form is mediated by the polymerization activity of *E. coli* DNA polymerase plus the four deoxyribonucleoside triphosphates. A dA residue will be incorporated opposite each deaminated cytosine, resulting in C–G to A–T transitions. Such a procedure usually results in the introduction of a large number of transitions throughout the target region.

Nucleotide misincorporation has also been successfully utilized to introduce alterations in the coding sequences of targeted regions [21,22]. By manipulations of in vitro polymerization reaction conditions, such as substitution of manganese for magnesium, or creation of large imbalances in the ratios of deoxyribonucleoside triphosphates, purified polymerases can be forced to misincorporate nucleotides at a high frequency. Gap misrepair has also been accomplished by forcing misincorporation of an alpha-thiophosphate deoxyribonucleotide [22]. In this procedure, a single alpha-thiophosphate nucleotide is incubated with DNA polymeraseI (Klenow), with or without one, two, or three of the deoxyribonucleotide triphosphate also present. Once the misincorporation of the alpha-thiophosphate deoxyribonucleotide has occurred, it cannot be edited out by the 3′-5′ exonucleolytic activity of the polymerase because of the sulphur atom in the covalent linkage. After an appropriate period of incubation in the presence of this analog, the gap is then completely filled in by completing the addition of the required deoxyribonucleoside triphosphates and the addition of DNA ligase to covalently close the gap. These mutagenesis procedures can be carried out on gapped duplex molecules where the target sequence has been rendered single stranded (see Fig. 5).

The important aspects of the misincorporation methodologies are that the DNA polymerase (*E. coli*, T-4 or *M. luteus*) will prime from the double-stranded region and repair the single stranded gap, and that the purified enzymes are quite error prone under the appropiate conditions. On a circular template, the filled-in gap region can be covalently closed to form a circle in the presence of DNA ligase. Thus, the misincorporated nucleotides will be sealed in a covalently closed circular molecule which is ready for transformation into the appropiate bacterial host.

The most difficult task utilizing these procedures is identifying the mutant transformants against an often large background of nonmutant clones. This task is simplified if there is a readily selectable phenotype for the mutant gene, but this is often not the case when eukaryotic genes are cloned in a bacterial host. To circumvent this problem, an ingenious blend of methodologies has recently been developed. The procedures to be described below take advantage of the fact that in a denaturing gradient gel electrophoresis system, a single point mutation within a DNA fragment can change the electrophoretic mobility such that the altered sequence is resolved from the unaltered strands of the same size [5,17]. The principle of the selection scheme is that DNA fragments of identical size, but differing by a single base change, will initially migrate through the polyacrylamide gel at a constant rate. When migration reaches certain critical concentrations of the denaturing gradient, there are specific domains which will "melt out" to produce partially denatured DNA. This melting retards the mobility. The position in the denaturing gradient gel at which the decrease in mobility occurs corresponds to the melting temperature (T_m) of the domain. A single-base difference within a particular domain will alter the T_m of that domain, retard the electrophoretic mobility, and result in separation of the mutant from nonmutant fragment.

Meyers et al. have developed new chemical mutagenesis strategies to complement the gel enrichment scheme described above [17]. Single-stranded DNA containing the DNA fragment to be mutagenized is obtained either via cloning the target sequence (usually 30–600 bp in length) in an M13 mp bacteriophage vector, or a GC clamp vector which contains both plasmid and M13 origins of replication. Single-stranded DNAs prepared from such vectors employed in the chemical sequencing of DNA. The treatment of the DNAs is performed under conditions in which approximately 10%–20% of the target DNA sequences contain one hit. One example of the type of mutagenesis which can be obtained by these procedures utilizes hydrazine, for which the primary mode of action is breakage of the pyrimidine rings of dC and dT residues. Following the mutagenesis protocol, the chemical mutagens are removed and a universal oligonucleotide primer is annealed to the DNA and used to prime AMV reverse transcriptase repair replication. Since this enzyme has a relatively high degree of infidelity, the primer extension products will frequently contain an incorrect base at positions corresponding to the damaged base on the single-stranded template DNA. The duplex target DNAs are excised by flanking restriction enzymes and inserted into a plasmid that can be used for direct genetic screening, or into a GC-clamp vector for further enrichment. This vector has a relatively thermal-stable, GC-rich sequence adjoining the cloning sites, which stabilizes the duplex, resulting in more dramatic melting and hence mobility differences between the mutant and nonmutant sequences in the adjoining target sequence.

Subsequent to insertion of the target sequences into this vector, they are propagated in an *E. coli* host, isolated, and digested with restriction enzymes to excise the target sequence and the GC-rich sequence. These DNAs are then electrophoresed in a preparative denaturing gradient gel to separate the mutant and wild-type target DNA sequences. The mutant sequences are isolated from the gel, recloned into a vector for DNA sequence analyses, and finally into an expression vector system. The primary advantages of this approach are: 1) it is efficient; 2) the mutagenesis is not confined to a precise location within a target fragment; 3) transitions and transversions for all four bases are possible; and 4) most importantly, it does not necessitate a phenotype for selection of the mutant alterations. This saturation mutagenesis procedure will certainly prove to be useful for studies of proteins for which little or no physical data is available.

SITE-DIRECTED MUTAGENESIS USING NONSENSE SUPPRESSING tRNAs

One of the most exciting innovations in site-directed mutagenesis of proteins is the use of suppressor transfer RNAs for amino acid substitutions. Until recently, such an approach was limited to the use of a half dozen genetically isolated suppressors. John Abelson, Jeffrey Miller, and their colleagues currently are extending this array of suppressor tRNA genes by designing and chemically synthesizing genes that will allow insertion of each of the 20 amino acids at a given nonsense codon within a protein-encoding messenger RNA. The synthetic genes are designed to be recognized by the cognate aminoacyl-tRNA synthetases but have alterated anticodons which enable them to base pair with UAG nonsense codons. With the completion of this technology, in a single set of experiments, it will be possible to examine the effects of every amino acid at a given codon within a protein.

The principle of their methodology is as follows. A nonsense codon (usually a UAG amber codon) is introduced at the desired site within the coding region of a protein-encoding gene. This is accomplished either by site-directed mutagenesis as described above, or by genetic means. A series of strains, each containing a different suppressor tRNA, is then transformed with the plasmid or viral vector harboring the nonsense mutant. Each of the synthetic tRNA genes is cloned in an M13mp vector and is subject to transcriptional regulation by the lactose-promoter, operator induction system. Upon induction, the modified tRNA is synthesized and charged with its cognate amino acid, which is then inserted into the peptide in response to the UAG codon in the messenger RNA.

The use of suppressor tRNA insertional mutagenesis is especially attractive for those proteins for which high-resolution structural information is available. Thus, at an amino acid position which is known to be important for either catalysis or folding of the protein, the effects of each of the other 19 amino acids can be examined without resorting to extensive synthetic DNA-mediated site-specific mutagenesis.

EXPRESSION VECTORS

It is well beyond the scope of this chapter to go into detail about the array of expression vectors now available for the large-scale production of proteins in bacteria. It should be pointed out that the primary elements required for large-scale production of any protein can be generalized into transcriptional and translational control sequences. The prokaryotic promoters which are most popular for expression are the *E. coli* Trp, bacteriophage λP_L, *E. coli* lac UV5, and the Trp-lacUV5 fusion promoter or so called Tac promoter [3,14,20]. Each of these promoters is complexed with an operator sequence for either induction or derepression of transcription. This is an extremely important feature, since high-level expression of some foreign proteins in *E. coli* is lethal to the host. Most of the vector systems now available for expression also have a ribosome binding sequence adjacent to a polylinker for cloning the foreign gene or cDNA.

Despite all the advances in expression technology, it is still not possible to predict whether or not a given protein will be highly expressed upon induction of transcription from these promoters. The major roadblocks to high-level expression are the efficiency of translation of the message and the stability of the protein itself. The efficiency of translation is dependent not only upon a strong ribosome-binding sequence, but is also influenced by features of the messenger structure and sequence itself. For physical studies of mutant proteins, the stability problem can be a concern, since improperly folded proteins are often quite unstable in *E. coli*. Despite these potential problems, numerous mutant proteins have been successfully expressed in *E. coli*.

Functional studies of mutant mammalian proteins can now be carried out in mammalian cell culture systems using a number of vector systems which have either viral, retroviral, or nuclear gene promoters and transcriptional enhancers. These systems are less well suited for large-scale production of the proteins but are extremely important for studies of intracellular protein localization and protein–protein interactions.

In summary, in choosing an expression vector system, the primary concerns should center around the end needs; that is, functional assays in an endogenous host system, or large-scale preparation for physical analyses. Monthly, the pages of many journals are devoted to descriptions of new or improved vectors for each of these needs.

CONCLUSION

Mutant analysis has been the hallmark for the study of gene function. With respect to protein structural–functional analyses, studies of modified proteins have often been confined to those proteins for which a good genetic selection system was available. This is no longer the case, since protein coding cDNAs and genes from any organism can now be easily cloned and modified. The studies of protein structure, function, folding, and design are now entering their golden age.

REFERENCES

1. Barany F (1985): Proc Nat Acad Sci USA 82:4202–4206.
2. Carter P, Hughes B, Winter G (1985): Nucl Acids Res 13:4431–4443.
3. DeBoer H, Comstock L, Vasser M (1983): Proc Nat Acad Sci USA 80:21–25.
4. Dente L, Cesareni G, Cortese R (1983): Nucl Acids Res 11:1645–1655.
5. Fischer S, Lerman L (1983): Proc Nat Acad Sci USA 80:1579–1583.
6. Gait M (1985): "Oligonucleotide Synthesis: A Practical Approach." UK:IRL Press.
7. Gillam S, Smith M (1979): Gene 8:99–106.
8. Gillam S, Smith M (1979): Gene 8:81–97.
9. Gray P, Leung D, Pennica D, Yelverton E, Najarian

R, Simonsen C, Derynck R, Sherwood P, Wallace D, Berger S, Levinson A, Goeddel D (1982): Nature 295:503–508.
10. Hutchinson C, Phillips S, Edgell M, Gillam S, Jahnke P, Smith M (1978): J Biol Chem 253:6551–6560.
11. Itakura K, Rossi J, Wallace R (1984): Ann Rev Biochem 53:323–356.
12. Kunkel T (1985): Proc Nat Acad USA 82:488–492.
13. Larson G, Itakura K, Itoh H, Rossi J (1983): Gene 22:31–39.
14. Maniatis T, Fritsch E, Sambrook J (1982): "Molecular Cloning—A Laboratory Manual." Cold Spring Harbor, NY:Publisher.
15. Messing J, Gronenborn B, Muller-Hill B, Hofschneider P (1977): Proc Nat Acad Sci USA 74:3642–3646.
16. Messing J, Vieira J (1982): Gene 19:1–10.
17. Meyers R, Lerman L, Maniatis T (1985): Science 229:242–247.
18. Norris K, Norris F, Christianson L, Fiil N (1983): Nucl Acids Res 11:5103–5112.
19. Razin A, Hirose T, Itakura K, Riggs A (1978): Proc Natl Acad Sci USA 75:4268–4270.
20. Remaut E, Stanssens P, Fiers W (1981): Gene 15:81–93.
21. Shortle D, DiMaio D, Nathans D (1981): Ann Rev Genet 15:265–294.
22. Shortle D, Grisafi P, Benkovic S, Botstein D (1982): Proc Nat Acad Sci USA 79:1588–1592.
23. Taniguchi T, Ohno S, Fujii-Kuriyama Y, Muramatsu M (1980): Gene 10:11–15.
24. Wallace R, Schold M, Johnson M, Dembek P, Itakura K (1981): Nucl Acids Res 9:3647–3656.
25. Wallace R, Shaffer J, Murphy R, Bonner J, Hirose T, Itakura, K (1979): Nucl Acids Res 6:3543–3557
26. Weinstock G, Rhys C, Berman M, Hampar B, Jackson D, Silhavy T (1983): Proc Nat Acad Sci USA 80:4432–4436.
27. Zagursky R, Berman M (1984): Gene 27:183–191.
28. Zinder N, Boeke J (1982): Gene 19:1–10.
29. Zoller M, Smith M (1983): Methods In Enzymology 100(B):468–500.
30. Zoller M, Smith M (1984): DNA 3:479–488.

Protein Engineering, pages 65–70

5

DNA Synthesis as an Aid to Protein Modification and Design

Marvin H. Caruthers

Department of Chemistry and Biochemistry, University of Colorado, Boulder, Colorado 80309

INTRODUCTION

Until recently, the chemical synthesis of deoxyoligonucleotides was a difficult and time-consuming task [1]. Now, however, methods are available for the rapid, manual synthesis of several DNA segments simultaneously [2]. The approach has been used successfully by nonchemists as well as those skilled in organo-chemical procedures. Moreover, these methods can be automated [3,4], which has led to the commercialization of so-called gene machines. The chemical methodology leading to these developments will be described.

Various directed mutagenesis procedures, as discussed elsewhere in this volume, are currently the most popular for modifying protein genes. Because DNA segments can now be prepared rapidly and in high yield, alternative methods for modifying proteins involving the total synthesis of protein genes have been developed and will be described in this chapter as well.

SYNTHESIS PROCEDURE

The general synthetic strategy involves adding mononucleotides sequentially to a nucleoside covalently attached to an insoluble polymer support (Fig. 1). Reagents, starting materials, and side products are removed simply by filtration. After various additional chemical steps, the next mononucleotide is joined to the growing, polymer-supported deoxyoligonucleotide. At the conclusion of the synthesis, the deoxyoligonucleotide is chemically freed of blocking groups, hydrolyzed from the support, and purified to homogeneity by polyacrylamide gel electrophoresis.

The Support

The most promising supports are inorganic, silica-based materials. Initially, high-performance liquid chromatography (HPLC) grade silica gel was used [5]. More recently, Fractosil [2] and controlled pore glass (CPG) [6] have proven to be superior because of improved yields (1–2% per cycle) and better handling properties. The chemistry used to covalently attach nucleosides to silica involves a two-step procedure [7]. Silica is first derivatized to contain a silylamine by reacting the support with 3-aminopropyl triethoxysilane. The next step is condensation of the amino silica with a nucleoside containing a 3′-*p*-ni-

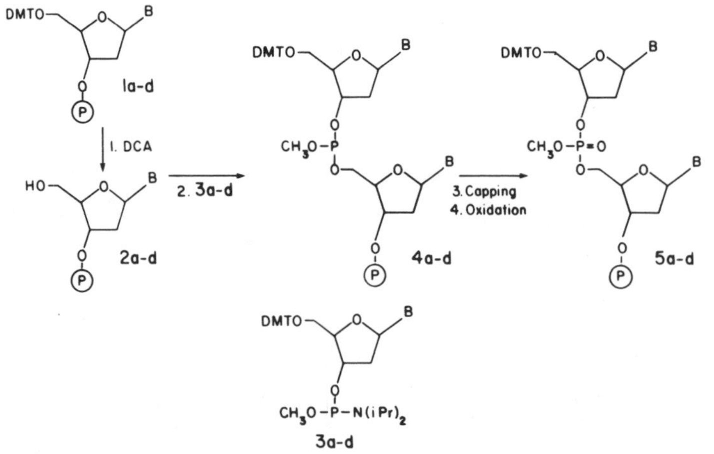

Fig. 1. *Steps in the synthesis of a dinucleotide. B:thymine or appropriately protected adenine, cytosine, or guanine; DMT:dimethoxytrityl; Ⓟ: silica or controlled pore glass support; DCA:dichloroacetic acid; iPr:isopropyl. Capping and oxidation steps are outlined in Table I.*

trophenyl succinate ester. The product having a nucleoside joined to the support through an ester linkage is compound 1a–d. Approximately 10 μmol nucleoside per gram support can easily be obtained using this procedure. Such capacity means that 25–50 mg silica is sufficient for preparing a DNA segment useful for biochemical or biological experiments. If, however, larger quantities are needed for biophysical studies, higher loadings can easily be obtained [8].

Synthesis Cycle

The addition of one mononucleotide to 1a, 1b, 1c, or 1d requires the following four steps: 1) removal of the dimethoxytrityl protecting group with acid to form 2a–d; 2) condensation with 3a–d; 3) acylation or capping of unreactive 5′-hydroxyl groups; 4) oxidation of the phosphite triester to the phosphate triester. Thus the synthesis proceeds in a 3′ to 5′ direction by stepwise addition of nucleotides. The individual steps for one synthesis cycle are summarized in Table I and discussed in the following paragraphs.

Removal of the dimethoxytrityl group can be completed with various protic acids or even Lewis acids [7]. Protic acids such as dichloroacetic acid are preferred because detritylation is very rapid (2–3 min) even as the DNA segment becomes quite large. This step, however, must be carefully controlled, since adenosine is very susceptible to depurination. Chain cleavage leading to an overall reduction in yield then occurs at the depurination site. Recently, new protecting groups called amidines which stabilize adenine toward depurination have been introduced [9]; these appear quite promising.

The key intermediates used for the sequential addition of mononucleotides to the support are appropriately protected deoxynucleoside phosphoramidites (3a–d). These compounds have several attractive features. They can be synthesized using standard organochemical procedures [10–11]. They are stable to moisture and oxidation conditions and can be stored for months. Consequently, these compounds can be handled much like any stable, organic reagent. Although compounds 3a–d are inert under neutral or basic conditions, they can be activated by weak acids such as tetrazole in acetonitrile and condensed in high yield to 4a–d. Tetrazole is the activating agent of choice, since it is a nonhygroscopic, commercially available material. Other mild acids, such as amine hydrochlorides, can also be used but are not recommended. Amine hydrochlorides

TABLE I. Chemical Steps for One Synthesis Cycle

Step	Reagent or solvent[a]	Purpose	Time (min)
1	Dichloroacetic acid in CH_2Cl_2 (2:100, v/v)	Detritylation	3
2	CH_2Cl_2	Wash	0.5
3	Acetonitrile	Wash	3 × 0.5
4	Dry acetonitrile	Wash	3 × 0.5
5	Activated nucleotide in acetonitrile[b]	Add one nucleotide	5
6	Acetonitrile	Wash	0.5
7	DMAP:THF:lutidine[c] (6:90:10, w/v/v) 0.1 ml acetic anhydride	Cap	2
8	THF:lutidine:H_2O (2:2:1, v/v/v)	Wash	1
9	I_2 solution[d]	Oxidation	1
10	Acetonitrile	Wash	0.5
11	CH_2Cl_2	Wash	0.5

[a]Multiple washes with the same solvent involve filtration between wash steps. Each step volume is 1 ml unless indicated.
[b]For each μmole of deoxynucleoside attached covalently to silica gel, 0.4 M tetrazole (0.2 ml) and 0.1 M deoxynucleoside phosphoramidite (0.2 ml) are premixed in acetonitrile.
[c]DMAP, dimethylaminopyridine; THF, tetrahydrofuran.
[d]THF:Lutidine:H_2O (2:2:1, v/v/v) containing 0.2 M iodine.

are hygroscopic and would introduce water into the condensation step. The synthesis procedure therefore involves dissolving the nucleoside phosphoramidite and tetrazole in acetonitrile, mixing, and adding the solution to the support. The condensation reaction is complete within a minute but is usually allowed to proceed for 5 minutes.

Based on careful analysis of various condensation reactions, approximately 1–2% of the deoxynucleoside or deoxyoligonucleotide bound to the support does not react with activated nucleotide. These unreacted oligomers are acylated or capped in order to guard against the formation of deoxyoligonucleotides having heterogeneous sequences. This capping step can best be accomplished using a tetrahydrofuran solution of acetic anhydride and dimethylaminopyridine. The reaction is complete within 1–2 minutes and does not lead to any detectable side products.

The last step in this cycle is oxidation of the phosphite triester to the phosphate triester using I_2 in 2,6-lutidine, water, and tetrahydrofuran. Oxidation is very rapid (1 minute or less) and side products are not generated. Attempts to postpone the oxidation until after all condensation steps have not been encouraging. Several uncharacterized side products are observed.

Deprotection of Synthetic DNA

Once a synthesis has been completed, the deoxyoligonucleotide is freed of protecting groups and isolated by polyacrylamide gel electrophoresis or reverse phase HPLC. Silica containing the DNA segment is first treated with triethylammonium thiophenoxide in dioxane to remove methyl groups from internucleotide phosphotriesters. This step is followed by treatment with concentrated ammonium hydroxide to hydrolyze the ester joining the deoxyoligonucleotide to the support and to remove protecting groups from deoxycytidine, deoxyguanosine, and deoxyadenosine. Alternatively, concentrated ammonium hydroxide can be used as a one-step procedure for removing the methyl [3] or other phosphorus-protecting groups [12], the base-protecting groups, and the deoxyoligonucleotide from the support. The reaction mixture containing deprotected deoxyoligonucleotides is then fractionated by electrophoresis on a polyacrylamide gel. When a slab gel is used, as many as eight compounds can be purified simultaneously. For various biochemical studies with synthetic DNA, reverse-phase HPLC is recommended. This procedure uses synthetic DNA which still contains the dimethoxytrityl ether. The reverse-phase column thus retards migration of the product through the column, whereas failure sequences, which lack the dimethoxytrityl group, pass through the column more quickly. Purification via reverse-phase HPLC is usually unnecessary for those using synthetic DNA for probing gene libraries, directed mutagenesis studies, or

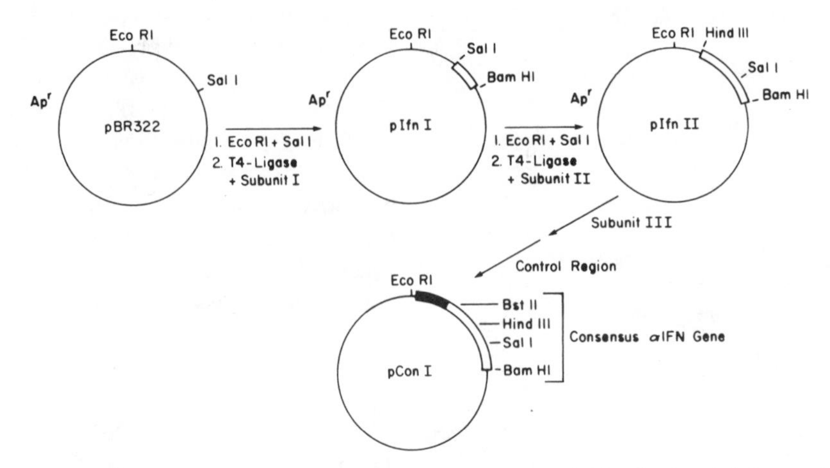

Fig. 2. *Steps for sequential cloning of synthetic consensus αIFN subunits. Ap^r:the ampicillin gene marker; the filled in region of the final plasmid construction, pConI, represents the control region including a promoter and ribosome-binding site.*

cloning experiments. Gel-purified segments are adequate for these experiments.

Synthesis Machines

Several types of machines have been used to synthesize DNA. Among the very simple are test tubes [13], sintered glass funnels [2], and syringes [14]. An intermediate level of sophistication has been achieved through semiautomatic machines. Numerous designs have recently been proposed which generally are based on earlier descriptions [15,16]. The solid support is placed in a column joined through a series of valves and tubing to a pump and an injector loop. Usually, reagents and solvents are attached to the machine and activated deoxynucleotide synthons are added through an injector loop. After one cycle, the machine shuts down and the operator then recycles the program. Completely automatic DNA synthesizers have also been described [3,4]. These automatic machines can be programmed to synthesize mixed probes containing more than one deoxymononucleotide at a predetermined position or deoxyoligonucleotides having defined sequences. In many cases, these machines have also been designed to deliver deprotected samples to the operator.

MODIFYING PROTEINS THROUGH GENE SYNTHESIS

When the first gene was chemically synthesized, it was described as "perhaps the greatest *tour de force* organic and biochemists have yet achieved. Like NASA with its Apollo programme, Khorana's group has shown it can be done, and both feats may never be repeated." [17]. Fortunately, we have scientists such as Khorana (and those at NASA) who will work at the very edge of what can be done and do not worry about such shallow criticism. Gene synthesis is a typical example. For several very good and logical reasons, the feat has been repeated so many times now that the reference list is beyond the scope of this article. Two examples will suffice, however, to illustrate the power of the approach.

Perhaps one of the most intriguing syntheses reported to date is that of an α-interferon called α-consensus interferon [18]. This protein is a hybrid molecule containing sequence elements from all the known subtypes. It has 10–20 -fold higher antiviral activity than any known α-interferon and also exhibits antiproliferative activity. Although the gene for this protein could have been synthesized by repeatedly and exhaustively submitting a natural α-interferon to many rounds of directed mutagenesis, it was much more conveniently synthesized chemically. The general plan is shown in Figure 2. The gene was synthesized and cloned in three sections. These sections were flanked by restriction sites generated within the gene by selecting the appropriate amino acid codons. Thus, the first section (the carboxy terminal region) was synthesized to contain a Bam HI site beyond the termination codon, a Sal I site approximately one-third the distance through the gene, and a linker attached to the Sal I site and terminating with an Eco RI site. Plasmid pBR322 was opened between Eco RI and Sal I, the synthetic Eco RI-Bam HI gene fragment was joined to the plasmid with T4 ligase, and the plasmid cloned. The plasmid derivative was then opened between Eco RI and Sal I and the next one-third of the gene introduced (an Eco RI-Sal I segment containing a Hind III site at one terminus of the gene fragment). A third repetition of this procedure with an Eco RI -Hind III segment having a Bst II site near the amino terminus completes the synthesis. Addition of a control region (again in modular form and synthetic so that it can be changed to enhance protein synthesis) leads to a gene system ready for expression. This procedure is not only very rapid for synthesizing a modified gene of any size, but the resulting gene can be further modified quickly. One needs only to break the gene between two designed, convenient restriction sites, synthesize the modified segments, and reclone the gene. This flexibility is extremely important for extensive studies on protein structure–function relationships and would be very difficult to build into a natural gene.

A synthetic cro gene has also proved to be very convenient for studying protein-DNA interactions [19]. Using the same strategy as outlined previously, a gene containing restriction sites throughout the sequence was prepared. Because the gene could therefore be opened precisely in a region of interest and then modified simply by inserting the desired DNA segments, a large number of mutant proteins were rapidly prepared and analyzed. An interesting set at the carboxy terminus involved sequential deletion of amino acids followed by rebuilding the terminus in altered form. The results demonstrate that the carboxy terminus is important for contacting the cro operator—an observation which could not be deduced from the X-ray crystal structure [20].

CONCLUSION

Because of recent advances in recombinant DNA technology, DNA sequencing, and DNA synthesis, the biochemist and molecular biologist for the first time have complete control over the information content of a DNA sequence. For example, any protein can now be modified in a selected fashion simply by redesigning its gene. This capability should usher in a new age of understanding on the dynamics of protein structure, enzyme catalytic activity, and recognition patterns among various proteins (e.g., hormones and receptors) or proteins and other macromolecules such as DNA or RNA.

REFERENCES

1. Amarnath V, Broom AD (1977): Chemical synthesis of oligonucleotides. Chem Rev 77:183–217.
2. Caruthers MH, Beaucage SL, Becker C, Efcavitch WE, Fisher EF, Galluppi G, Goldman R, deHaseth P, Martin F, Matteucci M, Stabinsky Y (1982): New methods for synthesizing deoxyoligonucleotides. In Setlow J, Hollaender A (eds): "Genetic Engineering." New York: Plenum, pp 1–16.
3. Alvarado-Urbina G, Sathe GM, Lin WC, Gillen MF, Duck PD, Bender R, Ogilvie KK (1981): Automated synthesis of gene fragments. Science 214:270–274.
4. Hunkapillar M, Kent S, Caruthers MH, Dreyer W, Firca J, Griffen C, Horvath S, Hunkapillar T, Tempst P, Hood L (1984): A microchemical facility for the analysis and synthesis of genes and proteins. Nature 310:105–111.
5. Matteucci MD, Caruthers MH (1980): The synthesis of oligodeoxypyrimidines on a polymer support. Tetrahedron Lett 21:719–722.
6. Adams SP, Kavka KS, Wykes EJ, Holder SB, Galluppi GR (1983): Hindered dialkylaminonucleoside phosphite reagents in the synthesis of two DNA 51-mers. J Am Chem Soc 105:661–663.

7. Caruthers MH (1982): Chemical synthesis of oligodeoxynucleotides using the phosphite triester intermediates. In Gassen HG, Lang A (eds): "Chemical and Enzymatic Synthesis of Gene Fragments." Weinheim: Verlag Chemie, pp 71–79.
8. Dorman MA, Noble SA, McBride LJ, Caruthers MH (1984): Synthesis of oligodeoxynucleotides and oligodeoxynucleotide analogs using phosphoramidite intermediates. Tetrahedron 40:95–102.
9. McBride LJ, Caruthers MH (1983): N^6(N-Methyl-2-pyrrolidine amidine) deoxyadenosine. A new deoxynucleoside protecting group. Tetrahedron Lett 24:2953–2956.
10. Beaucage SL, Caruthers MH (1981): Deoxynucleoside phosphoramidites-A new class of key intermediates for deoxypolynucleotide synthesis. Tetrahedron Lett 22:1859–1862.
11. McBride LJ, Caruthers MH (1983): An investigation of several deoxynucleoside phosphoramidites useful for synthesizing deoxyoligonucleotides. Tetrahedron Lett 24:245–248.
12. Sinha ND, Biernat J, Koster H (1983): β-Cyanoethyl N,N-dialkylamino/N-morpholinochloro phosphoamidites, new phosphitylating agents facilitating ease of deprotection and work-up of synthesized oligonucleotides. Tetrahedron Lett 24:5843–5846.
13. deHaseth PL, Goldman RA, Cech CL, Caruthers MH (1983): Chemical synthesis and biochemical reactivity of bacteriophage lambda P_R promoter. Nucleic Acids Res 11:773–787.
14. Tanaka T, Letsinger RL (1982): Syringe method for stepwise chemical synthesis of oligonucleotides. Nucleic Acids Res 10:3249–3260.
15. Matteucci MD, Caruthers MH (1981): Synthesis of deoxyoligonucleotides on a polymer support. J Am Chem Soc 103:3185–3191.
16. Gait MJ, Matthes HWD, Singh M, Sproat BS, Titmas RC (1982): Rapid synthesis of oligodeoxyribonucleotides VII. Solid phase synthesis of oligodeoxyribonucleotides by a continuous flow phosphotriester method on a kieselguhr-polyamide support. Nucleic Acids Res 10:6243–6254.
17. Cell Biology Correspondent (1973): Is enough too much? Nature New Biology 241:33.
18. Alton K, Stabinsky Y, Richards R, Ferguson B, Goldstein L, Altrock B, Miller L, Stebbing N (1983): Production, characterization and biological effects of recombinant DNA derived human IFN-α and IFN-γ analogs. In De Maeyer E, Schellekens H (eds) "The Biology of the Interferons." Amsterdam: Elsevier, pp 119–128.
19. Eisenbeis SJ, Nasoff MS, Noble SA, Bracco LP, Dodds DR, Caruthers MH (1985): Altered cro repressors from engineered mutagenesis of a synthetic cro gene. Proc Natl Acad Sci USA 82:1084–1088.
20. Ohlendorf DH, Anderson WF, Fisher RG, Takeda Y, Matthews BW (1982): The molecular basis of DNA-protein recognition inferred from the structure of cro repressor. Nature 298:718–723.

Protein Engineering, pages 71-82

6

Protein Purification

Richard R. Burgess

Biotechnology Center, and McArdle Laboratory for Cancer Research, University of Wisconsin, Madison, Wisconsin 53706

INTRODUCTION

Most biochemistry graduate students are faced at one time or another during their graduate career with the problem of purifying a protein or an enzyme. Sometimes this entails following a well-documented protocol, but more often it involves developing a new protocol or making modifications on an existing one. I personally love to purify enzymes. I find it an exciting and challenging job. What I would like to convey in this chapter is some sense of the process by which one decides on a purification procedure, and some of the new and exciting trends in protein purification. It will not be possible to go into details either of theory or of practice. The best source currently available to obtain these details is the book on protein purification by Robert Scopes [1], Volume 104 of Methods in Enzymology [2], and the review articles referenced in this chapter.

NATURE OF THE PROBLEM

The challenge of protein purification becomes more clear when one considers the mixture of macromolecules present in a cell extract. In addition to the protein of interest, several thousand other proteins with different properties are present in the extract, along with nucleic acids (DNA and RNA), polysaccharides, lipids, and small molecules. The proteins present in the bacterium, *E. coli*, may be dramatically visualized after resolution by two-dimensional gel electrophoresis [3], as shown in Figure 1. A given protein may be present at up to 10% or at less than 1/1000% of the total protein in the cell. The challenge, therefore, is to separate the protein of interest from all of the other components in the cell, especially the unwanted contaminating proteins, with reasonable efficiency, speed, yield, and purity.

EXTENT AND SCALE OF PURIFICATION SHOULD FIT THE PURPOSE

Just how extensively a protein is to be purified and the scale or size of the purification depend largely on how the protein, the final product, will be used. In general, one would like to obtain a protein product which is free of contaminating proteins and other materials. However, a number of uses of protein can be satisfied with less than completely pure protein. For example, when preparing monoclonal antibodies against a protein, microgram

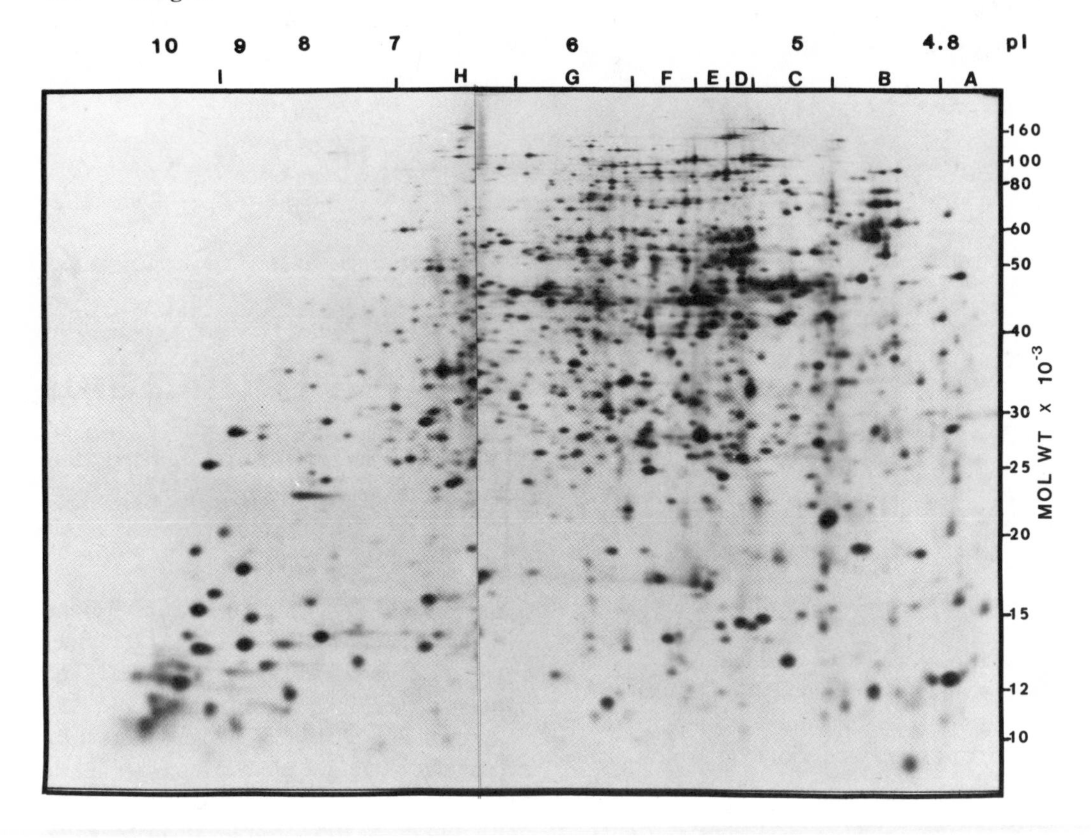

Fig. 1. E. coli *proteins resolved on two-dimensional gels. The approximate isoelectric point and molecular weight scales are indicated.* E. coli *K12, strain W3110, was labeled with* $^{35}SO_4$ *during growth in glucose minimal medium at 37°C. Composite autoradiogram made from nonequilibrium (left side) and pH 5–7 (right side) gels. (Adapted with permission from Neidhardt and Phillips [3]).*

amounts of protein are needed which need not be completely pure. Enzymatic studies (milligram amounts of enzyme) and protein chemistry studies (milligram to gram amounts) require highly purified protein preparations, since the basic physical and enzymatic properties of the enzyme are being measured. Protein structure work likewise requires highly purified protein at the gram level. When enzymes are being used to carry out routine enzymatic reactions, for example, restriction endonuclease reactions, it is not necessary that the protein be 100% pure, but merely that it be free of key contaminating enzyme activities, such as other nucleases, phosphatases, etc. The growing use of proteins as pharmaceuticals requires extraordinarily high levels of purity—greater than 99.99% purity in some cases—in order to remove materials which would be harmful when injected into a human patient. Finally, industrial-level purification, which may require purifications at the level of kilograms to tons, depends largely on the use, and in many cases quite impure protein preparations are suitable to carry out the desired enzymatic reaction. The scale of the purification again depends on the amount of material one desires. Micropurification produces nanogram to microgram amounts of material. A normal biochemical laboratory purification will produce milligram to 100-milligram amounts of protein. Larger-scale purifications may produce gram quantities of enzymes, while pilot plant purifications produce hun-

TABLE I. Amount and Purity Required for Different Uses

Use	Amount needed	Purity required
Immunology	μg–mg	
Polyclonal antibodies		High
Monoclonal antibodies		Medium
Enzymology	mg	High (>95%)
Physical properties	mg–g	High
Enzyme and protein chemistry	mg–g	High
Structure	g	High
Research enzyme	μg–mg	Variable; must be free of key contaminating enzymes
Pharmaceutical	mg–kg	Very high (>99.99%); free of pyrogens and bacterial endotoxins
Industrial enzyme	kg–ton	Variable, often low

dreds of grams to kilogram amounts, and industrial purifications in some cases are capable of producing ton quantities of enzymes. Table I summarizes the purity and amounts of proteins needed for different uses.

PROTEIN PROPERTIES AND HANDLES FOR PURIFICATIONS

The reason one is able to purify one protein from a mixture of thousands of proteins is that proteins vary tremendously in a number of their physical and chemical properties. By exploiting the difference in properties between the protein of interest and other proteins in the mixture, one can design a rational series of separation steps. These properties include:

Size and Shape

Proteins may vary in size from peptides of a few amino acids or a few hundred daltons to very large proteins containing over 3,000 amino acids with molecular weight over 300,000 daltons. Most proteins fall in the range 10,000–150,000 daltons (see Fig. 1). Proteins which are part of multisubunit aggregates may reach much larger sizes. The movement of a protein through a solution during centrifugation or through small pores in membranes, beads, or gels is influenced by the shape of the protein. Shapes range from spherical (globular) to quite assymetric.

Charge

The charge on a protein is determined by the sum of the positively and negatively charged amino acid residues. If a protein has many aspartic and glutamic acid residues, it will have a net negative charge and will be termed an acidic protein. If it has a preponderance of lysine, arginine, and in some cases histidine, it will be considered a basic protein. Obviously the charge on a protein will be determined by the pH of the solution.

Isoelectric Point

The isoelectric point, pI, is that pH at which the charge on a protein is zero and is determined by the number and titration curves of the positively and negatively charged amino acid residues on the protein. Protein pIs generally range from 4.5–8.5 (see Fig. 1).

Charge Distribution

The charged amino acid residues may be distributed uniformly on the surface of the protein, or they may be clustered such that one region is highly positive while another region is highly negative. Such nonrandom charge distribution can be used to discriminate between proteins.

Solubility

Proteins vary dramatically in their solubility in different solvents, all the way from being essentially insoluble to being soluble to a level of hundreds of milligrams per milliliter of solution. Key variables affecting solubility of a protein include pH, ionic strength, nature of the ions, and polarity of the solvent.

Density

The density of most proteins is approximately 1.4 g/cm^3 and is not generally a useful property for fractionating proteins. However, proteins containing large amounts of phosphate or lipid moieties are substantially more or less dense than average, respectively, and may be separated from the bulk of proteins using density gradient methods.

Hydrophobicity

The number and spatial distribution of hydrophobic amino acid residues present on the surface of the protein determine the ability of the protein to bind to other hydrophobic column materials, and therefore can be exploited in fractionation.

Ligand and Metal Binding

Many enzymes bind substrates, effector molecules, cofactors, DNA templates, and certain metal ions (e.g., Cu^{2+}, Zn^{2+}, Ca^{2+}, Co^{2+}) quite tightly. This binding can be used to bind that enzyme to a column to which the appropriate ligand, template, or metal ion has been immobilized.

Reversible Aggregation or Association

Under certain solution conditions, many enzymes will aggregate to form dimers, tetramers, etc. This ability to be a dimer under one condition, for example, and a monomer under another condition can be used if two fractionations based on size are carried out sequentially at those two different conditions [4].

Specific Sequence or Structure

The precise geometric presentation of amino acid residues on the surface of a protein can be used as the basis of a separation procedure if an antibody that recognizes only a particular site on a protein can be obtained. Such a column, with a monospecific antibody affixed that only binds to the protein of interest, is called an immunoaffinity column and can be highly specific.

Posttranslational Modifications

After protein synthesis many proteins are modified in the cells by the addition of carbohydrates to form glycoproteins, lipids to form lipoproteins, or phosphates to form phosphoproteins, or are otherwise modified. In many cases these modifications provide handles which can be used in fractionation. For example, the proteins containing carbohydrates on their surface often can be bound to columns containing lectins, which are molecules capable of binding tightly to carbohydrate moieties.

Unusual Properties

In addition to the types of properties mentioned above, certain proteins have unusual properties that can be exploited during their purification. An example is unusual thermostability. Most enzymes will be inactivated and will coagulate or precipitate if heated to 95°C. A protein that remains soluble and active after such a heat treatment can be separated easily from the bulk of the cellular proteins. Another such property is unusual resistance to proteases. These two properties often go hand in hand. An interesting example of a purification involving these properties is that of the *E. coli* alkaline phosphatase. The cellular extract is heated, and the insoluble coagulated proteins are removed by centrifugation. The supernatant that contains the phosphatase is then treated with a protease which digests the remaining contaminated proteins, leaving an essentially pure preparation of alkaline phosphatase.

TYPES OF PURIFICATION PROCESSES

All of the protein properties listed above are potentially usable in the design of an efficient protein purification protocol. The types of protein purification processes which can be used fall into several general categories. One type of process takes advantage of changes in the solubility of one protein with respect to other proteins. The second type of process involves the partitioning of a protein between two phases. The third category involves the adsorption of the protein of interest to solid matrix, often a column resin, and subsequent elution. The fourth category involves differential movement of the protein of interest, either in an electric field, in a centrifugal field, or through carefully size-controlled pores in a filter membrane or a gel filtration column.

The appropriate combination of a series of purification steps is sufficient in most cases to obtain a purified protein. Even if one only obtains a five-fold purification at each step, four successive steps will result in a $(5)^4$ or 625-fold purification. Table II summarizes some of the more common separation processes and the bases of the separations.

Figure 2 illustrates a hypothetical enzyme purification based on precipitation, adsorption chromatography, and gel filtration chromatography. The protein extract is first subject to addition of increasing amounts of ammonium sulfate. (The addition of ammonium sulfate to 30% of its saturating amount results in the precipitation of only a small amount of protein. This protein precipitate is removed by centrifugation. The supernatant is then adjusted to a higher ammonium sulfate concentration, stirred to allow precipitation, and centrifuged again. This process is repeated so that one takes a series of cuts between 0% and 100% ammonium sulfate saturation. Each precipitate is dissolved in buffer and assayed for total protein and enzyme activity.) That fraction containing the bulk of the enzyme activity is subject to subsequent purification. In this case, it is diluted to a salt concentration which allows the enzyme to bind to the positively charged resin of a DEAE cellulose column. The basic, neutral, or slightly acidic proteins in this sample will flow through this column, whereas the more acidic proteins will bind. As the salt concentration in the elution buffer (a salt gradient elution) is gradually raised, the proteins will elute based on their affinity for the ion exchange resin. Again the fractions are assayed for protein and enzyme activity and the fractions containing the bulk of the enzyme

TABLE II. Protein Separation Processes

Separation process	Basis of separation	Reference
Precipitation		
Ammonium sulfate	Solubility	[1]
Acetone	Solubility	[1,5]
Polyethyleneimine (Polymin P)	Solubility, charge	[6]
Isoelectric	Solubility, pI	[1]
Phase partitioning (e.g., polyethylene glycol, PEG)	Solubility	[2,7]
Chromatography		
Ion exchange	Charge, charge distribution	[1,2,8]
Hydrophobic	Hydrophobicity	[1,2]
Affinity	Ligand binding site	[1,2,8,9,10]
Immobilized metal affinity (IMAC)	Metal binding	[11]
Immunoaffinity	Specific antigenic site	[1,2,12,13]
Chromatofocusing	pI	[1,8]
Gel filtration	Size, shape	[1,2,8]
Electrophoresis		
Gelelectrophoresis	Charge, size	[1,2,3,5]
Isoelectric focusing	pI	[1,2,3,8]
Centrifugation	Size, shape, density	[4]
Ultrafiltration	Size, shape	[14]

START

100gm wet weight E. coli =20gm dry weight=12gm protein (100%enzyme,100% protein, purity factor=1)

Ammonium Sulfate Precipitation

(45-50% cut has 75% of enzyme, 15% of total protein, purity factor=5)

pooled

protein

enzyme

0 30 35 40 45 50 55 60 70 100

% saturation

Ion Exchange Chromatography

salt gradient elution (pooled fractions contain 60% of enzyme,2% of protein, purity factor=30)

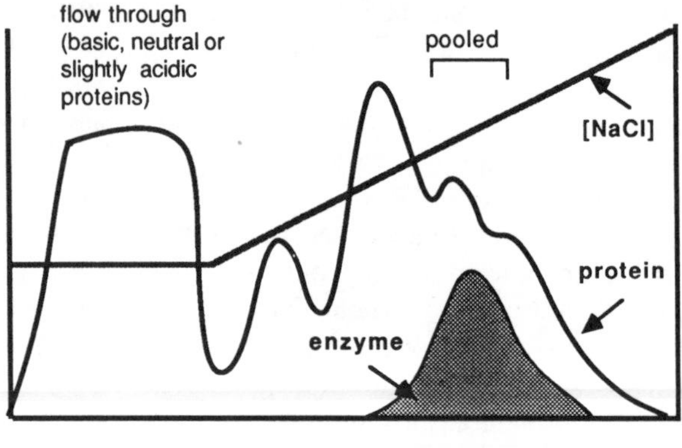

fraction number

Gel Filtration Chromatography

(pooled fractions contain 45% of enzyme, 0.3% of protein, purity factor=150)

void volume (large protein aggregates)

column volume

pooled

protein

enzyme

fraction number

Dialysis into storage buffer containing 50% glycerol for storage at -20°C or -70°C

FINAL PRODUCT

(36 mg)

Fig. 2. *Hypothetical enzyme purification involving ammonium sulfate precipitation, ion exchange chromatography, and gel filtration chromatography.*

activity pooled. The pooled and concentrated material is applied to a gel filtration column containing small, spherical beads with pores of defined size. The large proteins or aggregates in the sample cannot enter the pores, are excluded from the beads, and move through the column rapidly. They are found in the "void volume." Smaller proteins are able to penetrate into the porous beads that pack the column and move down the column at different rates, depending on the fraction of the time they spend inside the gel filtration bead out of the buffer flow. Last to elute are the very small proteins and salts in the original sample, at the "column volume." Fractions containing the enzyme activity are pooled; in this particular case, they are found to be free of contaminating enzymes by a variety of determinations. These fractions are dialyzed into a storage buffer often containing 50% glycerol and stored at −20° or −70°C.

GENERAL STRATEGY

Several principles are important in developing an efficient high-yield purification procedure for a given enzyme.

1) *One must develop an assay.* This assay will allow one to follow an enzyme through the various fractionation procedures. Classically this was an assay for the enzymatic activity of the enzyme of interest. Sodium dodecyl sulfate (SDS) polyacrylamide gel electrophoresis can also be used when the protein is a major band on the gel or migrates in a position on the gel that is unique [6]. Another method being used more recently is an immunological assay to detect the presence of a given protein in a fraction [15].

2) *One must choose a source as rich as possible in the desired product.* This often entails investigating the effects of growth conditions on the enzyme content of a bacterial cell or measuring the distribution of an enzyme in the various tissues of an organism. In those cases where the desired product is present at very low concentrations, large amounts of starting material must be used in order to obtain reasonable amounts of final product. More recently, the ability to clone genes and overproduce proteins in bacteria has vastly increased the amount of product present in a given amount of starting material (see below).

3) *One must use the minimal number of steps necessary to separate the desired products from the rest of the initial components.* The protein properties above provide more handles than one generally needs to purify a protein. One chooses a series of steps which result in the greatest factor of purification per step and eliminates steps which give only small purification factors. Since each step invariably leads to some losses, repeating a step rarely is productive. In those cases where a powerful fractionation step is available (such as in immunoaffinity chromatography, where a crude extract can be passed over a column containing an immobilized antibody and only the protein of interest retained), one can design purification procedures which involve very few steps, are very rapid, and often produce protein in very high yield.

4) *One must take precautions during a purification to minimize damage, irreversible inactivation, or loss of enzyme.* Generally this means working quickly, often at 4°C rather than at room temperature, and avoiding excess exposure to oxygen (for example, avoiding aeration or frothing of a protein solution). In addition, certain chemicals can be added to the buffer. These include protein stabilizers such as glycerol [16], a sulfhydryl reagent such as dithiothreitol to minimize oxidation of cysteine residues, and a metal chelater such as EDTA to prevent exposure of the enzyme to heavy metal ions that might inactivate the enzyme. In addition, loss of enzyme can occur due to degradation by proteases present during the preparation. Therefore, the presence of protease inhibitors is often beneficial. Finally, losses of enzyme due to nonspecific adsorption to the column material, the columns themselves, or the tubing can often be decreased or minimized by the presence of mild nonionic detergents, such as NP40, in the buffers [17]. Such detergents are required if

the proteins are, for example, membrane proteins and aggregates or are insoluble in buffers lacking detergents.

RECENT TRENDS IN PURIFICATION

The hypothetical enzyme purification in Figure 2 is an example of what would be considered a classic laboratory enzyme purification, and is typical of the sorts of purifications that have often been done in the past and still find wide usage. In addition, however, a number of more recent fractionation techniques have been developed which are excellent additions to the separation techniques available in designing a purification procedure. Some of these new methods result in higher resolution and greater specificity and therefore greater purification by a given step. The column procedures are often very fast (minutes instead of hours) and in many cases are capable of being scaled up to larger scale purifications, or are able to be automated or subject to continuous operation. Below are summarized some of these recent advances, with a brief summary and a reference to a recent review article where more references can be found.

Elution of Proteins From Polyacrylamide Gels [5]

For some time, the high resolution of SDS polyacrylamide gel electrophoresis has been apparent [3]. An example of the use of polyacrylamide gels as a micropurification step has been presented [5]. In the procedure, microgram amounts of material were separated from contaminating proteins on high-resolution SDS polyacrylamide gels. The band was located, and the protein of interest was eluted from the gel. The difficulty, of course, was that the eluted protein contained a detergent, SDS. It was found, however, that the SDS could be removed and the protein concentrated by addition of four volumes of cold acetone. The resulting precipitate could be dissolved in a small volume of a denaturant such as guanidinium hydrochloride (GuHCl). Upon dilution, the denatured protein was able to refold, forming an active enzymatic form. This procedure has been used widely to purify microgram amounts of proteins, often after minimal prior purification. In many cases enough protein can be obtained to produce antibodies or for protein microsequencing experiments. In several cases, new enzymatic activities have been identified by this method. This method will not work if the active enzyme contains two or more different-sized polypeptides or contains a noncovalently attached cofactor, because components necessary for activity will be separated from each other during electrophoresis.

General Affinity Columns [2,9–11]

The early affinity columns first described in 1968 by Cuatrecasas, Anfinsen, and colleagues involved the immobilization of a substrate or cofactor of an enzyme onto a column that would then bind the desired enzyme out of a crude enzyme mixture [9]. Each such affinity chromatography procedure required considerable knowledge about the nature and affinity of ligands that the enzyme of interest bound. More recently, several more general affinity chromatography methods have been described, one called the immobilized metal affinity chromatography or IMAC [11]. Metal ions, such as copper, cobalt, or zinc are attached to a column in such a way that they can form coordination bonds with unprotonated histidine and tryptophan residues on proteins. In this way proteins containing histidine and tryptophan residues oriented with the proper geometry on the surface of the enzyme can be bound to the column and eluted with a decreasing pH gradient.

A second general-affinity chromatography procedure involves the use of triazine dyes [10]. These dyes, such a Cibacron Blue and Procion Red, are attached to column resins, and the mixture of proteins is passed through the column. Only proteins with certain surface properties will bind to these resins and can be eluted, often with significant purification. These resins have the advantages of being relatively inexpensive compared to some more specific affinity resins and therefore it is possible that these purification procedures can be scaled up. Since a variety of such dyes exist, different dye-affinity columns can be tried until the appropriate one for a given enzyme is found.

Immunoaffinity Chromatography [12,13]

The advent of monoclonal antibodies has opened up a very exciting opportunity for highly specific purification of enzymes. A monoclonal antibody can be prepared that only reacts with the enzyme of interest in a very complex mixture of enzyme. This monoclonal antibody can be produced in large quantities, attached to a column resin, and be used to pull out the desired protein. The advantage of this procedure is the very high specificity and therefore the very high purification factor obtainable. In some cases greater than 10,000-fold purification can be achieved with this one step. The drawbacks include the fact that many antibody protein complexes are extraordinarily stable and require harsh elution conditions, such as pH 3. More recently antibodies have been isolated that allow the elution of bound enzymes by more mild conditions, thus opening up more general use of immunoaffinity chromatography for the purification of labile or easily damaged proteins [13; Burgess, unpublished results].

High-Performance Liquid Chromatography (HPLC) [2]

In the last few years, an explosion has occurred in the types of column resins available for protein purification. High-performance liquid chromatography involves highly uniform rigid resins with a variety of functional groups attached which allow either reverse phase fractionation, hydrophobic affinity chromatography, gel filtration chromatography, or ion exchange chromatography to be carried out at high pressure. As a result, the speed is greatly increased (often an entire protein fractionation on a HPLC column will take place in 20–60 minutes). In addition, in many cases the resolution is outstanding, leading to high purifications in very short times. Such columns are usually used with sophisticated equipment that allows precise and reproducible salt gradient production, excellent "plumbing" to avoid leaking and excessive dead volume in the system, sensitive detection systems, and often computer storage and manipulation of chromatographic data. Some skeptics refer to HPLC as "high-priced liquid chromatography."

Phase-Partitioning [2,7]

When appropriate mixtures of polyethylene glycol (PEG), salt, and water are mixed, a two-phase system occurs. The partitioning of a given protein between the two phases can be influenced by the salt and polymer concentrations, temperature, and pH used. Therefore, a protein can be fractionated by first being partitioned into one phase and then, after changing the conditions, being partitioned into the second phase. These procedures have been utilized in the large-scale purification of enzymes and seem to lend themselves well to scale-up in addition to being gentle procedures that often seem to stabilize enzymes.

Ultrafiltration [14]

In recent years, the ability to manufacture membranes containing very defined pore sizes has allowed one great latitude in separating proteins from solvents. It is possible, for example, to pass a protein mixture through an ultrafiltration unit in such a way that the water, buffer, salts, and the proteins smaller than 10,000 daltons will readily pass through the filter while larger proteins will be retained. If the retentate is now passed through a filter apparatus containing a 50,000-dalton cut-off filter, all the proteins 50,000 daltons or smaller will pass through the filter. In this way one can isolate a protein fraction containing proteins in the 10,000–50,000-dalton size range. This can be done on a large scale in a continuous fashion and is ideally suited for size fractionation at the pilot plant or the industrial scale. Ultrafiltration membranes are also extremely valuable for rapid concentration of protein solutions, removal of salts, and even for removal of cell debris at the early stages of purification.

Analysis [2,17]

In many of the micropurification procedures, if one is to analyze fractions for the presence of an enzyme, one needs highly sensitive assays. One such advance in analysis has come from the use of high-resolution polyacrylamide gels coupled with either highly sensitive staining methods [2] involving silver or gold, or in the use of monoclonal antibodies in detecting the presence of the protein [15]. Such sensitive procedures allow one to detect nanogram, (in some cases subnanogram), quantities of proteins, and to assay fractions even on a very small-scale preparation and still detect the presence of protein bands on gels.

PURIFYING PROTEINS THAT HAVE BEEN CLONED AND OVERPRODUCED IN *E. COLI*

A very exciting development in the last few years has been the ability to isolate the gene coding for an enzyme of interest and expression of that gene product in *E. coli* at high levels, often 10%–40% of the total cell protein. This has tremendously increased our ability to obtain gram quantities of interesting or important enzymes which previously had been extremely scarce and thus extremely expensive or difficult to purify. Table III outlines the process by which one can produce a large amount of a scarce protein, starting from microgram amounts of that protein. Once a strain of *E. coli* has been constructed in which the gene of interest can be expressed at very high levels, the problem, however, is not over. If the protein product is soluble, then conventional methods for purification can be used. Since one is starting with an extract that is highly enriched in the protein of interest, purification is often straightforward and rapid. However, in the majority of the cases a situation occurs that is both a problem and in some cases an advantage. This situation is the formation of inclusion bodies, insoluble precipitates, within the cell containing the overproduced protein. In more than 80% of the cases where proteins have been overproduced in *E. coli*, the proteins have proved to be largely insoluble and have formed these inclusion bodies. However, the high concentration of the desired protein in the inclusion body has now been used to advantage. An example of a purification is given below. When the sigma subunit of *E. coli* RNA polymerase is overproduced 200-fold in *E. coli*, this 70,000-dalton acidic protein is insoluble [18]. A purification has been developed, which is summarized below:

1) The cells are lysed on ice with Tris, lysozyme, and EDTA followed by 0.05% sodium deoxycholate. The crude extract contains about 10% sigma.
2) Cell debris and insoluble sigma are pelleted by low-speed centrifugation.
3) The pellet is washed with buffer to remove trapped soluble proteins. The washed pellet is about 70% sigma.
4) The pellet is dissolved in 6M GuHCl and 0.1 mM dithiothreitol to denature precipitated sigma, and then gradually diluted 60-fold to allow renaturation of sigma.
5) Materials that reprecipitate after diluting the GuHCl are removed by centrifugation.
6) The supernatant, containing about 90% sigma, is purified by rapid chromatography on a DEAE cellulose column or by gel filtration chromatography.

A comparison of this method with a more conventional procedure for purifying sigma is given in Table IV.

The renatured sigma was shown to be identical to conventionally prepared sigma that had never been denatured and renatured by showing that the two preparations had similar specific activities and identical kinetics of thermal inactivation and partial proteolysis. Many proteins have been purified from insoluble inclu-

TABLE III. How to Produce Large Amounts of Scarce Proteins

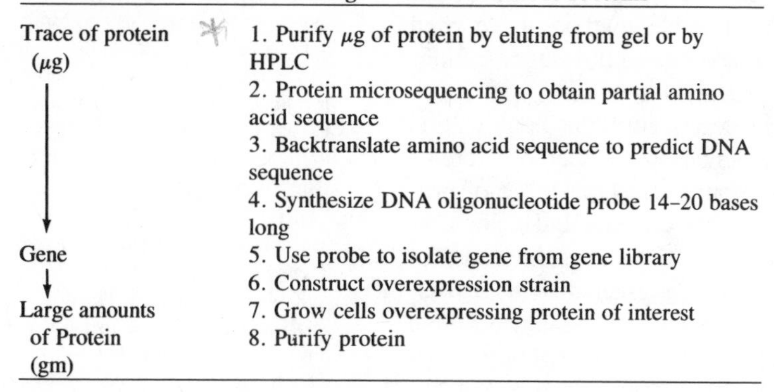

Trace of protein (μg)	1. Purify μg of protein by eluting from gel or by HPLC
↓	2. Protein microsequencing to obtain partial amino acid sequence
	3. Backtranslate amino acid sequence to predict DNA sequence
	4. Synthesize DNA oligonucleotide probe 14–20 bases long
Gene	5. Use probe to isolate gene from gene library
↓	6. Construct overexpression strain
Large amounts of Protein (gm)	7. Grow cells overexpressing protein of interest
	8. Purify protein

TABLE IV. Comparison of Old and New Methods for Purifying Sigma

	Old method [19]	New method [18]
Starting material	500 g *E. coli* (0.05% sigma)	50 g (10% sigma)
Time of preparation	2 weeks (5 columns)	1–2 days (1–2 columns)
Yield	1–2 mg	60–100 mg
Purity	96%	98–99%

sion bodies by this or similar methods of denaturation and renaturation. In certain cases, formation or reformation of correct disulfide bridges in nonbacterial proteins has been successfully completed after isolating the protein from the cell [20]. The problem of degradation of foreign proteins expressed in *E. coli* has been diminished by the use of special *E. coli* strains defective in cellular proteases [21].

An interesting strategy for purifying proteins expressed in *E. coli* has recently been presented [22]. Recombinant DNA technology has been used to produce a human enzyme, urogastrone, with five extra arginines fused to the carboxyterminus. The unusual basicity of the polyarginine fusion protein allowed substantial purification on a cation exchange column. The polyarginine was then removed by enzymatic cleavage with carboxypeptidase B, and the resulting urogastrone was purified by rechromatography.

This ability to produce large amounts of scarce proteins is not applicable in all cases. Certain proteins are coded by genomes containing introns, and therefore, in order for this above procedure to be used successfully, either cDNA clones must be prepared or the introns must be removed from genomic clones. In addition, certain proteins require for their enzymatic activity, posttranslational modifications which do not occur in *E. coli*. Such proteins may need to be expressed in other cells (such as yeast or mammalian cells) where such posttranslational modifications may be correctly carried out.

CONCLUSION

Protein purification is still largely an empirical science (or art?). Very rarely, except in the case of specific affinity columns, can one predict precisely whether a particular fractionation step will be effective. In an increasing number of cases the amino acid sequence of the protein to be purified is known from DNA sequencing of its gene. In these cases one can predict size, charge, and approximate isoelectric point. However, until methods for accurately predicting three-dimensional structure from sequence are available, properties such as shape, distribution of charge and hydrophobic residues, and ligand binding sites cannot be predicted. Even with complete structural information, no reliable methods have been developed to predict solubility, aggregation, modification, or stability.

The advances described in this book will be applied to an increasing number of proteins both to increase our basic knowledge of protein structure and function and to develop useful proteins for industrial and pharmaceutical purposes. Therefore a great demand exists, and will continue to exist, for scientists with a knowledge of how to use known protein properties, fractionation processes, empirical observations, and intuition to devise and carry out effective protein-purification schemes.

REFERENCES

1. Scopes R (1982): "Protein Purification: Principles and Practice." New York: Springer-Verlag, pp 1–277.
2. Jakoby WB (1984): "Enzyme Purification and Related Techniques." Methods in Enzymology, Vol 104 New York: Academic Press, Inc, pp 1–503.
3. Neidhardt FC, Phillips TA (1985): The protein catalog of *E. coli*. In Celis JE, Bravo R (eds): "Two-Dimensional Gel Electrophoresis of Proteins." New York: Academic Press, Inc, pp 417–444.
4. Burgess RR (1969): A new method for the large scale purification of *E. coli* DNA-dependent RNA polymerase. J Biol Chem 244:6160–6167.
5. Hager D, Burgess RR (1980): Elution of proteins from SDS-polyacrylamide gels, removal of SDS, and renaturation of enzymatic activity: results with sigma subunit of *E. coli* RNA polymerase, wheat germ topoisomerase and other enzymes. Anal Biochem 109:70–86.
6. Burgess RR, Jendrisak JJ (1975): A procedure for the rapid, large-scale purification of *E. coli* DNA-dependent RNA polymerase involving Polymin P precipitation and DNA cellulose chromatography. Biochemistry 14:4634–4638.
7. Hustedt H, Kroner KH, Menge U, and Kula MR (1985): Protein recovery using two-phase systems. Trends in Biotech 3:139–144.
8. Pharmacia Fine Chemicals AB Publications, Uppsala, Sweden: (1979) Affinity Chromatography: Principles and Methods; (1980) Ion Exchange Chromatography: Principles and Methods. (1981) Chromatofocusing with Polybuffer™ and PBE™. (1982) Gel Filtration: Theory and Practice. (1982) Isoelectric Focusing: Principles and Methods.
9. Parikh I, Cuatrecasas P (1985): Affinity Chromatography. Special Report, Chem and Eng News, August 26, 17–32.
10. Qadri F (1985): The reactive triazine dyes: their usefulness and limitations in protein purifications. Trends in Biotech 3:7–11.
11. Sulkowski E (1985): Purification of proteins by IMAC (immobilized metal-affinity chromatography). Trends in Biotech 3:1–6.
12. Chase HA (1984): Scale-up of immunoaffinity separation processes. J Biotech 1:67–80.
13. Hill CL, Bartholomew R, Beidler D, David GS (1983): "Switch" immunoaffinity chromatography with monoclonal antibodies. BioTechniques 1:14–17.
14. Tutunjian RS (1985): Scale-up considerations for membrane processes. Bio/Technology 3:615–626.
15. Howe JG, Hershey JWB (1981): A sensitive immunoblotting method for measuring protein synthesis initiation factor levels in lysates of *E. coli*. J Biol Chem 256:12836–12839.
16. Gekko K, Timasheff SN (1981): Mechanism of protein stabilization by glycerol: preferential hydration in glycerol-water mixtures. Biochemistry 20:4667–4676.
17. Faras AJ, Taylor JM, McDonnell JB, Levinson WE, Bishop JM (1972): Purification and characterization of the DNA polymerase associated with Rouse Sarcoma Virus. Biochemistry 11:2334–2342.
18. Gribskov M, Burgess RR (1983): Overexpression and purification of the sigma subunit of *E. coli* RNA polymerase. Gene 26:109–118.
19. Lowe P, Hager D, Burgess RR (1979): Purification and properties of the sigma subunit of *E. coli* RNA polymerase. Biochemistry 18:1344–1352.
20. Marston FA, Lowe PA, Doel MT, Schoemaker JM, White S, Angal S (1984): Purification of calf prochymosin (prorennin) synthesized in *E. coli*. Bio/Technology 2:800–804.
21. Baker TA, Grossman AD, Gross CA (1984): A gene regulating the heat shock response in *E. coli* also affects proteolysis. Proc Natl Acad Sci USA 81:6779–6783.
22. Sassenfeld HM, Brewer SJ (1984): A polypeptide fusion designed for the purification of recombinant proteins. Bio/Technology 2:76–81.

Protein Engineering, pages 83–90

7

Protein Folding

Thomas E. Creighton

MRC Laboratory of Molecular Biology, Hills Road, Cambridge CB2 2QH, United Kingdom

INTRODUCTION

A protein generally must be folded into a precise three-dimensional arrangement before it is biologically active [1]. This process usually occurs spontaneously after synthesis of the linear polypeptide chain, and natural proteins from biological sources are generally purified in their folded conformations, especially if they have been assayed by their biological activity. However, proteins that have been produced by chemical modification procedures or synthesized, either chemically or biologically, to high levels in the foreign environment of a different organism by recombinant DNA technology, are often isolated in unfolded forms and must be folded before a useful product is obtained. Fortunately, folding is usually a self-assembly process in which the folded conformation is specified solely by the covalent structure of the protein [2], but this is apparent only under the right conditions (and might be doubted by someone trying to purify a labile protein that spontaneously loses its biological activity). Also, the folding process need not occur efficiently, for it is very complex, probably too complex to occur randomly on a finite time scale [3].

FOLDED AND UNFOLDED PROTEINS

Polypeptide Conformational Flexibility

Protein three-dimensional structure arises because of the ability of the linear polypeptide chain to rotate about single bounds [4]. There are two such bonds of the backbone for each amino acid residue designated by the torsional angles ϕ and ψ:

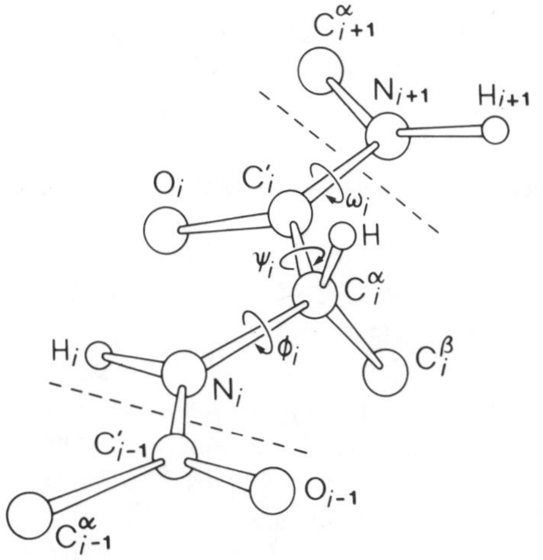

The third bond of the backbone, the peptide bond, is constrained to be planar, in either the *cis* or *trans* form ($\omega = 0°$ or $180°$); the *trans* form is the more stable, unless the next residue is Pro. Most bonds of the side chains are also rotatable.

Unfolded proteins have virtually complete freedom throughout the polypeptide chain [5], like a random coil, with only the limitation that atoms cannot occupy the same space. Steric constraints between atoms close in the covalent structure impose some limitations on the possible values of ϕ and ψ [4]. Unfolded proteins generally have average dimensions close to those expected for ransom coils, all their groups are fully exposed to solvent and other reagents, and bond rotations are occurring very rapidly. In contrast, folded proteins are very compact, with their interiors as closely packed as organic crystals [6], have many groups very protected from the solvent [7], and have greatly diminished flexibility due to steric interactions between neighboring atoms [8]. Any of these criteria may be used to distinguish between the folded and unfolded states [5].

A conformation can be specified closely for the folded state, with the positions of atoms well defined and often determined accurately by X-ray crystallography, but this is not possible for the unfolded state. It is rapidly interconverting between most, if not all, of the many conformations possible. The large number of combinations of different torsion angles all along the polypeptide chain results in there being an extremely large number of conformations possible. For example, if an amino acid residue can be considered to adopt an average of 10 conformations, a small polypeptide of 100 residues could exist in 10^{100} conformations, minus the number of those impossible because of steric overlap. Even a much more modest estimation with two conformations per residue would yield 2^{100}, or 10^{30}, conformations. Consequently, only the average properties of the unfolded state may usually be defined, and there is still considerable controversy and uncertainty about the extent to which intramolecular interactions cause departures from randomness under any particular set of conditions.

Requirements for a Folded Conformation

It is likely that only a very small fraction of all the possible polypeptides, with random amino acid sequences, are able to adopt folded conformations. Most polypeptides produced synthetically do not. Also, some small alterations of the covalent structures of folded proteins, such as removing a few residues from one end, cause them to unfold [9]; if the folded conformation is made impossible in some way, an unfolded protein results. No folded protein is known to adopt two or more substantially different folded conformations: conformational changes in proteins appear to involve primarily rearrangements of structural domains or subunits [1].

For a protein to fold, the folded state must be kinetically accessible and have a lower free energy than the unfolded state. The free energies of the two states are determined by all the physical interactions that take place within the protein molecule and with the solvent, plus all entropic considerations. The major factor favoring the unfolded state is believed to be its much greater conformational entropy, resulting from its occurrence in very many different conformations. The interactions responsible for the folded state are very weak individually [1]. They are able to stabilize a folded conformation only when very many of them are present simultaneously and cooperate. Furthermore, they contribute to the stability of the folded state only to the extent to which they are more favorable, primarily by being intramolecular, than the similar intermolecular interactions between the unfolded state and the solvent. The folded state also has the advantage of having less surface accessible to the solvent; nonpolar groups and water interact less favorably with each other than with themselves, so this hydrophobic effect favors the folded state. There is a rather fine energetic balance between the folded and unfolded states, so proteins usually adopt fold conformations only under a rather limited range of conditions, e.g., solvent, temperature, pH, and pressure.

Denaturants

Denaturants are substances that unfold proteins upon being added to the solvent; the best known are guanidinium chloride (GuHCl) and urea. All denaturants probably act by solvating more equally than water all parts of the protein, especially the nonpolar moieties. GuHCl and urea have been shown directly to diminish the hydrophobic interaction. Such solvation phenomena are often thought of as resulting from physical binding, and GuHCl and urea are often said to act by binding to the unfolded state preferentially, due to the greater number of binding sites. However, there is nothing special about such denaturants, and they are just extreme examples of the very many diverse compounds that affect the properties of water and protein solvation, and hence the relative stabilities of the folded and unfolded states [10]. Many increase protein stability, even the sulfate salt of the guanidinium ion.

Since the hydrophobic effect undoubtedly contributes to stability of the folded protein, it might be thought that the folded conformation would be dependent upon the presence of aqueous solvent. This is not the case, for dried proteins have increased stability to thermal unfolding [11]; apparently, removing the water has a greater effect upon the free energy of the unfolded state because of its greater number of polar groups that need to be solvated. This is just one example of how rationalizing protein stability requires careful consideration of the effects on the relative free energies of both the folded and unfolded states.

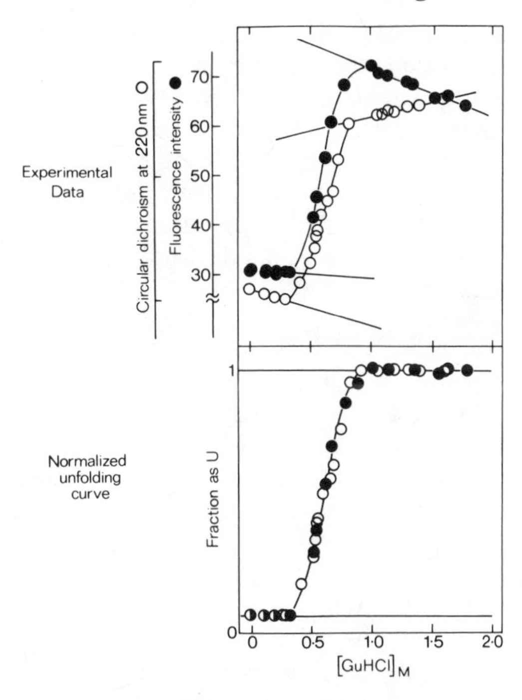

Fig. 1. *The GuHCl-induced unfolding transition of yeast phosphoglycerate kinase detected by the fluorescence intensity of 340 nm (●) and circular dichroism molar ellipticity at 220 nm (○). The experimental data are shown at the top; the straight lines show the effects of GuHCl on the spectral properties of the folded state at low concentrations, and on the unfolded state at high concentrations. The same effects are assumed throughout the transition region. Correcting for this, the fraction of unfolding indicated by the two spectral measurements is plotted in the lower half. The two curves coincide and are consistent with a two-state equilibrium unfolding transition. (From Nojima et al. [19].)*

FOLDING BY ALTERING THE ENVIRONMENT

Equilibrium Transitions

The equilibrium between the folded and unfolded states may be varied by gradually altering the environment, e.g., the solvent, temperature, or pressure (Fig. 1). The degree of folding may be followed by any of many methods sensitive to the protein conformation, but it should quantitatively measure the fractional degree of unfolding, f_u:

$$f_{\mathrm{u}} = \frac{\theta_N - \theta}{\theta_N - \theta_U}$$

where θ is the value of the parameter measured, θ_N and θ_U are the values of the same parameter for the folded and unfolded states, respectively. With many measurements, the unfolding transition is superimposed upon more gradual variations within the unfolded and folded states (Fig. 1). Therefore, the values of θ_N and θ_U must be extrapolated through the transition region to calculate the value of f_u.

With small proteins, the folding transitions detected by different measurements usually coincide, consistent with there being a simple two-state equilibrium transition between the folded and unfolded states

$$N \longleftrightarrow U$$

Consequently, $f_u = 0.5$ means that half the molecules are in state N; half in U. Intermediate, partially folded states are not populated substantially, because they are unstable relative to either *U* or *N* as a result of the cooperativity of protein folded structures.

More complex transitions can be observed with large proteins, particularly where individual domains fold independently, and with some small proteins, but the latter are not yet understood.

Energetics of Folding

Within the transition region of a two-state reaction, the values of the equilibrium constant K_U, between the *U* and *N* states, and their relative free energies, may be measured as follows (*ln* ≡ natural logarithm):

$$K_U = \frac{[U]}{[N]} = \frac{f_u}{1\text{-}f_u}$$

$$\Delta G_f = G_N - G_U = RT\, \ell n K_U$$

The value of ΔG may be obtained at other conditions, outside the transition region, only by extrapolation. Its value within the transition region usually varies linearly with denaturant concentration and is so extrapolated outside it, although there is no theoretical basis for doing so. Thermodynamic measurements of the folding transition, varying the temperature, are much more soundly based, and the elegant studies of Privalov [12] have supplied a plethora of accurate data. In spite of the complexity of proteins and their folding transitions, various small proteins show roughly similar behavior. In particular, they have maximal stability (i.e., most negative ΔG_f) at a temperature of about 5°C, with the stability decreasing at both higher and lower temperatures; hence, proteins may be unfolded by both cooling and heating. The pH is also important for stability of the folded state, because most proteins are ionized. The net charge of a protein is probably relevant to its stability, but the major pH-dependent factor is the inability of groups buried within folded proteins to ionize. In particular, proteins with buried His residues are common; these proteins tend to unfold under acidic conditions.

All measurements of protein stability indicate that folded conformations are only 5–15 kcal/mole more stable than the unfolded state under optimal conditions.

Disulfide Bonds

Of the many interactions within proteins, disulfide bonds between Cys residues are unique in that they may be varied independently of the others [13]. This is a result of the redox nature of the disulfide interaction, involving electron donors or acceptors

$$2\text{–SH} \longleftrightarrow \text{–S–S–} + 2H^+ + 2e^-$$

Consequently, proteins with disulfides may be unfolded with or without breaking the disulfides. Retaining the disulfide in the unfolded state decreases its conformational entropy, thereby destabilizing it. As a consequence, disulfide bonds tend to stabilize the folded state.

Some proteins are dependent for folding on the presence of their disulfide bonds, so merely reducing the disulfides is sufficient to unfold these proteins, with no requirements for other denaturants. All proteins are synthesized with Cys residues in the thiol form, so folding accompanies disulfide bond formation.

Practical Considerations

If protein folding transitions occurred as described above, proteins could be folded and unfolded simply by altering the conditions. However, complications often occur, usually as a result of insolubility of the unfolded protein under conditions required for folding. The nature of these intermolecular interactions is not known; they may be nonspecific, or they may be due to more specific interactions between partly folded molecules. They appear to be more prevalent among larger, more complex proteins, those composed of multiple domains or subunits, where specific interactions between folded domains on different molecules could be imagined. Folding is then an intramolecular process competing with an intermolecular one and is favored by lowering the protein concentration. It can also be favored by adsorbing the unfolded protein reversibly to a solid support, such as an ion-exchange resin, in such a way as not to interfere with folding.

Other complications can occur. Unfolded proteins are very susceptible to proteolytic cleavage; this is especially disastrous if the protein has proteolytic activity, for the first few molecules to fold can rapidly degrade the remaining unfolded molecules. No standard protocol can be given for folding proteins in general, for each protein is unique, with its own folding and solubility properties.

RATES OF FOLDING

The rates of folding and unfolding vary enormously, depending upon the protein and the conditions. The fastest reported folding reactions occur on the millisecond time scale, as for the fast-refolding form of ribonuclease A with the four disulfides intact [14], but the rate depends upon the conditions. Other proteins, generally larger and more complex, fold more slowly, on the minute time scale, even under optimal conditions. The rates of unfolding of proteins are generally more dependent on the conditions (e.g., denaturant concentration) than are the rates of folding.

There have been many detailed studies of protein-folding kinetics in the hope of detecting and identifying intermediates in the process. Unstable intermediates might be populated kinetically, since the kinetic observations can be carried out under conditions favoring folding, by rapidly changing from unfolding conditions to those where the folded state is most stable. However, such intermediates will be transient, and only certain intermediates could accumulate: those that occur before the rate-limiting step of the pathway and have free energies comparable to or less than that of the starting state.

The kinetics of unfolding are usually simple, with unfolding occurring in a single step, and no partially unfolded intermediates detectable. In contrast, the kinetics of refolding are generally observed to be complex, with two or more kinetic phases apparent. Such complexities were initially ascribed to the presence of intermediates, but more thorough studies showed that they are usually due to the presence of different populations of the unfolded protein that are only slowly interconverted. It must be remembered that the unfolded protein is a very heterogeneous population, with each molecule probably having at each instant of time a distinct conformation that is rapidly changing. A single rate of folding could only be observed if all these conformations could interconvert more rapidly than folding occured, which would be prima facie evidence for a limited number of intermediate states. Rather than intermediates, the different kinetic phases are primarily due to the unfolded protein having different *cis–trans* isomers of peptide bonds preceding Pro residues. The two isomers have comparable free energies in the unfolded state but are only slowly interconverted. Each peptide bond is always either *cis* or *trans* in the folded state, and only a fraction of the unfolded molecules have the right isomers of all the bonds. The effect of having wrong *cis* or *trans* isomers is still under study, but it certainly varies for each bond. In some cases, folding may be blocked, whereas there is evidence for a transient accumulation of folded states with certain incorrent isomers [14,15].

The observed rates of folding are very many orders of magnitude greater than those predicted for a random search of all possible conformations [3], implying that there are a limited number of intermediate conformational states. However, many such states are still possible, since single-bond rotations in an unfolded protein occur approximately 10^{13} times per second. Also, diffusion will bring most pairs of groups into contact about 10^5 times per second [1].

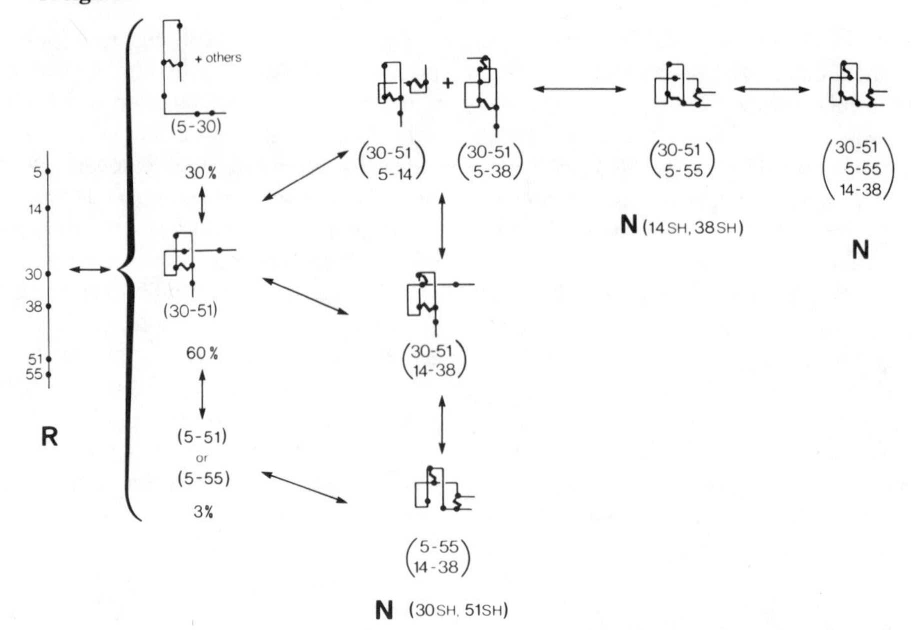

Fig. 2. *The pathway of folding and disulfide formation in BPTI. The polypeptide chain of 58 amino acid residues is shown schematically as a solid black line, with the positions of the six Cys residues indicated. The residue numbers of the Cys residues paired in disulfide bonds are indicated below each species. Those that adopt a stable native-like conformation are indicated as N with any free thiol groups indicated.*

Reduced BPTI (R) is very unfolded and consequently forms initial disulfides nearly randomly. However, these disulfides are interchanged rapidly, so a single arrow is depicted and a bracket encloses the non-random spectrum of one-disulfide intermediates, with their relative levels of accumulation indicated. The predominant intermediate (30–51) readily forms 3 second disulfides. The species (5–55, 14–38) arises from a minor one-disulfide intermediate, probably either (5–51) or (5–55). Intermediates (30–51, 5–14) and (30–51, 5–38) have comparable kinetic roles in the rearrangements of these intermediates to (30–51, 5–55) and are indicated together, with a "+" between them. The resulting native-like intermediate (30–51, 5–55) very rapidly forms the third disulfide bond to complete refolding. (From Creighton and Goldberg [20].)

MECHANISMS OF FOLDING

Intermediates

Elucidating the pathway and mechanism of protein folding requires knowledge of the intermediate conformations that are encountered, their kinetic roles, and how they are interconverted. In spite of their presumed importance, relatively few folding intermediates have been detected and characterized. It appears that unfolded proteins very rapidly adopt relatively compact conformations when placed under refolding conditions. Their hydrodynamic volume decreases [16], and many of their amide hydrogen atoms are protected from exchange with the solvent [15]. There are also quasi-native species, such as ribonuclease A with an incorrect peptide bond *cis–trans* isomer [15].

The unique properties of the disulfide interaction permit kinetic intermediates with different disulfides to be accumulated, trapped, isolated, and characterized [13]. With proteins that require disulfide bonds for folding, the intermediates that occur during both unfolding and refolding may be analyzed under a single set of conditions, with no need for denaturants. A very nonrandom spectrum of intermediates accumulates with bovine pancreatic trypsin inhibitor (BPTI) (Fig. 2).

Pathways of Folding

The folding pathway, plus the rates of the intramolecular transitions, can be elucidated from the kinetics of disulfide bond formation and breakage. The disulfides are then probes of the folding transitions, indicating what Cys residues come into proximity. Of course, a single disulfide intermediate need not signify a single protein conformation. What conformations account for the occurrence of any disulfide intermediate should be apparent from its conformational properties when trapped. That the BPTI pathway (Fig. 2) is determined by the conformational properties of the protein is demonstrated by the effects on the pathway of modifying the protein and the environment. Disulfide formation is cooperative, and the kinetic intermediates do not accumulate substantially at equilibrium [13].

The most suprising aspect of the BPTI pathway was that disulfide bond rearrangements are involved in both unfolding and refolding, not simply a sequential breaking or making of the disulfides (Fig. 2). Essentially the same pathway has been conserved in two evolutionary related proteins [17], but the rates of the various steps differ. In particular, the direct, sequential pathways are more favorable energetically than the rearrangement pathways. This is believed to be due to the lower stabilities of the BPTI-like folded conformations of these two proteins.

Transition States

The transition state with the highest free energy in the BPTI pathway is that involved in the disulfide rearrangements into and out of the folded conformation. It probably is a distorted form of the folded conformation and would have a high energy due to the cooperativity of folded structures. The disulfide rearrangements may be a special means of getting into and out of BPTI's folded conformation, which is remarkably stable. The related proteins with lower stabilities would not require this special folding mechanism [18].

That the rate-limiting transition state in both folding and unfolding of most proteins (including *cis–trans* Pro peptide bond isomerization) is a distorted form of the native conformation is suggested by the observation that partially folded intermediates generally are absent in unfolding but accumulate during refolding. The nature of the folding transition state has also been studied by determining the effects on the rates of folding and unfolding of altering the conditions and the protein covalent structure. The few results obtained thus far suggest that the transition states are native-like in some respects but not in others. Such studies may be the most productive way of determining the mechanism of protein folding in cases where intermediates cannot be detected or trapped. Much remains to be done!

REFERENCES

1. Creighton TE (1984): "Proteins: Structures and Molecular Properties." New York: WH Freeman.
2. Anfinsen CB (1973): Principles that govern the folding of protein chains. Science 181:223–230.
3. Levinthal C (1968): Are there pathways for protein folding? J Chim Phys 65:44–45.
4. Ramachandran GN, Sasisekharan V (1968): Conformation of polypeptides and proteins. Adv Protein Chem 23:283–437.
5. Tanford C (1968): Protein denaturation. Adv Protein Chem 23:121–282.
6. Richards FM (1977): Areas, volumes, packing, and protein structure. Ann Rev Biophys Bioeng 6:151–176.
7. Englander SW, Kallenbach NR (1984): Hydrogen exchange and structural dynamics of proteins and nucleic acids. Quart Rev Biophys 16:521–655.
8. Karplus M, McCammon JA (1981): The internal dynamics of globular proteins. CRC Crit Rev Biochem 9:293–349.
9. Baldwin RL, Creighton TE (1980): Recent experimental work on the pathway and mechanism of protein folding. In Jaenicke R (ed): "Protein Folding." Amsterdam: Elsevier/North Holland Biomedical Press, pp 217–260.
10. Arakawa T, Timasheff S (1984): Protein stabilization and destabilization by guanidinium salts. Biochem 23:5924–5929.
11. Fujita Y, Noda Y (1981): Effect of hydration on the thermal stability of protein as measured by differential scanning calorimetry. Int J Peptide Protein Res 18:12–17.
12. Privalov PL (1979): Stability of proteins. Small globular proteins. Adv Protein Chem 33:167–241.

13. Creighton TE (1978): Experimental studies of protein folding and unfolding. Prog Biophys Mol Biol 33:231–297.
14. Baldwin RL (1975): Intermediates in protein folding reactions and the mechanism of protein folding. Ann Rev Biochem 44:453–475.
15. Kim PS, Baldwin RL (1982): Specific intermediates in the folding reactions of small proteins and the mechanism of protein folding. Ann Rev Biochem 51:459–489.
16. Creighton TE (1980): Kinetic study of protein unfolding and refolding using urea gradient electrophoresis. J Mol Biol 137:61–80.
17. Hollecker M, Creighton TE (1983): Evolutionary conservation and variation of protein folding pathways. Two protease inhibitor analogues from black mamba venom. J Mol Biol 168:409–437.
18. Goldenberg DP, Creighton TE (1984): Energetics of protein structure and folding. Biopolymers 24:167–182.
19. Nojima H, Ikai I, Oshima T, Noda H (1977): Reversible thermal unfolding of thermostable phosphoglycerate kinase. Thermostability associated with mean zero enthalpy charge. J Mol Biol 116:429–442.
20. Creighton TE, Goldenberg DP (1984): Kinetic role of a meta-stable native-like two-disulphide species in the folding transition of bovine pancreatic trypsin inhibitor. J Mol Biol 179:497–526.

Protein Engineering, pages 91–102

8

Mutagenesis as a Probe of Protein Folding and Stability

Anne M. Beasty, Mark Hurle, Joanna T. Manz, Thomas Stackhouse and C. Robert Matthews

Department of Chemistry, Lehigh University, Allentown, Pennsylvania 18104 (A.M.B.), Department of Chemistry, The Pennsylvania State University, University Park, Pennsylvania 16802 (M.H., C.R.M.), Department of Immunology, University of Washington, Seattle, Washington 98195 (J.T.M.), Department of Biochemistry, University of California, Davis, Davis, California 95616 (T.S.)

INTRODUCTION

The pioneering studies by Anfinsen and his colleagues nearly three decades ago on the recovery of enzymatic activity following the folding of denatured and disulfide bond-reduced ribonuclease A demonstrated that the amino acid sequence of a protein determines its unique three-dimensional structure [1]. This discovery led to a flurry of activity whose goal was to elucidate the mechanism by which the complex conformational change between the unfolded random coil and the native conformation, i.e., protein folding, occurs [2–4].

An implicit assumption in these early experiments was that the folding process is not a random search of all possible conformations with the native conformation favored because it represents the global minimum in free energy. Crude attempts to calculate the time required to sample randomly the conformations accessible suggested that, for a protein of 100 amino acids, around 10^{50} years would be necessary for a random search. These results reinforced the idea that folding proceeds along a limited number of pathways and, therefore, that discrete intermediates might be detectable.

Although the concept of folding intermediates has been in vogue for quite some time, it has proven to be exceedingly difficult to observe such species and to characterize their structures. The principal reason for the lack of data is the high cooperativity of the protein-folding reaction. The cooperative nature of this process means that only the native and unfolded conformations are highly populated at equilibrium during unfolding. Intermediates are sufficiently unstable that they are not significantly populated. The best evidence for this point was provided by calorimetric studies of the thermal unfolding of five small globular proteins: cytochrome C, lysozyme, ribonuclease A, myoglobin, and chymotrypsin [5]. The near equality of the van't Hoff enthalpy and the calorimetric enthalpy proved conclusively that stable intermediates comprise less than 5% of the total population during the thermal unfolding transition. For this reason, equilibrium studies of protein folding have provided little information on the mechanism.

The absence of stable intermediates in folding led to studies of the transient response of the system to rapid changes in environmental conditions designed to cause unfolding of the native conformation or refolding of the unfolded conformation. The hope was that kinetic studies might reveal transient intermediates in folding that can be highly populated, at least for a short time. This approach has proven to be successful in a number of small globular proteins [6–8]. Unfortunately, the short lifetimes of the intermediates, often less than 10 seconds, makes their structural characterization exceedingly difficult.

An alternative way of solving the protein-folding problem is to determine not what are the structures of intermediate, partially folded species but rather which amino acids play key roles in directing the process and which do not. The concept of a limited search implies not only that discrete intermediates may exist but also that certain amino acids or groups of amino acids, e.g., those in an α helix or β turn, may be required to direct the folding along a productive pathway. Replacement or modification of these essential amino acids would be expected to have an observable effect on the kinetics of the folding process and/or the stability of the native conformation.

Early approaches towards the modification of the amino acid sequences of proteins involved comparison of the folding properties of homologous proteins containing naturally occurring replacements [9] and of chemically modified proteins in which a single amino acid was derivatized [10]. Although homologous proteins were indeed found to have altered folding properties, the fact that such proteins usually differ by more than a single amino acid has made the assessment of the role of an individual amino acid impossible. Another problem is the lack of control on the replacement of specific amino acids; one studies what evolution has provided. The chemical modification of a specific amino acid has also been achieved; however, this method is applicable only to certain types of amino acids and is usually not specific to amino acid position.

The spectacular developments in recombinant DNA technology in the past few years now make it possible to achieve the ultimate level in amino acid sequence modification: the replacement of a specific amino acid by any other naturally occurring amino acid [11]. Because this set contains residues whose side chains vary in volume, charge, hydrogen-bonding capability, and hydrophobicity, the influence of each of these properties on the folding and stability can be tested. If an X-ray structure of the protein is available, these effects may be able to be interpreted in terms of local or long-range interactions. It is a reasonable expectation that the development of a data base that includes the results of many such replacements will eventually lead to general rules that can be used to predict the folding of any globular protein. This chapter will describe the results of initial studies on the effects of amino acid replacements on protein folding and stability and attempt to provide a view of future studies in this field.

CURRENT STUDIES

Equilibrium Studies

Methods. Equilibrium studies of the reversible unfolding process can yield quantitative estimates of the stabilities of the wild-type protein and various point mutants. The two most commonly used methods are thermal unfolding and chemical denaturant–induced unfolding. Each approach has certain advantages and disadvantages. The thermal unfolding can be monitored by differential scanning calorimetry, which is capable of providing a wealth of quantitative thermodynamic data [5]. The van't Hoff and calorimetric enthalpies, the entropy, the free energy, the melting temperature, and the difference in heat capacities between the native and denatured forms can all be obtained. The van't Hoff and calorimetric enthalpies are particuarly useful in determining whether stable intermediates are present (see above). The enthalpy, entropy, and heat capacity difference all have structural interpretations in terms of the protein itself or its interactions with the solvent and therefore have great potential value for interpreting the effects of mutations at the molecular level. A major drawback of the calorimetric approach for a number of proteins is the lack of reversibility following exposure to high temperatures for extended periods of time. The validity of the thermodynamic parameters obtained in an irreversible system is uncertain. Although the reasons for such a lack of reversibility have not been systematically studied, oxidation of sulfhydryls is one possibility, since the exclusion of molecular oxygen from the solvent significantly improves reversibility in the case of the α subunit of tryptophan synthase (C.R. Matthews, unpublished results)

The second commonly used procedure to study protein folding employs chemical denaturants such as urea and guanidine hydrochloride. In this case, the unfolding transition is usually monitored by a spectroscopic method such as difference ultraviolet, fluorescence, or far UV circular dichroism spectroscopy. The former two techniques are sensitive to changes in tertiary structure, and the latter to changes in secondary structure. The free energy of folding for wild-type and mutant proteins can be extracted from this data; however, the estimates of these values in the absence of denaturant must be obtained by extrapolation. Two different mathematical models have been proposed for the form of this extrapolation, each leading to a different estimate of the stability [12]. Since one is often more interested in the change in stability induced by a mutation than in the absolute stability, consistent application of one or the other extrapolation method minimizes this problem. An advantage of the chemical denaturation approach is the high reversibility that is normally observed. A major disadvantage is the difficulty in obtaining the thermodynamic parameters other than free energy change. Unfolding studies at a series of temperatures are required to determine the corresponding enthalpy and entropy changes.

Results. The effects of single amino acid replacements on protein stability have been reported for two proteins: the α subunit of tryptophan synthase and the lysozyme from the bacteriophage T4. The replacement of Gly by Glu or Arg at position 211 in the α subunit of tryptophan synthase [13] has rather small effects on the melting temperature at pH 7.8; however, the enthalpy and entropy change by 20%–25% (Table I). These changes are in opposing directions so that the net effect on free energy is rather small. The magnitude of the enthalpy change, 15–20 Kcal mol^{-1}, indicates that more than one noncovalent interaction must be affected by the mutation. The replacement of Glu by Gln or Ser at position 49 can either increase or decrease the stability of the α subunit, depending upon pH [14]. This effect has been attributed to the ionization of a buried carboxyl moiety from Glu 49. A significant feature of the thermal transition is its two-state behavior; the equality of the van't Hoff and calorimetric enthalpies demonstrates that no stable intermediates are present.

In contrast, both the urea-and guanidine hydrochloride-induced unfolding of the α subunit involve at least one stable intermediate. Spectroscopic and hydrogen exchange experiments suggest that the principal intermediate has a folded amino domain and an unfolded carboxyl domain [15–17]. Limited tryptic digestion had previously shown that the

TABLE I. Thermodynamic Parameters for the Unfolding of the Wild-Type and Mutant α Subunits at pH 7.8, 57.8°C

Protein	Tm(°C)	ΔH(kcal mol^{-1})	ΔS(kcal mol^{-1})	ΔG(kcal mol^{-1})
Wild type	57.8 ± 0.5	93.8 ± 5.9	283 ± 18	0.0
Gly 211→Arg	58.0 ± 0.2	111 ± 4	337 ± 12	0.1 ± 0.1
Gly 211→Glu	59.6 ± 0.6	109 ± 6	342 ± 19	0.6 ± 0.2

α subunit is comprised of a larger, more stable amino domain, residues 1–188, and a smaller, less stable carboxyl domain, residues 189–268 [18]. For the Gly 211→Glu mutation, the major effect is to increase the stability of the native conformation with respect to the intermediate. In other words, a mutation in the carboxyl domain has altered the free energy of unfolding of the carboxyl domain. The replacement of Glu at position 49 with a variety of amino acids has been reported to selectively alter the stability of the intermediate with respect to the unfolded form [19], although recent results in our laboratory show that both the transition between native and intermediate forms and the transition between intermediate and unfolded forms are affected by the Glu 49→Met mutation. A mutation in the amino domain could influence the stability of the carboxyl domain by altering the binding constant of the two folded domains. The absence of an X-ray structure for the α subunit precludes the possibility of interpreting these results at the molecular level.

It is interesting to note that one or more of the replacements at both positions increase the stability of the α subunit compared to the wild-type protein. These mutants, however, are enzymatically inactive. Apparently, the evolutionary effort to maximize both stability and activity results in less than optimal stability. One wonders whether the same situation might be true for activity.

The second protein system in which the effect of point mutations on protein stability has been reported is phage T4 lysozyme [20]. In this case, a high-resolution X-ray structure is available that will eventually assist in efforts to interpret the results at the molecular level. Evaluation of the thermal unfolding reaction for a series of mutants at different positions came to the same conclusion as that for the α subunit: namely, that while the changes in enthalpy and entropy are large they compensate each other and result in small changes in the free energy of folding. Comparison of a refined structure for the Arg 96→His mutant with that of wild-type lysozyme did not yield a ready explanation for the 13.6° decrease in melting temperature at pH 3 [21].

The introduction of an intraprotein disulfide bond is expected to increase protein stability by decreasing the conformations available to the unfolded form and, thereby, the entropy change for unfolding. Construction of the Ile 3→Cys 97 mutant in T4 lysozyme and subsequent oxidation to form the Cys 3–Cys disulfide bond decreased the rate of heat inactivation [23]. This result is consistent with increased thermal stability and demonstrates the success of the approach.

Limitations. The inability to correlate structural and stability changes in T4 lysozyme points out a current limitation in being able to predict the effect of an amino acid replacement on stability: protein structures are sufficiently complex, and the relevant potential functions are not known to sufficient accuracy to allow the calculation of the stability difference between wild-type and mutant proteins at the accuracy required. One of the goals of current studies must be the development of a data base which can be used to refine and improve such calculations.

Kinetic Studies

Methods. Protein-folding reactions can span a time range from submilliseconds to hundreds of seconds. The breadth of this range requires several different techniques. For many fast reactions, i.e., those in the 10^{-5}–10^{0}-second time range, capacitive temperature jump instrumentation is employed. For reactions in the intermediate time range, i.e., 10^{-2}–10^{1} seconds, stopped-flow drive trains interfaced to absorbance or fluorescence spectrophotometers are used. Finally, reactions that occur over longer time ranges, i.e., those whose relaxation times are 10^{1} to 10^{3} seconds, are more accurately measured with a double beam spectrophotometer and manual mixing procedures.

For single-subunit proteins, the folding reactions will be first order, i.e., the rates will not depend on the protein concentration. Therefore, the data can be computer fit to one or more exponentials and the relaxation times and associated amplitudes determined. The relaxation times, whose inverse are the apparent rate constants of folding, depend on the *final* conditions in any unfolding or refolding jump. Under certain conditions, these relaxation times can be simply related to microscopic rate constants for individual steps in a folding model. The amplitudes depend on both the *initial* conditions which determine the populations of various stable species and the *final* conditions.

Although kinetic studies are essential for developing a folding model, it has proven to be very difficult to determine the structural basis for a given kinetic phase. Initial efforts in the study of the effects of single amino acid replacements on the kinetics of protein folding strongly suggest that mutagenesis will play an important role in advancing our knowledge of these complex conformational changes.

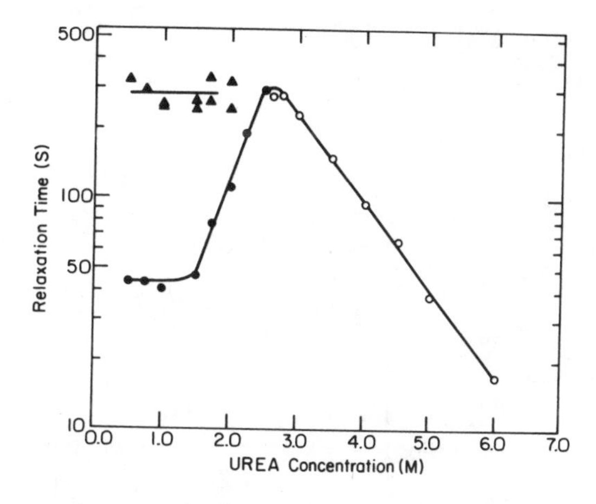

Fig. 1. *The semilog plot of the urea dependence of the relaxation times for the single slow phase in unfolding (○) and the two slow phases in refolding (●, ▲) of the α subunit of Trp synthase at pH 7.8, 25°C. The unfolding relaxation time is believed to reflect the $N \rightarrow I_3$ reaction (Fig. 2), while the urea dependence segment of the faster relaxation time in refolding reflects the $I_3 \rightarrow N$ reaction. The urea-independent relaxation time whose average value is 282 sec is believed to reflect the $I_1 \rightarrow I_2$ reaction, and that whose average value is 44 sec reflects the $I_2 \rightarrow I_3$ reaction.*

Results. Published work from our laboratory thus far has involved studies on the folding of the α subunit of Trp synthase from *E. coli* [23]. Studies are also in progress in our laboratory on dihydrofolate reductase from *E. coli* [K. Perry, N. Touchette, C.R. Matthews; manuscript in preparation]. A semilog plot of the relaxation times observed for the slow phases in unfolding and refolding as a function of the final urea concentration is shown in Figure 1. Unfolding is governed by a single slow reaction whose relaxation time decreases at higher urea concentration. This is the behavior expected for a protein-folding-type reaction whose equilibrium constant is shifting to favor the unfolded form at high urea concentrations. Refolding is more complex, with a fast phase that cannot be measured by manual mixing techniques, followed by two slower phases whose relaxation times vary as a function of urea concentration, as shown in Figure 1. The relaxation time of the faster of these two slow phases decreases as the final urea concentration decreases down to ~1.5 M urea. Below that point, the relaxation time becomes independent of urea concentration. The slower phase is independent of the urea concentration over the entire range observed.

This and other data has been used to propose a folding model for the α subunit (Figure 2). The model involves two unfolded forms, three intermediate forms, and one native form. Unfolding is governed by the slow unfolding of N to I_3. Refolding involves first the rapid collapse of U_1 and U_2 to I_1 and I_2, respectively. These species are then converted to the native form by the slow reactions whose time constants are plotted in Figure 1. The $I_1 \rightarrow I_2$ reaction has a relaxation time of 282 seconds that is independent of urea concentration, while the $I_2 \rightarrow I_3$ and $I_3 \rightarrow N$ reactions both contribute to the faster phase. At urea concentrations between 1.5 and 3.0 M, the $I_3 \rightarrow N$ reaction is rate limiting. Since this reaction involves protein folding, the relaxation time decreases as the equilibrium constant shifts to favor the native form. Below 1.5 M urea, the $I_2 \rightarrow I_3$ reaction whose relaxation time is 44 seconds becomes rate-limiting in the conversion of I_2 to N. The urea independence of both the $I_1 \rightarrow I_2$ and the $I_2 \rightarrow I_3$ reactions suggests

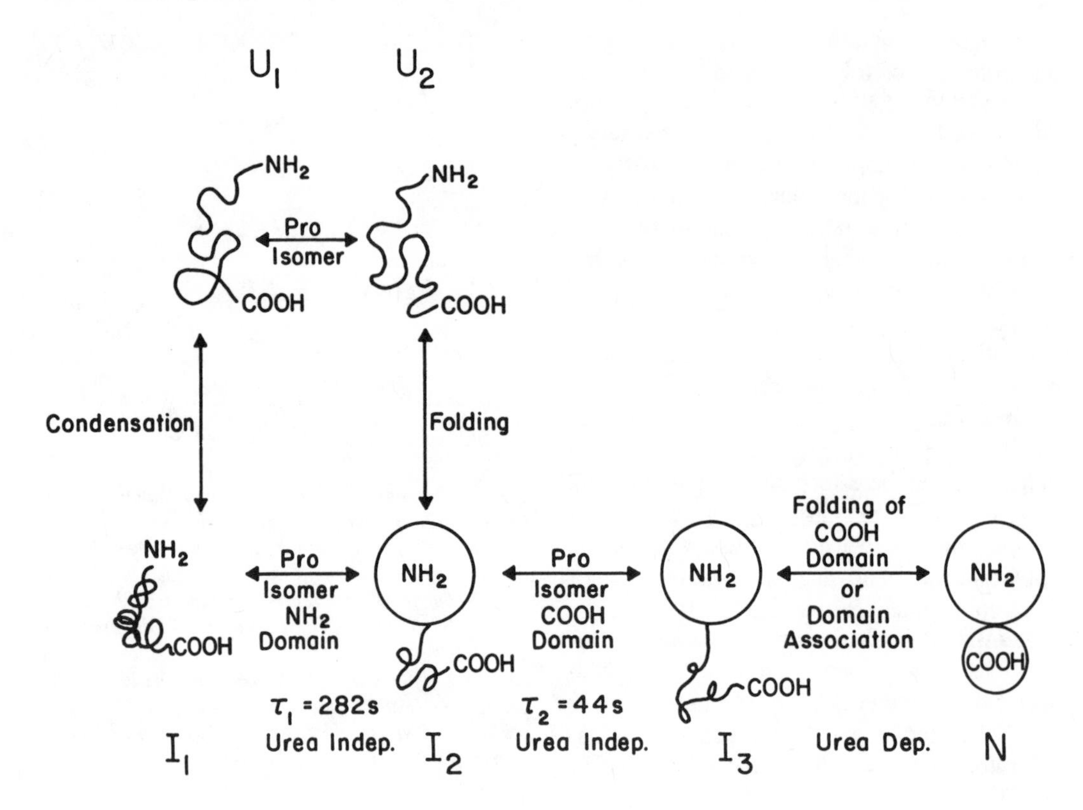

Fig. 2. *The folding model for the α subunit of Trp synthase. The cartoon structures are meant to represent current views on the actual structures of the intermediates; detailed descriptions are provided in the text. The relaxation times for the $I_1 \rightarrow I_2$ and $I_2 \rightarrow I_3$ reactions are designated τ_1 and τ_2, respectively. The values shown for τ_1 and τ_2 are those observed in refolding at pH 7.8, 25°C. At urea concentrations between 1.5 and 3 M, the $I_3 \rightarrow N$ reaction is rate limiting in the conversion of I_2 to N. This reaction accelerates as the urea concentration is lowered, making the $I_2 \rightarrow I_3$ reaction rate-limiting below 1.5 M urea.*

that they reflect not protein folding but rather cis/trans isomerizations of the polypeptide backbone at X-Pro peptide bonds. It has been suggested previously that the incorrect isomeric form at an X-Pro peptide bond could give rise to slow phases in folding [24], and data now exist that support this hypothesis for several proteins. The rates of such isomerization reactions are expected to be independent of the denaturant concentration.

The structural interpretation of the intermediates in the kinetic model is also shown in Figure 2. Hydrodynamic, spectroscopic, and, more recently, hydrogen-exchange results show that: I_1 has a compact but not correctly folded structure with a nonnative X-Pro isomer in the amino domain; I_2 has a folded amino domain and an unfolded carboxyl domain with a nonnative X-Pro isomer; and I_3 is similar to I_2, except that the X-Pro isomer in the carboxyl domain has now adopted the native isomeric form.

The effort required to define a folding model and identify the structures of the intermediates was important because it provided the framework for asking specific questions about the rate-limiting steps in folding. In terms of the mutagenic approach, one can ask if single amino acid replacements selectively alter the rates of protein-folding and the Pro isomerization reactions. Because the latter reactions involve very localized events in the polypeptide backbone in unfolded protein, one would not expect them to be sensitive to amino acid replacements at distant sites. In contrast, the protein-folding reaction could be affected by the replacement of an amino acid at a key position anywhere in the protein. For the above folding model, one can also ask whether the $I_3 \rightarrow N$ step is rate-limited by the folding of the carboxyl domain or by the association of the folded amino and carboxyl domains.

These two questions were answered by examining the effect of two point mutations in the α subunit: Gly 211→Glu and Phe 22→Leu. The former resides in the carboxyl domain and might be expected to alter the rate of the $I_3 \rightarrow N$ reaction for either potential rate-limiting step. The latter mutation, in the amino domain, would only affect the rate if domain association is rate-limiting. Also, since neither replacement is on the amino side of a Pro, one would not expect the rates of the Pro isomerization steps observed at low urea concentrations in refolding to be affected.

Figure 3 shows the relaxation times observed for the unfolding and refolding of the Gly 211→Glu and Phe 22→Leu mutant proteins, as well as those of the wild-type α subunit. The patterns are qualitatively similar, a demonstration that the folding mechanisms are the same. However, certain aspects are quantitatively different, and it is these differences that illustrate the potential of the mutagenic approach for unraveling protein-folding mechanisms.

The major difference in the kinetics of folding caused by both mutations is to increase the relaxation times for *both* unfolding and refolding in the interconversion of I_3 and N. For the Gly 211→Glu mutant protein, unfolding slows down by a factor of ~10 while refolding slows by a factor of 4. For the Phe 22→Leu mutant protein, unfolding slows by a factor of 2.5 while refolding slows by a factor of 4. In contrast, the phases assigned to Pro isomerization are much less dramatically affected if at all. Both of the relaxation times for refolding to low urea concentrations of the Gly 211→Glu mutant protein are virtually identical to those for the wild-type protein. The data for the slowest phase in refolding for the Phe 22→Leu mutant is rather scattered; however, the relaxation time is probably within experimental error for that of the wild-type protein. The second Pro isomerization phase for the Phe 22→Leu mutant is also within experimental error of the value for the wild-type protein.

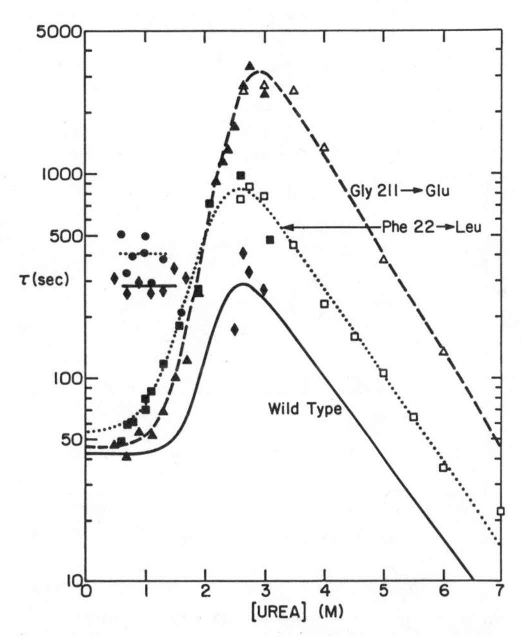

Fig. 3. *Semilog plots of the urea dependences of the relaxation times for the slow phases in unfolding (open symbols) and refolding (closed symbols) for the Phe 22→Leu (□, ■, ●, dotted lines) and Gly 211→Glu (△, ▲, ♦, dashed lines) mutant α subunits at pH 7.8, 25°C. The wild type α subunit data (Fig. 1) is shown for comparison as the solid line. The average value for the τ_1 phase for the Gly 211→Glu mutant protein (♦), 282 seconds, is identical to that for wild type protein.*

The interpretation of this selective effect of the amino acid replacements on the protein-folding phase is done most easily in terms of a reaction coordinate diagram (Figure 4). The stable conformations N and I_3 interconvert by

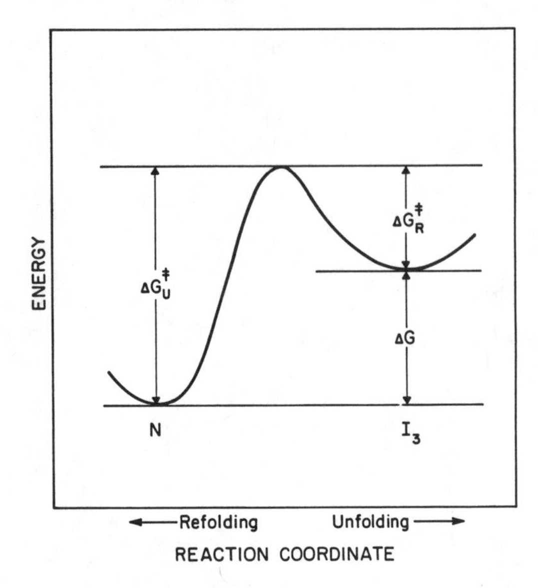

Fig. 4. *Hypothetical reaction coordinate diagram for the N→I_3 reaction in the absence of denaturant. The symbols $\Delta G_u{}^{\ddagger}$, $\Delta G_r{}^{\ddagger}$, and ΔG correspond to the unfolding activation free energy, the refolding activation free energy, and the free energy difference between the stable N and I_3 states, respectively.*

way of a hypothetical transition state. The equilibrium constant relating N and I_3 is determined by the difference between the free energies of these states, designated ΔG. The unfolding rate constant, k_u, is related to the difference in energies between the native and transition state, $\Delta G_u{}^{\ddagger}$. In a similar fashion, the rate constant for refolding, k_r, is related to $\Delta G_u{}^{\ddagger}$. The link between this diagram and the relaxation time data is provided by the well-known kinetic equation for a simple two-species system:

$$\tau^{-1} = k_u + k_r$$

where k_u and k_r are the unfolding and refolding rate constants. For unfolding jumps ending at high urea concentrations, where the protein will be unfolded when equilibrium is established, $k_u >> k_r$ and $\tau^{-1} \cong k_u$. For refolding jumps ending at low urea concentrations, $k_r >> k_u$ and $\tau^{-1} \cong k_r$. Therefore, the effect of the mutations on the relaxation time at low and high urea concentrations can be interpreted in terms of the effects on the microscopic rates constants and their associated reaction coordinate diagram.

The most important effect of both the Phe 22→Leu and Gly 211→Glu mutations is to increase the relaxation times or equivalently decrease the rate constants for both unfolding and refolding. In terms of the reaction coordinate diagram, the energy of the transition state has increased with respect to the energies of both stable states (Fig. 5). A mutant protein whose folding kinetics have been altered in this way can be labeled as a kinetic folding mutant. This designation implies that the amino acid replacement has had a real effect on this rate-limiting step in folding, i.e., it selectively alters the energy of the transition state relative to the energies of the stable states. Amino acid positions that display this behavior play key roles in the folding process. The observation that the increases in the relaxation times for unfolding and refolding are not precisely equal means that the relative energies of the stable states have also been affected. This implies a small change in the equilibrium constant for N and I_3 which is not shown in Figure 5.

This behavior can be contrasted to that of mutants which only alter the rate of unfolding or refolding but not both (Beasty et al.; manuscript in preparation). In the latter case, the effect can be described in terms of a shift of the energy of one of the stable states with respect to the energies of the transition state and the other stable state; the relative energies of the transition state and second stable state are unchanged (Fig. 6). In this case, the principal effect is on the free energy difference between N and I_3, ΔG, and the associated equilibrium constant, K, where $\Delta G = -RT \ln K$. Such a mutation would not alter the rate-limiting step in the sense described above and therefore would not play a key role in the rate-limiting step.

Because mutations in both the amino and carboxyl domains have a real effect on the

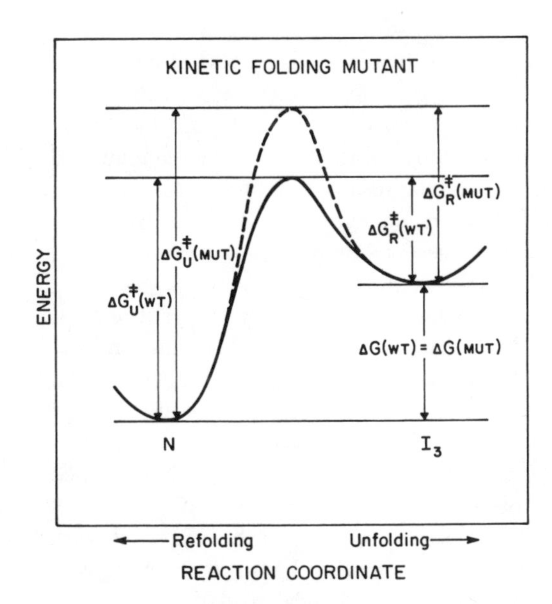

Fig. 5. *Hypothetical reaction coordinate diagrams for the N→I_3 reaction for a wild-type protein (----) and a kinetic folding mutant (----). The diagrams are only qualitatively correct and are representative of the relative energies of the various species in the absence of denaturant.*

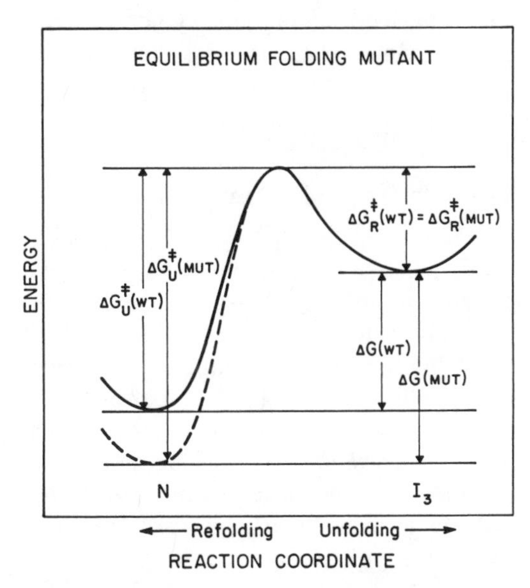

Fig. 6. *Hypothetical reaction coordinate diagrams for the N←→I_3 reaction for a wild-type protein (-----) and an equilibrium folding mutant (----). The diagrams are only qualitatively correct and are representative of the relative energies of the various species in the absence of denaturant.*

rates of interconversion of I_3 and N, the reaction must correspond to either the domain association reaction suggested above or to some other molecule-wide phenomenon. It cannot be limited by the folding of the carboxyl domain. This information would have been exceedingly difficult to obtain in any other fashion and emphasizes the power of the mutagenic approach in identifying the structural basis of kinetic phases in protein folding.

Limitations. A practical limitation of kinetic studies on the folding of mutant proteins is the requirement for relatively large quantities of protein (i.e., ~100 mg), for a complete analysis of the type shown above. This problem can be minimized by improving production through genetic engineering, by using where possible the inherently more sensitive fluorescence methods in place of absorbance methods, and by limiting the study to a few urea concentrations for both unfolding and refolding once the mechanism has been established.

Another problem involves the necessity of measuring both the unfolding and refolding rate constants to make a correct interpretation of the effect of the mutation. This requirement can be circumvented if one of the rate constants and the equilibrium constant are available. For the α subunit of Trp synthase, the kinetic scheme is such that the unfolding of the intermediates to the unfolded forms cannot be detected because it is preceded by the slower I_3→N step. Unfortunately, the folding of the unfolded forms to the intermediates is in the submillisecond time scale [C. Froebe and C.R. Matthews; manuscript in preparation], precluding the possibility of obtaining either of the rate constants for this step. Therefore, mutagenesis will probably not be useful in identifying the amino acids that play key roles in these particular folding reactions.

FUTURE DIRECTIONS

The Role of Specific Amino Acids in Determining the Mechanism of Folding

The published studies described above and those in progress on staphylococcal nuclease [25] and cytochrome C from yeast suggest that the ability to make single-amino-acid replacements will be a powerful new tool in efforts to unravel the mechanism of folding. Given the opportunity to make specific amino acid changes, one must have specific testable hypotheses to focus the mutagenic efforts along productive lines. An X-ray structure provides the basis for such hypotheses. One of the hazards with this approach is that one may expend a great deal of effort synthesizing mutant proteins that are insufficiently stable to be purified from the bacterial hosts. A second potential problem is that the mutation may change the structure to such an extent that comparison with the wild-type protein is invalid.

Another approach which partially circumvents these latter two problems is to return to more classical genetics and use random mutagenesis combined with an appropriate selection or screen to identify the mutants. As an example, temperature-sensitive mutants are likely to form a class which will have altered stabilities and therefore be of particular interest in their effects on the folding properties. Our own experience with the collection of mutants in the α subunit of tryptophan synthase developed by Yanofsky and his colleagues [26] is that mutants selected for a loss in enzymatic activity have a good chance of altering the folding or stability as well. Such a collection of random mutations could form the basis for specific hypotheses that could then be tested by site-directed mutagenesis.

Role of Secondary Structure in Determining Folding Mechanism

One of the proposed folding mechanisms involves the initial formation of recognized elements of secondary structure such as β-turns or α helices [27]. These isolated elements are sufficiently unstable that they only exist for microseconds before melting. Occasionally, they collide with another element and dock, and the resulting, more-stable aggregate persists for a longer time. Eventually, a progressive association of these elements leads to the native conformation.

The potential role in folding of a specific α helix can be tested by the systematic replacement of its constituent amino acids. One might imagine that residues involved in docking the helix against other elements of structure would play a more critical role than those which protrude into the solvent. Another approach towards testing the role of the helix is to vary the amino acids according to their propensity to be involved in helical structures, e.g., according to their Chou-Fasman parameters [28]. Correlations would support the idea that the appropriate segment of the polypeptide indeed behaved as a helix in the absence of the remaining protein framework.

Effect of Loop Insertions on Folding and Stability

Oligonucleotide mutagenesis can also be used to insert peptide segments of varying length at any point in the sequence of the protein. Although one could imagine a variety of uses, in terms of the folding problem one would like to know if such insertions would decrease the rate of folding by decreasing the likelihood that given segments of the protein would have a productive encounter. By varying the site of the insertion, one could also determine if such effects are uniform or selectively distributed.

Potential of Shortened Polypeptides to Fold

One can also use oligonucleotide-directed mutagenesis to decrease the length of a polypeptide either by removing an internal segment or by placing a terminating codon at an appropriate site. The questions to be answered might focus on the role of what would appear from an X-ray structure to be superfluous structures (e.g., loops) or the role of the folding of structural domains. Although in some cases domains can be recognized by visual inspection of an X-ray structure, algorithms have been developed to assist in such structural dissections [29]. For larger, multifunctional enzymes, it may be possible to separate the activities into discrete polypeptides.

Interactions Between Amino Acid Residues That Influence Folding

Mutagenesis also provides an opportunity to test interactions between pairs of amino acids that play key roles in the folding process [30]. The approach involves the measurement of the effects of both single mutants and the double mutant on the folding and stability. If the residues act independently, the results for the double mutant should be equal to the sum of the effects of the two single mutants, in terms of free energy changes. If the residues interact, either directly or indirectly, this equality will not hold. This experiment can demonstrate how energy flows through the protein framework and, thereby, improve our understanding of the relationship between structure, folding and stability.

SPECULATIONS

Expectations and Limitations of the Mutagenic Approach

It is a generally held expectation that mutagenesis will provide an avenue to new biomaterials with improved properties, and there is little doubt that this expectation will be realized eventually. However, the results of current experiments suggest that the realization of this goal will require a significant effort to develop the data base required for intelligent protein engineering. Our understanding of the forces that stabilize proteins is sufficiently primitive that we are not yet in a position to interpret in any quantitative way the effects of mutations on stability. Similarly, the calculation of the binding constants of small molecules to proteins is still at a stage where one has difficulty in developing confidence in the present methods. Although part of this problem will be alleviated by more powerful computational methods, a data base of the effects of mutations on folding, stability, and activity will be required to calibrate and test these methods. The ultimate advantage of the computational approach will be to provide a rational way to design multiple amino acid changes required to optimize particular properties of naturally occurring proteins and, eventually, to allow ab initio design of new biomaterials.

Extension to the Study of Structure in Small Peptides

The complexities involved in attempting to quantitate the effects of single-amino-acid replacements on protein stability mentioned above suggest that a simpler approach towards understanding the relevant forces is to study organized structure in small peptides [31]. Although it was previously thought that peptides of less than ~ 100 amino acids could not form secondary structure, recent experiments on the S-peptide, a 20-residue fragment from ribonuclease A, show that it forms a marginally stable α helix under appropriate conditions. Analogs of this peptide and a shorter version obtained by cyanogen bromide cleavage can be synthesized, and the role of specific amino acids in stabilizing the helix can be assessed. Although the interactions involved are still being determined, it seems clear that the information obtained from this "model" system will be required to understand the forces stabilizing proteins. Thus, the synthetic approach provides a variation on mutagenesis that will probably play an important role in understanding the relationship between structure, folding, stability, and function in proteins.

REFERENCES

1. Anfinsen CB (1973): Principles that govern the folding of protein chains. Science *181*:223–230.
2. Tanford C (1968): Protein denaturation. Adv Prot Chem 23:121–282.
3. Tanford C (1970): Protein denaturation. Adv Prot Chem *24*:1–95.
4. Anfinsen CB, Scheraga HA (1975): Experimental and theoretical aspects of protein folding. Adv Prot Chem *29*:205–300.
5. Privalov PL (1979): Stability of proteins: small globular proteins. Adv Prot Chem *33*:167–241.
6. Wetlaufer DB (1973): Acquisition of three-dimensional structure of proteins. Ann Rev Biochem

*42:*135–158.
7. Baldwin RL (1975): Intermediates in protein folding reactions and the mechanism of protein folding. Ann Rev Biochem *44:*453–475.
8. Kim PS, Baldwin RL (1982): Specific intermediates in the folding reactions of small proteins and the mechanism of protein folding. Ann Rev Biochem *51:*459–489.
9. Knapp JA, Pace CN (1974): Guanidine hydrochloride and acid denaturation of horse, cow and candida krusei cytochromes C. Biochemistry *13:*1289–1294.
10. Cupo JF, Pace CN (1983): Conformational stability of mixed disulfide derivatives of β-lactoglobulin B. Biochemistry *22:*2654–2658.
11. Itakura K, Rossi JJ, Wallace RB (1984): Synthesis and use of synthetic oligonucleotides. Ann Rev Biochem *53:*323–356.
12. Pace CN, Vanderburg KE (1979): Determining globular protein stability: guanidine hydrochloride denaturation of myoglobin. Biochemistry *18:*288–292.
13. Matthews CR, Crisanti MM, Gepner GL, Velcelebi G, Sturtevant J (1980): Effect of single amino acid substitutions on the thermal stability of the α subunit of tryptophan synthase. Biochemistry *19:*1290–1293.
14. Yutani K, Khechinashvili NN, Lapshina EA, Privalov PL, Sugino Y (1980): Int J Peptide Protein Res *20:*331–336.
15. Crisanti MM, Matthews CR (1981): Characterization of the slow steps in the folding of the α subunit of tryptophan synthase. Biochemistry *20:*2700–2706.
16. Miles EW, Yutani, Ogasahara K (1982): Guanidine hydrochloride induced unfolding of the α subunit of tryptophan synthase and of the two proteolytic fragments: evidence for stepwise unfolding of the two domains. Biochemistry *21:*2586–2592.
17. Beasty AM, Matthews CR (1985): Characterization of an early intermediate in the folding of the α subunit of tryptophan synthase by hydrogen exchange measurement. Biochemistry *24:*3547–3553.
18. Higgins W, Fairwell T, Miles EW (1979): An active proteolytic derivative of the α subunit of tryptophan synthase. Identification of the site of cleavage and characterization of the fragments. Biochemistry *22:*4827–4835.
19. Yutani K, Ogasahara K, Sugino Y (1982): pH dependence of stability of the wild-type tryptophan synthase α subunit and two mutant proteins (Glu 49→Met or Gln). J Mol Biol *144:*455–465.
20. Hawkes R, Grutter MG, Schellman J (1984): Thermodynamic stability and point mutations of bacteriophage T4 lysozyme. J Mol Biol *175:*195–212.
21. Grutter MG, Hawkes RB, Matthews BW (1979): Molecular basis of thermal stability in the lysozyme from bacteriophage T4. Nature *227:*667–669.
22. Perry LJ, Wetzel R (1984): Disulfide bone engineered into T4 lysozyme: stabilization of the protein toward thermal inactivation. Science *226:*555–557.
23. Matthews CR, Crisanti MM, Manz JT, Gepner GL (1983): Effect of a single amino acid substitution on the folding of the α subunit of tryptophan synthase. Biochemistry *22:*1445–1452.
24. Brandts JF, Halvorson HR, Brennan M (1975): Consideration of the slow step in protein denaturation reactions is due to cis-trans isomerism of proline residues. Biochemistry *14:*4953–4963.
25. Shortle D, Lin B (1985): Genetic analysis of staphylococcal nuclease: identification of three intragenic (global) suppressors of nuclease minus mutations. Genetics *110:*539–555.
26. Yanofsky C (1967): Gene structure and protein structure. Harvey Lect *61:*145–168.
27. Ptitsyn OB, Finkelstein AV (1980): Self-organization of proteins and the problem of their three-dimensional structure prediction. In Jaenicke R, (ed): "Protein Folding." New York: Elsevier/North-Holland Biomedical Press, pp 101–115.
28. Chou PY, Fasman GD (1978): Prediction of the secondary structure of proteins from their amino acid sequence. Adv Enzymol *47:*45–148.
29. Rose GD (1979): Hierarchic organization of domains in globular proteins. J Mol Biol *134:*447–470.
30. Ackers GK, Smith FR (1985): Effects of site-specific amino acid modification on protein interactions and biological function. Ann Rev Biochem *54:*597–629.
31. Shoemaker KR, Kim PS, Brems DN, Marqusee S, York EJ, Chaiken IM, Stewart JM and Baldwin RL (1985): Nature of the charged-group effect on the stability of the C-peptide helix. Proc Natl Acad Sci USA *82:*2349–2353.

Protein Engineering, pages 103–108

9

Genetic Strategies for Analyzing Proteins

David Shortle

Department of Biological Chemistry, The Johns Hopkins University School of Medicine, Baltimore, Maryland 21205

All of the important biological properties of a protein stem from the specific three-dimensional conformation it assumes spontaneously in an aqueous environment. Although ample evidence exists that this "native" conformation is determined by the linear sequence of amino acid residues in the one or more polypeptide chains of the protein, the process by which this one-dimensional sequence information is converted into precise three-dimensional structural information is understood in only a rough, qualitative way. Even with the help of the largest computers available, one cannot yet transform amino acid sequence data into meaningful predictions about where a given amino acid in a protein will be positioned in the native conformation or what its structural and/or functional roles will be. Discovery of the quantitative rules that govern this complex physical transformation stands as one of the central goals of protein chemistry.

For a particular protein, X-ray diffraction analysis can often provide a detailed model of the native conformation, one which describes the exact spatial position of each amino acid residue and may provide clues as to which regions of the molecule are involved in biological functions, such as enzyme active sites, allosteric sites, regions of subunit–subunit interactions, etc. Hypotheses about the structural or functional roles of individual amino acids suggested by examination of the X-ray model can now be put to rigorous tests by changing the naturally occurring amino acid at a given position in the chain to a different one and then examining the properties of the altered protein. Such site-specific mutagenesis (see chapter 4, by Rossi and Zoller, for details) represents the most important genetic approach that can be applied to the analysis of how a protein's properties are precisely determined by its amino acid sequence. In addition, there are three other genetic strategies that have been extensively used by molecular geneticists prior to the development of recombinant DNA technology and that provide valuable alternative ways of confronting this general problem. These strategies are briefly discussed in this chapter.

The amino acid residues responsible for a particular property (e.g., catalytic activity, a stable native conformation, a rapid forward kinetic constant of protein folding) can, in principle, be identified directly in the absence of X-ray structural or biochemical information by isolating and characterizing *phenotypic mutations* (Fig. 1). The genetic strategy of concentrating on phenotypic mutations can best be viewed as a kind of systems analysis: one in which a large number of stable perturbations (mutations) are introduced into a complex system (the protein), but only those

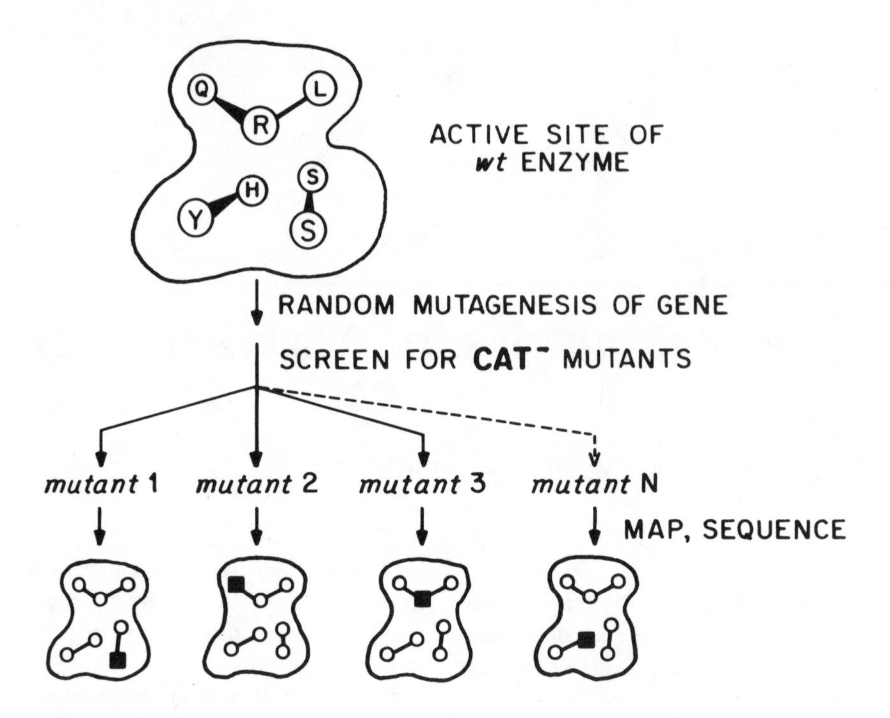

Fig. 1. *Identification of active site residues by isolating and characterizing mutations with a CAT-phenotype.*

perturbations that produce a specific modification in the behavior of the system (the mutant phenotype) are identified and studied. All of the perturbations that modify some other behavior of the system, or that do not affect the system at all, are ignored (at least in the early stages of analysis).

The only requirements for applying this genetic strategy are: 1) a large collection of randomly induced mutations within the gene that encodes the protein, and 2) an assay which will detect the subset of mutations that significantly alter the property of the protein under investigation. Once the gene for a protein has been cloned by recombinant DNA methods, single-base substitution mutations can be efficiently induced at essentially every nucleotide position [1], making it possible to generate pools of random mutations containing one third of all possible single amino acid substitutions. As a result, the first requirement usually can be met for any gene of interest by applying more-or-less standard in vitro manipulations of DNA. It is the requirement for an assay capable of detecting specific abnormal properties of mutant proteins that potentially limits the feasibility of employing this genetic strategy. Once a workable assay has been developed, it can be turned into an instrument analogous to a microscope for examining how the various properties of a protein are encoded in its amino acid sequence. The greater the specificity and sensitivity of the assay, the higher the level of resolution attainable.

In order to systematically search through a pool of randomly induced mutations, the assay employed must permit isolation of viable organisms (usually bacteria) expressing the mutant phenotype, either by *selection* (inhibition of growth of all organisms with nonmutant genes) or by *screening* (visual discrimination between organisms with mutant versus nonmutant genes). Since a selection eliminates all wild-type genes, it allows the isolation of very rare mutations; one in a billion is not uncommon. However, it is seldom possible to devise a situation where the growth of an organism is contingent on the presence of a mutant protein. Fortunately, the new methods of inducing mutations in recombinant DNA molecules are very efficient, yielding single-base changes at random sites across a gene at frequencies as high as 50%. Consequently, since only a relatively small number of organisms need be assayed (hundreds to thousands) in order to find a variety of mutations, screening for mutations with a particular mutant phenotype offers many opportunities in those cases when no selection can be devised. In fact, any in vitro biochemical assay that can be carried out on the protein in an unpurified state potentially can be developed into a mutant screen. As long as more than a few assays can be carried out per day, it may be feasible to screen mutagenized collections of genes one by one for the occasional mutation that displays an interesting mutant phenotype. While invariably laborious, this *brute force* approach has the real advantage that all it takes is perseverance.

The protein that has been most exhaustively analyzed through the isolation and characterization of phenotypic mutations is the repressor of the *lac* operon of *Escherichia coli*. This oligomeric protein, which consists of four identical chains 360 amino acids in length, is the principal regulatory protein for a cluster of genes involved in the catabolism of the sugar lactose. Because very powerful genetic selections exist for mutations that either enhance or reduce the ability of this protein to turn off expression of the *lac* operon, many hundreds of mutations in the repressor gene have been recovered and partially analyzed without the aid of recombinant DNA techniques [2]. This work, which has extended over many years in a number of laboratories and best demonstrates the versatility of phenotypic mutations, has idenified single amino acid substitutions that: 1) reduce the protein's binding affinity for the operator, a unique regulatory site in the DNA adjacent to the promoter; 2) enhance nonspecific affinity for DNA; 3) eliminate the allosteric site where galactosides bind and mediate changes in affinity for the operator; 4) block oligomerization to form tetramers; and 5) alter the allosteric changes that transmit the signal of galactoside binding to the DNA binding region. Unfortunately, the X-ray crystal structure of this protein has not yet been solved, so that the molecular mechanisms responsible for these mutant phenotypes remain an exciting but unwritten chapter in the long book of contributions of the *lac* operon to molecular biology.

As detailed in the last section of this book, many proteins which were extensively studied in the past by biochemical and physical techniques are now being analyzed by genetic methods. Since a majority of these proteins are enzymes, a simple assay for enzymatic activity will usually provide a ready screen for phenotypic mutants. One example of how straightforward mutant isolation and mapping becomes once a simple screen for activity has been devised is provided by staphylococcal nuclease. By replica-plating *E. coli* colonies expressing the cloned gene for this enzyme onto an indicator plate that provides a semiquantitative measure of total enzyme activity, mutations have been identified that alter practically every one of the amino acid residues previously determined to play a significant role in the active site [3,4]. In addition, more than 50 mutations have been recovered that alter residues outside the active site. Since a stable native conformation is requisite for biological and enzymatic activity, it is not surprising that many of these mutations destabilize and/or alter the structure of the active conformation. In some cases the perturbation of stability caused by the mutation results in a higher rate of degradation by cellular proteases, increasing the mutation's effect on assayable activity.

It is important to emphasize that phenotypic mutations can contribute to the analysis of proteins in several quite different ways. First of all, the isolation of a mutation that alters a single amino acid residue and thereby significantly alters a specific property of the protein immediately implicates that residue, and perhaps others nearby in the native conformation, as having an important role in that property. Thus the relevant information responsible for different properties can be traced to unique amino acids in the polypeptide chain. Secondly, the phenotypic mutation can be introduced into an overexpression vector that directs a microorganism to synthesize large quantities of the mutant protein. For many mutant proteins, especially those that are unstable to proteolysis, this step enormously simplifies the task of purifying enough material to determine how, in physical chemical terms, the amino acid substitution has changed the protein's behavior. Since an amino acid position can play more than one role in determining the structure and function of a protein, such biochemical studies are essential before conclusions can be confidently drawn about the molecular basis of an observed alteration in a specific property.

A genetic strategy known as *second-site suppression analysis* or *reversion analysis* represents another way of using phenotypic mutations to extend the study of proteins. If a particular mutation causes a mutant phenotype by destabilizing a specific interaction that normally involves the amino acid residue changed by the mutation, it is reasonable to expect that a second mutation which eliminates the mutant phenotype (i.e., restores the wild-type phenotype) might do so by exerting an opposite effect on the same interaction. For example, if an amino acid substitution in the DNA binding site of a repressor protein reduces its affinity for the operator sequence and thereby gives rise to a mutant phenotype, mutagenesis of this mutant gene followed by phenotypic screening for wild-type may lead to the isolation of a variety of *revertants*. In all cases, it should be possible to mutate the original mutant amino acid back to the wild type, restoring the interaction to its original level. In addition to such *true revertants*, it may be possible to change the mutant amino acid residue to some other residue that permits a more-or-less normal interaction with DNA to occur. This type of *pseudorevertant* sheds light on the range of side chains at a specific residue position that can engage in the interaction.

A potentially more interesting type of pseudorevertant is one in which the second, or suppressor, mutation occurs at a site in the gene some distance from the original mutation. If it can be demonstrated that such a second-site suppressor exerts its effect on the mutant phenotype by restoring the perturbed interaction of the original mutant protein to more normal levels, then a second site important in the interaction has been identified. In the process of determining how the suppressor mutation works, it is helpful to ascertain if the suppressor is *allele-specific* or *global* in its action. As its name implies, an allele-specific suppressor is only able to restore the mutation with which it was isolated to a wild-type phenotype. On combination with other mutations, this type of suppressor has either no effect or, in some cases, may have the opposite effect of making the mutant phenotype more severe. At the opposite extreme is the global suppressor that is able to suppress the mutant phenotype of a number of different mutations that alter amino acids at different positions in the polypeptide chain. From the pattern of mutations that are or are not suppressed and from the behavior of the suppressor in an otherwise wild-type gene, hypotheses can be formed about the mechanism of action of individual suppressors.

As described by Nelson and Sauer [5], second-site suppression analysis of several different mutations that diminish the affinity of lambda repressor for its binding site (operator) on DNA has led to the identification of two global suppressors. The amino acid substitutions produced by these two mutations increase the DNA binding affinity when recombined with several low-affinity mutants (to give a doubly mutant protein), and not surprisingly, they also increase the affinity of wild-type repressor for its operator sequence. From the proposed model of repressor–operator complex, both amino acid changes occur at sites that are near enough to the DNA helix to create new noncovalent protein–DNA bonds.

Second-site suppression analysis of four mutations in the gene for staphylococcal nuclease that involve amino acids outside the active site led to the identification of three global suppressor mutations, one of which was able to suppress at least 15 different *nuc*-mutations [4]. In vitro characterization of mutant proteins, with and without these suppressors, suggests that the initial *nuc*-mutation destabilizes the native conformation and that the suppressors exert at least part of their effect by adding an increment of stability to the mutant protein. And again not surprisingly, proteins with the suppressor changes alone are more stable to unfolding than is the wild-type protein. Unlike the case with lambda repressor, however, examination of the model of staph nuclease derived from X-ray crystallography did not suggest a reason why these particular amino acid changes should confer a higher than wild-type level of stability. This result illustrates the power of working with mutations isolated on the basis of phenotype for identifying important positions in a protein, in a way that is not based on hypotheses, nor on preconceived ideas as to what one might find.

A third genetic strategy that holds considerable promise for analyzing how amino acid sequence information transforms into structure and function is the detailed characterization of doubly mutant proteins. Once two interesting mutant proteins harboring single amino acid substitutions at different positions have been isolated and the effects of each amino acid change quantitated, both mutations can be put together in the same gene and a doubly mutant protein isolated (Fig. 2). The important question can then be addressed: How does the effect of one substitution depend on the context in which it occurs? In other words, does the quantitative effect of amino acid substitution A on a particular property of the protein change when substitution B is present in the same polypeptide chain? If the effects of A and B are independent of each other, the quantitative change observed in the double mutant should be the arithmetic sum of the change observed with A alone plus the change observed with B alone.

However, if the effects are not independent, the properties of the double mutant will not result from the simple addition of effects of the single mutations. In this case, a mechanism must be inferred by which the two amino acid positions are able to "communicate" what residues are present, since substitution A produces a quantitatively different effect when the wild-type amino acid is at B as opposed to the mutant amino acid, and likewise with substitution B. Presumably, such a mechanism of communication or coupling between two sites in a protein must involve some sort of structural modification or conformational change.

As described in more detail in chapter 24 by Fersht and Leatherbarrow, several pairs of amino acid substitutions in the active site of tyrosinyl-tRNA synthetase (*Bacillus stearothermophilus*) have been introduced into the same polypeptide chain and found to exhibit nonindependent effects on the free energy of stabilization of the transition state [6]. The most pronounced coupling of effects was seen between amino acid positions only three residues apart, a reasonable distance over which a conformational perturbation might be propagated.

A conceptually similar application of doubly mutant proteins is in the analysis of the interaction between amino acid positions responsible for a known conformational change. Probably the most thoroughly studied conformational transition of any protein is the change in the quaternary structure of hemoglobin that occurs on binding four molecules of oxygen. The increase in free energy of the protein on shifting from the deoxy to the oxy conformation in effect lowers its affinity for oxygen and gives rise to the cooperativity of oxygen binding. In an elegant attempt to define the "pathway" by which the heme group communicates its state of oxygenation to the interfaces between subunits (the sites where most of the conformational changes occur), Ackers and

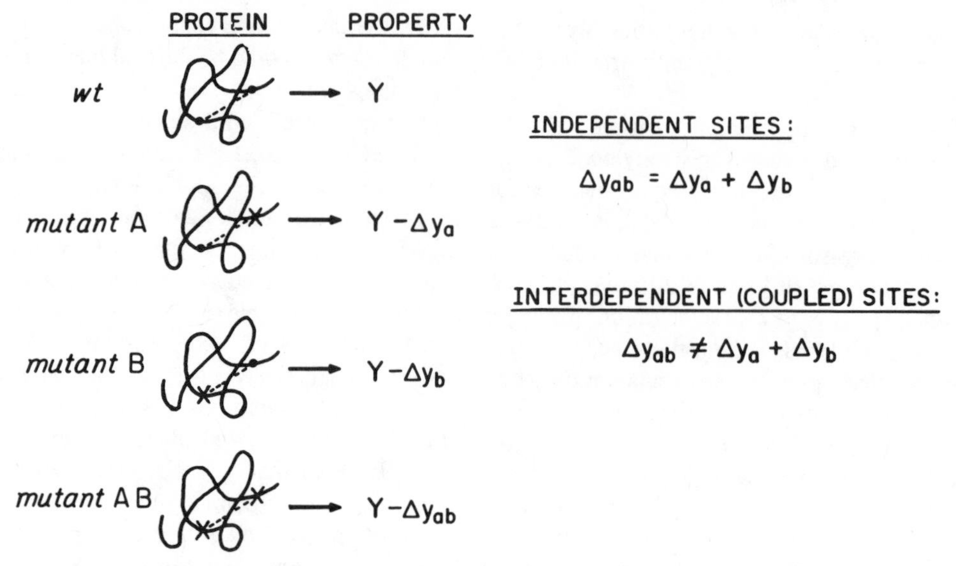

Fig. 2. *Testing for interactions between positions A and B with respect to the quantitative property Y by characterization of singly and doubly mutant proteins.*

Smith have generated and characterized a number of doubly mutant human hemoglobins [7]. These doubly mutant proteins were not constructed by manipulation of genes, but rather by taking naturally occurring mutant hemoglobins found in the human population, disassociating the alpha and beta chains, and then reassociating the peptide chains in vitro to form complete tetramers. Therefore, instead of being in the same polypeptide chains, the two mutations are in different chains, so that only long-range coupling interactions that can be propagated between subunits are studied with these types of hybrid molecules. The interested reader is encouraged to see reference 7 for a more complete discussion of the logic and methodology behind this approach.

In the years ahead, site-directed mutagenesis with oligonucleotides and the more classical genetic strategies described in this chapter will contribute in complementary ways to our understanding of how amino acid sequence is transformed into the wide diversity of properties exhibited by the many thousands of proteins found in nature. A genetic approach based on phenotypes points to residue positions that are important with respect to the protein property being assayed through the mutant phenotype. Once these important positions have been identified and their interactions defined, quantitative hypotheses can be formulated and submitted to rigorous testing by the study of very precise single and double mutations constructed with oligonucleotides.

REFERENCES

1. Botstein D, Shortle D (1985): Strategies and applications of in vitro mutagenesis. Science 229:1193–1201.
2. Miller JH (1978): The *lacI* gene: its role in *lac* operon control and its use as a genetic system. In Miller JH, Reznikoff WS (eds) "The Operon." Cold Spring Harbor Laboratory, New York pp 31–88.
3. Shortle D (1983): A genetic system for analysis of staphylococcal nuclease. Gene 22:181–189.
4. Shortle D, Lin B (1985): Genetic analysis of staphylococcal nuclease: identification of three intragenic "global" suppressors of nuclease-minus mutations. Genetics 110:539–555.
5. Nelson HCM, Sauer RT (1985): Lambda repressor mutations that increase the affinity and specificity of operator binding. Cell 42:549–558.
6. Carter PJ, Winter G, Wilkinson AJ, Fersht AR (1984): The use of double mutants to detect structural changes in the active site of the tyrosyl-tRNA synthetase (*Bacillus stearothermophilus*). Cell 38:835–840.
7. Ackers GK, Smith FR (1985): Effects of site-specific amino acid modification on protein interactions and biological functions. Ann Rev Biochem. 54:597–629.

Protein Engineering, pages 109–121

10

Temperature-Sensitive Mutations Affecting Kinetic Steps in Protein-Folding Pathways

Jonathan King, Cameron Haase, and Myeong-hee Yu

Department of Biology, Massachusetts Institute of Technology, Cambridge, Massachusetts

GENETIC ASPECTS OF THE PROTEIN-FOLDING PROBLEM

The spatial organization of the polypeptide chains of over 100 proteins is known to atomic dimensions, and the sequence of amino acids forming these polypeptide chains has been completely determined. Yet it is still not possible to deduce the spatial conformation of a protein of unknown function from inspection of its amino acid sequence. Since for at least a subset of proteins the amino acid sequence can fully determine the conformation in the absence of cellular components [1], there must be some code, grammar, or set of rules that relate sequence to three=dimensional conformation. The nature of these instructions represents an undeciphered aspect of the genetic code.

The difficulty in solving this problem stems in large part from the lack of two kinds of information: 1) the actual pathways that polypeptide chains follow in achieving their native conformation, and 2) the identity and organization of those amino acid residues within polypeptide chains that direct the folding process.

The Importance of Intermediates and Pathways

In the classical view of protein folding, the native conformation represents the lowest energy state available to the polypeptide chain under a given set of conditions, and the chain finds this conformation through a relative random search. This view does not explain how the chain avoids getting trapped in local energy minima different from the native conformation. In fact, the direct experimental analysis of protein refolding in vitro has revealed the existence of distinct pathways and intermediates [5,17,].

The concept of a folding pathway has emerged particularly clearly, from the analysis of bovine pancreatic trypsin inhibitor by Thomas Creighton. An intermediate in this pathway contains a nonnative disulfide bond, and therefore almost certainly a nonnative conformation. This band is not an "incorrect" disulfide, as it is often refered to, but rather a step in the pathway toward the final native conformation.

Role of Amino Acid Sequence in Determining the Folding Pathway

Given the existence of intermediates in polypeptide chain folding pathways, the conformation of these intermediates must be determined by the amino acid sequence. In cases such as BPTI, where intermediates have conformations which may be absent from the native protein, it would be difficult to decipher the instructions in the sequence by inspection of only the native conformation. The native conformation is likely to be the product of both kinetic steps in the folding pathway, stabilized by interactions in the native protein, both of which are specified by the amino acid sequence.

Once one recognizes the importance of the pathway, we are faced with the experimental problem of determining the structure of these intermediates [17]. Approaches to this difficult problem are described in the contributions to this volume by Matthews, Creighton and Goldenberg, and Kim and Baldwin.

Which Residues Carry the Folding Instructions?

Most algorithms that attempt to predict or describe chain folding from amino acid sequence implicitly assume that every residue carries information with respect to conformation; each residue is assigned some property or propensity, and these are added or multiplied to generate the prediction/description [3]. In fact this is not the case, except perhaps for very small polypeptide chains. For example, comparison of hemoglobin sequences from different species reveals clearly that a large number of residues can be altered without altering the folding pathway.

These observations suggest that the folding instructions are dispersed in the coding sequence, just as control signals in DNA are dispersed through the nucleotide sequence. Until recently, little experimental data has been available as to which amino acid sequences, or which aspects of the sequence, carry the folding pathway instructions [19]. The studies of Baldwin and co-workers with ribonuclease S-peptide [18] and the mutational analysis reported here are attempts to fill this gap.

In taking a genetic approach to the problem of how the folding and conformational information is distributed through a polypeptide chain, we have addressed the following questions:

1. Which residues in a polypeptide chain drive the folding pathway?

2. How are these residues organized? As words—for example, 'start helix' and 'stop helix'?

3. What is the size and general character of these words? How many residues are in them? Are they contiguous or dispersed?

Our approach has been to isolate and characterize mutants which specifically alter the folding and subunit association pathway of a polypeptide chain, without altering the native protein. Such mutants distinguish residues involved in the kinetic control of conformation from residues involved in the stability and activity of the native protein [20]. This approach is complementary to the efforts to characterize mutations which alter the stability of the mature protein [14,23]. It is likely that many residues will have roles in both aspects of the functioning of the polypeptide chain. We thought it likely, however, that at least with large proteins, these aspects might be segregated in different local sequences.

TEMPERATURE-SENSITIVE MUTATIONS

Temperature-sensitive mutations were developed initially by Horowitz and Leupold, [29] and later by Edgar and Lielausis [7] in order to identify genes whose proteins were essential for the reproduction of the organisms. The general phenotype of ts mutations is that a functional protein carrying a single amino acid substitution is formed at low (permissive) temperature, whereas active protein does not accumulate at a higher restrictive temperature. The reverse class, cold-sensitive mutations, though much rarer, have been isolated for some proteins [4,16,22].

The ts mutations were early divided into two classes: TL, for thermolabile, and TSS, for temperature-sensitive synthesis. In the TL class, the active protein produced at permissive temperature is inactivated on exposure to the higher restrictive temperature. The best-studied examples are temperature-sensitive mutations which decrease the melting temperature of T4 lysozyme [13,14].

On the other hand, an analysis of mutants of T4 DNA polymerase revealed that, though some of the proteins purified from ts mutant-infected cells were thermolabile, others were not [6,26]. Many of these represent the TSS class in which the defect is only expressed if the mutant protein is synthesized at restrictive temperature. Edgar and Lielausis [8] suspected that these were folding defects but were unable to prove it. As we show below, many of these are indeed defective in polypeptide chain folding or subunit association [24].

Thermostable Tail Spike Protein of Phage P22

Sorting out the effect of mutations on the native protein from effects on intermediates required a protein whose folding intermediates were easily distinguished from the native protein. The system that has turned out to be most useful is the tail spike endorhamnosidase of *Salmonella* phage P22 [2]. This is a large structural protein, which forms the attachment apparatus of the phage. The tail spike is an elongated trimer of polypeptide chains coded for by gene 9 of P22 [2, 11]. The gene 9 polypeptide chain is 666 amino acids long, without unusual features of amino acid composition or sequence [21]. Its secondary structure is predominantly β-sheet [27].

As with most structural proteins, it performs multiple functions: One end of the protein is involved in the formation of an irreversible but noncovalent bond with the portal vertex of the phage particle, the other end the of the protein has an endorhamnosidase activity which cleaves the O-antigen of the *Salmonella* host cell [15]. The protein is a strong antigen and is the major target of neutralizing antibody made by immunized rabbits.

However, the most distinctive aspect of the protein is its thermal and physical stability: Heating to above 80°C is required for loss of activity. The native spike is also resistant to detergents and proteases, unless denatured [11]. The SDS resistance of the native spike has allowed easy resolution of unfolded and misfolded states from the native states. As a result, it has been possible to follow the folding kinetics in vivo of newly synthesized gene 9 polypeptide chains maturing into the native spike.

Tail Spike Chain Folding and Assembly Pathway

Figure 1 shows the in vivo folding and subunit association pathway for the tail spike pro-

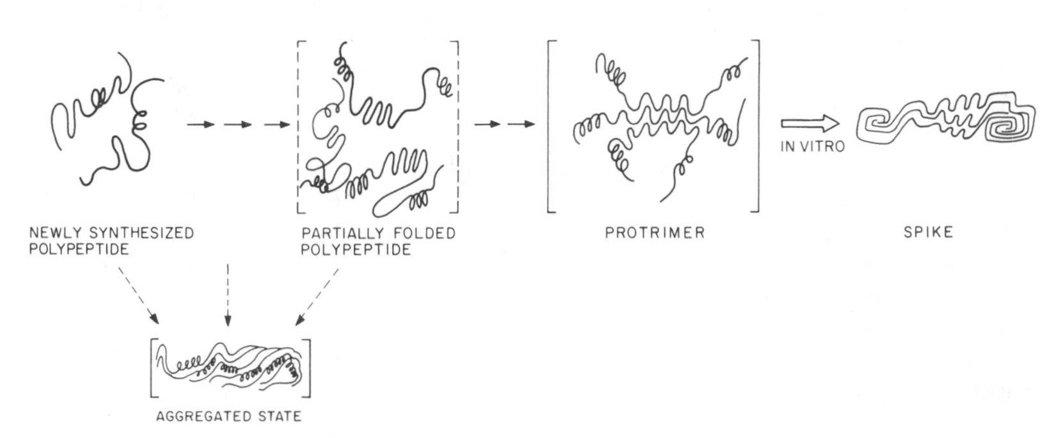

Fig. 1. *Pathway of tail spike maturation. Newly released polypeptide chains fold partially into a conformation that has enough structure for specific chain-chain recognition. These species associate into the protrimer, in which the chains are associated but not fully folded. Both the unassociated and protrimer forms are sensitive to proteases and SDS. The protrimer folds further to form the highly stable native spike, which is SDS and protease resistant, and is stable to 80°C. This last reaction proceeds in vitro [10]. Though the native spike is heat stable, an early intermediate is heat labile [9,11]. At the high end of the temperature range for phage growth, a substantial fraction of newly synthesized polypeptide chains end up as an aggregated dead - end. This aggregate is formed from folding intermediates and not from the native protein.*

tein [10]. The chains are released from the ribosome as single polypeptide chains. They then partially fold into a conformation that has a site for chain–chain recognition. This yields an intermediate, the protrimer, which has the chains associated but only partially folded. This intermediate is metastable and much longer lived than the refolding intermediates found with smaller proteins [17], about 1 hour on ice. It can be resolved from the native protein by electrophoresis through native gels, presumably due to a larger radius of gyration compared to the more compact native spike.

This intermediate folds further into the native spike, and this reaction proceeds in vitro upon warming the material [10]. In this last reaction, the thermostability, resistance to proteases, and detergents develops.

Note that this pathway is not like hemoglobin, with subunits folding, then associating. It is more like collagen, in which chain association is not separable from chain folding. The maturation pathway is purely conformational, with no proteolytic cleavage, glycosylation, phosphorylation, or other covalent modification known.

As with myosin or collagen, there is no species corresponding to a native monomer. The individual polypeptide chains require interaction with their sibs in order to achieve the native conformation [10].

An early intermediate step in this pathway is thermolabile. As the temperature is increased, the fraction of chains successfully passing through the folding pathway and achieving the native conformation decreases [11, 28]. The incompletely folded chains accumulate as large aggregates. This thermolabile early intermediate in the wild-type pathway appears to be the locus of action of the temperature-sensitive folding mutations.

The last step in the pathway, the protrimer to native spike transition, has the opposite character. It is inhibited in the cold, permitting trapping of the protrimer. The conformational transition proceeds at optimal efficiency in the 37 to 40°C temperature range [10].

Temperature-Sensitive Folding (TSF) Mutations

To isolate mutants blocked in the folding pathway, we focused on temperature-sensitive mutations, since a number of lines of experimentation indicated that many TSS mutants were indeed defective in maturation [25]. We have isolated over 100 temperature-sensitive mutations in gene 9, which encodes the thermostable tail spike protein. Figure 2 shows a map of 31 sites of such mutations. The mapping is not yet complete, but the mutations are distributed over a minimum of 55 sites. They occur over the whole chain but are more concentrated in the central region.

These are conventional ts mutants: They form a plaque at low temperature but not at high temperature [7]. To study the mutant proteins, we introduce an additional mutation-blocking capsid shell assembly, so that tail spikes accumulate as soluble precursors and also are overproduced.

In analyzing the temperature-sensitive mutants, two protein preparations must be characterized for each mutant protein; polypeptide chains released from the ribosome at restrictive temperature, and the protein synthesized at permissive temperature. The character of the TSF mutants is such that these have very different spatial conformations.

At permissive temperature, the polypeptide chains carrying TSF substitutions proceed through the folding and subunit assembly pathway and form native tail spikes. This can be seen in the pulse/chase experiment shown in the top part of Figure 3. Newly synthesized, but incompletely folded or associated polypeptide chains are dissociated by SDS and migrate to the position expected of a 71,000-dalton chain. As these pass through the folding and association pathway, they appear higher up in the gel as native spikes. The mature spikes do not bind SDS and migrate more slowly [11].

At restrictive temperature, the results are very different. The polypeptide chains carrying the mutant amino acid are synthesized at

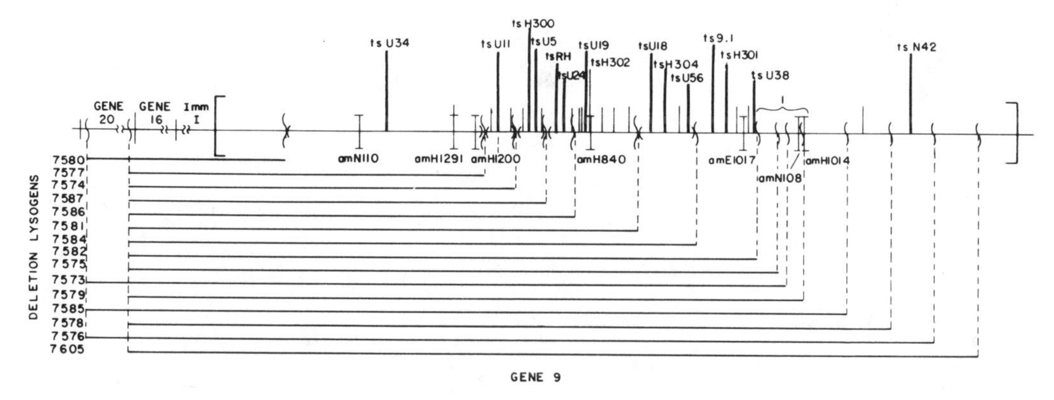

Fig. 2. *Map of temperature-sensitive folding mutations in gene 9. The 31 ticks above the line represent sites of TSF mutations [12;24;25;28], some of which are labeled for further reference. The sites of seven amber mutations are shown below the line.*

normal rates and are stable within the cell. However, they are blocked relatively early in the folding pathway and accumulate as SDS-sensitive polypeptide chains (Fig. 3, lower gel). As summarized in Figure 5, these chains display none of the activities of the native protein—head binding, endorhamnosidase, absorption, or antigenicity—and do not form the protrimer [12,24]. In fact, they are precipitated by antibody made against guanidine-denatured tail spikes, indicating that their conformation is closer to the unfolded [12]. On cell lysis, and possibly before, these incompletely folded chains associate into large aggregates.

The high-temperature state of the tail spike polypeptide chain is not simply a misfolded species. As shown in Figure 6, on shifting infected cells to permissive temperature, active tail spikes are recovered [12,24]. Thus, the mutant polypeptide chains accumulating are reversibly related to intermediates in the folding pathway.

The TSF proteins matured at permissive temperature carry the TSF amino acid substitution. However, they are functionally indistinguishable from the wild type, displaying wild-type levels of head binding, absorption, and endorhamnosidase activity [9]. In particular, they are as thermostable as the wild type, requiring temperatures above 80°C for heat denaturation [9]. Figure 4 shows the thermal inactivation of mutant and wild-type tail spikes at 90°C. The substitutions clearly have little effect on the mature state of the protein formed at permissive temperature. In a sense, they represent residues that are not involved in the activities of the mature protein.

Though indistinguishable from the wild-type native protein with respect to function and heat stability, some of the mutant tail spikes are distinguishable from the wild type in that they have an altered electrophoretic mobility [11]. When electrophoresed through gels of different pore size, the protein mobilities extrapolate back to different free mobilities, indicating that the mobility differences are due to charge rather than shape changes. Of the 33 TS mutant proteins we have studied, 15 have altered electrophoretic mobility, even when synthesized under permissive conditions.

The mutant proteins coded by TSF mutations at 33 sites have been studied in some detail. The great majority of them have the features summarized in Figure 5; they do not affect the native protein formed at permissive temperature, but they block the formation of the native protein if the polypeptide chain was released from the ribosome at restrictive temperature.

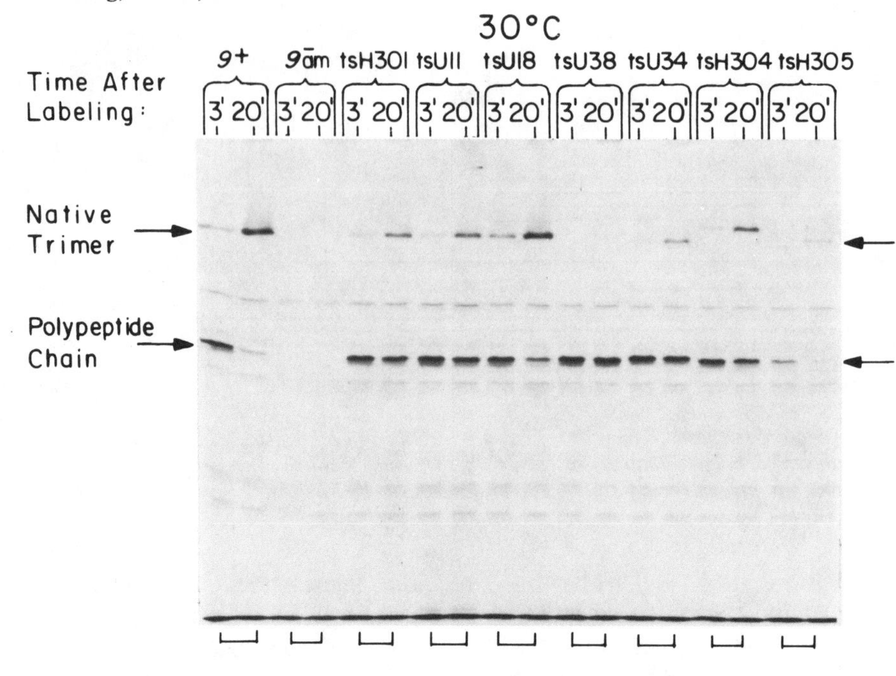

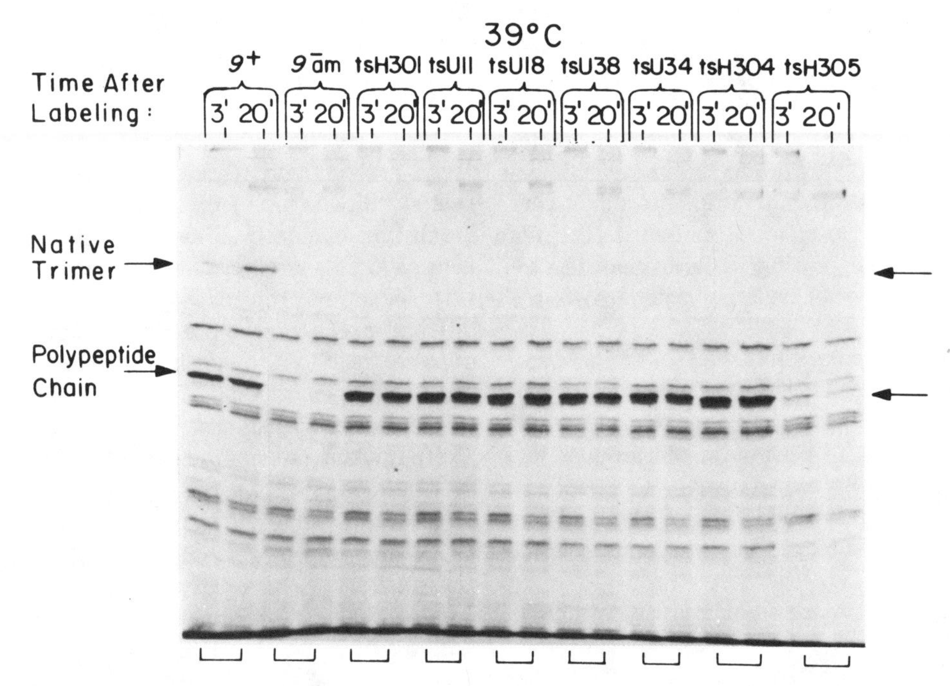

Fig. 3. *The state of newly synthesized TSF gene 9 polypeptide chains at different temperatures. Incompletely folded and native forms of the tail spike are resolved in these gels. Both categories are missing from cells infected with a nonsense mutant of gene 9, as can be run in the second set of lanes. The native spike is only formed by the mutants at 30°C; at 39°C the mutant polypeptide chains accumulate as SDS-polypeptide chain complexes. Cell cultures infected with reference and various TSF mutants of the tail spike were divided and incubated at permissive and restrictive temperatures [24]. The cultures were exposed to a short pulse of* ^{14}*C-labeled amino acids. Samples were withdrawn at 3 minutes and 20 minutes, and the reaction was stopped by freezing. Samples were thawed and electrophoresed through an SDS-acrylamide gel, without prior heating.*

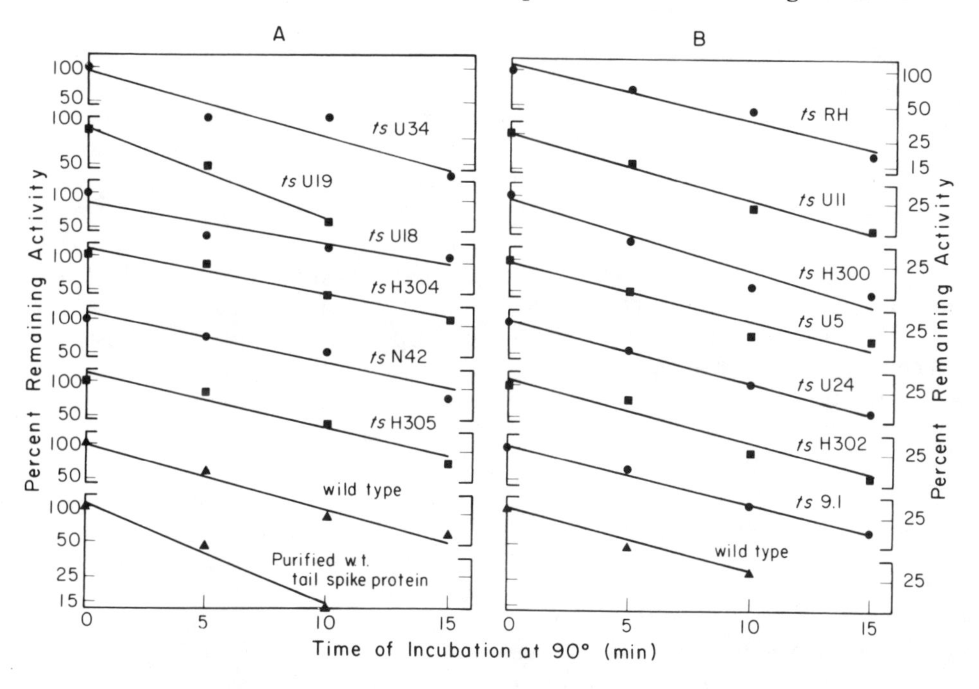

Fig. 4. *Thermostability of TSF mature tail spikes. Mutant and wild-type tail spikes which had been synthesized at permissive termperature were incubated at 90°C. At various times samples were withdrawn, cooled quickly to 30°C, and assayed for in vitro assembly activity [9].*

Thus, these mutations affect kinetic steps in the folding and subunit association pathway at restrictive temperature. The genetic map in Figure 2 is in fact a map of sites in the polypeptide chain which play a critical role in the folding pathway, at least at elevated temperature.

As noted above, an early step in the wild-type pathway is thermolabile. The reduction in maturation efficiency with increased temperature observed with the mutants (Fig. 7) is an intensification of this process. The TSF mutations probably act on to further destabilize the intermediate that is thermolabile in

COMPARISONS OF THE HIGH TEMPERATURE AND LOW TEMPERATURE FORMS OF THE MUTANT POLYPEPTIDE CHAINS

	39° C ts POLYPEPTIDE CHAINS	30° C MATURE PROTEIN
SYNTHESIS	YES	YES
HEAD BINDING	NO	YES
ENDORHAMNOSIDASE	NO	YES
ANTIGENICITY	NO	YES
TRYPSIN RESISTANCE	NO	YES
SDS RESISTANCE	NO	YES

CONCLUSION: ts CHAINS ARE ACCUMULATING IN AN INCOMPLETELY OR INCORRECTLY FOLDED FORM.

Fig. 5. *Summary of properties of the gene 9 TSF mutations.*

wild type, or as considered below, to speed up an off-pathway step.

Are TSF Sites Involved in Chain Folding or Chain Association?

At restrictive temperature, all of the TSF mutant polypeptide chains we have studied are blocked prior to the formation of the protrimer intermediate [12]. In addition to the biochemical characterization, we have characterized the mutants with respect to intragenic complementation and dominance [25; A. Marra and J. King, unpublished results]. Despite the wide distribution of the TSF mutations through the gene, we have never found evidence for intragenic complementation [Denhardt et al., 1964]. The inability of mutant chains to rescue each other suggests that the mutant chains are blocked prior to the chain interaction steps.

We have interpreted these and other results as indicating that the mutant amino acid substitutions prevent the chains from achieving the conformation needed for specific chain–chain recognition and association. This kinetic explanation (Fig 8) explains why the mutant proteins, once native, are as thermostable as wild type.

An alternative model is that the mutations do not act kinetically, but destabilize a specific domain needed for association. However, the distribution of the mutations throughout the polypeptide chain indicates that a very large fraction of the polypeptide chain participates in the "domain"— in fact, the entire chain. Furthermore, since amino acid substitutions at over 30 different sites destabilize this special domain, each one cannot make a very large contribution to its stability. Yet the different mutants all sharply reduce protrimer formation at restrictive temperature.

These observations are better accounted for by a model in which the mutations interfere with kinetic steps in the pathway, such that the chain never achieves the correct conformation for association. The substitutions could also affect kinetic steps by speeding up off-pathway interactions.

Role of Aggregation

Characterization of the physical state of the inactive TSF mutant polypeptide chains synthesized at restrictive temperature indicates that they are aggregated. After lysis of infected cells, the mutant gene 9 polypeptide chains are in a rapidly sedimenting form. The complex bands with a density of 1.3g/ml in a cesium chloride density gradient. indicating that it is aggregated polypeptide chains.

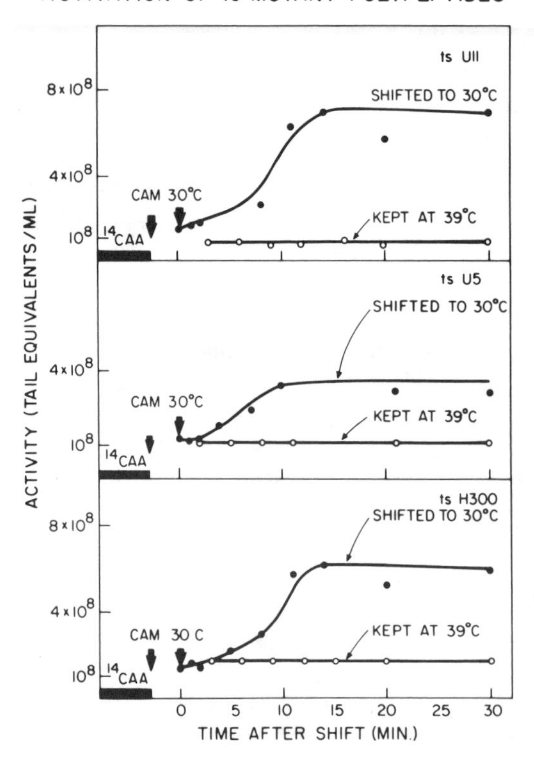

Fig. 6. *Reversability of the ts polypeptide chains. Mutant infected cells were incubated at restrictive temperture with a short pulse of radioactive amino acids. Chloramphenicol was added to stop protein synthesis, and the cells were shifted to permissive temperature. Samples were withdrawn periodicially and assayed both for active tails and for SDS-resistant spikes [24]. After the shift a substantial fraction of the inactive chains formed at high temperature were recovered as active tail spikes. The appearance of SDS resistant spikes paralleled the recovery of activity (data not shown).*

These aggregates represent dead ends formed from incompletely folded polypeptide chains and are also present in wild-type infected cells at high temperature. However, the aggregation step could be a primary effect of the TSF mutations. Increasing the rate of an off-pathway step would have the same overall effect as decreasing the stability of an on-pathway intermediate (Fig. 8).

Measurements of the overall rate of the tail spike maturation pathway as a function of temperature reveal that the mutations do no alter the rates, but rather reduce the yield with increasing temperature [28]. This is consistent with an increase in the efficiency of an off-pathway step, such as aggregation.

Amino Acid Substitutions in TSF Mutants

We have determined the amino acid substitutions of a number of these mutant proteins by isolating restriction fragments from the phage DNA, and sequencing the fragments. The genetic map was used to select the fragments to isolate [28]. A number of mutations have also been sequenced by Javed Siddiqui by cloning restriction fragments into M13 and sequencing the insert. In all cases where we found nucleotide substitutions, they represented a single amino acid substitution. These are shown in Table I.

We initially thought that many of these mutations would occur at the sites of prolines. The restricted rotation around the peptide bond would be more important at high temperature than at low temperature in controlling the correct trajectory during the folding process. In fact, only one proline residue has been identifed as a site of TSF mutations, though we have a number of independent occurrences at that site. The substitutions are quite diverse, as shown in Table I. However, two classes stand out: replacements of glycines and replacements of threonines.

Do TSF Mutations Mark Turns?

Four glycine residues have been identified as sites of TSF mutations. For all four, the substitutions are by bulky charged amino

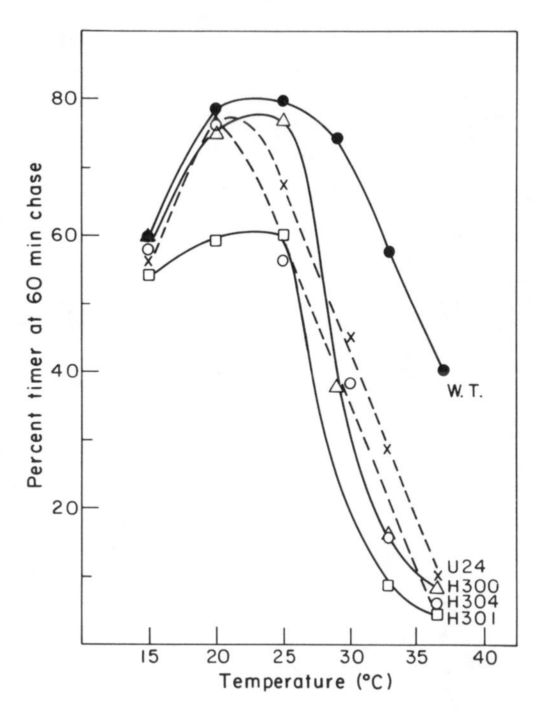

Fig. 7. *Yield versus temperature for wild-type and TSF mutant proteins. Infected cells were exposed to a short pulse of amino acids; then portions were incubatd at various temperatures. Samples were withdrawn and assayed by gel electrophoresis for SDS-sensitive incompletely folded chains, and SDS-resistant native spikes. This figure shows the yield of native spikes as fraction of total gene 9 polypeptide chains synthesized, at 60 minutes after the pulse of radioactive amino acids [28].*

acids. Though it is not surprising that these substitutions would cause problems, it was difficult to understand how they could be accommodated into the native protein at permissive temperature. In fact, the native mutant proteins specified by these mutants have altered electrophoretic mobilities and isoelectric points. This indicates that they represent amino acid substitutions at the protein surface, changing the effective charge. The location of the glycines at the protein surface allows bulky polar residues to be accommodated in place of the glycines.

The low temperature result shows that there is no steric limitation to the accommodation of the glycine replacements. Since the high tem-

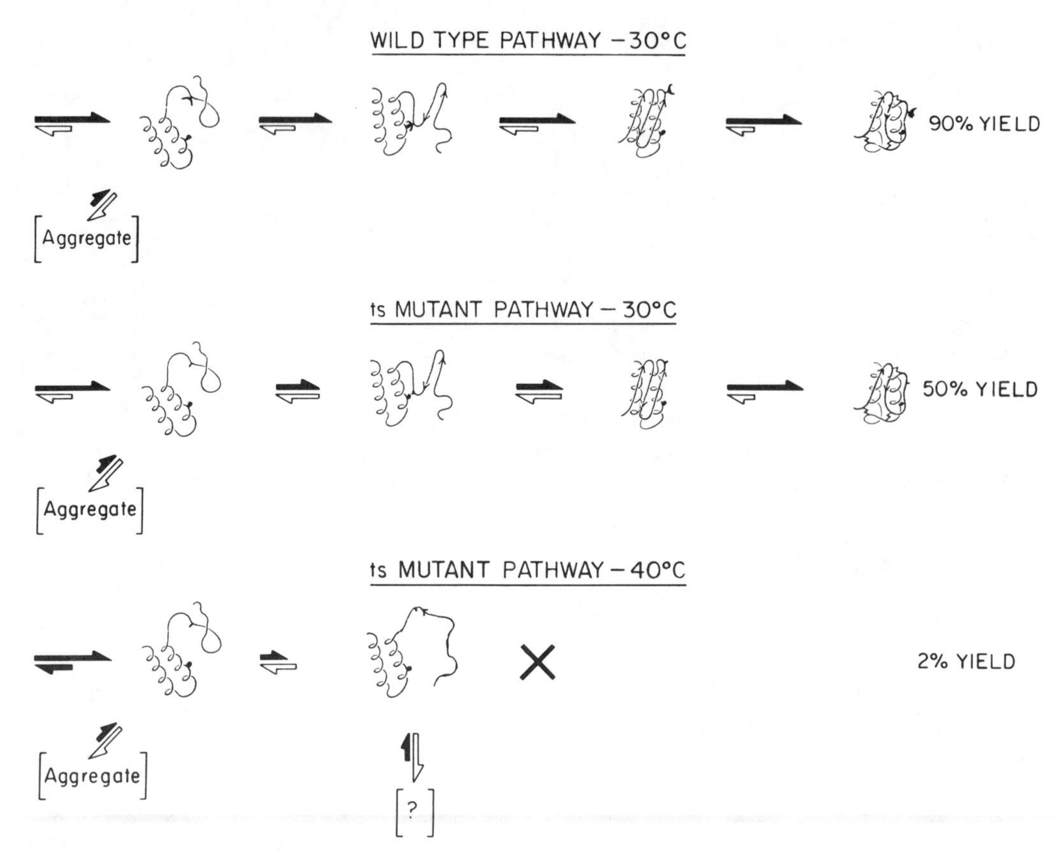

Fig. 8. *Model for the molecular basis of a TSF defect. This model shows an interaction between a segment of polypeptide chain forming β-strands, with a nucleating pair of α-helices. The interaction stabilizes the complex, ensuring the correct pathway at elevated temperature. At lower temperature the reaction still proceeds with lower efficiency. This interaction is only present in the intermediate and is not found in the final structure, as indicated in the pathway. Thus the substitution does not destabilize the native structure, but only the intermediate. Modified from a drawing by Jane Richardson [30].*

perature defect is due to an effect on the folding pathway, as indicated by the mutant phenotypes, the polar replacements of glycines are presumably the source of the kinetic block.

Since turns are concentrated at the surface of proteins, and glycines are common at turns, we interpret these sites as representing glycines at surface turns. The charged residue would alter the interaction with the solvent, or perhaps introduce interactions with other parts of the chain, which at elevated temperature interferes with the folding pathway.

Three of the sequenced TSF mutations are replacements of threonines. The mutant amino acids lack the side chain hydroxyl group. The side-chain hydroxyl groups of threonines are most frequently involved in internal hydrogen bonds with an amido or carbonyl group of a nearby backbone. If we imagine a bend or turn in which the threonine hydroxyl is hydrogen bonded to the backbone of the opposite side of the turn, it is easy to understand the temperature effect. At permissive temperature a turn could be stable enough to continue through the pathway. However, at elevated temperature the turn would be thermolabile without the stabilizing bond (Figure 8).

Possibility of "Foldons"

Examination of the local sequence surrounding the two threonine> isoleucine replacements (shown in Table II) reveals distinctive homology. At three positions the residues are identical. Tyrosine and tryptophan are related, and at the 2 position of a beta-turn, glycine and proline are equivalent amino acids. Since the replacements of threonine by isoleucine cause similar functional defects, the local sequences may well specify similar steps in the folding pathway. Note that decrease in native yield as a function of temperature is similar for both threonine > isoleucine substitutions (Fig. 7). In light of the previous discussion, these sequences might be specifying turns in the folding pathway. We term such a sequence a "foldon" [28]. We were able to identify the sequence relationship because contiguous amino acids were homologous. However, if the relationship was between residues dispersed somewhat, the pattern would probably go unrecognized.

These sites could be essential sites for folding, regardless of temperature. In that case, our substitutions are a selected subset which would interfere only at elevated temperature. Presumably other amino acid substitions would prevent correct folding at all temperatures. Alternatively, they could mark residues which are only needed at higher temperatures—for example, the model for the threonine turns.

These findings indicate that the distribution of temperature-sensitive folding mutations in the genes of at least prokaryotes probably marks particular features of protein conformation, or particular kinetic steps on or off the maturation pathway. As we obtain better data on the actual conformation of intermediates, aspects of the rules through which the amino acid sequence controls protein conformation should emerge more clearly.

ACKNOWLEDGMENTS

This work supported by Grant PCM-8402546 from the National Science Foundation and Grant GM17,980 from the National Institutes of Health. We thank David Goldenberg and Donna Smith for valuable discussions. Ruth Martin provided expert assistance in preparation of the manuscript.

TABLE I. Amino Acid Substitutions of Gene 9 ts Folding Mutants

Mutation	Residue substitution	Local sequence
tsU9	177 Gly > Arg	Phe · Ile · Gly · Asp · *Gly* · Asn · Leu · Ile · Phe
tsH304	244 Gly > Arg	Val · Lys · Phe · Pro · *Gly* · Ile · Glu · Thr · Leu
TsH302	323 Gly > Asp	Asn · Tyr · Val · Ile · *Gly* · Gly · Arg · Thr · Ser
tsU38	435 Gly > Glu	Leu · Leu · Val · Arg · *Gly* · Ala · Leu · Gly · Val
tsH300	235 Thr > Ile	Gly · Tyr · Gln · Pro · *Thr* · Val · Ser · Asp · Tyr
tsU18	307 Thr > Ala	Asp · Gly · Ile · Ile · *Thr* · Phe · Glu · Asn · Leu
tsH301	368 Thr > Ile	Thr · Trp · Gln · Gly · *Thr* · Val · Gly · Ser · Thr
tsU5	227 Ser > Phe	Thr · Leu · Lys · Gln · *Ser* · Lys · Thr · Asp · Gly
tsN48	333 Ser > Asn	Gly · Ser · Val · Ser · *Ser* · Ala · Gln · Phe · Leu
tsU19	285 Arg > Lys	Gly · Phe · Leu · Phe · *Arg* · Gly · Cys · His · Phe
tsU53	382 Arg > Ser	Asn · Leu · Gln · Phe · *Arg* · Asp · Ser · Val · Val
tsH303; tsU11	250 Pro > Ser	Glu · Thr · Leu · Leu · *Pro* · Pro · Asn · Ala · Lys
tsU24	258 Ile > Leu	Lys · Gly · Gln · Asn · *Ile* · Thr · Ser · Thr · Leu
tsRAF; tsRH	270 Val · Gly	Glu · Cys · Ile · Gly · *Val* · Glu · Val · His · Arg
ts9.1	334 Ala > Val	Ser · Val · Ser · Ser · *Ala* · Gln · Phe · Leu · Arg

TABLE II. Potential Foldon for a Stable Turn in the Tail Spike

Residues	Allele	Sequence[a]
231–239	Wild type	Gly · **Tyr** · **Gln** · **Pro** · **Thr** · **Val** · Ser · Asp · Tyr
231–239	tsH300	**Ile**
364–372	Wild type	Thr · **Trp** · **Gln** · **Gly** · **Thr** · **Val** · Gly · Ser · Thr
364–372	tsH301	**Ile**

[a]Homologous sequences shown in boldface.

REFERENCES

1. Anfinsen CB (1978): Principles that govern the folding of protein chains. Science 181:223–230
2. Berget PB, Poteete AR (1980): Structure and Functions of the phage P22 tail protein J Virol 34:234–243.
3. Chou PY, Fasman GD, (1978): Empirical predictions of protein conformation. Ann Rev Biochem 47:251–276.
4. Cox JH, Strack HB (1971): Cold-sensitive mutants of bacteriophage Lambda. Genetics 67:5–17.
5. Creighton TE (1981): Experimental elucidation of pathways of protein unfolding and refolding. In Jaenicke R. (ed): "Protein Folding." Elsevier/North-Holland Biomedical Press, pp 427–441.
6. DeWaard PA, Paul AV, Lehman IR (1965): The structural gene for deoxyribonucleic acid polymerase in bacteriophage T4 and T5. Proc. Natl Acad Sci USA 54:1241–1248.
7. Edgar RS, Lielausis I (1964): Temperature-sensitive mutants of bacteeriophage T4D: Their isolation and genetic characterization. Genetics 49:649–662.
8. Edgar RS, Lielausis I (1965): Serological studies with mutants of phage T4D defective in genes determining tail fiber structure. Genetics 52:1187–1200.
9. Goldenberg D, King J (1981): Temperature-sensitive Mutants Blocked in the Folding of Subunit Assembly of the Bacteriophage P22 Tail Spike Protein. II. Active Mutant Proteins Matured at 30°C. J Mol Biol 145:633–651.
10. Goldenberg D, King J (1982): Trimeric intermediate in the in vivo folding and subunit assembly of the tail spike endhorhamnosidase of bacteriophage P22. Proc Natl Acad Sci USA 79:3403–3407.
11. Goldenberg DP, Berget PB, King J (1982): Maturation of the Tail Spike Endhorhamnosidase of Salmonella Phage P22. J of Biol Chem 257:7864–7871.
12. Goldenberg DP, Smith DH, King J (1983): Genetic Analysis of the Folding Pathway for the Tail Spike Protein of Phage P22. Proc Nat Acad Sci USA 80:7060–7064.
13. Grutter MG, Hawkes RB, Matthews BW (1979): Molecular basis of thermostability in the lysozyme from bacteriophage T4. Nature (London) 277:667–669.
14. Hawkes R, Grutter MG, Schellman J (1984): Thermodynamic stability and point mutations of bacteriophage T4 lysozyme. J Mol Biol 175:195–212.
15. Iwashita S, Kanegasaki (1976): Enzymatic and molecular properties of base-plate parts of bacteriophage P22. Eur J Biochem 65: 87–94.
16. Jarvik J, Botstein D (1973): A genetic method for determining the order of events in a biological pathway. Proc Natl Acad Sci USA 70:2046–2050.
17. Kim PS, Baldwin RL (1982): Specific intermediates in the folding reactions of small proteins and the mechanism of protein folding. Ann Rev Biochem 51:459–489.
18. Kim PS, Baldwin RL (1984): A helix stop signal in the isolated S-peptide of ribonuclease A. Nature (London) 307:329–334.
19. King J, Yu M-H (1986): Mutational analysis of protein folding pathways. "In Methods in Enzymology." New York: Academic Press (in press).
20. Matthews CR, Chrisanti MM, Manz JT, Gepner GI (1983): Effect of a single amino acid substitution on the folding of the alpha subunit of tryptophan synthase. Biochemistry 22:1445–1452.
21. Sauer, RT, Krovatin W, Poteete AR, Berget PB, (1982): Phage P22 tail protein: gene and amino acid sequence. Biochemistry 21:5811–5815.
22. Scotti PD (1968): A new class of temperature conditional lethal mutants of bacteriophage T4D. Mutation Res 6:1–14.
23. Shortle D Genetic analysis of Staphyloccal nuclease. In Inouye M (ed): "Protein Engineering." New York: Academic Press (in press).
24. Smith DH, King J (1981): Temperature-sensitive mutants blocked in the folding or subunit assembly of the bacteriophage P22 tail spike protein. III. Inactive polypeptide chains synthesized at 39°C. J Mol Biol 145:653–676.
25. Smith DH, Berget PB, King J (1980): Temperature-

sensitive mutants blocked in the folding or subunit assembly of the bacteriophage P22 tail-spike protein. I. Fine-structure mapping. Genetics 96:331–352.

26. Swartz MN, Nakamura H, Lehman IR (1972): Activiation of defective DNA polymerase in ts mutants in T4. Virology 47:338–353.
27. Thomas GL Jr, Li Y, Fuller MT, King J (1982): Structural studies of P22 phage, precursor particles and proteins by laser raman spectroscopoy. Biochem 80:3866–3878.
28. Yu M-H, King J (1984): Single amino acid substitutions influencing the folding pathways of the phage P22 tail spike endorhamnosidase. Proc Natl Acad Sci USA 81:6584-6588.
29. Horowitz NH, Leupold U (1951): Some recent studies on the one gene-one enzyme hypothesis. Cold Spring Harbor Symp Quant Biol 16:65–72.
30. Richardson JS (1981): The anatomy and taxonomy of protein structure. Adv Prot Chem 34:167–330.

III
ENERGETICS AND PROTEIN DESIGN

Protein Engineering, pages 125–126

OVERVIEW: SECTION III

David Eisenberg and Robert L. Baldwin

Parts I and II of this book have focused on how protein structures are determined and how proteins of known structure can be modified. Part III focuses on how we can design new proteins. Protein design is a new science, and its principles are being formulated now. This is the origin of the sense of excitement in the chapters in Part III.

Several chapters in Part III emphasize intermolecular forces. The reason for this emphasis is clear if you consider the following: Natural polypeptides have been selected to fold into functioning proteins; proteins designed by humans have not. There is no reason to suppose that a random string of amino acids will fold into a compact molecule of definite three-dimensional structure. Thus, coupled to the problem of designing functioning proteins, there is the even more basic problem of designing properly folding proteins. To do this, it is necessary to understand the forces acting between different parts of the protein chain.

General ideas on protein stability are given in the first paper of Part III by Baldwin and Eisenberg. These include: the central idea that natural proteins fold spontaneously, the concepts used in measuring protein stability, and a discussion of the main forces that determine stability. The following two chapters (by Richardson and Richardson, and by Ohlendorf, Finzel, Weber, and Salemme) are on design principles. The first of these articles examines how strands of extended polypeptide chains can be linked by tight hairpin turns and packed together to form hydrogen-bonded β-sheets. The result is a proposal for a synthetic protein, "betabellin." The second of these two articles considers how α-helices can be linked by turns and then packed together to form stable tetrameric bundles.

The next three articles consider the effect on protein stability of specific component forces. Two papers by Lesser, Lee, Zehfus, and Rose; and by Eisenberg, Wilcox, and McLachlan describe hydrophobic forces. The hydrophobic interaction has long been thought to stabilize proteins, but finding quantitative values has been difficult. These papers address this problem. The paper by Dill discusses protein stability in general. It also considers the entropy of protein folding, another elusive quantity. All of these forces will have to be understood more thoroughly before protein design becomes routine.

The final three chapters of Part III describe synthesis and stability of polypeptides. Peptides that bind to the natural protein calmodulin have been designed by Erickson-Viitanen, O'Neil and DeGrado. These are amphiphilic α-helices, having one polar and one apolar face. The design of this class of α-helix, as models for apolipoproteins and peptide toxins, is considered in the article by Kaiser. Kaiser also considers the design of amphiphilic β-strands. The final chapter of Part III, by Klibanov and Ahern, discusses the basis of thermal stability of proteins.

Taken together, the chapters of Part III demonstrate that we have entered the era of protein design. Although the general principles of protein design are only now being discovered, it is already clear that they include a mixture of elements: some theory of intermolecular forces, some homology of the design with structures of natural proteins, and some trial-and-error synthesis followed by characterization of structure and function of the designed protein.

Protein Engineering, pages 127–148

11

Protein Stability

Robert L. Baldwin and David Eisenberg

Department of Biochemistry, Stanford University Medical Center, Stanford, California 94305 (R.L.B.), Department of Chemistry, and Molecular Biology Institute, University of California, Los Angeles, California 90024 (D.E.)

SPONTANEOUS FOLDING OF PROTEINS

In the 1950s it was widely believed that accessory factors in addition to the unfolded polypeptide chain were needed to fold a protein correctly or to form the correct S–S bonds of a disulfide-containing protein. The dilemma is as follows. Most disulfide-containing proteins unfold completely when their S–S bonds are reduced by a thiol. Once the protein is reduced and unfolded, a large number of combinations of –SH groups can be obtained by reoxidation: $7 \times 5 \times 3 \times 1 = 105$ combinations for a reduced protein with 8 –SH groups. At the start of reoxidation, a given –SH group can react with any of the other 7 –SH groups, and so on. To form a correct S–S bond requires positioning of the –SH groups through folding, but the folded structure isn't stable until the S–S bonds are made.

Anfinsen's Experiment

Anfinsen resolved the dilemma with the following experiments. He chose a small protein, ribonuclease A (RNaseA), with 4 S–S bonds. RNaseA was a particularly well-studied protein, and many of the properties needed to design and interpret these experiments were known. When its S–S bonds are left intact, RNaseA undergoes a readily reversible unfolding/refolding reaction; Figure 1 shows thermal unfolding at two pHs. When the 4 S–S bonds are broken, RNaseA unfolds completely. S–S bonds can be remade by oxidation but, following procedures in use before Anfinsen's work, the product is unfolded, scrambled RNaseA containing incorrectly paired S–S bonds.

Anfinsen and co-workers [1] discovered that the S–S bonds of RNaseA can be paired correctly in the following conditions:

1) A small amount of a reducing agent (a thiol) is added in addition to the disulfide used to obtain oxidation of the S–S bonds. The thiol causes the S–S bonds that are accessible to be broken and the disulfide remakes the S–S bonds, but in new pairings. The result is to reshuffle the S–S bonds. As the correct bonds are made, the protein folds and the S–S bonds are no longer accessible to thiol, and the reshuffling stops. The protein itself can take part in causing the reshuffling to occur: –SH groups attack S–S bonds in the same protein molecule via the disulfide interchange reaction.

2) The second condition for forming correct S–S bonds is that reoxidation should take place in conditions in which folded RNaseA is stable. If 8M urea is used as a solvent for reoxi-

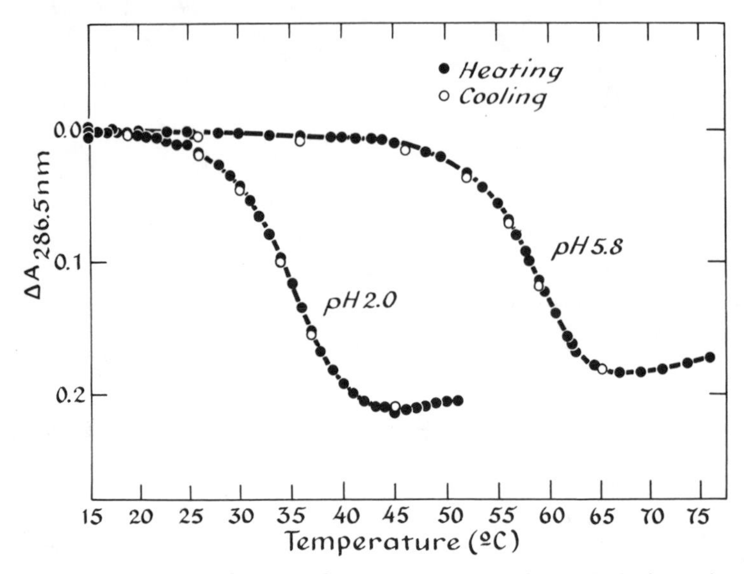

Fig. 1. *Equilibrium transition curves for thermal unfolding of ribonuclease A at two pH values. Tyrosine absorbance (287 nm) is used to monitor unfolding. The same transition curve is obtained either by heating or by cooling, which shows that equilibrium has been reached between the native and unfolded protein. (From Garel and Baldwin [40].)*

dation, scrambled RNaseA is obtained. The scrambled protein can be converted to native RNaseA by diluting out the urea and then adding thiol plus an oxidant. An enzyme has been found, protein disulfide isomerase (PDI), that speeds up the reshuffling reaction.

The conclusion drawn by Anfinsen was that the free energy of folding provides the driving force not only for correct folding but also for correct pairing of the S–S bonds, which may be regarded as fasteners that reinforce the correctly folded structure. No accessory factors connected with the protein synthetic machinery are necessary, although, as pointed out above, protein disulfide isomerase speeds up the formation of correct S–S bonds. Consequently, the information needed for correct folding must be contained entirely in the amino sequence of the protein. For possible caveats, see below.

Thermodynamic vs. Kinetic Control of Folding

Two views of the folding process have developed over the years. In the first view, only thermodynamic stability determines the conformation of a folded protein. If this view is correct, it follows that the kinetic pathway of folding must be such that the most stable product is formed. Moreover, it should be possible in principle to predict the folding of a protein from its amino acid sequence by energy minimization. Prediction of structure involves expressing the potential energy of a protein as a function of dihedral angles, bond distances, etc., and of the forces acting between all pairs of atoms in the protein. A major problem in prediction is that the protein can be trapped in a local energy minimum without reaching the global minimum.

The alternative view of folding is that the conformation of the final product is determined by the kinetic pathway of folding. Folding occurs by the fastest route available, and the final folded conformation could be determined by the folding pathway without necessarily reaching the thermodynamically most stable structure. According to the kinetic view of folding, a protein can be trapped in a local energy minimum in real life and not only in computer simulations of folding.

Experimental approaches to the folding problem are sometimes chosen that can succeed as planned only if one or the other of these views is correct. If the kinetic view is correct, it may be possible to find mutations that block folding at an intermediate stage even though the mutation has little effect on the stability of the folded conformation (for a review, see Reference 2). On the other hand, if the thermodynamic view of folding is correct, such mutants do not exist, and mutations that have been interpreted as kinetic-block mutations actually act by other mechanisms. Before considering this problem in more detail, we discuss two developments that have had a major impact on the controversy.

Levinthal's Calculation

Equilibrium studies of the unfolding/refolding reaction of small proteins can often be represented by the 2-state model N⇌U (see below). U is the unfolded protein and N is the native conformation. The 2-state model says that only N and U are present inside the unfolding transition zone where both species are measurable (cf. Figure 1): Folding intermediates are not detectable.

Levinthal [3] took the 2-state model to its logical conclusion by asking how slowly folding would occur if there were no folding intermediates and folding takes place by a random search. By twisting motions about all bonds, the unfolded protein tests all possible conformations and when, by chance, it finds the correct final conformation it remains folded. Estimates of the time required for folding by a random search are usually made by considering only the conformation of the polypeptide backbone. Suppose the protein has 100 peptide bonds. On the two sides of a peptide bond there are two rotatable bonds (ϕ,Ψ). Suppose there are three possible conformations about each rotatable bond: then there are 3^{200} possible conformations of the polypeptide backbone. Suppose that bond rotation occurs at the rate 10^{+13} sec^{-1} and let n be the number of peptide bonds. Then the time t required to sample each polypeptide conformation is [4] $t = 3^{2n}/(2n \times 10^{13}) = 1.3 \times 10^{80}$ sec or 4×10^{72} years.

Since folding of small proteins usually takes place in seconds or less at 37°C, folding cannot occur by a random search process. Instead, there must be folding intermediates. The calculation given above is badly oversimplified. When the conformations are rejected that are impossible because the chain must pass through itself, the time needed for folding is drastically reduced [5]. Nevertheless, folding by a random search process is still impossibly slow.

Levinthal's calculation has sometimes been used as an argument that folding is kinetically determined, because: 1) There isn't time to test all possible conformations, so there is no guarantee of reaching the thermodynamically most stable state, and 2) since there must be folding intermediates, there probably is a defined folding pathway and it need not yield as product the most stable conformation. The basic counterargument is given in the section below titled "Thermodynamic Control of Conformation in Single-Domain Proteins."

Covalent Processing After Folding

An enzyme may be activated by covalent processing after folding is complete. The pancreatic zymogens chymotrypsinogen and trypsinogen are classic examples of inactive proenzymes that are activated after folding and S–S bond formation have taken place. In each proenzyme, the peptide bond between Arg15 and Ile16 is broken by cleavage with trypsin, the newly created α-NH_3^+ group of Ile16 swings inward to form a salt bridge with the COO^- group of Asp194, and changes in folding create a binding site for substrate. In the case of chymotrypsin, a further proteolytic cleavage produces α-chymotrypsin, whose three polypeptide chains are held together by 5 S–S bonds. When the S–S bonds of chymotrypsinogen are allowed to be reshuffled by the enzyme PDI, no changes in folding or S–S pairing are observed. But when PDI is allowed to reshuffle the S–S bonds of α-chymotrypsin, inactive enzyme is obtained [6]. A similar result was obtained by these authors with insulin, which contains two polypeptide chains held together by its three S–S bonds. The probable reason why PDI inactivates α-

chymotrypsin and insulin but not chymotrypsinogen is that the three chains of α-chymotrypsin, or the two chains of insulin, dissociate as the protein unfolds when the S–S bonds are broken. Since folding is cooperative, the separated chains must be reunited before folding can occur, but this is an improbable event when S–S bonds no longer hold the chains together.

This study is interesting not only for its physiological significance (it led to the discovery of proinsulin), but because it provides an example of a folding process which is pathway dependent. The active enzyme α-chymotrypsin is obtained only via synthesis and folding of the precursor chymotrypsinogen. Active enzyme is not obtained in good yield from the unfolded, reduced form of α-chymotrypsin by reoxidation procedures that are successful when the protein is a single-polypeptide chain. Since these examples of pathway-dependent folding are limited to proteins with two or more polypeptide chains connected by S–S bonds, they do not answer the question of whether the folded conformations of single-chain proteins depend on the folding pathway.

Thermodynamic Control of Conformation in Single-Domain Proteins

Chemical equilibrium between N and U is reached inside the transition zone for the folding reactions of typical single-domain proteins. Failure to reach equilibrium, when it occurs, can usually be attributed to side reactions such as aggregation of the unfolded protein or chemical reaction of –SH groups that are exposed on unfolding. Standard tests for reaching equilibrium have been applied: The same transition curve is obtained beginning with either native or unfolded protein (cf. Figure 1). A kinetic test for reaching equilibrium is also satisfied: The apparent rate constant for the slowest kinetic phase in refolding is the same as that measured for unfolding in the same conditions, inside the transition zone [7]. Provided that the S–S bonds are left intact in such experiments, the same result is obtained for proteins with S–S bonds (RNaseA) or without S–S bonds (horseheart cytochrome *c* [7]).

Consequently, if some other, more stable, folded conformation is possible, it must be inaccessible starting either from the native or from the unfolded conformation in the conditions used for applying these tests, inside the folding transition zone. Small changes in conformation of single-domain proteins are usually found to occur on a time scale of minutes or faster. Thus, if a more stable conformation can exist, it should differ from the native structure in a major way; otherwise, it should be possible to reach this hypothetical structure from the native state in a reasonable time span.

This argument leaves open the question of whether or not there is a unique, sequential pathway of folding. The answer to this question is not known; finding the answer is the object of current research [8]. It should be stressed that the discussion above applies only to the folding of single-domain proteins. The situation is quite different as regards the folding and assembly of oligomeric proteins or large assemblies such as microtubules. Concerted changes in conformation among the subunits of allosteric enzymes are sometimes extremely slow. Mutations have been found that block the folding and assembly of a trimeric tailspike protein of phage P22 [2].

MEASUREMENT OF PROTEIN STABILITY

At present, there is wide interest in the problem of protein stability, for several reasons. The improvement of enzyme stability is one of the main targets in genetic engineering of proteins. To gain control of folding, it is essential to understand how stability is determined. When measured in kcal/mole, the net stability of a folded protein is small compared to the input factors that affect stability (see the section below "Determinants of Stability"). A modest change, such as Ala → Val in a single residue, may tip the balance and make a mutant protein unstable. The ready availability of mutant proteins obtained by genetic engineering makes it attractive to study the problem of protein stability. We consider in this section techniques of measuring the difference in stability between wild-type and mutant protein, and of measuring the stability of the wild-type protein.

In many cases one can measure the equilibrium transition curve for the unfolding of a small protein. In the older literature, rates of denaturation were often used to characterize stability. The rate of denaturation is measured in conditions where unfolding is irreversible, usually because the unfolded protein aggregates and precipitates. Consequently, denaturation rates do not give direct information about the equilibrium stability of the protein; it is better to characterize stability by the equilibrium transition curve.

We discuss below the practical problems of measuring protein stability. The analysis is straightforward for small proteins that obey the two-state model, and much more complex for larger proteins that do not.

Tests for Two-State Unfolding

To find out if the two-state model is applicable, one begins by examining the X-ray structure, when it is available, to see if there is evidence for more than one domain. A domain is a continuous segment of polypeptide chain that is folded upon itself and makes only minimal contacts with other parts of the chain. Since "minimal contacts" is a subjective term, this definition of a domain is not always unambiguous. If a protein has two domains that unfold independently, or nearly so, the two-state model is not applicable.

There is a quantitative calorimetric test for two-state unfolding (reviewed by Privalov [9]). It can be expressed as $\Delta H_{cal}/\Delta H_{vH} \geqq 1$, where ΔH_{cal} is the calorimetrically measured heat of unfolding and ΔH_{vH} is the apparent enthalpy change found from the temperature dependence of the apparent equilibrium constant for two-state unfolding (the van't Hoff relation). If unfolding is two-state, $\Delta H_{cal} = \Delta H_{vH}$; otherwise, ΔH_{cal} is greater. Extensive studies of single-domain proteins have given values of $\Delta H_{cal}/\Delta H_{vH}$ close to 1.00, whereas papain (a small protein whose X-ray structure has two well-defined domains) shows $\Delta H_{cal}/\Delta H_{vH} = 1.8$ [10].

A standard qualitative test for two-state unfolding is to compare the transition curves obtained by two different probes; the two-state model is used to express the data given by each probe as the fraction of unfolded protein. If the two curves do not coincide, and if the noncoincidence is not an artifact caused by improperly chosen baselines, then the two-state model is not applicable. Figure 2 shows such a comparison applied to the thermal unfolding of RNaseA in the presence of a specific inhibitor, 2′CMP. The two probes are tyrosine absorbance, which measures integrity of the tertiary structure, and nucleotide absorbance, which measures binding of 2′CMP. The two unfolding transition curves do coincide, and consequently the 2′CMP binding site is not disrupted before the entire tertiary structure unfolds. A more common choice of

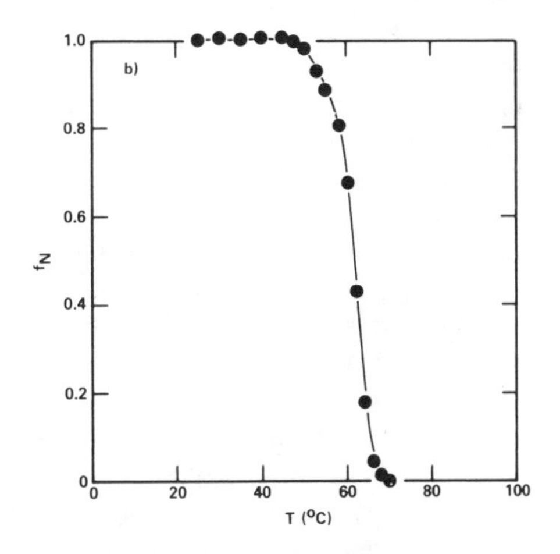

Fig. 2. *Comparison of the unfolding transition curves of ribonuclease A at pH 5.8 measured by two different probes of unfolding: tyrosine absorbance at 287 nm (solid line) and the ability to bind the specific inhibitor 2′ CMP (●). (From Nall and Baldwin [41].)*

two probes is to compare the transition curve obtained with a probe of secondary structure (usually circular dichroism in the far UV, 200–240 nm) with the curve given by a probe of tertiary structure.

Measurement of Stability

One should decide if the aim is to place wild-type and mutant proteins in a series based on relative stability or to measure the thermodynamic stability in kcal/mole in standard conditions (e.g., at 25°, pH 7.0, 0 M denaturant). To give the rank order of stability for a series of mutant proteins, it is sufficient to give the T_m for thermal unfolding or the denaturant concentration at the transition midpoint; this can be done whether or not the two-state model is applicable. Considerably more effort and caution are required to determine the Gibbs free energy stabilizing the native protein, and to extrapolate this to fixed conditions.

In either case, the first step is to choose the unfolding agent and conditions. Typical choices are thermal unfolding at low pH (pH 2–pH 4) or unfolding by GuHCl at 25°C. Low pH is used for thermal unfolding to reduce the T_m and also to minimize the risk of S–S interchange when the protein if unfolded. Urea is used in place of GuHCl if the unfolded protein behaves properly in urea but aggregates in GuHCl; this sometimes happens at low GuHCl concentrations, probably because GuHCl is a salt as well as a denaturant. For more detail, see the review by Pace [11]. The second step is to study the reversibility of unfolding by comparing the transition curves obtained starting either with folded or unfolded protein, and also to determine the time needed to reach equilibrium. To rank order a set of mutant proteins, one can simply record the T_m or GuHCl concentration at the transition midpoint. To determine the thermodynamic stability, one must investigate whether the two-state model is applicable, as indicated above. It is also necessary to be certain that the sample is homogeneous before following the procedure outlined below for measuring thermodynamic stability.

The calculations are simple when the two-state model is applicable. The fraction native, f_N, is

$$f_N = (y - y_U)/(y_N - y_U) \qquad (1)$$

where y is the value of the probe (e.g., absorbance at 287 nm) at some point in the transition and y_N, y_U are the values for the native and unfolded species extrapolated to this point in the transition zone [7]. It is important to measure y as a function of temperature or GuHCl concentration above and below the transition zone, to have the correct baselines for extrapolating y_N and y_U into the transition zone. Then the equilibrium constant K is

$$K = f_N/(1 - f_N) \qquad (2)$$

and the standard-state difference in Gibbs free energy between N and U is

$$\Delta G^\circ = -RT \, \ell n \, K \qquad (3)$$

To obtain ΔG° in standard conditions, it is customary to measure K at different positions in the transition zone and to extrapolate these values to 0 M. Linear extrapolations of ΔG° vs M (GuHCl) or (urea) are often used [11,12] but the problem of possible curvature in this plot at low GuHCl or urea concentrations is not resolved (see the section describing denaturants and stabilizers under the main heading "Variables Affecting Protein Stability").

Urea-Gradient Gel Electrophoresis

An ingenious method has been devised for doing these things in a single experiment. The experiment has, moreover, two other advantages: 1) It tests the homogeneity of the protein being studied, and 2) it can be used to measure directly the difference in stability between wild-type and mutant protein. The experiment gives the urea concentration at the midpoint of the urea-induced unfolding transition [13–15]. Electrophoresis takes place in a polyacrylamide gel containing a 0 → 8 M gradient of urea perpendicular to the direction of migration. Provided that the U ⇌ N reaction is fast compared to the rate of separation of U and N in the gel, then the entire protein sample at a given urea concentration moves in a band whose mobility is a weighted average of U and N, and the gel traces out a curve that reflects the unfolding transition (Figure 3). Reversibility of unfolding is checked by comparing two gels, one with N and the other with U layered on top of the gel. A procedure for obtaining the curve of ΔG° vs M (urea), and for extrapolating it to 0 M, has been given [15].

VARIABLES AFFECTING PROTEIN STABILITY

In experiments on protein stability, the key variables are temperature, pH, ionic strength, and denaturant concentration. In some experiments the protein concentration is also important. Ionic strength effects on stability are studied at ionic strengths below 1 M; specific ion effects that depend on the particular salt used generally become dominant above 1 M. The factors controlling stability of water-soluble proteins are discussed briefly here. Again, the focus is on small, single-chain proteins.

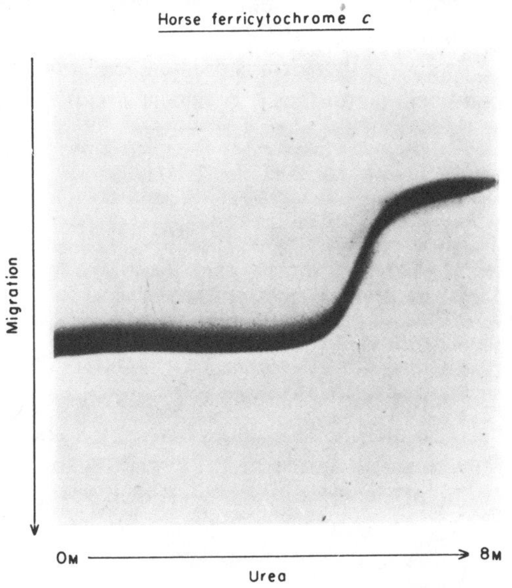

Fig. 3. *The unfolding transition curve of horseheart cytochrome* c *(Fe^{3+}) measured by polyacrylamide gel electrophoresis in the presence of a 0 → 8 M urea gradient, perpendicular to the direction of migration. The native protein (at the left) is more compact and migrates more rapidly than the unfolded protein on the right. (From Creighton [13].)*

Temperature

Almost all water-soluble proteins unfold when heated. The transition curves shown in Figure 1 for RNaseA are typical. Although protein stability often decreases below 25°C (see Figure 4) and the possibility of cold denaturation has been discussed [16,17], there are few, if any, examples of proteins that actually become unfolded by reducing the temperature to 0°C in H_2O at atmospheric pressure (a clearcut example has been given [17A]). On the other hand, several examples are known of oligomeric proteins and subunit assemblies that dissociate as the temperature is reduced close to 0°C.

Since proteins unfold as the temperature is increased, a favorable enthalpy change helps to drive folding. The enthalpy change that occurs on unfolding has been measured by

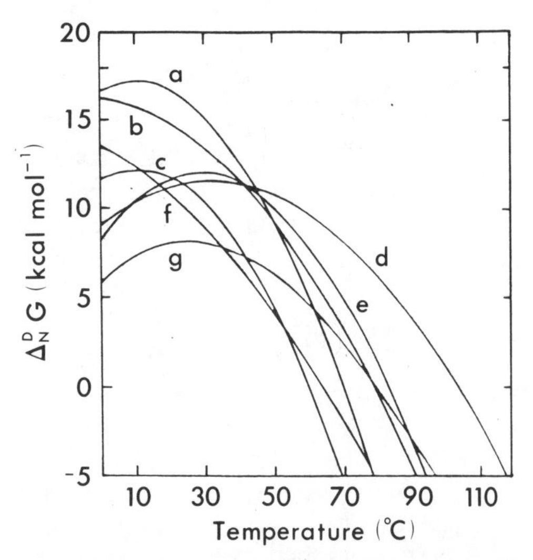

Fig. 4. *The Gibbs free energy of unfolding is plotted as a function of temperature for several small proteins. The data have been obtained by differential scanning calorimetry. (a) trypsin, (b) hen lysozyme, (c) α-chymotrypsin, (d) pancreatic trypsin inhibitor (dimer), (e) metmyoglobin, (f) ribonuclease A, (g) ferricytochrome* c. *(From Privalov [9].)*

differential scanning calorimetry in a wide range of conditions and for several proteins. The results have been summarized by Privalov [9]. There are three findings of particular interest here: 1) The change in partial molar enthalpy on unfolding, ΔH°, is far from being independent of temperature: ΔH° is small at 25°C but becomes large at high temperatures, where ΔH° is positive for unfolding. 2) The change in ΔH° with temperature is linear and ΔCp (the difference in heat capacity between the native and unfolded species) is constant up to 80°C. ΔCp governs the temperature dependence of ΔH° according to $\Delta H°\ (T_2) = \Delta H°\ (T_1) + \Delta Cp(T_2 - T_1)$ when ΔCp = constant. For all proteins studied, ΔCp is large and positive. 3) ΔH° depends only on temperature, as judged by experiments in which pH is used to vary the temperature at which unfolding occurs.

As a result, the change in partial molar Gibbs free energy on unfolding (ΔG°) is a complex function of temperature. It can, however, be expressed in terms of three parameters [12]: T_m (in °K), ΔH° at $T_m = \Delta H_m$, and ΔCp.

$$\Delta G° = \Delta H_m(1 - T/T_m) - \Delta Cp(T_m - T) + T\ell n(T/T_m) \quad (4)$$

Figure 4, taken from Privalov [9], shows ΔG° vs temperature curves for several small proteins. ΔG° is 0 at T_m and, for several of the proteins shown, passes through a maximum around 30°C.

The origins of the enthalpy change on folding are still under study. Groups inside the protein are close-packed [18] and make better van der Waals contacts than when the protein is unfolded. Thus, improved van der Waals contacts may contribute to the enthalpy change [19] and so may peptide H bonds. As the peptide H bond –NH···O = C– is formed in H_2O, H bonds to water of the –NH and –CO groups are broken. Whether or not the net ΔH for the overall reaction differs significantly from 0 is still under study. Analysis of the thermodynamics of reactions in which urea [20] and diketopiperazine [21] dimers are formed in H_2O indicate that the enthalpy of the peptide H bond probably makes an important contribution to the enthalpy of protein folding. Although the hydrophobic interaction is entropy-driven at 25°C [22], it becomes enthalpy-driven at high temperature [23] and the hydrophobic interaction makes an important contribution to the enthalpy of protein folding. The large positive value of ΔCp is usually attributed to reordering of the water structure around hydrophobic sidechains and "melting out" of this structure with increasing temperature [22,24].

pH

Protein stability typically decreases sharply in acidic and basic pH ranges. Figure 5, taken from Privalov [9], shows the typical behavior. There is a simple explanation, whose basis has been pointed out by several authors (see Reference 25, equation A-11). Protons participate in the unfolding process because the pKs of some groups are different in the native and unfolded species. Consequently, the equilibrium constant for folding contains a term in (H^+) and, if this term is omitted, the apparent equilibrium constant K^{app} depends on pH.

$$N + rH^+ \rightleftharpoons UH_r \quad (5)$$

$$K = (N)\,(H^+)^r/(U) \quad (5a)$$

$$K^{app} = (N)/(U) = K/(H^+)^r \quad (5b)$$

The number of protons, r, is not an integer because the pK differences between N and U may be small; furthermore, r varies with pH.

The pH vs stability curves shown in Figure 5 are curves of T_m vs pH. Nevertheless, the variation of K^{app} with pH at constant temperature provides a satisfactory qualitative explanation for understanding the curve of T_m vs pH.

A group such as a sidechain $-COO^-$ or $-NH_3^+$ group in a protein shows an observed pK that depends on the local charge distribu-

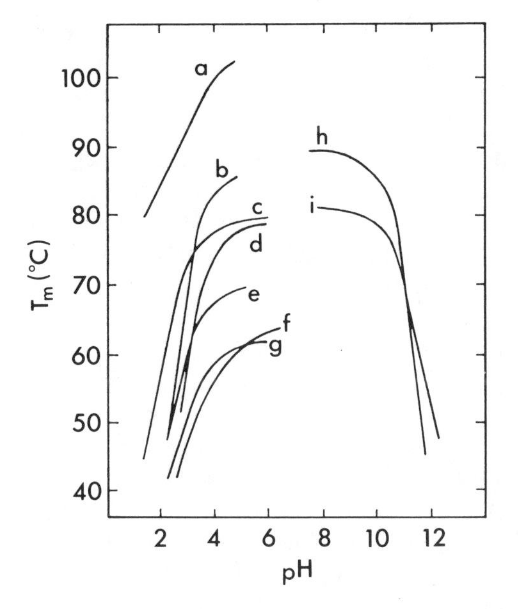

Fig. 5. *Curve of T_m versus pH for several small proteins: (a) bovine pancreatic trypsin inhibitor, (b) papain, (c) hen lysozyme, (d) ferricytochrome* c, *(e) trypsin, (f) ribonuclease A, (g) α-chymotrypsin, (h) parvalbumin (in the presence of Ca^{2+}), (i) metmyoglobin. (From Privalov [9].)*

tion in much the same manner as the pKs of the two $-COO^-$ groups of a dibasic acid. (The term pK refers here to the observed or apparent ionization constant.) Nearby positively charged groups repel H^+ ions in solution and cause the pK to decrease whereas nearby negative groups attract H^+ and cause the pK to increase. For a protein with many charged groups, the problem is mathematically and physically complex. Nevertheless, the availability of powerful computers makes tackling the problem feasible and there is much current interest in finding a solution [26].

We point out a simple qualitative fact: charge interactions usually cause the pK of a group to be lower in N than in U at acidic pH (below the isoelectric pH) but higher in N than in U at alkaline pH (above the isoelectric pH). The reason is that the protein has a net negative charge above the isoelectric pH but a net positive charge below the isoelectric pH, and a protein molecule expands to occupy a larger volume when it unfolds. Thus, the charge density decreases on unfolding, whether the net charge is positive or negative. Since the isoelectric pH is usually in the vicinity of pH 7, these considerations explain the general shape of the T_m vs pH curves (Figure 5) as follows. The T_m falls with decreasing pH below pH 7 because U binds more protons than N (the pKs of some groups are higher in U) whereas the T_m falls with increasing pH above pH 7 because N binds more protons than U (the pKs of some groups are higher in N). The pK differences between N and U are explained by the higher charge density in N, since N has a smaller molecular volume than U but has the same number of ionizing groups. Note that the value of r in equation 5 is determined only by groups whose pKs are close to the pH under consideration. A Lys side chain whose pK is 10.5 in N and 10.0 in U is fully protonated at pH 4 both in N and in U. Note also that the charge distribution may be asymmetric and most of the local charges within 7 Å of a given group may be positive even though the pH is above the isolectric pH and the overall net charge is negative.

Groups in proteins may have abnormal pKs, and show unusual pK changes in pK on unfolding, either because of an unusual local charge distribution, or because the ionizing group is partly buried out of contact with H_2O [26], or because the group participates in a salt bridge or other H bond. Data for these effects have been obtained by using proton NMR to measure the pKs of individual His residues in a protein. There are often large pK differences (~1 pK unit) between two His residues in the same protein. When a group participates in a salt bridge, the sign of the pK shift is predictable because the proximity of an oppositely charged partner always lowers the pK of a $-COO^-$ group and raises the pK of a $-NH_3^+$ group, and the magnitude of the shift gives the stability of the salt bridge. A good example is provided by the salt bridge between the $-\alpha-NH_3^+$ and α-COO^- groups of BPTI, for which one pK shift has been measured by NMR, using a transaminated derivative of BPTI to provide a reference molecule without the salt bridge [27]. The salt bridge lowers the pK of the α–COO^- group from 4.0 to 2.9; the change in pK of the α–$NH_3{}^+$ group was not measured for technical reasons. The pK change of the α–COO^- group indicates that the salt bridge contributes 1.5 kcal/mole to the stability of BPTI. This agrees satisfactorily with the estimate of 1.0 kcal/mole measured from the change in thermal stability in different conditions (in 6 M GuHCl).

Ionic Strength

There exist few studies of the effect of ionic strength on the stability of proteins. Increasing ionic strength gives a slight increase in the T_m of RNaseA at pH 6.5, where the curve of T_m versus pH (Figure 5) indicates little difference in the binding of protons by N and by U, but there is a large effect of ionic strength on T_m at pH 0.7 where T_m is strongly dependent on pH [25]. There are, however, systematic studies in the older literature of the effect of ionic strength on the solubility of proteins (see review, Reference 28). Because solubility and stability are closely related thermodynamic properties, the effect of ionic strength on stability can be calculated from measurements of solubility as a function of ionic strength for both N and U, provided that the 2-state model $U \rightleftharpoons N$ satisfactorily describes the folding reaction. Studies of ionic strength effects are complicated if binding of specific ions occurs (see below). Even at low ionic strengths, Scatchard observed specific binding of Cl^- and I^- to bovine serum albumin (see Reference 29 and the references therein).

The effect of ionic strength on protein solubility at low ionic strength ("salting-in") varies from one protein to the next and depends also on conditions (pH, temperature). Large effects are observed in some cases and sizable effects are expected from the effects of ionic strength on the solubility of amino acids and peptides [28]. Bacterial mutants that contain a thermosensitive protein can be grown at the nonpermissive temperature, in several cases, by adding neutral salts [30].

In principle, the electrostatic interactions that contribute to protein stability should be screened by ionic strength [31] and, from this point of view, one expects to find a dependence of protein stability on ionic strength. In practice, the magnitude of these interactions is still under study and little is known about how effectively they can be screened by ionic strength.

Denaturants and Stabilizers

Denaturants and stabilizers may change the stability of a protein in the same manner described above for H^+ (See the section above titled "pH"): If X binds preferentially to N or to U, it becomes a participant in the unfolding reaction and must be included in writing the equilibrium constant K. If the binding of X is omitted, then the apparent constant K^{app} depends on the concentration of X. This description of denaturant action has been developed systematically by Tanford and applied particularly to GuHCl as a denaturant (see review in reference 32).

Unlike protons, denaturants and stabilizers are effective at high molar concentrations: above 1 M. Therefore Schellman [33] developed a different thermodynamic description for the action of denaturants. It uses the standard formalism for describing the thermodynamic properties of a mixed solvent. One important difference between the two treatments is that, in the simplest case for each treatment, $\Delta G°$ for folding is a linear function of ln M in Tanford's description but a linear function of M in Schellman's description. For the experimentalist who wants to obtain $\Delta G°$ at 0 M by extrapolation, it is important to know whether to extrapolate versus M or ln M. A thorough study of $\Delta G°$ vs. M (GuHCl) for the denaturation of myoglobin [34] shows that some curvature at low M (GuHCl) is observable when a linear plot is used, but the curvature is less than expected if the ln M plot were correct. It should be noted that both the binding and mixed-solvent mechanisms may be operative for a given denaturant, and also that the simplest case may be too simple to describe correctly either mechanism. Thus, curvature in a plot of $\Delta G°$ vs M (GuHCl) may be a standard property of the mixed-solvent mechanism. The problem is being studied actively and should be resolved in the near future.

Detailed thermodynamic studies of solvent additives that change protein stability have been made by Timasheff and co-workers for several classes of compounds, including glycerol and sugars which act as stabilizers (see References 35 and 36 and the references therein). In several systems, they find that both specific binding and mixed solvent (or preferential hydration) mechanisms are operative.

Their work shows that a form of the cavity model is useful for describing the mixed solvent mechanism. The Gibbs free energy of making a spherical cavity in a solvent of hard spheres has been calculated from scaled particle theory, and the result has been used in predicting successfully the solubility of nonpolar solutes in H_2O [37]. The expression shows that the Gibbs free energy of the cavity depends principally on its surface area [38]. In a mixed solvent (water plus an additive), there is an excess or deficiency of the additive in the domain of the protein. It is a thermodynamic quantity and may be measured by dialysis equilibrium. It is sometimes expressed as preferential hydration, which may be either positive or negative, or it can be expressed as preferential interaction with the additive. The sign of this term (whether there is an excess or a deficiency of the additive) determines whether it increases or decreases the solubility of a protein. The effect of an additive in stabilizing or destabilizing a protein depends on the difference between the terms for N and U. According to the cavity model, the term depends in each case on the surface area of the cavity (which is larger for U than for N) and on the excess of the additive at the cavity surface. In a solution of an additive in H_2O there is a surface excess (or deficiency) of the additive at the air–water interface, and this produces a change in surface tension at the air–H_2O interface. In the simplest application of the cavity model, the protein is treated as inert and the excess or deficiency of the additive in the domain of the protein is identified with the surface excess (or deficiency) of the additive at an air-water interface in the absence of the protein. Gibbs' equation relates the surface excess of an additive to the increment it produces in surface tension at the interface. For salts that do not show specific binding to proteins, there is a good correlation [35,36] between the surface tension increment of the salt and its tendency to stabilize or destabilize proteins. Earlier, a similar correlation between surface tension increment and protein solubility had been found [39] for Hofmeister salts at high salt concentrations (>1 M).

DETERMINANTS OF STABILITY

Proteins fold spontaneously under conditions of constant temperature and pressure (see section titled "Anfinsen's Experiment"), showing that the Gibbs free energy of the system decreases during folding. This has led scientists to hope that the structures of folded proteins can be predicted, starting from knowledge of their amino acid sequences. However, this goal has not yet been achieved. The reasons include the following:

1) Proteins are too large to be treated by the fundamental methods of quantum and statistical mechanics, without making massive approximations.

2) The stabilizing energies of proteins (typically about 10–20 kcal/mol) are miniscule compared to the total molecular energy of a protein (of the order of 10^7 kcal/mol for a small protein of 100 amino acid residues). This means that the quantum mechanical energy would have to be computed to an accuracy of about one part in a million to make useful predictions.

3) Water, salts, and sometimes other small molecules strongly affect how proteins fold, so it is necessary to consider the solvent in any calculation, as well as the protein. Because the entropy of the solvent is important,

the energies of many different configurations of the solvent need to be considered. In other words, the free energy G, as distinct from the internal energy E, is needed to predict protein stability.

4) As discussed earlier, there may be kinetic as well as thermodynamic control of folding, placing the problem beyond the power of time-independent quantum and statistical mechanics.

In the face of these barriers to applying fundamental theoretical chemistry to the problem, biophysical chemists have concentrated on defining the forces that are most important in stabilizing proteins, and on estimating their contributions to the overall stability. This section summarizes current lines of work on the problem.

Hydrogen Bonds

The most common secondary structures of folded proteins are the hydrogen bonded alpha helix and beta sheet. These two structures are discussed and illustrated in this book [42,43]. They were predicted by Pauling and Corey [44–46] on the assumption that hydrogen bonds are one of the chief stabilizing forces in proteins. The question considered in this section is whether the ubiquity of hydrogen bonds in folded proteins means that they stabilize the folded forms of proteins relative to denatured, unfolded forms.

The hydrogen bonds in 15 proteins, whose structures are known accurately from x-ray crystallography, were classified by Baker and Hubbard [47]. They found that "Almost all groups capable of hydrogen bonding are, in fact, hydrogen bonded," either to other protein groups, or to water. Of main chain C=O and N–H groups, the proportion hydrogen bonded to each other varies from protein to protein, in the range of 40%–70%. The larger figure is for all α-helical proteins. Amino acid side chains can form up to five (for Arg) hydrogen bonds. Internal side chains with the capacity for hydrogen bonding form short bonds, with bond angles near to ideal. This indicates that these hydrogen bonds "provide a directing influence on the protein structure." Hydrogen bonds formed by external side chains, both to water and to other side chains, are less ideal, suggesting that they are weaker.

What is the value of the free energy change for formation of a hydrogen bond? This depends on the exact process. It is of the order of −6 kcal/mol-of-hydrogen-bond for formation of the bond from groups in a vacuum [48]:

$$\text{R–OH (vacuum)} + \text{R–OH (vacuum)} \rightarrow \text{R–O–H} \cdots \underset{\text{H}}{\text{O}}\text{–R} \qquad (1)$$

In enzyme–substrate complexes, however, experimental free energies of hydrogen bonds are much lower. From protein engineering studies [49] it was found that deletion of a side chain between enzyme and substrate that leaves an unpaired, uncharged hydrogen-bond donor or acceptor weakens binding energy by about 0.5–1.5 kcal/mol. Leaving an unpaired, charged donor or acceptor weakens binding by about 3 kcal/mol more. The energies are smaller than the vacuum energies because, as explained below, when protein–substrate hydrogen bonds are lost, protein–water bonds are gained.

The value of the hydrogen bond free energy in protein folding may be similarly small, or even smaller, as suggested by studies on model systems [50–52]. The reason is that during folding, protein donor and acceptor groups *replace* hydrogen bonds to water with hydrogen bonds to other protein (P) groups; there is no net increase in the number of hydrogen bonds, and probably not a great change in their strength:

$$\text{P–C=O} \cdots \text{H}_2\text{O} + \text{H}_2\text{O} \cdots \text{H–N–P} \rightarrow \text{P–C=O} \cdots \text{H–N–P} + \text{H}_2\text{O} \cdots \text{H}_2\text{O} \qquad (2)$$

The free energy of this reaction is negative, only if the hydrogen bonds between protein groups and between water molecules are more stable than those between protein and water. It is not yet certain that this is so. At present we can say that hydrogen bonds are crucial in determining the pattern of secondary and tertiary structure of proteins, but that each protein hydrogen bond probably contributes at most a small amount to the overall stability. However, because there are many such bonds, and because they tend to form cooperatively, their overall effect can be significant. We return to this point in the section below titled "Configurational Entropy."

S–S Bonds

The disulfide bridge between Cys residues is the only common type of covalent bond that is formed during folding of proteins. The reaction may be represented by

$$\mathrm{P(SH)_2} + \mathrm{RSSR} \rightarrow \mathrm{P(S{-}S)} + 2\,\mathrm{RSH} \quad (3)$$

in which RSH is a thiol-containing molecule.

Some interesting questions about the role of disulfide bridges in protein folding are: 1) When do they form in the biosynthetic history of a protein? 2) How much do they contribute to the free energy of folding? 3) Do disulfide bonds force the protein into its folded form, or do other forces guide the protein into a conformation in which the disulfide bonds can easily form?

The first question was investigated by Bergmann and Kuehl [53]. They found that intrachain disulfide bonds are completely formed on nascent immunoglobulin chains within 1 s of the chain passing through the membrane into the cisterna of the endoplasmic reticulum, that is, the disulfide bond is probably formed while the protein chain is still folding.

Despite this formation during the early biosynthetic stages, disulfide bonds in many (but not all) proteins can be cleaved in vitro, and then properly reformed. The pathway involved in disulfide bond reformation has been followed in elegant detail by Creighton [54] for the bovine pancreatic trypsin inhibitor protein (BPTI). He has found that this protein does not refold by simple, sequential formation of its three disulfides. Instead it goes through complex disulfide rearrangements at the stage when two bridges are formed.

The energies of disulfide bridges have also been investigated by Creighton in BPTI [55]. The energy of a disulfide bridge relative to two thiol groups depends upon the *redox potential* of the environment, and so fixed values cannot be cited, without defining other conditions. However, it is possible to say that the free energy of disulfide formation varies for different bonds. In BPTI the bridge between Cys 5 and Cys 55 is buried in the interior of the molecule and is characterized by a large equilibrium constant for formation. In contrast the bridge between Cys 14 and Cys 30 is on the surface, where these side chains have more flexibility. Its equilibrium constant for disulfide formation is lower.

These measurements demonstrate that other interactions influence the extent to which two Cys residues are held in the proper positions to form a S–S bridge. Nevertheless, the disulfide bridges themselves can stabilize a folded protein conformation that would not be otherwise stable. A classic example is that of insulin, which is biosynthesized as an inactive precursor that folds and forms disulfide bonds. The precursor is then converted to insulin by proteolytic cleavage. If disulfide interchange is then permitted, insulin then spontaneously unfolds and loses activity. Thus, in some proteins disulfide bonds stabilize conformations that would otherwise be unstable.

In summary, disulfide bridges stabilize some proteins, but their energies of formation are influenced by the overall folding of the protein. The role of disulfide bonds in the stability of proteins is discussed in greater detail in this book in the chapter by Klibanov and Ahern [56].

Electrostatic Forces

Two charges q_1 and q_2 interact with energy U_{qq}, when separated by a distance r in a medium of dielectric constant ϵ,

$$U_{qq} = \frac{q_1 q_2}{\epsilon \, r} \quad (4)$$

Also a charge q interacts with a dipole of moment m with energy U_{qm} when the two are separated by r, and the moment forms an angle theta with the line from the dipole to the charge,

$$U_{qm} = \frac{qm \cos \Theta}{\epsilon r^2} \quad (5)$$

The interaction energy of two dipoles can be expressed in a similar fashion, but is proportional to the cube of the separation of the dipoles, and depends on their mutual orientation [48].

To apply these equations we must answer two questions: 1) What values should be used for the charges and dipole moments of chemical groupings on the protein? 2) What value should be used for the dielectric constant? This second question is especially troublesome: some calculations (for example, [57]) have been done with a value for the dielectric constant of unity, characteristic of a vacuum and others with a values nearer to 80, characteristic of water. Rees [58] argued that a value near that of water may be appropriate for groups interacting at the surface of a protein.

In 1986 there is little agreement about what quantitative conclusions can be reached on the electrostatic energies in proteins. There is, however, fairly widespread acceptance of a number of qualitative conclusions on the role of electrostatic forces in protein stability. The remainder of this section summarizes these conclusions and the evidence for them.

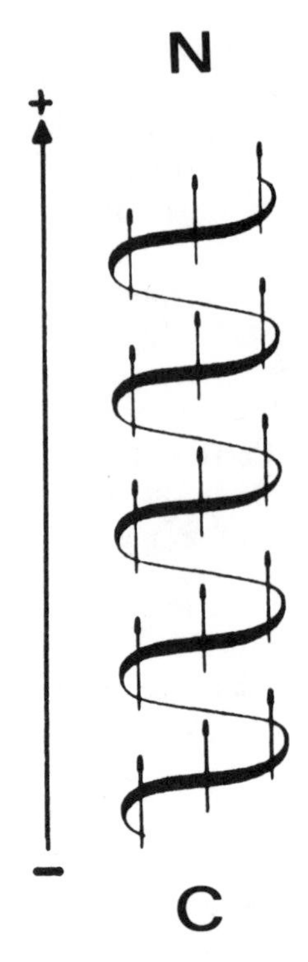

Fig. 6. *A schematic drawing of the peptide dipole moments in an α-helix. The entire helix has a dipole moment with the positive pole at the N-terminus and the negative end at the C-terminus. (Adapted from Hol [63].)*

Ion pairs. Ion pairs in proteins have been studied by Barlow and Thornton [59] and Rashin and Honig [60]. Barlow and Thornton defined a pair as two charged groups within 4 Å of each other. A total of 229 such ion pairs were found in 38 proteins of known structure, only 85 of which involved groups of the same charge (repulsive interaction). Ion pairs are formed by 38% of Arg residues, 29% of His, 16% of Asp, and 20% of Lys. Most pairs occur between residues that are distant in the sequence and in different segments of secondary structure. Thus, they stabilize tertiary structure more than secondary structure. About 85% of ion pairs are at the protein surface, and they are poorly conserved among different members of protein evolutionary families. This suggests that ion pairs do not exert a crucial influence on the stability of structures or on a pathway of folding.

The amide and helix dipoles [61–63]. The amide group in protein backbone is well known to have a partial negative charge on the oxygen atom and a partial positive charge on the N-H group. These result in an effective amide dipole moment of about 3.5 Debye (1.2 $\times 10^{-29}$ Coulomb-meter), pointing parallel to the C=O and N–H bonds, from the former toward the latter. This dipole moment is about twice the magnitude of the dipole moment of water.

When a polypeptide is coiled as an alpha helix, the dipoles are parallel to the helix axis pointing towards the N-terminus (Figure 6). The effect of these combined dipoles is that of 0.5–0.75 unit positive charge at the N-terminus of the helix, and 0.5–0.75 unit negative charge at the C-terminus. A strand of parallel beta sheet also has an effective dipole, with a positive charge at the N-terminus and negative at the C-terminus, but the effective charge is considerably smaller.

Interactions of helix dipoles [62,63]. Many proteins fold with their α-helices aligned in an antiparallel fashion. This is a favorable alignment of dipoles, just as for antiparallel bar magnets. The most striking examples are the bundles of antiparallel α-helices discussed by Ohlendorf et al. in this volume [42]. Other proteins contain antiparallel beta sheets, and antiparallel arrangements of strands of sheet and α-helices, both of which may confer electrostatic stabilization. Whether the electrostatic energy of antiparallel structure elements is sufficient to account for the observed modes of tertiary folding is uncertain at this time.

Interactions of helix dipoles with ions. The ion–dipole force, described by Equation 5, seems to stabilize α-helices and to position negative ions towards the N-termini of α-helices [63]. You will recall from above that the α-helix has the electrical properties of a dipole, with positive pole at the N-terminus. From this you might expect that negative ions might often be positioned near the N-termini of α-helices, and conversely that negative ions placed near the N-terminus would tend to stabilize the helix. Both these expectations have been confirmed by experiments.

Twenty-one examples were found by Hol [63] of phosphate moieties in low-molecular-weight ligands bound to proteins at or near the N-terminus of alpha helical dipoles. In the majority of cases, the phosphate-containing molecules bound between the N-terminus of a helix and a positively charged side chain. An even more striking example of ion–dipole interactions was reported by Pflugrath and Quiocho [64], who found a sulfate ion sequestered from solvent in the interior of a sulfate-binding protein. It is bound solely by hydrogen bonds and is near the N-termini of three α-helices.

You may wonder if positive ions have been found at the C-termini of α-helices. In general, they have not, which may suggest that there is a reason for the association of phosphate ions at the N-termini of helices other than the helix dipole.

The stabilizing effect of negative charges towards the N-termini of helices was suggested [65] as being the explanation that the negatively charged Asp and Glu residues are found more frequently toward the N-termini of α-helices, and that positively charged residues are found more frequently toward the C-termini [66]. The importance of such ion–helix dipole interactions have been confirmed in a simpler, better defined system: a 13 residue α-helix isolated from the N-terminus of the enzyme ribonuclease [67]. When the histidine at position 2 is deprotonated, the helix is destabilized, even though His^+ is a strong helix-breaker in polypeptides of random sequence.

Further evidence for interactions of the helix dipole with ions comes from a study of the pK_a values of histidine residues in hemoglobin [68]. In carbonmonoxy hemoglobin A, a particular histidine residue was found by Perutz et al. to have a pK_a value of 7.8, compared to the pK_a value of about 6.6 characteristic of free histidine at the surface of proteins. This shift can be accounted for by its interaction with the negative pole of the C terminus of helices F and FG of the protein.

Hydrophobic Interactions

Evidence for the hydrophobic interaction and its origin. The strongest generalization that can be stated about the 250 or so known three-dimensional structures of folded proteins is that apolar side chains (especially Ile, Leu, Val, Trp, Tyr, and Phe) tend to be largely buried away from solvent, and that the charged and polar side chains tend to contact the solvent, especially with their side chain N and O atoms. Exactly this distribution was predicted by Kauzmann in 1959 [69], before the three-dimensional structure of any protein was yet known in detail. Kauzmann termed the interaction of the apolar groups in the protein interior "hydrophobic".

Kauzmann's arguments for the hydrophobic effect were thermodynamic, based on studies of the distribution of small apolar molecules, such as methane and benzene, between an apolar solvent and water. These small apolar molecules are taken as models of the apolar side chains of the protein, and the apolar solvent is taken as a model for the protein interior. The solubility of the small solute is generally lower in water than in the apolar solute, meaning that the free energy of transfer to water is positive. For studies near room temperature, Kauzmann explained the low solubility in water by an ordering of water molecules around the apolar solvent. This ordering lowers the entropy of the water. A lower entropy means a greater free energy [$\Delta G = \Delta H - T \Delta S$]. A recent interpretation of the solubility of hydrocarbons data suggest that at high temperatures the enthalpy also plays a role in the low solubility of apolar molecules in water [23]. In short, the hydrophobic interaction is caused by the avoidance of water by the apolar side chains of the protein.

Describing the hydrophobic interaction in terms of residue properties. Several simple methods have been devised to estimate the stability conferred on proteins by the hydrophobic interaction. One is to assign to each amino acid side chain a value for its "hydrophobicity." This is the free energy that is gained by moving the residue from the protein interior to water. (Hydrophobicities are usually taken to be positive for apolar residues; thus, moving a residue from water to the protein interior has a negative contribution to the free energy). Residue hydrophobicities have been estimated by several methods [70–73]. One of the most common is to measure the distribution of the side chain, or some close chemical analog, between water and an apolar solvent that mimics the interior of the protein. This distribution is closely related to the free energy of transfer, which is one measure of hydrophobicity.

The significance and uses of residue hydrophobicities have been discussed at length [74–80]. Some uses to which they have been put include the following:

1) Correlation of residue type with its position in the protein: either buried inside or at the surface [72,74,78,79]. In this book, Lesser et al. [80] describe a set of residue hydrophobicities that correlate well with the extent that each side chain is buried.

2) Detection from the amino acid sequence of hydrophobic, membrane-spanning α-helices [78,81–84].

3) Detection from the amino acid sequence of antigenic sites at the protein surface [85]. We note that this is presently a hotly discussed question, with answers still far from certain.

4) Detection of turns in the polypeptide backbone [86].

5) Description of the amphiphilicity (asymmetry of hydrophobicity) with the aid of hydrophobic moments [87–90]. The hydrophobic moment is a vector assigned to a protein segment which points in the direction of increasing hydrophobicity, and whose length is a measure of the asymmetric distribution of hydrophobic and hydrophilic residues. This method is illustrated in Figure 7, which shows that the hydrophobic moments of each of the four peptide chains of melittin point inwards toward the hydrophobic center.

6) Estimation of the hydrophobic contribution to the free energy of folding, when the accessibility to solvent of residues is known [91].

All of these methods are attempts to develop semiquantitative, semiempirical descriptions of protein–solvent interactions. Some of es-

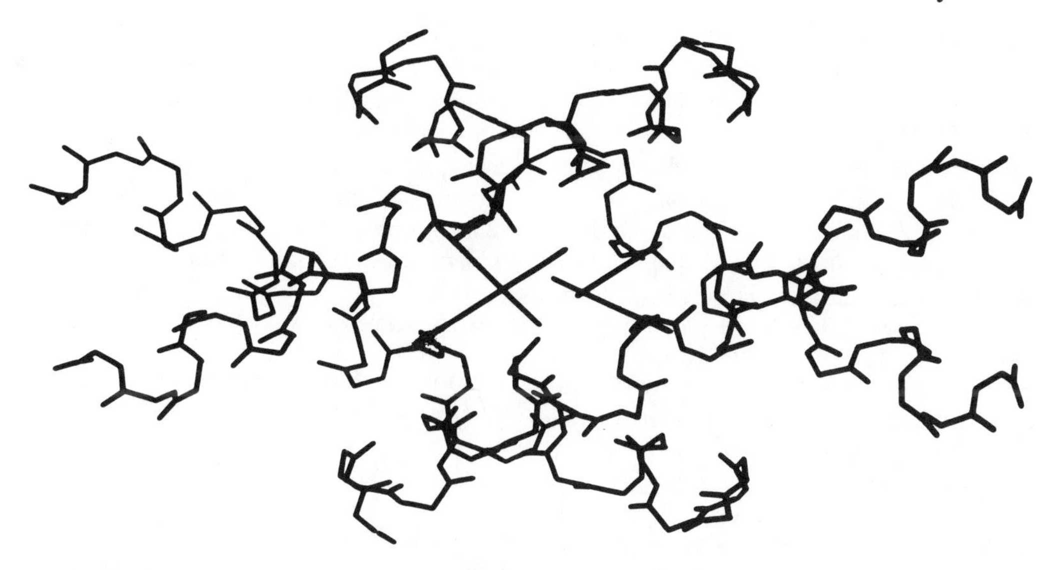

Fig. 7. *The hydrophobic moments of the melittin molecule. Melittin is a tetrameric protein, each polypeptide chain of which contains 26 amino acid residues, coiled as an α-helix. Each of the four α-helices is amphiphilic, having a polar face and an apolar face. The apolar faces of four helices pack together in a hydrophobic interaction. The amphiphilicity of each helix is shown here by a line, representing the hydrophobic moment of the helix. This is a vector which points from the center of the helix towards the direction of greater hydrophobicity. Notice that the hydrophobic moments of the four helices oppose each other in direction.*

sentially the same ideas are used in qualitative form in the design of peptides. These approaches are summarized in this book in the articles by Kaiser [92] and Erickson-Vitanen and DeGrado [93].

Describing hydrophobic interactions and protein-solvent interactions in terms of atomic properties. Any description of protein stability in terms of the hydrophobicities of whole amino acid residues must be crude. The reason is that most amino acid residues (such as Lys) have *both* water-attracting polar parts (the ammonium ion) and apolar parts (the methylene groups). Any property of the entire residue (such as hydrophobicity) is really a summation of atomic contributions. What is needed is an atomic-level description of solvent interactions with each atom of the residue.

A measure of the exposure of any protein atom to solvent was suggested by Lee and Richards [94] and has been extended by others [95–97]. A sphere the size of a water molecule is moved across the surface of the protein. Any atom it can touch is considered to be accessible to solvent, and the measure of accessibility is the area swept out by the center of probe sphere.

An estimate of the contribution of any protein atom to the solvent free energy of stabilization can be made by multiplying its solvent-accessible area with an Atomic Solvation Parameter, $\Delta\sigma$, which gives the tendency of the atom to be in water. These $\Delta\sigma$ values for five types of protein atom have been derived from data on the free energy of transfer of amino acids [98]. These five atom types, with increasing attraction for water, are: sulfur, carbon, uncharged nitrogen/oxygen, charged oxygen, and charged nitrogen. When this method of estimating free energies was applied to two-protein models, it was found that the calculated free energy for the folded protein is lower than that for the unfolded polypeptide chain, and is lower than that for an incorrectly folded structure [98]. The folded protein is more stable than the extended chain by about

1 kcal/mole-of-residue, as estimated by this method.

Other Forces

For the sake of completeness, three other interatomic forces that are important in protein structure should be mentioned. All three forces are at work in both folded and unfolded protein chains, and there is no reason to be certain that any of these three forces tends to stabilize one type of tertiary structure over others. Nevertheless in any complete computation they must be considered (for an example see Reference 99).

The three forces are: 1) valence forces that result in the covalent bonds that hold the polypeptide together; 2) dispersion forces, the universal weak, attractive force between noncovalently bonded atoms (this energy falls off with the inverse sixth power of the atomic separation); 3) the repulsive force, the repulsive interaction experienced by all noncovalently bonded atoms when their electron clouds overlap as they are brought within the sum of their van der Waals radii.

Conformational Entropy

Meaning of the conformational entropy. There is another important determinant of protein stability, the conformational entropy of the protein. Suppose that the number of arrangements of the folded protein is F, and the number of arrangements in the same protein in the unfolded state is U. Then the entropy change for unfolding is given by

$$\Delta S_c = k \, \ell n \, [U/F] \tag{6}$$

in which k is Boltzmann's constant. Because there are many more arrangements of the flexible, thermally agitated, unfolded protein chain than for the folded, compact molecule, ΔS_c is invariably positive. Thus the unfolded state has greater entropy, and the conformational entropy is a destabilizing component of the free energy. As the temperature is raised, the entropy change contributed more strongly (recall that $\Delta G° = \Delta H° - T\,\Delta S°$) and the product of T with the conformational entropy change eventually overwhelms the stabilizing forces, and the protein undergoes thermal denaturation.

Estimating the conformational entropy—even its order of magnitude—is a frontier problem. It is discussed by Dill in a chapter in this volume [100], and in greater detail in a research article [5].

Relating conformational entropy to stabilizing terms in the free energy. In a folded protein, the unfavorable conformational entropy is compensated by many small, stabilizing interactions. These include hydrogen bonds, hydrophobic bonds, electrostatic forces, and so forth. The relationship between the stabilizing forces and the destabilizing configurational entropy has been considered by Creighton [101], Jencks [102], and others.

As an oversimplified example, let us consider the folding of a protein chain into a U (Figure 8) with hydrogen bonds and hydrophobic interactions stabilizing the U. The overall standard free energy of folding $\Delta G°_{net}$ for this model can be expressed:

$$\Delta G°_{net} = \Delta G°_U + \Delta G°_1 + \Delta G°_2 + \cdots \tag{7}$$

In this equation $\Delta G°_U$ is the free energy associated with the conformational entropy of forming the U at temperature T. $\Delta G°_1$ is the free energy of forming the first interaction, $\Delta G°_2$ is the free energy of forming the second interaction, and so forth. These interactions may be hydrophobic or may be hydrogen bonds or disulfide bonds. $\Delta G°_U$ is positive, and the other terms are negative.

Imagine that the protein chain can fold into the U, with the bonding groups in their proper positions to bond, but with the bonds not yet formed. At this point, the free energy change is positive, with only the positive $\Delta G°_U$ term contributing. However, once the entropic investment has been made, the price can be

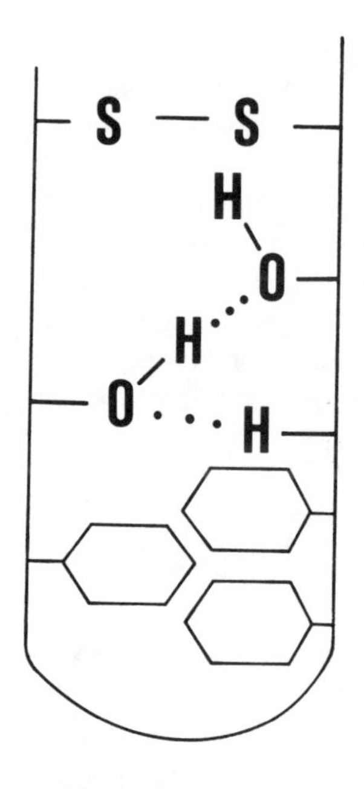

Fig. 8. *A schematic drawing of the folding of a U-shaped protein. The protein is stabilized in its folded conformation by a S–S bond, some hydrogen bonds, and a hydrophobic interaction, represented by adjacent rings. When the protein has assumed the U conformation, all interactions can form.*

repaid by all the favorable bonds, now in the proper positions to form.

Proteins have evolved their networks of hydrogen bonds, disulfide bonds, and hydrophobic interactions so that they can be satisfied simultaneously. This is illustrated in Figure 8: When one of the bonds in the U is in the proper conformation to form, all the others care, too. In a more realistic example, not all bonds would be positioned to form at once, but still the formation of one bond would bring other groups in proximity with their partners. If we now imagine all the bonds forming, all the small negative $\Delta G°$'s on the right-hand side of Equation (7) now contribute. For any stable protein structure, $\Delta G°$ net must, of course, be negative.

This cooperative character of interactions in folding proteins has an important effect. Even though each interaction provides marginal binding energy, when all act together they provide the required stability. Moreover, the addition of one more small stabilizing interaction can have a significant effect. The reason is that the price of the conformational entropy is paid only once as the protein assumes its native fold ($\Delta G°_U$), while each new bond adds another negative $\Delta G°_i$. Also because the equilibrium constant between folded and unfolded forms depends *exponentially* on the free energy, any negative increment in the free energy has a significant effect on stability. Deeper and more quantitative discussions of the relationship of the net stabilizing energy to its components are given by Jencks [102] and Creighton [101].

CONCLUSIONS

Current research on protein folding and stability is vigorous. Knowing how proteins fold into stable, active structures is a question of importance for most fields of molecular biology and molecular medicine. This is partly a consequence of the development of rapid gene sequencing during the past decade. There is now a bottleneck in biochemical progress: The sequences of genes are readily determined, and amino acid sequences can be predicted from the gene sequences. But how are the protein chains folded into functional structures? Also understanding how proteins fold is a bottleneck in biotechnology: Cloned proteins can be expressed in bacteria, but often they form insoluble aggregates, rather than folded, active molecules.

Although our understanding of protein folding is far from complete, there has been much progress since 1950. It is now believed that most proteins fold spontaneously, guided by the decrease of free energy of the system. Many proteins can be denatured and then refolded into active structures. Known exceptions are proteins that undergo covalent modification after initial folding. There has been much discussion on whether proteins fold to the structure of lowest free energy, or whether there is "kinetic control" of folding. Kinetic control would mean that the decrease in free energy guides the folding to a structure which is at a local minimum of free energy, but there is some other (inaccessible) folded structure of lower free energy. At least for single-domain proteins, a structure of lowest free energy is consistent with observations.

It is not known whether the pathway of folding to this low-energy structure is a unique pathway, or if there are several or many pathways. A recent article [8] likens protein folding to solving a jigsaw puzzle. There are faster methods for solving jigsaw puzzles than a random search for pieces, and yet there are many paths. Once several pieces have been correctly attached to form one local patch, extension can be rapid.

The size of the free energy that stabilizes a folded protein relative to its denatured conformation is small, often only 10–20 kcal/mol. It is readily possible to determine a value for this energy when the protein contains a single domain. In this case, it is often valid to consider a protein as being in either of two states, folded or unfolded. This approximation is probably too simple for more complex proteins, or even for a single-domain protein in the many different environments of a living cell. The stability of a protein is affected by temperature, pH, and ionic strength, and these experimental parameters can be altered to study the free energy and range of stability.

The small free energy of stabilization of a folded protein arises from many interatomic forces, some stabilizing and some destabilizing. Hydrogen bonds are obvious stabilizing forces in folded proteins but are also present between the unfolded protein and water. Hence, they may not strongly promote stability. Nevertheless, intraprotein hydrogen bonds form cooperatively, and so are important for proper folding. For example, in an α-helix all amide hydrogen bonds can be satisfied simultaneously (other than those at the ends).

The hydrophobic interaction is a stabilizing force that is present in the folded protein but is not in the unfolded protein. This suggests that it is a major factor in stabilizing folded proteins. However, evaluating the hydrophobic interaction presents a challenge, because it is not a simple pair-wise force like the hydrogen bond. This is the reason for recent attention to hydrophobicities, hydrophobic moments, and other methods that attempt to describe hydrophobic interactions.

A major destabilizing effect on folded proteins is the greater conformational entropy of the unfolded protein chain. The folded form can be stable only if the stabilizing forces, including hydrophobic interactions, S-S bonds, and hydrogen bonds, outweigh this effect.

REFERENCES

1. Anfinsen CB, Haber E, Sela M, & White, FH Jr. (1961): Proc Natl Acad Sci USA 47:1309–1314.
2. King J (1986): in "Enzyme Structure", a volume of Methods in Enzymology, ed. by C.H.W. Hirs & S.N. Timasheff, in press. Yu M-H, King J (1984): Proc Natl Acad Sci USA 81:6584–6588.
3. Levinthal, C (1968) J. Chim. Phys. 65, 44–45.
4. Wetlaufer DB (1973): Proc Natl Acad Sci USA 70:697–701.
5. Dill KA (1985): Biochemistry 24:1501–1509.
6. Givol D, DeLorenzo F, Goldberger RF, Anfinsen, CB (1965): Proc Natl Acad Sci USA 53:676–684.
7. Ikai A, Fish WW, Tanford C (1973): J Mol Biol 73:165–184.
8. Harrison SC, Durbin R (1985): Proc Natl Acad Sci USA 82:4028–4030.
9. Privalov PL (1979): Adv Protein Chem 33:167–241.
10. Tiktopulo EI, Privalov PL (1978): FEBS Lett 91: 57–58.
11. Pace CN (1986): In "Enzyme Structure," a volume of Methods in Enzymology, ed. by Hirs CHW, Timasheff SN (in press).
12. Schellman JA (1980): In "Protein Folding" ed. Jaenicke R, Elsevier-North Holland, p 331–343.
13. Creighton TE (1979): J Mol Biol 129:235–264.
14. Creighton TE (1986): "Enzyme Structure," a volume of Methods in Enzymology ed. by Hirs CHW, Timasheff SN (in press).
15. Hollecker M, Creighton TE (1982): Biochim et Biophys Acta 701:395–404.
16. Brandts JF, Hunt L (1967): J Am Chem Soc 89:4826–4838.
17. Hawley SA (1971): Biochemistry 10:2436–2442.

17a. Privalov PL, Griko YV, Venjaminov SY, Kutyshenko VP (1986) J Mol Biol 190:487–498.

18. Richards FM (1977): Ann Rev Biophys Bioeng 6:151–176.
19. Bello J (1978): Intl J Prot Res 12:38–41.
20. Schellman JA (1955): C R Trav Lab Carlsberg Ser Chim 29:223–229.
21. Gill SJ, Noll L (1972): J Phys Chem 76:3056–3068.

22. Kauzmann W (1959): Adv Protein Chem 14:1–63.
23. Baldwin RL (1986): Proc Natl Acad Sci USA (in press).
24. Frank HS, Evans MW (1945): J Chem Phys 13:507–532.
25. Hermans J Jr, Scheraga HA (1961): J Am Chem Soc 83:3283–3292.
26. Matthew JB (1985): Ann Rev Biophys Chem 14:387–417.
27. Brown LR, DeMarco A, Richarz R, et al (1978): Eur J Biochem 88:87–95.
28. Cohn EJ, Edsall JT (1943): "Proteins, Amino Acids and Peptides," New York: Reinhold.
29. Scatchard G, Wu YV, Shen AL (1959): J Am Chem Soc 81:6104–6109.
30. Kohno T, Roth J (1979): Biochemistry 18:1386–1392.
31. Matthew JB, Richards FM (1982): Biochemistry 21:4989–4999.
32. Tanford C (1968): Adv Protein Chem 23:122–282.
33. Schellman JA (1978): Biopolymers 17:1305–1322.
34. Paçe CN, Vanderburg KE (1979): Biochemistry 18: 288–292.
35. Arakawa T, Timasheff SN (1984): Biochemistry 23:5912–5923.
36. Arakawa T, Timasheff SN (1984): Biochemistry 23:5924–5929.
37. Pierotti RA (1965): J Phys Chem 67:1840–1845.
38. Lee B (1985): In "Mathematics and Computers in Biomedical Applications," Eisenfeld J, DeLisi C (eds.) Elsevier-North Holland (Amsterdam), p 3–10.
39. Melander W, Horvath C (1977): Arch Biochem Biophys 183:200–215.
40. Garel J-R, Baldwin RL (1973): Proc Natl Acad Sci USA 70:3347–3351.
41. Nall BT, Baldwin RL (1977): Biochemistry 16:3572–3576.
42. Ohlendorf DH, Finzel BC, Weber PC, Salemme FR (1987): This volume, chapter 13.
43. Richardson JS, Richardson DC (1987): This volume, chapter 12.
44. Pauling L, Corey RB, Branson HR (1951): Proc Nat Acad Sci USA 37:205.
45. Pauling L, Corey RB (1951): Proc Nat Acad Sci USA 37:729.
46. Pauling L, Corey RB (1953): Proc Nat Acad Sci USA 39:253.
47. Baker EN, Hubbard RE (1984): Prog Biophys Molec Biol 44:97–179.
48. Eisenberg D, Kauzmann W (1969): The Structure and Properties of Water," Clarendon Press, Oxford.
49. Fersht AR, Shi J-P, Knill-Jones J, Lowe DM, Wilkinson AJ, Blow DM, Brick P, Carter P, Waye MMY, Winter G (1985): Nature 314:253.
50. Klotz IM, Franzen JS (1962): J Am Chem Soc 84:3461–3466.
51. Schellman JA (1955): Compt Rendu Trav Lab Carlsberg Ser Chim 29: 223–229.
52. Susi H, Timasheff SN, Ard JS (1964): J Biol Chem 239:3051–3054.
53. Bergmann LW, Kuehl WM (1979): J Biol Chem 254:8869–8876.
54. Creighton TE (1980): J Mol Biol 144:521–550.
55. Creighton TE (1984): "Proteins." WH Freeman & Co, New York.
56. Klibanov AM, Ahern TJ (1987): This volume, chapter 19.
57. Hol WGJ, van Duijnen PT, Berendsen HJC (1978): Nature 273:443–446.
58. Rees DC (1980): J Mol Biol 141:323–326.
59. Barlow DJ, Thornton JM (1983): J Mol Biol 168:867–885.
60. Rashin AA, Honig B (1984): J Mol Biol 173:515–521.
61. Wada A (1976): Adv Biophys 9:1–63.
62. Sheridan RP, Levy RM, Salemme FR (1982): Proc Nat Acad Sci USA 79:4545–4549.
63. Hol WGJ (1985): Prog Biophys Molec Biol 45:149–195.
64. Pflugrath JW, Quiocho FA (1985): Nature 314:257–260.
65. Blagdon DE, Goodman M (1975): Biopolymers 14:241–245.
66. Chou PY, Fasman GD (1974): Biochemistry 13:211–221.
67. Shoemaker KR, Kim PS, Brems DN, Marqusee S, York EJ, Chaiken IM Stewart JM, Baldwin RL (1985): Proc Nat Acad Sci USA 82:2349–2353.
68. Perutz MF, Gronenborn AM, Clore GM, Fogg JH, Shih T-b (1985): J Mol Biol 183:491–498.
69. Kauzmann W (1959): Adv Protein Chem 14:1–63.
70. Nozaki Y, Tanford C (1971): J Biol Chem 246:2211–2217.
71. Wolfenden R, Anderson L, Cullis PM, Southgate CCB (1981): Biochemistry 20:849–855.
72. Janin J (1979): Nature (London) 277:491–492.
73. Chothia C (1976): J Mol Biol 105:1–14.
74. Rose GD, Geselowitz AR, Lesser GJ, et al (1985): Science 229:834–838.
75. Edsall JT, McKenzie WA (1983): Adv Biophys16:53–183.
76. Eisenberg D, Weiss RM, Terwilliger TC, et al (1982): Faraday Symp Chem Soc 17:109–120.
77. Guy HR (1985): Biophys J 47:61–70.
78. Kyte S, Doolittle RF (1982): J Mol Biol 157:105–132.
79. Rose GD, Roy S (1980): Proc Nat Acad Sci USA 77:4643–4647.
80. Lesser GJ, Lee RH, Zehfus MH, Rose GD (1987): This volume, chapter 14.
81. Eisenberg D, Schwarz E, Komaromy M, et al (1984): J Mol Biol 179:125–142.
82. Argos P, Rao JKM, Hargrave PA (1982): Eur J Biochem 128:565–575.
83. Kuhn LA, Keigh JS Jr (1985): Biochim Biophys Acta 828:351–361.

84. Engelman DM, Goldman A, Steitz TA (1981): Meth Enzymol 88:81–88.
85. Hopp TP, Woods KR (1981): Proc Nat Acad Sci USA 78: 3824–3828.
86. Rose GD (1978): Nature 272:586–591.
87. Eisenberg D, Weiss RM, Terwilliger TC (1982): Nature (London) 299:371–374.
88. Eisenberg D, Weiss RM, Terwilliger TC (1984): Proc Nat Acad Sci USA 81:140–144.
89. Finer-Moore J, Stroud RM (1984): Proc Nat Acad Sci USA 81:155–159.
90. Pownall HJ, Knapp RD, Gotto AM Jr, Massey JB, (1983): FEBS Lett 159:17–23.
91. Eisenberg D, Wilcox W, McLachlan AD (1986): J Cell Biochem 31:11–17.
92. Kaiser ET (1985): This volume, chapter 17.
93. Erickson-Vitanen S, DeGrado WF (1987): This volume, chapter 18.
94. Lee B, Richards FM (1971): J Mol Biol 55:379–400.
95. Shrake A, Rupley JA (1973): J Mol Biol 79:351–371.
96. Richmond TJ, Richards FM (1978): J Mol Biol 119:537–555.
97. Richmond TJ (1984) J Mol Biol 178:63–89.
98. Eisenberg D, McLachlan AD (1986): Nature 319:199–203.
99. Novotny J, Bruccoleri R, Karplus M (1984): J Mol Biol 177:787–818.
100. Dill KA (1987): This volume, chapter 16.
101. Creighton TE (1983): Biopolymers 22:49–58.
102. Jencks WP (1981): Proc Nat Acad Sci USA 78:4046–4050.

Protein Engineering, pages 149–163

12

Some Design Principles: Betabellin

Jane S. Richardson and David C. Richardson

Department of Biochemistry, Duke University Medical Center, Durham, North Carolina 27710

INTRODUCTION

We have undertaken the design and synthesis of novel small proteins "from scratch" in hopes of illuminating the natural design principles by which amino acid sequence determines three-dimensional structure in proteins. The old aim of predicting tertiary structures from sequences has made considerable progress in recent years, but any summary must conclude that the process still does not work well. However, in tune with the more active meddling encouraged by the recent revolution in genetic techniques, we have been gambling that the inverse process of going from structure to sequence can in favorable cases be much easier.

The idea is to pick a specific, simple tertiary structure described at about the level of detail in Figure 1, and then to choose an amino acid sequence which should fold up to form that tertiary structure. As symbolized by Figure 2, the sequence decorates the backbone with side chains and all their folding interactions and determines the surface properties where the interactions with solvent and other molecules take place. From the viewpoint of design, the critical question is why a particular collection of side chains is suitable for one specific tertiary structure and disfavors alternative arrangements. Many of the factors involved are known or guessed at, and in a de novo design all of those can be arranged to work cooperatively toward one common structure, within the limits of compromising their occasional conflicts.

This sort of project has much in common with other current efforts described in this book, such as peptide syntheses aimed at producing secondary structures (Kaiser; Erickson-Viitanen, O'Neil, and DeGrado; Baldwin and Eisenberg, this volume) and the site-directed mutagenesis projects of section IV. Our approach is complementary to those because, on the one hand, it makes the whole sequence, rather than one residue at a time, new and nonhomologous; this lets us ask fundamental, really "stupid" questions as opposed to more local and sophisticated ones. On the other hand, it specifically emphasizes tertiary structure and packing in nonrepeating, fairly long sequences. The closest other work to ours is Gutte's design and synthesis of a 24-residue hydrophobic polypeptide that binds DDT [1]. Presumably all of these approaches, and others as well, will need to interact and contribute results in order to build a useful understanding of the determinants of protein folding and structure.

As a specific example to illustrate how this design process works, we will use betabellin, an "invented" protein of molecular weight

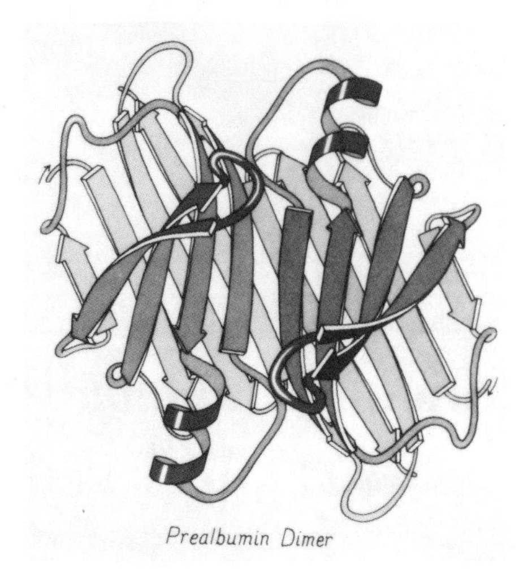

Fig. 1. *Schematic backbone structure of prealbumin, to symbolize the degree of detail to which one specifies a tertiary structure for the design process. (Reproduced in color on p. 340).*

7000 that is intended to form a sandwich of two identical, four-stranded, antiparallel β-sheets. In this collaboration, the design, and the eventual crystallography, are done at Duke, while the amino acid synthesis and purification are done by Bruce Erickson and his colleagues at The Rockefeller.

CHOICE OF A TERTIARY STRUCTURE TO AIM AT

Methods for both peptide and nucleic acid synthesis have made qualitative advances recently, but it is still very important that the target structure be small—certainly less than 100 residues. On the other hand, it will probably take on the order of 50 or 60 residues to make a soluble protein that has a unique, stable tertiary structure. For de novo design, rather than modification of existing native proteins, it is essential to choose a tertiary structure type that is well understood. We may be wrong about our level of understanding (and that is one of the main things we hope to find out, one way or the other), but at any rate it would be foolish to start with something obviously inscrutable.

The main criteria, then, are small size coupled with comprehensible organization. Many

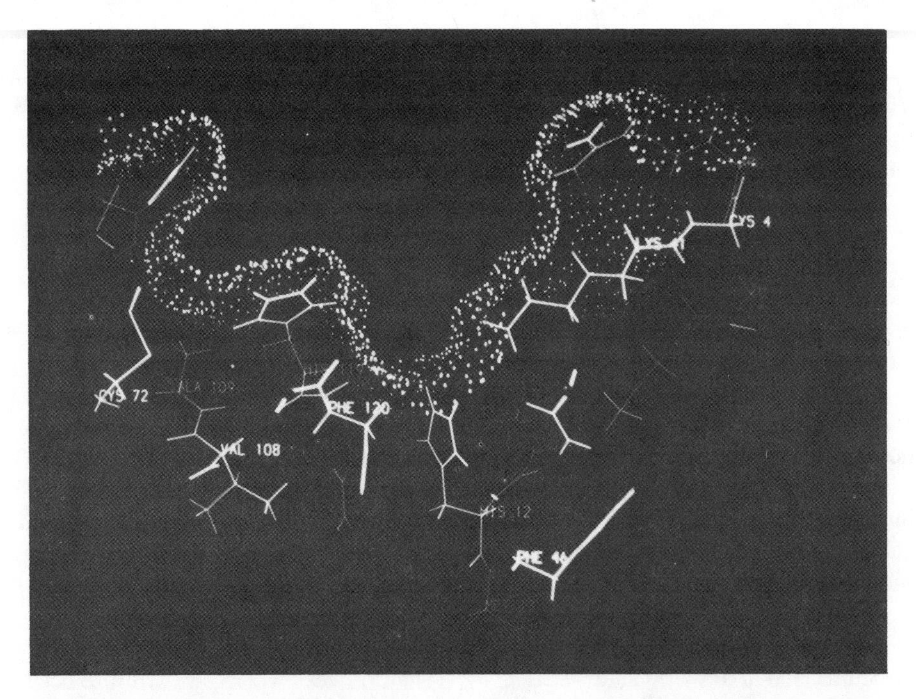

Fig. 2. *An all-atom model of ribonuclease A, with dots showing the accessible surface, to symbolize the process of decorating the schematic with side chains, which contribute essentially all the interactions both internally and with solvent and other molecules. (Reproduced in color on p. 340.)*

well-known protein structures fail on one count or the other, or on both. Small disulfide-rich proteins, for instance (such as the basic pancreatic trypsin inhibitor in Figure 3a), have been widely and productively used for both experimental and theoretical studies because of their small size, but their variability and irregularity have prevented any global understanding of their construction principles. Parallel α/β structures, however, fail the test mainly on the ground of size. They are so common and so similar to one another that there are plentiful empirical rules for their construction (such as the righthandedness of all $\beta\alpha\beta$ connections) even if theoretical understanding is tenuous. However, it would probably require at least 140 residues to construct a stable, classic nucleotide-binding domain, and at least 200 for a parallel β-barrel like the one in triose phosphate isomerase. Figure 3b illustrates both the large size, and the occasional unpredictable complexity, of parallel α/β structures.

The three tertiary structure types that emerge as the best candidates for de novo design are the up-and-down β-barrel, the Greek key β-barrel, and the up-and-down four helix cluster (all shown in Figure 4). They probably require only 60–80 residues for stability, and they are simple and common enough to have acquired a large body of empirical rules about their construction. Even if some of the "rules" turn out to be wrong, they give a rational basis for initial design and a set of hypotheses to test when making later variants.

Internal symmetry provides another opportunity for both simplification and effective size reduction. Many tertiary structures can show

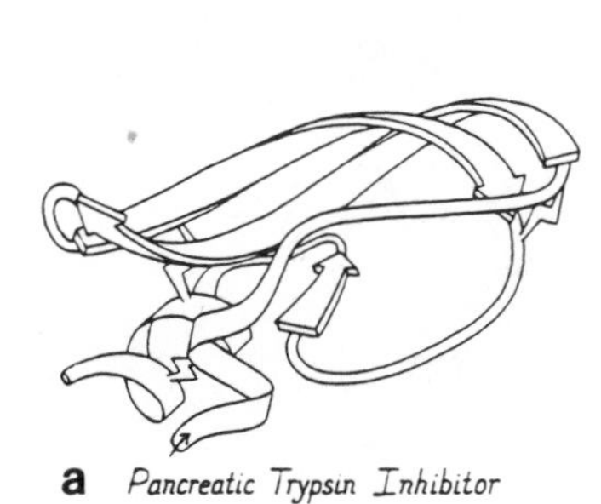

b Carboxypeptidase

Fig. 3. *Tertiary structure types that are unsuitable for de novo design and synthesis: a) Pancreatic trypsin inhibitor, a small SS-rich irregular protein, which is a good size but not well enough understood; b) Carboxypeptidase A, a parallel alpha/beta protein, for which many structural principles are known but which is much too large and complex. (3b reproduced in color on p. 341. Drawing by Duncan McRee.)*

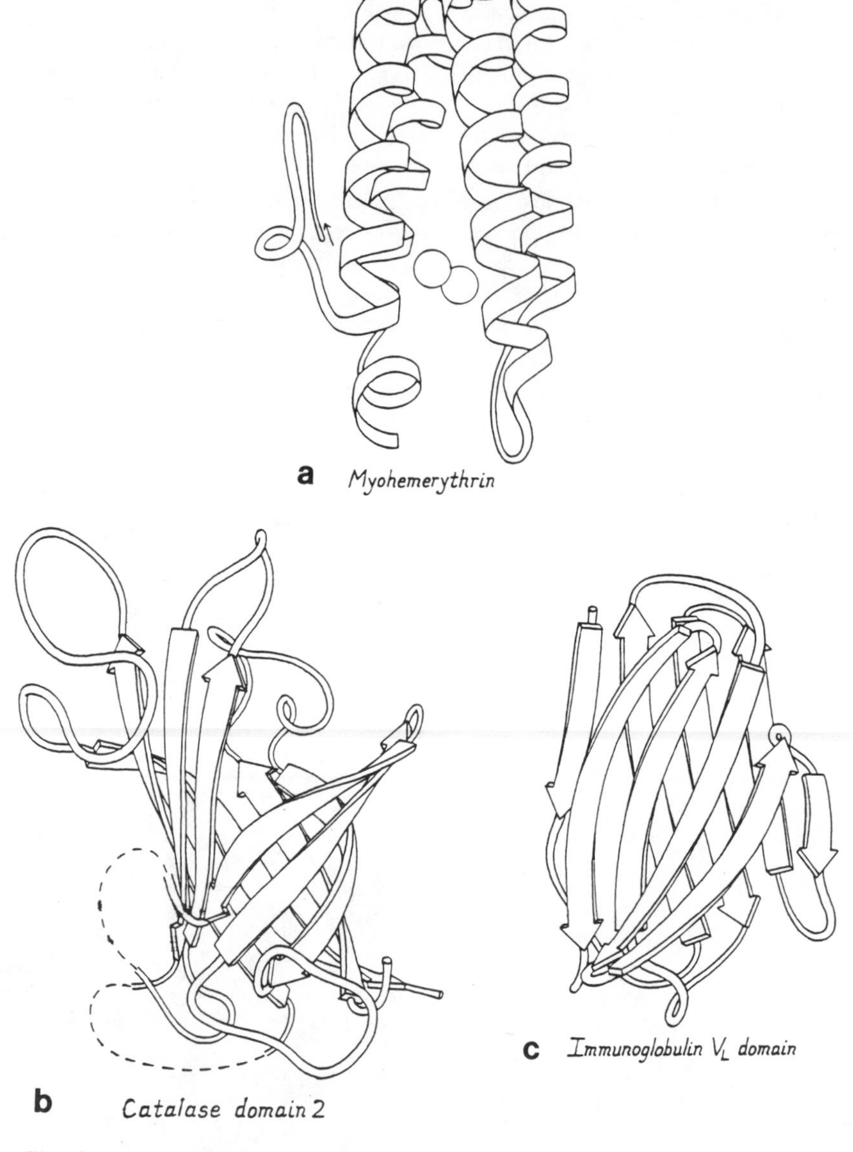

Fig. 4. *Tertiary structure types that are suitable for de novo design and synthesis: a) Myohemerythrin, a four-helix cluster; b) Catalase domain 2, an up-and-down antiparallel β-barrel; c) An immunoglobulin variable domain, a Greek key β-barrel.*

approximate internal symmetry, including the three types chosen above. Although duplication of the amino acid sequence is very bad for doing NMR and not ideal for X-ray crystallography, it is an enormous advantage for the initial attempts at peptide synthesis, especially since the syntheses may need to be repeated and modified many times before obtaining a usable quantity of pure product with a suitably well-defined tertiary structure. To take advantage of internal twofold repeats, Bruce Erickson has designed a crosslinker (called DAB) [2] with a tail joined to the solid-phase resin and two arms on which the two identical half-chains are simultaneously synthesized. Although Greek key β-barrels are much more common than up-and-down ones, they are not suitable for use with the DAB crosslinker because they are thought to fold by curling up a long, two-stranded antiparallel β ribbon (Fig. 5), whereas the two strands attached to DAB are made in parallel from their C termini.

In accordance with all the above requirements, betabellin is an up-and down antiparallel β barrel with four strands (31 residues) forming each of the two identical β sheets. Of the two common shapes for β-barrels, the flatter, lower-twist, seven to ten-stranded form was chosen rather than the high-twist, five- or six-stranded form, because the internal packing is somewhat easier to visualize. Figure 6 is a schematic drawing of the tertiary structure toward which betabellin is targetted.

CHOICE OF A SEQUENCE TO FORM THAT TERTIARY STRUCTURE

First we must lay out a blank template for the sequence, as in Figure 7, with a proposed number of residues and a specific hydrogen-bonding arrangement for the β-sheet. This level of detail immediately exposes several simple constraints, such as the fact that an up-and-down sheet with successive tight turns cannot have an odd number of residues in any internal strand. (If you try it, you will find either that the turn cannot H-bond or that it will turn in the wrong direction.) Six residues per strand would be minimal; betabellin has eight per strand to allow a margin of error and later, we hope, to let us change a few side chains at one end to make a binding site without destroying the β-sheets. Although the side chains alternate direction along a β-strand, at a turn there are four in a row that point more or less to the same side (see Fig. 7). To make the side chains at the top end (the potential binding pocket) point inward for eight-residue strands, the N-terminus must be at the lower left rather than the lower right; combined with the universal righthanded twist of β-sheet structures, this means that the crosslink between the C-termini must be rather long. DAB was chosen to fit the length requirements, and it is designed to tuck in between the bottoms of the two sheets where outward-pointing side chains leave room. The fourth strand is only seven residues long to pull the crosslinker up into the desired position. Thus, there are $8 + 8 + 8 + 7 = 31$ residues in each of the identical sheets.

Along the β-strands, side chains alternately stick out into the solvent and in toward the hydrophobic core. A very basic design rule for antiparallel β-sheets, then, is an alternation of hydrophilic and hydrophobic residues. Native β-sheet proteins show this hydrophilic–

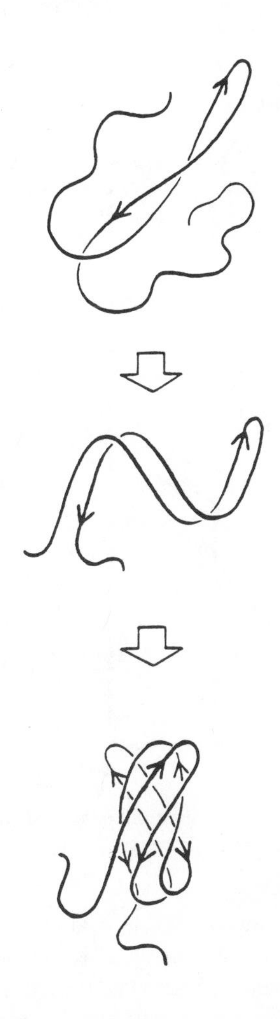

Fig. 5. *A proposed foldling scheme by which Greek key β-barrels form from a long, twisted ribbon of two antiparallel β-strands.*

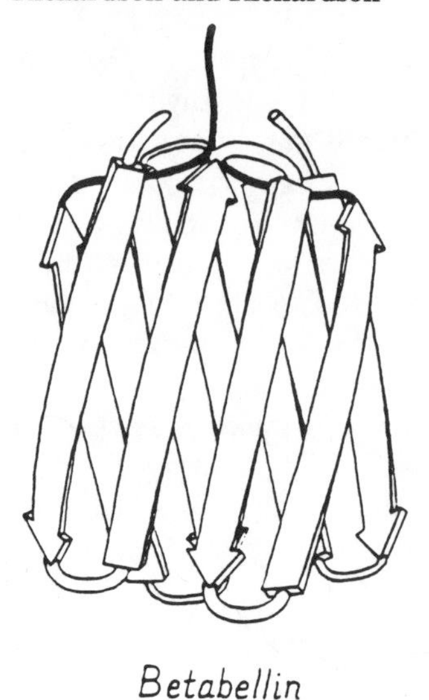

Fig. 6. *A schematic drawing of the tertiary structure proposed for betabellin.*

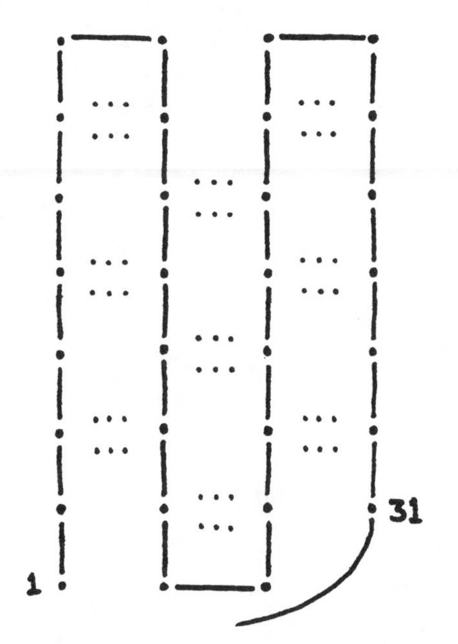

Fig. 7. *A template showing the proposed strand and hydrogen-bond layout for betabellin.*

hydrophobic alternation as a very strong but far from perfect regularity. In betabellin, as in other "designed" proteins so far, this sort of regularity is greatly exaggerated as part of the effort to maximize stability. We do not yet know, however, whether this sort of overdesign is in fact a good thing; we hope to find out.

Turns are crucial in a β-sheet protein, particularly if the exact topology is of concern. On a β-strand progressing upwards, a turn with residue 1 (out of 4) pointing toward you will turn to the right, while a turn with residue 1 pointing away from you will turn to the left [3]; these two different cases are separated by an offset of only one in the sequence. In betabellin we are trying to engineer a specific β-sheet topology; therefore the amino acids in the turns are chosen even more for their position-specificity than for their overall turn preference. Figure 8 shows position-specific turn preferences for several potentially useful amino acids [from reference 4]. Pro is the most specific, favoring position 2, and Asn the next best, favoring positions 1 or 3. Two thirds of the Pro–Asn sequences in proteins of known structure are the central pair of a turn [5]. Gly and Ser, both excellent turn formers, are dangerous because they like several adjacent positions almost equally. Betabellin, therefore, has Pro-Asn at each of its six tight hairpin turns.[1]

Another obvious design criterion is secondary structure prediction, preferably by several different methods. As an example, Figure 9 shows a Chou-Fasman [6] prediction for the initial betabellin sequence. We have also used Robson [7] and Ptitsyn-Finkelstein [8] methods, and the parallel vs. antiparallel preferences from Lifson and Sander [9]. One cannot maximize the prediction values, because β-strands made solely from the one, or even two, best residues would not pack into a stable, unique tertiary structure. However, the chosen strand sequences must predict acceptably high for β and low for helix. After compromise with the other necessary criteria, the sequence ended up with four strong β-residues per strand, as shown on the betabellin tem-

[1]Although Pro-Asn seems the obvious best choice and was also used in the only other β-sheet designed so far [1], it may after all have been a mistake (see below).

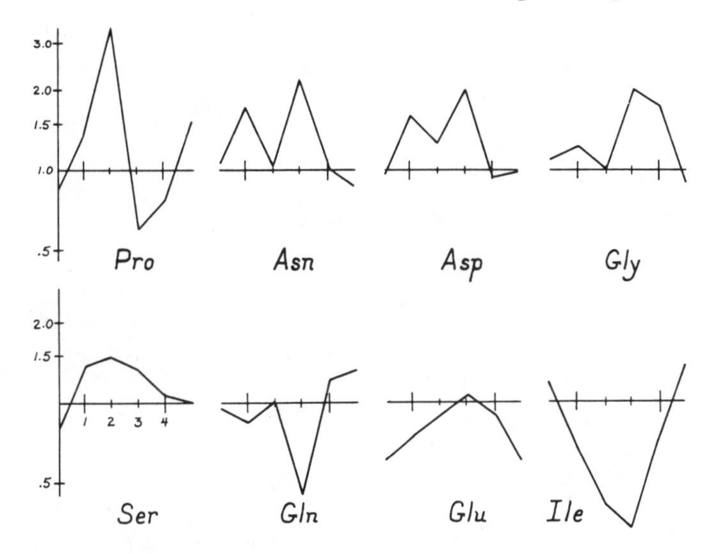

Fig. 8. *Plots of position-specific turn preference for several potentially useful amino acids.*

```
S T V T A R Q P N V T Y S I S P N T A T V R L P N Y T L S I G

    h h h h                         h h h
  B B B b     b b B B B B         b B B B B   b b B B B B          (C-F)
          t T T T t t   T T T T T t         T T T t t   t

C B B B T B B T T B B B B B C C T B B B B B B B T T B B B B B      (G-O-R)

  B B B B B T T   B B B B B B T T B B B B B B T T B B B B B        (P-F)
```

Fig. 9. *Secondary-structure predictions for the initial betabellin sequence (the sequence is across the top in one-letter code), by the Chou-Fasman, the Garnier-Osguthorpe-Robson, and the Ptitsyn-Finkelstein methods. H: helix; B: beta; T: turn; C: coil.*

plate in Figure 10. The betabellin sequence was also checked for *lack* of homology with any native protein sequence in the PIR-NBRF database [10].

As an example of the necessary compromises, let us consider the conflicts between predicted secondary structure potentials and the requirements within specific tertiary structures. On the inside of antiparallel β-sandwich proteins over 70% of the residues are accounted for by just five amino acids: Val, Leu, Ile, Phe, and Ala [11]. The first four are good β-formers, but Ala most definitively is not. Those out-of-character alanines are presumably needed to provide very short side chains to make the internal packing work. Betabellin was designed with a sprinkling of alanines on the inside (especially where the surrounding side chains are bulky, such as next to the Tyr), but there is never more than one Ala per strand. An isolated Ala presumably would not nucleate an α-helix, while an isolated Gly could conceivably form a turn.

Side chains next to each other on adjacent β-strands are in very close contact, and these positions show moderately strong pairing preferences—as much as 2:1 above or below frequencies predicted from overall β-sheet composition [3]. Many of these pair preferences are readily understandable, such as the bias in favor of unlike rather than like charges and like rather than unlike hydrophobicities.

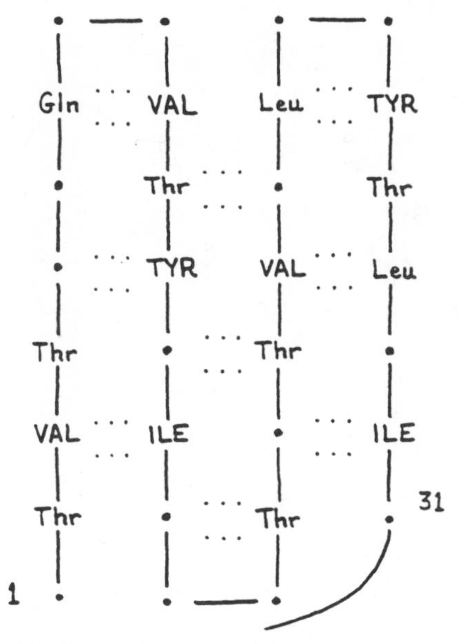

Fig. 10. *β-sheet-forming residues shown on the betabellin template (capital letters are the strongest and small letters the next best).*

The particular type of pair preference that influenced the betabellin sequence most prevasively is the preference (in antiparallel, but *not* in parallel, β-sheet) for a side chain branched at Cβ (i.e., Val, Ile, or Thr) to pair with an unbranched one. Those branched-chain residues are among the strongest β-formers, so they occur often in the sequence; however, it is presumably important for good packing to stagger them so that they are not near neighbors, as seen in Figure 10.

So far we have only considered interactions within a single β-sheet. Good steric fit between the two sheet surfaces is relatively subtle and difficult to describe, but it is certainly vital for producing a stable tertiary structure. This is especially true in betabellin, which has almost no hydrogen bonds between the two halves and only one covalent connection, through the DAB crosslinker. The easiest way to study internal packing is with dot-surface computer graphics. The original type of dot surfaces developed by Mike Connolly [12] show the smoothed outer surface of a molecular or a substructure. They are especially effective for seeing how knobs and cavities complement each other, and where there may be small mismatches, as illustrated in Figure 11.

For even more detailed study of the patterns of specific steric contacts, we have developed a modified type of dot-surface display. The probe spheres are only .25 Å in radius rather than the usual water-sized 1.4 Å radius, and dots are kept for display only where the probe touches two different atoms rather than keeping dots on the accessible surface where the probe touches a single atom. These small-probe contact surfaces show isolated patches of dots that represent individual van der Waals contacts or hydrogen bonds (see Figure 12). They allow very detailed evaluation of internal packing, but they can only be used for structures with quite accurate coordinates. In fact, they give sensible results only if the hydrogen atoms are represented explicitly on aromatic rings and NH's, rather than by united-atom spherical approximations. Since our betabellin models are only approximate guesses and not refined X-ray structures, we cannot use these techniques directly on our designed models.

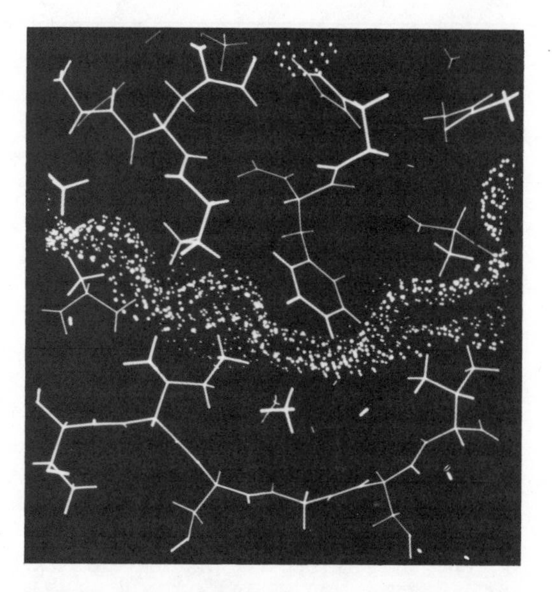

Fig. 11. *Connolly dot surfaces showing the fit of the top beta sheet (green dots) to the bottom β-sheet (purple dots) in Cu, Zn superoxide dismutase. (Reproduced in color on p. 341.)*

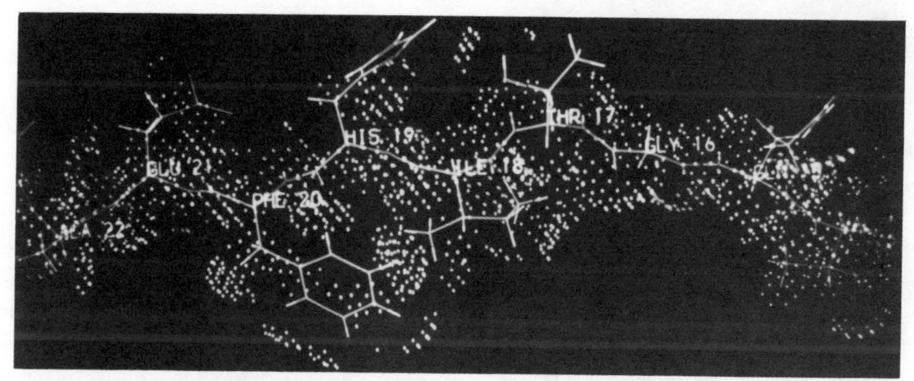

Fig. 12. *Modified dot surfaces that show internal contacts between atoms in a β-strand. (Reproduced in color on p. 341.)*

Instead, we use them to study known structures similar to what we are designing, and then try to apply what we have observed to the new designs.

The results of that process are, first of all, to aim for a twist angle between the two-β-sheets of 30–40° (see Figure 13), since that is usual for β-barrels with about eight strands (3,11). More contacts are observed between adjacent strands than across the β barrels, which produces relatively smooth sheet surfaces [11]. However, there are significant ripples in those surfaces that match impressively well across the sheet contacts, as we have already seen in Figure 11. There is great variety in how these complementary ripples are produced, but in betabellin we have tried to use two of the commoner types of interaction [13]: long diagonal ridges of side chains that include branched Cβ's in their strongly preferred conformation [14], and long side chains such as Tyr to provide interactions on the two diagonal sheet corners exposed by the β sheet twist.

We also found it useful to build CPK models of the pair of sheets and to modify side chains and their conformations conservatively until they settled into a snug fit when put together and jiggled. The CPK model also suggested several other changes. Residue 12 was intially a Phe, but in the model the end of its ring reached about to the outside edge of the β-sheet sandwich; it was therefore changed to a Tyr so that the OH group could interact with solvent. One of the hairpin turns initially had a Thr in position 1 and a Pro in position 2; the CPK model showed this to produce disturbingly close contacts, and so position 1 was changed to Ser, which does indeed have a greater statistical preference for that position [4]. The CPK model also confirmed the possibility of building a sterically reasonable disulfide bond between residues 21 and 21′, which we have used in several of the later versions of betabellin.

The computer model of betabellin shown in Figure 13 was built using Richard Feldmann's graphics system at NIH. The β-sheets were modeled on four adjacent strands from Concanavalin A (which is very regular, but unfortunately has a very low twist), while the turn and side-chain conformations and the sheet-to-sheet spacing were adjusted by hand. These coordinates were used by Jiri Novotny as a starting point for energy minimization, using CHARMM. The starting model had relatively few bad contacts, the SS geometry was good, and nothing moved very far during minimization, as shown in Figure 14. The hydrophilic/hydrophobic ratio for the outer surface of the minimized model was higher than for natural proteins because of the overdesign, whereas

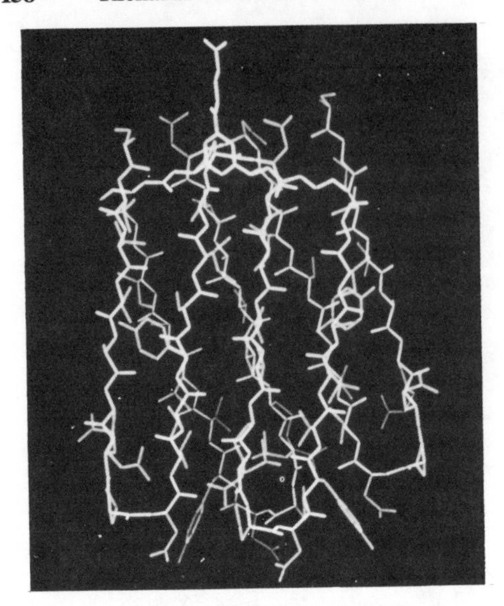

Fig. 13. *An atomic model of betabellin built on Richard Feldmann's computer graphpics system at NIH. (Reproduced in color on p. 341.)*

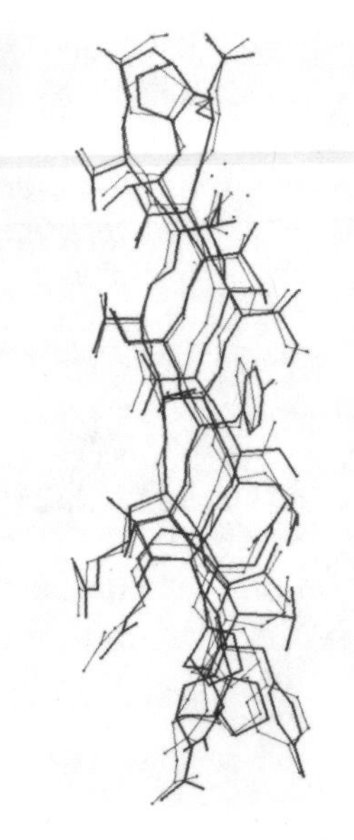

Fig. 14. *Superposition of betabellin models before and after energy minimization (Novotny.)*

for deliberately "misfolded" proteins [15] it is smaller.

The design of betabellin also included considerations based on ease of synthesis, ability to substitute and sequence residues at a potential future binding site, and ease of structure determination by X-ray crystallography or by 2-D NMR. For example, each tripeptide sequence is unique to aid in sequential assignment of NMR resonances, and residue 26 is iodo-Phe rather than Tyr to provide a built-in heavy-atom derivative for crystallography. In order to make a try at designing betabellin to crystallize, we examined the crystal packing contacts in a number of the most accurately determined protein structures. All the contacts looked remarkably similar, and remarkably unlike what we had expected. There are relatively few contacts between protein atoms, while on the average there is one layer of highly ordered water molecules bridging the contacts. Figure 15 shows an example from the electron density map of crambin [16]. As might be expected from the fact that proteins have not evolved to crystallize, such contacts are rather forgiving: The two surfaces need not have complementary shapes, and hydrogen bonds need not match a donor on one side to an acceptor on the other. As a first-order model for the crystallization process, then, one can imagine two molecular surfaces, each with a partial layer of ordered water. If those two water structures are nearly the same, then the two surfaces can come together, discarding half of the waters. The surface side chains of betabellin, therefore, are predominantly Ser and Thr to give a large and flexible capacity for hydrogen bonding with water, while keeping the side chains short and thus more likely to form well-ordered crystals if they crystallize at all.

SYNTHESIS AND CHARACTERIZATION

The end product of this entire design process is just a single word: STVTARQPNVTYSISPNTATVRLPNYTLSIG! If this were a magical incantation, we could expect immediate feedback: Either the demon would come to do our bidding or the earth would swallow us up. Unfortunately, things are slower than that nowadays. Five different versions of betabellin have been made so far by solid-phase peptide synthesis, and consider-

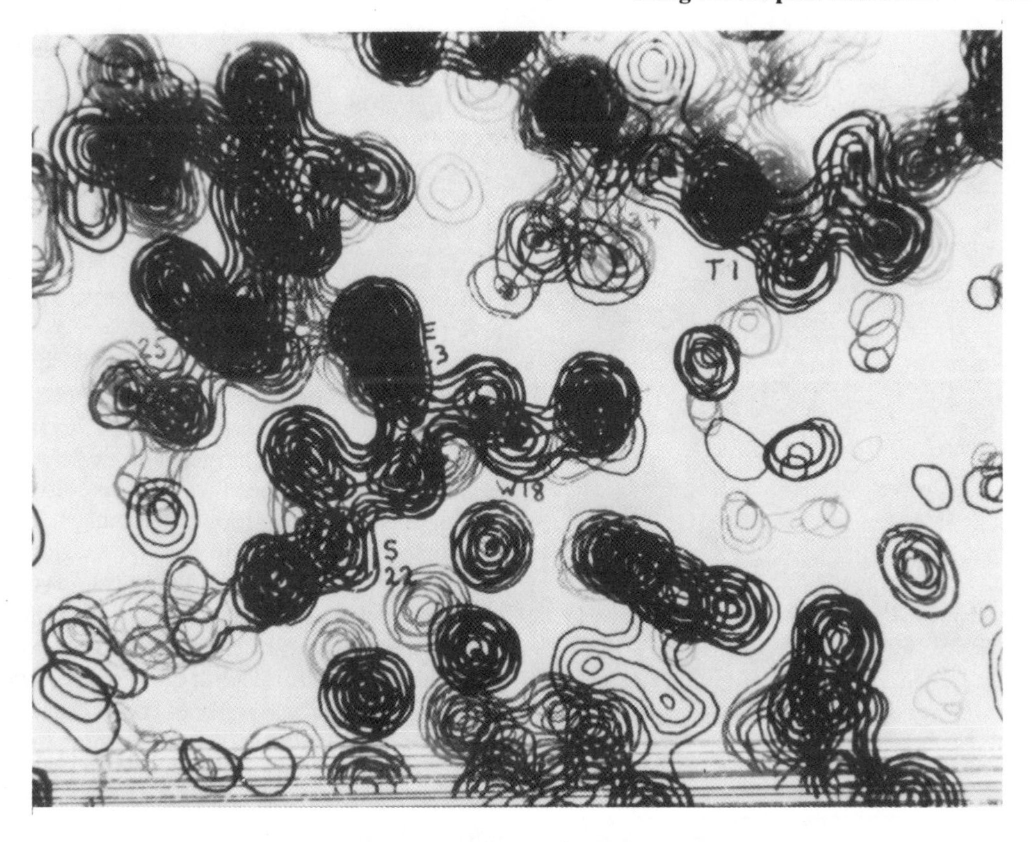

Fig. 15. *Part of the electron density map of crambin, with circles in the lower center representing well-ordered water molecules in a crystal packing contact.*

able progress has been made on their purification and characterization, but we do not yet know anything firm about their three-dimensional structures.

The raw material from the first synthesis was soluble up to about 20 mg/ml in water, but it showed eight or nine peaks on HPLC. Circular dichroism (CD) spectra of the various components were all extremely similar. They showed definitively that there was no α-helix at all (see Figure 16), which was very reassuring. They indicated something like 40% β structure and the rest coil [17]. CD spectra for antiparallel-β proteins are extremely variable, but betabellin is within the observed range (rather like spectra of serine proteases). It seems that betabellin indeed has considerable β-sheet structure, but probably a bit less than was intended.

Amino acid composition showed that several of the betabellin 1 components were missing a Val or a Thr or both. Successive amino acids with branched β-carbons sometimes couple poorly, so that the culprit was almost certainly the Thr-Val-Thr sequence in the first strand. That central Val was changed to Leu, and also the arginines were changed to lysines, to allow milder cleavage and deprotection chemistry. Particularly in β-sheet, each residue interacts with almost everything else, so that after making those two changes it was necessary to include five additional changes in order to return to approximate agreement with all the design criteria (see Fig. 17). Two related sequences were synthesized in parallel: one with and one without an internal disulfide at residue 21, which we hoped would act as a structure probe and stabilizer. The synthesis was indeed cleaner, and it was also discovered that purification worked much better on gels and partition columns than on HPLC. The disulfide formed spontaneously, and the SS-

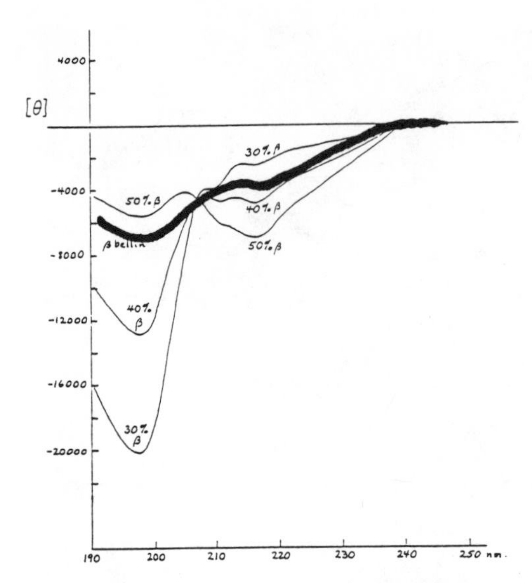

Fig. 16. *CD spectrum of betabellin 2 (heavy line) superimposed on reference spectra representing no helix and varying mixtures of β and coil (Greenfield and Fasman).*

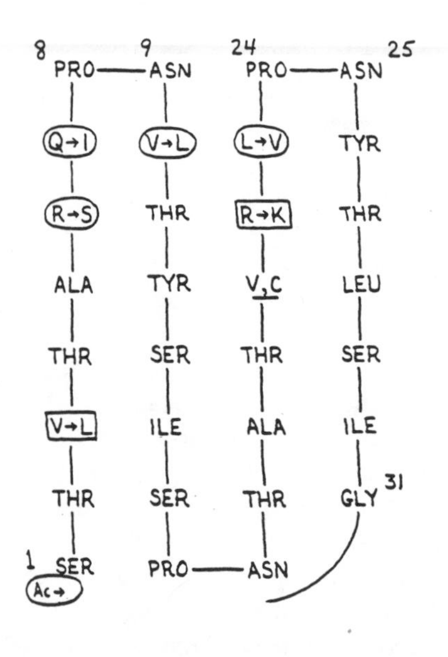

Fig. 17. *Rectangles show the initially required changes from betabellin 1 to 2, while circles show the additional changes necessary to preserve the proposed type of tertiary structure.*

linked material ran cleanly as a monomer of about 7000 molecularweight, so that there was apparently no crosslinking between separate molecules.

THE IMPORTANCE OF MAKING GOOD MISTAKES

A major aim of this project is to learn more about the principles of protein folding. Since there are good reasons to think that our understanding is far from perfect in this area, it would actually be a disappointment if everything about betabellin turned out exactly as planned. We hope, therefore, to make useful mistakes and to learn from them. Our guiding conceptual analogy is with the immunoglobulin variable domain, those β-barrel framework is so overdetermined that it can accommodate an enormous range of changes in the hypervariable loops and still form the same tertiary structure. Similarly, we hope to overdetermine the well-understood features of betabellin's structure, so that the inevitable mistakes will only somewhat modify or destabilize its folding rather than destroying it altogether.

Recently there have been two discoveries about antiparallel β structure, of the sort that we would have used in the design of betabellin had they been known at that time. One of them happens to be done right in betabellin, while the other one is done wrong.

The first of these discoveries, by Ray Salemme [18], is that in addition to the well-known twist preference, a two-stranded antiparallel β-ribbon has a preferred side toward which it will bend or curl (the effect disappears for three strands or more). Side chains bracketed by a closely spaced pair of hydrogen bonds stick out on one side of the β-ribbon, while side chains between a wide pair of H-bonds stick out on the other side. The narrow-pair side is the one that will preferentially end up on the inner, concave side of the bend or curl. Stable proteins should have their hydrophobic side chains on the concave, narrow-pair side of such β-ribbons; this is in fact true, both for isolated two-strand ribbons such as in trypsin inhibitor (Fig. 3a) and also for the strand pairs in Greek key β-barrels (e.g., Fig. 18) from which they are thought to fold (Fig. 5). In betabellin we wished to avoid Greek key folding, and in fact the sequence is arranged correctly to achieve that, with hydrophobic residues on the wide-pair side around the central tight turn.

The second discovery, by Janet Thornton and Lynn Sibanda [19], is that the turns in tight β-hairpins in fact behave very differently from turns in general, because of conformational constraints imposed by the twist of the surrounding pair of β-strands. The four Cα's of the residues in a turn are not all in one plane, but form a dihedral angle which for the Type I turn (by far the commonest, in general) is about +45°. The strong righthanded twist of a β-ribbon also imposes a dihedral angle at the end of the hairpin, but of about −45° (see Fig. 19). Therefore the Type I turns works very poorly for a tight β-hairpin, whereas the otherwise very rare mirror-image version, Type I′, works perfectly (shown schematically in Fig. 20) and occurs most often in those locations. Unfortunately, our choice of Pro in position 2 of the betabellin turns locks them into the Type I conformation, so that in the folded structure the two sorts of twist would be fighting one another. The probable result would be the breakage of at least one hydrogen bond next to each turn, making the β-sheet shorter. Type I′ turns are very unfavorable with normal l-amino acids, so that the β-hairpins in native proteins contain at least one and sometimes two glycines. Using peptide synthesis, however, we can put d-amino acids at those positions, encouraging Type I′ turns and directly testing Thornton and Sibanda's hypothesis. The most recent syntheses (betabellins 4 and 5) being done at The Rockefeller are a pair of sequences identical elsewhere but one with Pro-Asn turns and the other with d-Pro-d-Asn turns. Since there are six hairpin turns in the structure, this difference should produce a large effect.

The third "mistake" has been found experimentally. Betabellin 3 was the first version to

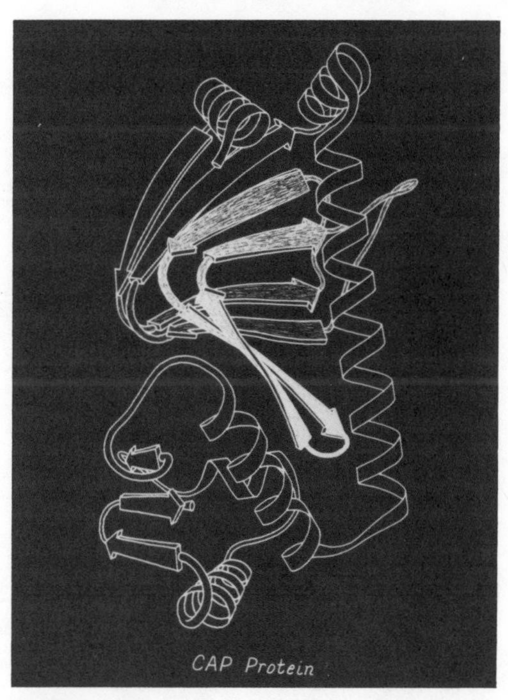

Fig. 18. *CAP protein, an examle of an eight-stranded Greek key β barrel in which a pair of strands win along next to each other from a dividing point in the center (darkest shading) all the way to their ends (lightest shading).*

incorporate the internal disulfide. The monomeric, SS-linked material turned out to have two nonexchanging components with vastly different partition ratios: One was very hydrophilic and the other was strongly hydrophobic. Apparently, betabellin was capable of unfolding or turning inside out, and each conformation could be locked in place by the disulfide. The β-sheets would at least have to be distorted to form the disulfide if their hydrophobic sides were facing out, and indeed the CD of the hydrophobic form shows less β content. Characterization of these two components is still in progress, but we do know that when the SS of the hydrophobic form is reduced and its solvent conditions changed, its CD slowly changes, whereas the hydrophilic form is very stable. There are two things we would like to do with this phenomenon. One is to get rid of it entirely, so that we can produce only hydrophilic material as our original design intended;

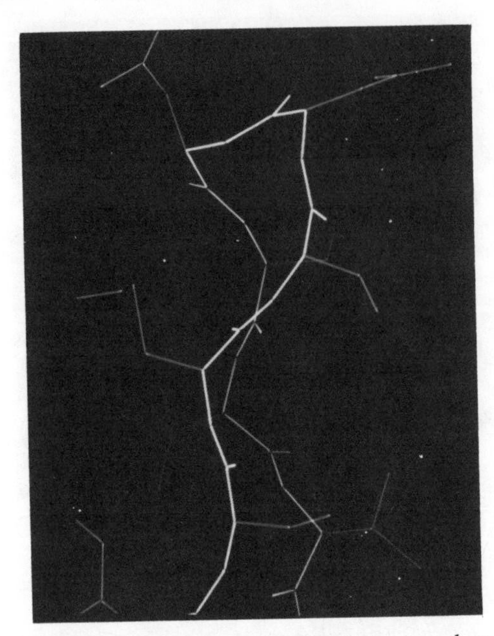

Fig. 19. *An example of a Type I′ hairpin turn, showing how its dihedral angle matches the twist of the surrounding β strands.*

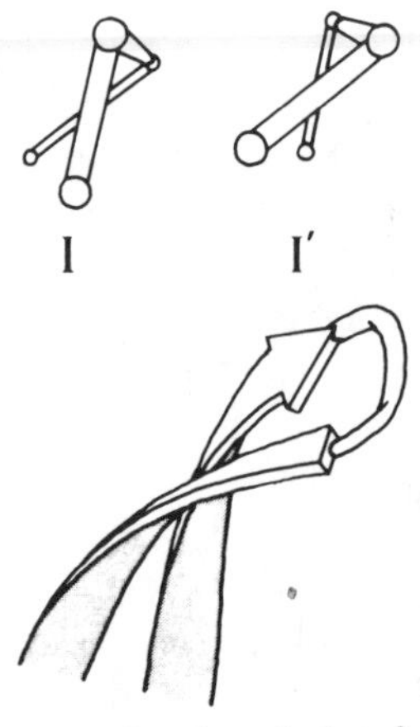

Fig. 20. *Schematic drawing of a beta hairpin, showing how the Type I′ matches and the Type I fights the β sheet twist.*

this may be possible by some gentle refolding process, or it may require sequence changes in addition. The second is to find out which feature(s) of the sequence are mainly responsible for this behavior, by varying the amount and distribution of charged, hydroxyl group, and hydrophobic side chains and perhaps even the strand connectivity. Betabellin 6 will have eight more charged residues, to test their effect on hydrophobic partitioning.

IMPLICATIONS

Current betabellin syntheses are aimed at understanding and improving particular structural features of the molecule, such as the β-hairpin turns and the inside-out partitioning behavior. This can probably be done by studying the solubility, CD spectra, and stability of different sequence variants.

The original goal of testing the feasibility of protein design still requires determination of a complete three-dimensional structure of some one of the betabellins. This now seems much closer, because of greatly improved purification procedures. With tens of milligrams of purified product we can begin extensive crystallization trials and other physical measurements such as NMR.

Another possible way of producing "invented" proteins is the route of synthesizing DNA fragments, ligating them into a gene, cloning it, and expressing the protein. In a collaboration with Richard Ogden at the Agouron Institute, a four-helix cluster protein of 81 residues called Felix has been designed and is in the process of being made. Protein synthesis and nucleic acid synthesis have somewhat different advantages and limitations; it should be productive and interesting to try both.

If it does turn out to be possible to make a novel, synthetic protein with a well-defined tertiary structure, then the next logical step would be to design a binding site on its surface. At present we would be working from our models of the possible betabellin structure, and later, one hopes, from an experimentally determined structure. The end of betabellin opposite from the crosslinker was designed to have six potentially substitutable side chains pointing inward around a possible binding site. Affinity columns on multiply substituted material could be used to pull out and then sequence the best variants. If tight binding to transition-state intermediates were used as a probe, then perhaps it is not too large a step from ligand binding to actual catalysis. We are exploring these ideas, in collaboration with Bruce Erickson and with Richard Wolfendon at the University of North Carolina.

This project has already rewarded our efforts by stimulating new and unanticipated ways of thinking about protein structure, and we are very optimistic about achieving more concrete results soon. It is useful as a final caution, however, to keep in mind that all the high hopes and the pretty pictures do not really prove that we know what we are doing yet, as is symbolized by Figure 21.

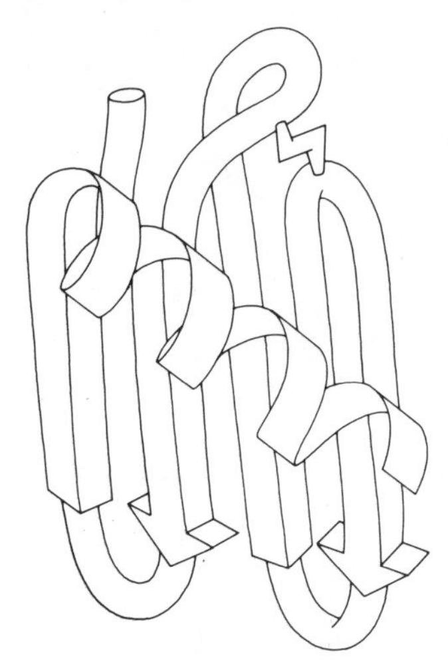

Fig. 21. *Just a cautionary reminder that not all apparently plausible protein structures will turn out to work.*

REFERENCES

1. Moser R, Thomas RM, Gutte B (1983): FEBS Lett. 157:247–251.
2. Unson CG, Erickson BW, Richardson DC, Richardson JS (1984): Fed Proc 43:1837.
3. Richardson J (1981): Advances Prot Chem 34:167–339.
4. Chou PY, Fasman GD (1977): J Mol Biol 115:135–175.
5. Zimmerman SS, Scheraga HA (1977): PNAS 74:4126–9.
6. Chou PY, Fasman GD (1974): Biochemistry 13:222–245.
7. Garnier J, Osguthorpe D, Robson B (1978): J Mol Biol 120:97–120.
8. Ptitsyn OB, Finkelstein AV (1983): Biopolymer 22: 15–25, or program ALB available from the Brookhaven Protein Data Bank, Upton NY 11973.
9. Lifson S, Sander C (1980): J Mol Biol 139:627–639
10. Protein Identification Resource, National Biomedical Research Foundation, Georgetown U Med Center, Washington, DC 20007.
11. Chothia C, Janin J (1981): Proc Nat Acad Sci USA 78:4146–4150.
12. Connolly ML (1983): Science 221:709–713.
13. Cohen FE, Sternberg MJE, Taylor WR (1981): J Mol Biol 148:253–272.
14. Janin J, Wodak S, Levitt M, Maigret B (1978) J Mol Biol 125:357–386.
15. Novotny J, Bruccoleri R, Karplus M (1984): 177:787–818.
16. Hendrickson WA, Teeter MM (1981): Nature 290:107–113.
17. Greenfield N, Fasman GD (1969): Biochemistry 8:4108–4115.
18. Salemme FR, (1983): Prog Biophys Molec Biol 42:95–133.
19. Sibanda BL, Thornton JM (1985): Nature 316:170–174.

Protein Engineering, pages 165–173

13

Some Design Principles for Structurally Stable Proteins

Douglas H. Ohlendorf, Barry C. Finzel, Patricia C. Weber, and F. Raymond Salemme

Protein Engineering Division, Genex Corporation, 16020 Industrial Drive, Gaithersburg, Maryland 20877

INTRODUCTION

Emerging capabilities in synthetic and recombinant DNA methods make feasible the site-specific modification or de novo synthesis of designer protein molecules with novel functions. Whatever the functional objective in protein design, a necessary requirement is the production of a folded molecule that retains its three-dimensional structural stability in the dynamically active solution environment. Indeed, in many cases of industrial relevance, the objective of engineering a protein may focus simply on improving its stability under conditions of pH or temperature that favor rapid catalysis but otherwise cause enzyme denaturation. The purpose of what follows is to describe some of the factors that govern protein structural stability in a context that relates both to de novo protein design and to chemical or recombinant-DNA modification of existing enzymes to enhance stability.

STRUCTURAL ORIGINS OF STABILITY IN 4-α-HELICAL PROTEINS: AN EXAMPLE OF MOLECULAR ENGINEERING PRINCIPLES

Crystal structure determinations of proteins have revealed many occurrences of remarkable similarity in tertiary folding. Although local structural features are frequently determined by characteristic conformational properties of particular sequences, overall similarities in tertiary folding are commonly observed in the absence of detailed amino acid sequence homology. These recurrences in functionally different proteins presumably reflect convergent evolutionary solutions to basic requirements for protein structural stability. For example, recent studies of recurrent β-sheet motifs in proteins have shown how structural similarities reflect requirements for both facile folding pathways and extended stabilization of the molecule as a whole [1].

Here we examine factors stabilizing a recurrent protein motif that is organized as a sequentially connected array of four α-helices (Figure 1). Naturally occurring 4-α-helical proteins serve a variety of functions, incorporate several different prosthetic groups, and occur both in monomeric form and a variety of oligomeric assemblies [2,3]. The combination of basic structural simplicity and functional diversity makes the 4-α-helical motif a prime candidate for attempts at ab initio design of proteins of novel function. The objective of the following is to describe the physical principles that stabilize the 4-α-helical structural motif.

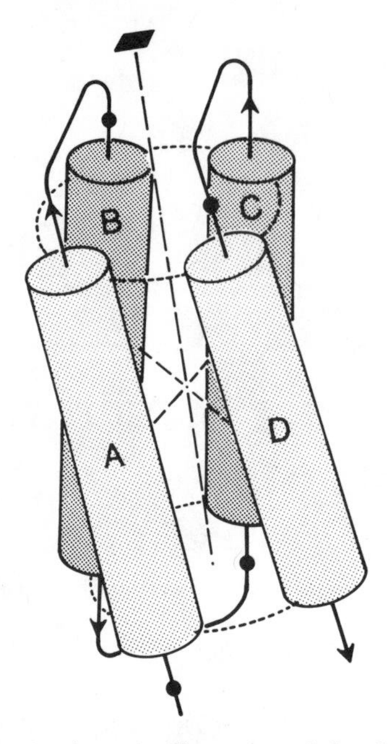

Fig. 1. *A schematic illustration of the 4-α-helical array common to a number of sequentially and functionally unrelated proteins. The structure is organized as a sequentially connected array of four α-helices (represented as cylinders) that pack together with adjacent antiparallel N to C senses, to give a bundle with approximately square cross-section.*

Previous studies [2] have shown that the subunits of cytochrome b562, cytochrome c′, myohemerythrin, hemerythrin, tobacco mosaic virus coat protein, and a subdomain of T4 phage lysozyme [4] share a common core structural organization schematically illustrated in Figure 1. Here four, sequentially connected α-helices pack together so that adjacent helices align with antiparallel N to C senses at angles of about 18°. This arrangement produces a 4-α-helical bundle of approximately square cross-section with an overall left-handed twist. Several factors can be enumerated that contribute to the stability of this arrangement among molecules otherwise having little overall similarity in amino acid sequence.

The Structural Rigidity of α-Helices

α-helices show the least conformational variability among the hydrogen-bonded secondary structures commonly found in globular proteins. This is a reflection of both the depth and localization of the potential energy minimum that characterizes the α-helical conformation for individual L-amino acids. How-

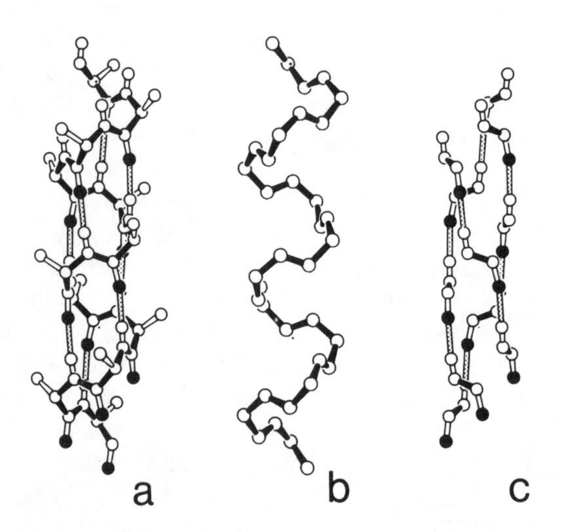

Fig. 2. *α-helices (a) are rigid structures because they are constructed from interconnected bonded helices of opposite hand. Part b illustrates the right-handed polypeptide backbone helix, and c the three contiguous hydrogen bond networks that form left-handed helices.*

ever, many of the important packing properties of α-helices that determine their organization in proteins depend upon the regular and extended spacing of the pendant amino acid side chains on the helix surface. Figure 2 shows some alternative views of the bonding interactions in an α-helix that illustrate how the extended regularity of the α-helix is stabilized. The extended structure can be viewed as an interconnected set of bonded interactions that follow helices of opposite senses. The covalent polyptide backbone follows a right-handed helix, while the three contiguous hydrogen-bond networks form left-handed helices. It is clear that extensive deformation or unraveling of an α-helix will generally involve an alteration of the helical twist of the constituent polypeptide chain. Viewed from this perspective, it can be seen that the hydrogen-bond helical networks are arranged to prevent untwisting of the covalent helix, so that the covalent and hydrogen-bonded sets of helical interactions act in opposition to make the α-helix structurally rigid [5]. This simple principle has been exploited in manufacturing ropes and cables (Figure 3) where the strands are laid with one helical sense, and the fibers within the strands with the opposite helical sense. Unwinding of the cable when stretched is prevented, since the strands and fibers cannot simultaneously unwind in opposite directions.

Helix Dipole Interactions

Each peptide group in a protein possesses a local dipole moment. This results from the electronic distribution in the peptide bond that produces partial positive and negative charges on the carbonyl oxygen and peptide amide nitrogen, respectively. Since the backbone conformation of an α-helix produces an arrangement where the CO and NH peptide groups are oriented with the same sense and aligned parallel to the α-helix axis, the summation of the local dipoles produces a macroscopic dipole in the helix as a whole. Computational studies [3] suggest that the invariant antiparallel packing sense of adjacent

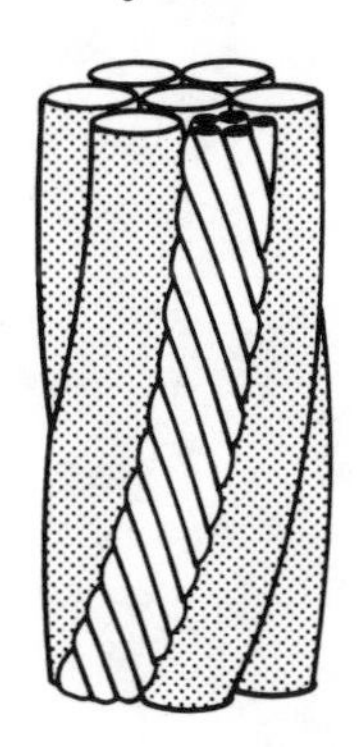

Fig. 3. *The cable illustrated is constructed to resist disassembly by untwisting. This is achieved by arranging the lay of fibers in the strands with the opposite helical sense of the lay of the strands in the cable. Four-α-helical structures can resist dissembly both because individual α-helices are built up of helices with opposite hands (Fig. 2), and because the right-handed helices pack together to form a left-handed bundle (Fig. 1).*

helices in 4-α-helical proteins is a result of favorable helix dipole interactions.

Helix-Packing Geometry in 4-α-Helical Bundles

A predominant geometrical feature of the 4-α-helical proteins is the packing arrangement of adjacent helices with an angle of approximately 18° between their helix axes. A variety of models have been proposed to account for the common occurrence of this interhelical packing arrangement in both fibrous [6] and globular proteins [7,8]. However, detailed comparison of the pairwise helical packing in known 4-α-helical proteins, both among themselves and with published models, showed unexpected variability. This observation motivated further investigations of the interhelical packing in the 4-α-helical proteins. The fundamental result to emerge [2] was the observation that, independent of the detailed pairwise interactions between helices relatively inclined at 18° (Figure 4), square arrays of four helices with this interaction angle tend naturally to produce structures where all pairwise helix interactions are symmetry related and those of individual residues pseudoequivalent. This relationship tends to

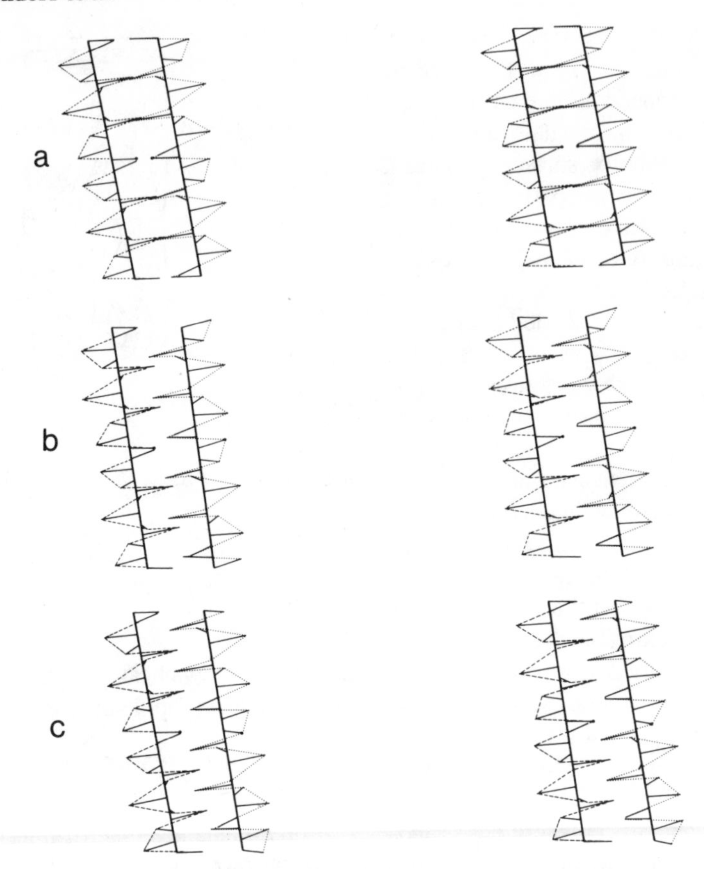

Fig. 4. *Stereoviews of pairwise α-helix packing interactions at about 18°. Individual 23-residue helices are represented as helix axes with projecting radial vectors [6,8]. Parts a,b, and c show alternative side-chain packing patterns that in each case produce nearly equivalent (pseudoequivalent) local packing interactions among the residues along the axes of each helix. As illustrated in Figure 2 of Reference 2, these patterns also repeat cyclically between all pairs of helices when arranged as in Figure 1.*

make the packing forces holding the pairs of helices that form four faces of the 4-α-helical bundle equivalent, and so results in a regular structure. The symmetric packing of four essentially straight α-helices produces an array where the individual helices diverge from a common point of closest approach. The point of closest interhelix approach frequently occurs near the end of the 4-α-helical bundle. The resulting divergence between the α-helices provides an internal binding pocket that typically incorporates a bound prosthetic group in many of the 4-α-helical proteins [2].

Helix Interconnections

The observed 4-α-helical proteins exhibit a substantial diversity in both the lengths and local conformations of their interconnecting polypeptide segments. This observation suggests that, apart from a generalized tendency for extended polypeptide chains to assume statistical solution conformations that have right-handed supercoiled character [1], α-helix packing effects are the determinative factors in the folding [3] and final organization of 4-α-helical proteins. However, the attractive-

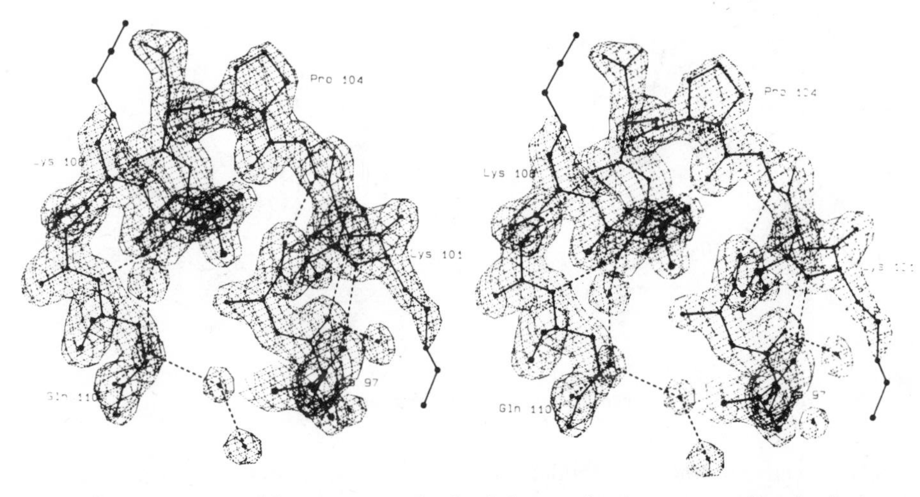

Fig. 5. *A stereoview of the structure and refined electron density corresponding the single residue connection between the penultimate and terminal α-helices in cytochrome c' [9].*

ness of the 4-α-helical motif as a candidate for ab initio protein design experiments raises the question of designing a minimal connection between α-helices in a 4-α-helical array. The recent 1.67 Å resolution refinement of cytochrome c′ from R. molischianum [9] provides an example where a single glycine residue interconnects α-helices terminating at alanine 102 and commencing at proline 104 (Figure 5). This efficient chain reversal serves to fix the relative rotational orientations of the connected helices, which are tightly packed owing to a number of close alanine interactions between the two helices. Thus, the incorporation of this minimal connection into structures of novel design necessitates a consideration of the more extended packing interactions between the helices. The observation of the cytochrome c′ structure clearly shows how the requirements of local conformation and more extended packing can be met simultaneously in the formation of a tight interhelical bend. While this is directly applicable to the de novo design of α-helical bundles, this example points out the necessity to consider both local conformational and more extended packing effects in ab initio protein design experiments.

In addition to the foregoing factors, there are considerations described by others that may be useful constituents of a design protocol for 4-α-helical arrays. These include both local features of protein sequence, such as the situation of charged α-helix amino acid side chains to form favorable charge–dipole or dipole–dipole interactions with the helix macrodipole (described above), or more extended sequence periodicities required to ensure stable hydrophobic associations between the helices [10]. Taken together, these approaches suggest a rational approach to de novo protein design that will no doubt be subjected to experimental test in the near future.

ENGINEERING THERMAL STABILITY IN PROTEINS

The preceding discussion of 4-α-helical proteins dealt with some fundamental issues relevant to the long-range objective of ab initio protein design. More immediate objectives of protein engineering involve situations where it may be desirable to enhance the stability of a naturally occurring protein through chemical or site-specific modification of its structure. Since the folding free energy of proteins is typically on the order of 10 Kcal/M (only a small fraction of a single covalent bond formation free energy), and reflects compensa-

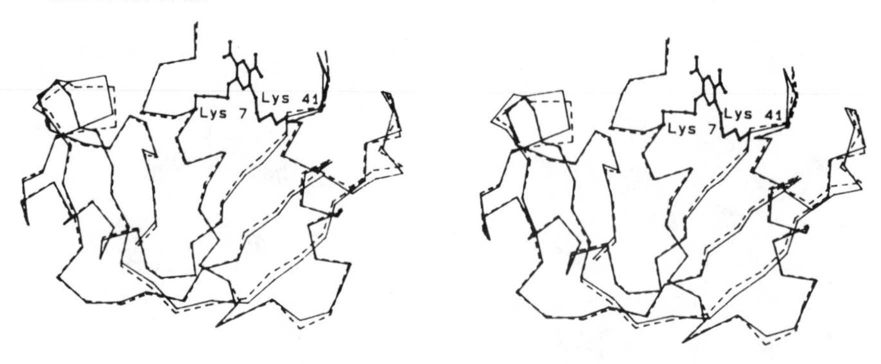

Fig. 6. *A stereoview showing results of a least-squares superposition of DNP-crosslinked RNase-A (backbone solid) with the unmodified RNase-A structure (backbone dashed).*

tory enthalpic and entropic contributions of large magnitude, apparently subtle alterations in protein composition or structure can produce significant differences in protein thermal stability. In what follows, we describe results of some recently completed studies aimed at elucidating the origins of thermal stability in some well studied proteins.

Since it is the totality of pairwise interatomic interactions that contributes enthalpic components of a protein's stabilization free energy, there exist a multiplicity of potential strategies for increasing stability based on alterations of packing (i.e., van der Waals), electrostatic, or hydrogen-bonding interactions to enhance the stability of proteins of known structure. Alternatively, alterations of the protein's structure can be envisioned that affect entropic components of the stabilization free energy. However, owing to the complex and interactive nature of bonding and entropy contributions, both within the protein and with surrounding solvent, it is presently difficult either to explain the origins of thermal stability in well-studied systems such as thermolabile mutants of T4 phage lysozyme [11] or accurately predict them computationally. Nevertheless, numerous studies suggest that the introduction of covalent crosslinks can enhance protein thermal stability. This is a potentially attractive approach, given available capabilities in site-specific modification of proteins. For example, the structure of a known protein can be examined computationally or with computer graphics to find sites geometrically and sterically compatible with the introduction of a disulfide crosslink. Cysteine residues can then be specifically substituted for preexisting amino acids at the appropriate sites in the native protein by modification of the encoding DNA [12]. In fact, experiments carried out in Genex laboratories have used this approach to introduce several disulfide crosslinks into the secreted enzyme subtilisin. Contrary to expectation, analysis by differential scanning calorimetry of the thermal denaturation properties of these modified proteins showed, in all cases examined, that the crosslinked protein was less stable than the native enzyme. Results of this sort clearly motivate a more detailed physical understanding of the origins of thermal stabilization by crosslink introduction. To investigate this phenomenon, we undertook the X-ray crystallographic structure determination of a derivative of Ribonuclease-A (RNase-A) that incorporates a 2,4-dinitrophenylene (DNP) group covalently crosslinked at the 1 and 5 positions to the side-chain amino nitrogens of RNase-A lysine residues 7 and 41. In thermal unfolding experiments carried out at pH 2.0, the crosslinked protein (RNase-X) melted at a temperature some 25°C higher than the native unmodified enzyme [13]. The stabilization of the crosslinked molecule was attributed to entropic effects that altered the relative free energies of folding the native and crosslinked proteins. Specifically, the stabilization of RNase-X was ascribed to a reduction in conformational entropy that predominantly mani-

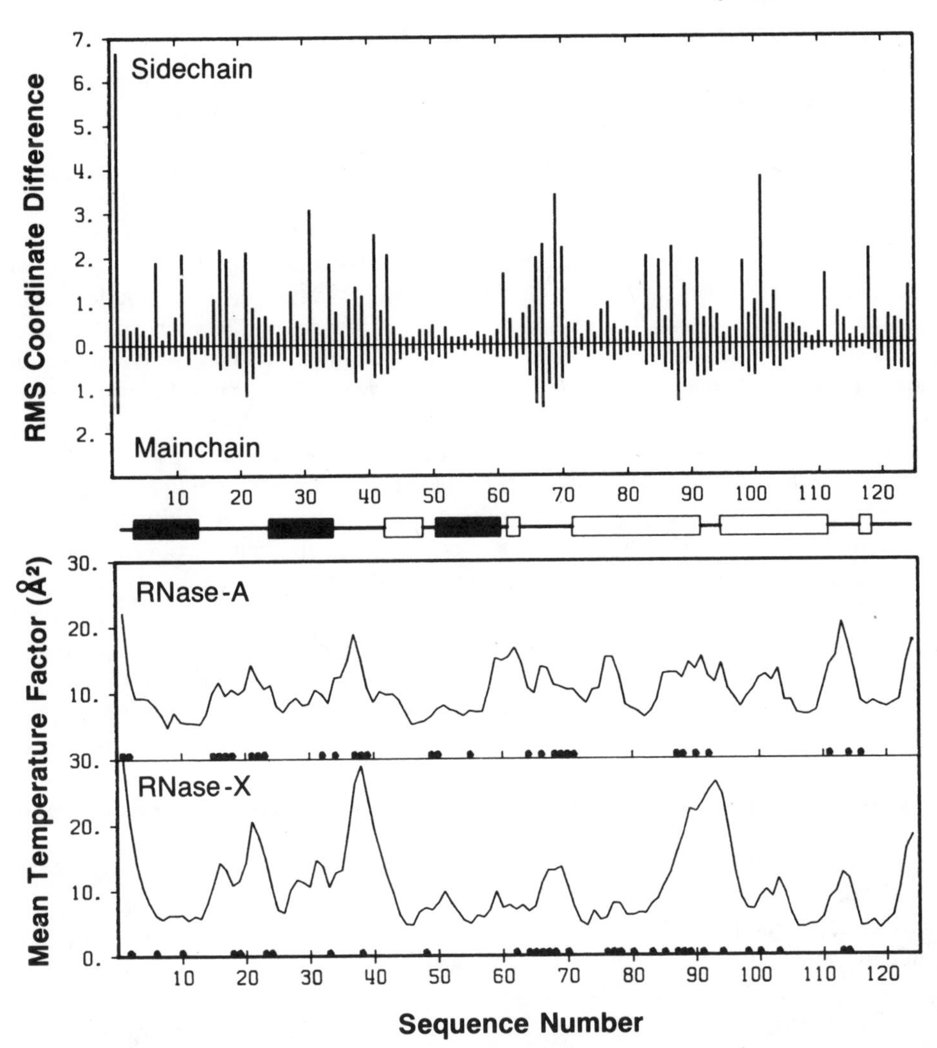

Fig. 7. *Comparison of the refined RNase-A and DNP-crosslinked (RNase-X) X-ray structures [14]. Top panel shows r.m.s. positional differences averaged over backbone (0.52 Å) and side-chain (1.34 Å) atoms after least-squares superposition of the structures. Positional differences of the observed magnitude are typical of structure determinations in different crystal forms of the same protein. The schematic shows the molecular secondary structure with α-helices as black bars, β-strands as open bars, and connecting loops as lines. Bottom panel compares average backbone B-value behavior for the molecules (connected curves), together with positions of residues involved in intermolecular contacts (dots) in the different crystal cells. Relatively little alteration in local backbone B-value is attributable to crosslink introduction.*

fests itself in the protein's unfolded state. This results in a destabilization of the unfolded state of the crosslinked molecule relative to the folded states of both RNase-A and RNase-X. Because it is implicit in this explanation that the folded free energies of the native and modified proteins be essentially equivalent, we determined and crystallographically refined the RNase-X structure to an R-factor of 0.18 at 2 Å resolution [14] and compared its structure to published data on the native RNase-A molecule [15]. Figure 6 shows a least squares superposition of the RNase-A and RNase-X α-carbon backbone structures, together with the situation of the DNP crosslinking group. As evident from Figure 7, the native and derivatized structures are very similar. Differences that do occur primarily involve alterations in side-chain positions. These differences include a) alterations in the lysine 7 and 41 side-chain conformations resulting from crosslink introduction; b) some small shifts in adjacent residues caused by DNP steric interactions; c) some shifts in adjacent active site residues caused by counterion concentration and pH differences between the two crystal forms; and d) some differences in side-chain ordering due to formation of alternative contacts in the crystal lattices. Although these are in some sense indistinguishable in type and magnitude from differences observed between native and thermolabile mutants of T4 phage lysozyme [11], they are nevertheless typical of variations in structure observed when the same protein is examined in different crystal lattice environments [16,17]. Figure 7 also compares the refined B-value behavior (a measure of molecular flexibility of the protein in the crystal lattice) of RNase-A and RNase-X and shows no major alterations in the dynamic properties of the molecule as a consequence of crosslink introduction.

Taken together, these results suggest that both enthalpic and entropic components of the molecules free energies are similar in native and crosslinked enzymes, a result consistent with the proposal that crosslink introduction alters thermal stability by altering relative entropy contributions in the molecules' unfolded states. Assuming this to be the case, the question remains as to why the introduction of disulfide bonds into subtilisin described above should have destabilized the enzyme. One possibility is that disulfide introduction into subtilisin unfavorably altered the native enzyme structure to an extent that offset favorable entropic differences manifest in the native and modified unfolded states. Alternately, it is possible that crosslink introduction may have unfavorably altered folded-state entropic contributions to the molecule's free energy. This proposal is based on theoretical considerations [18,19], dynamics simulations [20], and interpretation of thermodynamic data [21] that collectively suggest that proteins may be stabilized in part by entropy effects associated with collective vibrations of the structure as a whole [21]. It seems clear that "improperly" situated crosslinks could potentially reduce cooperative flexibility in the protein, with a consequent reduction in vibrational entropy. This unfavorable alteration in the entropic contribution to the folded proteins free energy could more than offset favorable enthalpy changes associated with covalent bond formation. In this context, it is useful to emphasize the observation that the introduction of the chemical crosslink into RNase-A notably did not alter the apparent flexibility of the structure, although it will obviously be interesting to examine the structures and B-value behavior of the thermolabile crosslinked subtilisins when they are completed. Nevertheless, because recent computational experiments suggest that protein normal-mode behavior constitute a useful basis set for describing protein motion, it would seem worthwhile to test this hypothesis by introducing crosslinks that respectively inhibited or enhanced the molecules' cooperative vibrational properties proposed to entropically stabilize the folded protein structure.

ACKNOWLEDGMENT

This work supported in part by NIH research grants GM 33325 and 30393.

REFERENCES

1. Salemme FR (1983): Structural properties of protein β-sheets. Prog Biophys Molec Biol 42:95–133.
2. Weber PC, Salemme FR (1980): Structural and functional diversity in 4-α-helical proteins. Nature 287:82–84.
3. Sheridan RP, Levy RM, Salemme FR (1982): α-helix dipole model and electrostatic stabilization of 4-α-helical proteins. Proc Nat Acad Sci USA 79:4545–4549.
4. Bernstein FC, Koetzle TF, Williams GJB, Meyer EF Jr., Brice MD, Rodgers JR, Kennard O, Schimanouchi T, Tasumi M (1977): The protein data

bank: a computer-based archival file for macromolecular structures. J Mol Biol 112:535–542.

5. Salemme FR (1985): Engineering aspects of protein structure. Ann New York Acad Sci 439:97–106.
6. Crick FHC (1953): The packing of α-helices: simple coiled-coils. Acta Cryst 6:689–697.
7. Chothia C, Levitt M, Richardson D (1977): Structure of proteins: packing of α-helices and pleated sheets. Proc Nat Acad Sci USA 74:4130–4134.
8. Richmond TJ, Richards FM (1978): Packing of α-helices: geometrical constraints and contact areas. J Mol Biol 119:537–555.
9. Finzel BC, Weber PC, Hardman KD, Salemme FR (1986): The structure of ferricytochrome c′ from Rhodospirillum molischianum at 1.67 Å resolution. J Mol Biol 186:627–643.
10. Eisenberg D, Weiss RM, Terwilliger TC (1984): The hydrophobic moment detects periodicity in protein hydrophobicity. Proc Nat Acad Sci USA 81:140–144.
11. Grutter MG, Hawkes RB, Matthews BW (1979): Molecular basis of thermostability in the lysozyme from bacteriophage T4. Nature 277:667–669.
12. Perry LJ, Wetzel R (1984): Disulfide bond engineered into T4 lysozyme: stabilization of the protein toward thermal inactivation. Science 226:555–557.
13. Lin SH, Konishi Y, Denton ME, Scheraga HA (1984): Influence of an extrinsic cross-link on the folding pathway of ribonuclease A. Conformational and thermodynamic analysis of cross-linked (lysine 7-lysine 41)-ribonuclease A. Biochemistry 23:5504–5512.
14. Weber PC, Sheriff S, Ohlendorf DH, Finzel BC, Salemme FR (1985): The 2 Å resolution structure of a thermostable ribonuclease-A chemically cross-linked between lysine residues 7 and 41. Proc Nat Acad Sci USA 82:8473–8477.
15. Wlodawer A, Sjolin L (1983): Structure of Ribonuclease-A: Results of joint neutron and x-ray refinement at 2.0 Å resolution. Biochemistry 22:2720–2728.
16. Finzel BC, Salemme FR (1985): Lattice mobility and anomalous temperature factor behavior in cytochrome c′. Nature 315:686–688.
17. Sheriff S, Hendrickson WA, Stenkamp RE, Sieker LC (1985): Influence of solvent accessibility and intermolecular contacts on atomic mobilities in hemerythrins. Proc Nat Acad Sci USA 82:1104–1107.
18. Peticolas WL (1979): Low frequency vibrations and the dynamics of proteins and polypeptides. Meth Enzymol 61:425–456.
19. Karplus M, McCammon JA (1983): Dynamics of proteins: elements and function. Ann Rev Biochem 53:263–300.
20. Irikura KK, Brooks BR, Karplus M (1985): Transition from B to Z DNA: Contribution of internal fluctuations to the configurational entropy difference. Science 229:571–572.
21. Sturtevant JM (1977): Heat capacity and entropy changes in processes involving proteins. Proc Nat Acad Sci USA 74:2236–2240.
22. Salemme FR (1982): Cooperative motion and hydrogen exchange stability in protein β-sheets. Nature 299:754–756.

Protein Engineering, pages 175–179

14

Hydrophobic Interactions in Proteins

Glenn J. Lesser, Richard H. Lee, Michael H. Zehfus, and George D. Rose

Department of Biological Chemistry, Hershey Medical Center, Pennsylvania State University, Hershey, Pennsylvania 17033

INTRODUCTION

The hydrophobic effect has been a topic of chemical interest for more than a century. This phenomenon, in its most general terms, is concerned with the solubility of molecules in water and in nonpolar solvents [1]. Liquid water dissolves polar substances readily but nonpolar substances only sparingly. Upon mixing, water "squeezes out" hydrophobic molecules, resulting in a separation into polar and nonpolar phases. The spontaneous segregation of oil after mixing with water serves as a familiar example.

Hydrophobicity is believed to play a major role in organizing the self-assembly of protein molecules, because some amino acid residues are abundantly water soluble while others are only sparingly so [2]. The 20 commonly occurring residues can be classified with respect to their side chains as polar, nonpolar, or amphiphilic. However, even residues with nonpolar side chains should be viewed as amphiphilic, because backbone N–H and C=O groups are themselves highly polar.

According to the popular "oil drop" model, the protein interior is expected to be enriched in nonpolar ("oily") residues which cluster together to form, in effect, an organic phase [3]. The effect is analogous to the spontaneous segregation of oil in water, with the important distinction that residues in proteins are covalently bound to their chain neighbors and cannot partition independently. This appealing model, if correct, should predict the relative tendencies of residues to partition between the inside and outside of water-soluble proteins. The model is complicated by the fact that the protein interior is more heterogeneous than a simple hydrocarbon solvent, and the solvation of groups on the surface may be hindered by intramolecular neighbors.

The oil drop concept of a protein molecule stems from the solution thermodynamics of simple model compounds [3,4]. The solubility of hydrocarbons and amino acids has been studied extensively both in water and in organic solvents such as ethanol [4,5]. Adoption of these measurements to explain larger composites, such as proteins, would seem to be a valid extrapolation.

However, quantitative assessment of the actual situation in proteins of known structure reveals unsuspected complexity. Richards [6] noted that

> Of the accessible areas of native structures, roughly half represents polar atoms and half nonpolar atoms. Thus the "grease" is by no means all "buried." In the folding process there are roughly equivalent decreases in the accessibility of both the polar and nonpolar groups. The relevant forces and

the final structure require more careful definition than is implied by the common feeling that inside equals nonpolar and outside equals polar.

Can the thermodynamics of simple model compounds account for the observed heterogeneity of actual molecules? Ideally, the distribution of residues between the inside and outside of a protein molecule would be governed solely by side-chain solubility, since backbone atoms remain invariant from residue to residue. Practically, significant sources of nonideality arise because, as mentioned previously, the residues are covalently linked and cannot distribute independently. In particular, a chain site with a polar residue adjacent in sequence to a nonpolar residue may not be able to satisfy both tendencies simultaneously.

Despite these complications, it is conceivable that segments of the polypeptide chain distribute according to their aggregate solubility. In the simplest case, the hydrophobicity of a composite segment would be reckoned by summing the contributions of constituent chemical groups. Such group additivity obtains, for example, in the case of straight-chain aliphatic hydrocarbons [1].

To test the idea that chain segments behave as predictible composites, the hydrophobicity of individual residues must be known. In such studies, hydrophobicity is measured as the free energy required to transfer a standard quantity of substance from water to a nonaqueous solvent. This standard-state free energy, $\Delta G^\circ_{water \rightarrow organic}$, is derived by measuring the solubility in either phase. The free energy of transfer, $\Delta G^\circ_{p \rightarrow p'}$, between phases p and p′ is related to the partition coefficient between these phases, $K_{p \rightarrow p'}$, by the equation

$$\Delta G^\circ_{p \rightarrow p'} = RT\ 1\text{–}K_{p \rightarrow p'}\ [3\text{–}5].$$

Many scales of hydrophobicity have been determined for the amino acids. Surprisingly, there is often little agreement between the scales of different investigators. In the next section, a novel scale is introduced.

SCALES OF HYDROPHOBICITY

Many scales of hydrophobicity have been determined for the amino acids, their residues, and their analogs. Such scales can be divided into two general classes. *Solution scales* are based upon the solubility of amino acid solutes in aqueous and nonaqueous solvents. These solubilities are determined and then used to calculate partition coefficients, and, in turn, free energies of transfer. *Empirical scales* are based upon observed distributions of the residues between the solvent-accessible surface (aqueous phase) and the buried interior (nonaqueous phase) in proteins of known structure.

Examination of scales in the recent literature often reveals qualitative discrepancies. Choosing an example from solution studies, tryptophan and tyrosine are very hydrophobic in the scale of Nozaki and Tanford [5], but quite hydrophilic in the scale of Wolfenden et al. [7]. Qualitative disagreement is also evident in empirical studies where different investigators have used differing criteria to determine residue "buriedness." These issues are analyzed in another place [8].

In response to this dilemma, we have developed a new empirical scale [9] by measuring the mean solvent accessibility [10] of the residues in a data base of X-ray–elucidated proteins. The mean solvent accessibility, $\langle A_{res} \rangle$, is the residue surface area that remains accessible to solvent, on average, within folded proteins. In work of this sort, it is presumed that surface area is related to hydrophobicity in a linear way and, in particular, that the more hydrophobic the residue, the more completely buried it will be.

Mean accessibilities must be normalized before valid comparisons can be made. For example, although $\langle A_{val} \rangle$ and $\langle A_{gly} \rangle$ are approximately equal, an isolated valyl residue has almost twice the area of an isolated glycl residue, and hence it must bury a larger fraction of its available surface upon folding.

To normalize $\langle A_{res} \rangle$, a standard state accessible surface area is needed. For a given residue type, X, this standard state accessibility is defined as the average surface area that residue has in Gly-X-Gly tripeptides. The tripeptide surface area depends upon the values of its dihedral angles. Values are chosen to reflect the actual ensemble of dihedral angles observed in the data base, and these are used to construct a representative set of tripeptides. The standard state accessibility, A°_{res}, is taken as the average over this representative set. For example, if there are N phenylalanine residues in the data base, then N tripeptides, Gly-Phe-Gly, are constructured with angles

$$(\phi_i, \psi_i, \chi_i^1) \quad (i=1, 2, \ldots, N)$$

A°_{phe} is the average of this ensemble.

The fractional accessibility of a residue can now be calculated as its mean accessibility in proteins divided by the standard state accessibility, $\langle A_{res} \rangle / A^\circ_{res}$. This fractional accessibility is an intrinsic measure of residue hydrophobicity. In addition, the difference between the standard state accessibility and the mean accessibility, $A^\circ_{res} - \langle A_{res} \rangle$, measures the area the residue buries, on average, upon folding. This difference is proportional to the residue's hydrophobic contribution to the conformational free energy of the protein.

Table I lists standard-state, mean, and fractional accessibilities for the 20 residues.

PREDICTIVE METHODS—HYDROPHOBICITY PROFILES

The distribution of hydrophobic groups along the linear sequence can be used in a predictive manner. A *hydrophobicity profile* for a protein graphs the average hydrophobicity per residue against the sequence number. The plotted curve reveals the loci of minima and maxima in hydrophobicity along the linear polypeptide chain, as illustrated in Figure 1. The method has been applied to proteins of known sequence but unknown structure in order to predict: 1) peptide chain turns [11]; 2) interior/exterior regions [12,13]; 3) antigenic sites [14–16]; and 4) membrane-spanning segments [13].

TABLE I. Standard-State, Mean, and Fractional Accessibilities for the Amino Acid Residues[a]

Residue	A° ($Å^2$)	$\langle A \rangle$ ($Å^2$)	$A^\circ - \langle A \rangle$ $(Å)^2$	$(A^\circ - \langle A \rangle)/A^\circ$
Ala	118.1	31.5	86.6	.74
Arg	256.0	93.8	162.2	.64
Asn	165.5	62.2	103.3	.63
Asp	158.7	60.9	97.8	.62
Cys	146.1	13.9	132.3	.91
Gln	193.2	74.0	119.2	.62
Glu	186.2	72.3	113.9	.62
Gly	88.1	25.2	62.9	.72
His	202.5	46.7	155.8	.78
Ile	181.0	23.0	158.0	.88
Leu	193.1	29.0	164.1	.85
Lys	225.8	110.3	115.5	.52
Met	203.4	30.5	172.9	.85
Phe	222.8	28.7	194.1	.88
Pro	146.8	53.7	92.9	.64
Ser	129.8	44.2	85.6	.66
Thr	152.5	46.0	106.5	.70
Trp	266.3	41.7	224.6	.85
Tyr	236.8	59.1	177.7	.76
Val	164.5	23.5	141.0	.86

A° = area in the standard state. For a given residue type, X, the standard state accessibility is defined as the average surface area that residue has in a representative ensemble of Gly-X-Gly tripeptides.

$\langle A \rangle$ = average accessible area in proteins.

$A^\circ - \langle A \rangle$ = average area buried upon transfer from the standard state to the folded protein.

$(A^\circ - \langle A \rangle)/A^\circ$ = mean fractional area loss, equal to the average area buried normalized by the standard state area.

[a]Proteins used in the study were obtained from the Brookhaven data base [17]. These include: actinidin (2ACT), concanavalin A (3CNA), carboxypeptidase A (1CPA), calcium-binding parvalbumin B (3CPV), cytochrome c (3CYT), elastase (1EST), ferredoxin (3FXC), flavodoxin (3FXN), D-glyceraldehyde-3-phosphate dehydrogenase (1GPD), high potential iron protein (1HIP), lactate dehydrogenase (1LDX), lysozyme (6LYZ), myoglobin (3MBN), papain (8PAP), pancreatic trypsin inhibitor (3PTI), trypsin (3PTP), ribonuclease A (3RSA), rubredoxin (3RXN), subtilisin (1SBT), staphylococcal nuclease (1SNS), superoxide dismutase (2SOD), triose phosphate isomerase (1TIM), and thermolysin (1TLN).

The hydrophobicity profile for an amino acid sequence is simple to construct, with only two significant parameters: 1) the choice of hydrophobicity scale and 2) the degree of av-

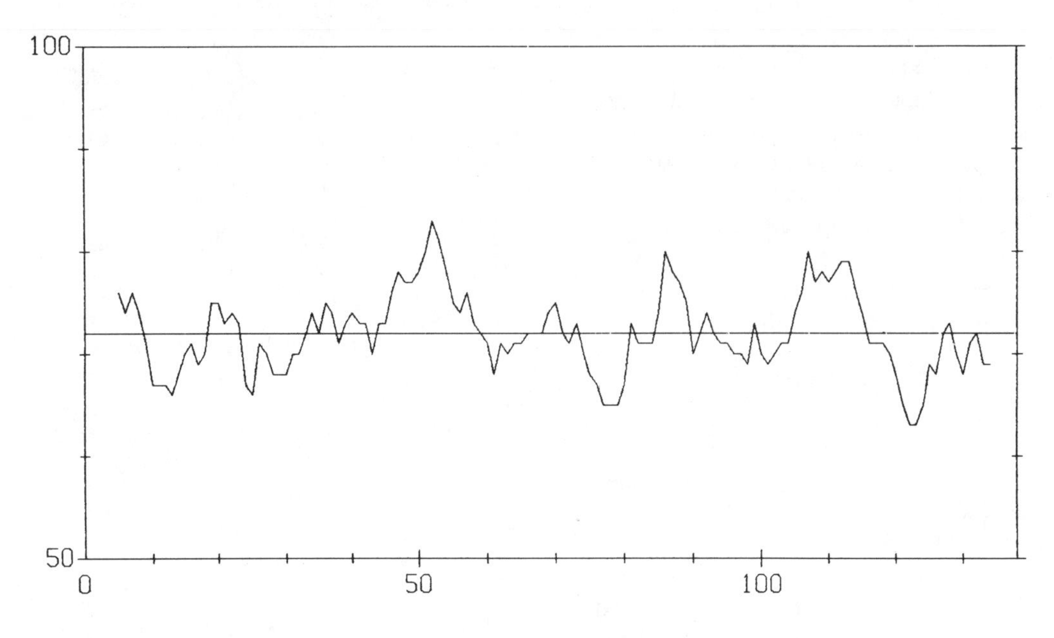

Fig. 1. *Hydrophobicity profile for flavodoxin using a 9-residue window and the scale of fractional area loss listed in Table I. At each residue, N, the ordinate specifies the percent area lost, on average, for the 9-residue segment N ± 4 residues. A 9-residue window is used to distinguish interior and exterior regions. (A smaller window would be used to identify turns or antigenic sites, and a larger window to identify membrane-spanning segments.)*

Local minima in hydrophobicity predict regions on the outside, while local maxima predict the buried interior. Two of the major peaks in hydrophobicity—near residues 52 and 86—correspond to inner strands of the beta sheet. The remaining large peak, residues 107–113, corresponds to an outer strand.

eraging. The chosen scale is used to establish a table of hydrophobicities for each of the 20 residue types. Averaging is then achieved by evaluating the mean hydrophobicity within a moving window that is stepped along the sequence from each residue to the next. To resolve structural components, the size of the averaging window must be larger than a single turn (i.e., greater than four residues) but smaller than a complex segment such as a turn-helix (i.e., less than approximately 12 residues). To identify membrane-spanning segments, the window should approximate the expected length of a segment (i.e., about 19 residues). Smaller windows yield profiles with greater detail, but most of the minor fluctuations in hydrophobicity appear to have little relation to overall segmentation of the molecule.

ACKNOWLEDGMENT

This work was supported by National Research Service Awards to MHZ and RHL, an award for short-term training of medical scientists to GJL, a Research Career Development Award to GDR, and by GM29458.

REFERENCES

1. Tanford C (1980): "The Hydrophobic Effect." New York: John Wiley & Sons.

2. Tanford C (1978): The hydrophobic effect and the organization of living matter. Science 200:1012–1018.
3. Kauzmann W (1959): Some factors in the interpretation of protein denaturation. In Anfinsen CB, Edsall JT, Richards FM (eds): "Advances in Protein Chemistry." Orlando: Academic Press Inc., pp 1–63.
4. Cohn EJ, Edsall JT (1943): "Proteins, Amino Acids, and Peptides as Ions and Dipolar Ions" New York: Reinhold.
5. Nozaki Y, Tanford C (1971): The solubility of amino acids and two glycine peptides in aqueous ethanol and dioxane solutions—establishment of a hydrophobicity scale. J Biol Chem 246:2211–2217.
6. Richards FM (1977): Areas, volumes, packing, and protein structure. Annu Rev Biophys Bioeng Vol 6:151–176.
7. Wolfenden R, Andersson L, Cullis PM, Southgate CCB (1981): Affinities of amino acid side chains for solvent water. Biochem 20:849–855.
8. Rose GD, Gierasch LM, Smith JA (1985): Turns in peptides and proteins. In Anfinsen CB, Edsall JT, Richards FM "Adv. Protein Chem." Orlando: Academic Press Inc., pp 1-109.
9. Rose GD, Geselowitz AR, Lesser GJ, Lee RH, Zehfus MH (1985): Hydrophobicity of amino acid residues in globular proteins. Science 229:834–838.
10. Lee BK, Richards FM (1971): The interpretation of protein structures: estimation of static accessibility. J Mol Biol 55:379–400.
11. Rose GD (1978): Precition of chain turns in globular proteins on a hydrophobic basis. Nature 272:586–590.
12. Rose GD, Roy S (1980): Hydrophobic basis of packing in globular proteins. Proc Nat Acad Sci, USA 77:4643–4647.
13. Kyte J, Doolittle RF (1982): A simple method for displaying the hydropathic character of a protein. J Mol Biol 157:105–132.
14. Both GW, Sleigh MJ (1980): Complete nucleotide sequence of the haemagglutinin gene from a human influenza virus of the Hong Kong subtype. Nucleic Acids Res 8:2561–2575.
15. Hopp TP, Woods KR (1981): Prediction of protein antigenic determinants from amino acid sequences. Proc Nat Acad Sci, USA 78:3824–3838.
16. Novotny J, Auffray C (1984): A program for prediction of protein secondary structure from nucleotide sequence data: application to histocompatibility antigens. Nucleic Acids Res 12:243–255.
17. Bernstein FC, Koetzle TG, Williams GJB, Meyer EF Jr., Brice MD, Rogers JR, Kennard O, Shimanouchi T, Tasumi M (1977): The protein data bank: a computer-based archival file for macromolecular structures. J Mol Biol 112:535–542.

Protein Engineering, pages 181–185

15

Hydrophobicity and Amphiphilicity in Protein Structure

David Eisenberg, William Wilcox, and Andrew D. McLachlan

Molecular Biology Institute and Department of Chemistry and Biochemistry, University of California, Los Angeles, Los Angeles, California 90024 (D.E., W.W.), and MRC Laboratory of Molecular Biology, Hills Road, Cambridge CB2 2QH, England (A.D.M.)

INTRODUCTION

The energy of folding a protein from a disorganized coil to a native structure is often considered to be a sum of terms, involving various covalent and noncovalent forces [1–4]. Of these terms, the hydrophobic interaction has been the most elusive for treatment with standard potential energy functions. The reason is that this contribution depends in a complex way on the arrangements of water about both the native and the unfolded structures [5–7]. Because of this, the hydrophobic contribution to the free energy cannot be expressed simply as an interaction between pairs of protein atoms.

The importance of the hydrophobic interaction in maintaining the stabilities of proteins was first clearly stated by Kauzmann [5]. Since then, there have been many contributions to understanding this force, among others [6–11]. An important step in providing a quantitative link between protein structure and energy was made by Richards and others [12–15], who emphasized that the accessibility to aqueous solvent of protein atoms can be expressed in terms of the areas of atoms adjacent to the protein surface. As a protein folds, carbon and sulfur atoms lose a greater fraction of their area than do oxygen and nitrogen atoms. Yet the problem remains of establishing a relationship between accessible area and energy.

This link can be provided by considering the hydrophobicity of exposed and buried amino acid residues. The hydrophobicity, *H*, of a residue is a measure of its free energy of transfer from a relatively apolar phase, such as the protein interior, to aqueous solution. In a pioneering paper, Nozaki and Tanford [16] measured hydrophobicities for several amino residues, and their work has been followed by many useful studies, which have gradually provided values of hydrophobicities for all the 20 coded amino acid residues of proteins [e.g., 17–20].

By combining measurements of accessible surface areas of proteins with numerical hydrophobicities, we present here a method for estimating the hydrophobic energy of protein folding.

A second method for assessing the hydrophobic interaction in protein folding combines residue hydrophobicities with known features of protein structure. In this method, the directions and magnitudes of the hydrophobic moments of the various segments of secondary structure of a protein are considered. It can be illustrated with the incorrect protein folds devised by Novotný et al. [1] and their correctly folded analogs.

METHODS

The Hydrophobic Residue Method

It has been suggested that a rough estimate for the free energy of a protein may be given by

$$G_{\mathrm{H}} = \sum_{\text{residues},i} H_{\mathrm{i}}\, M(x_{\mathrm{i}}) \qquad (1)$$

in which H_i is the hydrophobicity of the *i*th residue and $M(x_i)$ represents the hydrophobicity of the environment. Suppose that some fraction A/A° of the surface of a residue is available to solvent, and that the rest of the residue [1−(A/A°)] is buried within the protein, and hence is shielded from solvent. Further suppose that we can represent the hydrophobicity of the environment of the aqueous solvent by +½ and the hydrophobicity of the environment of the interior of the protein by −½. The justification for the values of ± ½ is implicit in Equation 1, if G_H is regarded as the free energy of transfer of a single residue of hydrophobicity H_i. Then we can write

$$G_{\mathrm{H}} = \sum_{\text{residues}} H_{\mathrm{i}} \frac{[A_{\mathrm{i}} - ½]}{A_{\mathrm{i}}{}^{\circ}} \qquad (2)$$

For the calculations of G_{H} by Equation 2, we have used values of $A_{\mathrm{i}}/A_{\mathrm{i}}{}^{\circ}$ tabulated by Novotný et al. [1]. For values of H, we have adopted the consensus scale of Eisenberg et al. [20]. It is also possible to extend Equation 2 to use atomic accessibilities, rather than residue accessibilities [22].

Hydrophobic Moments

Structural hydrophobic moments were calculated for segments of secondary structure in the example proteins, using formula 2 from [20]:

$$\mu_{\mathrm{H}} = \sum_{\text{residues } i} H_{\mathrm{i}}\, \mathbf{s}_i$$

in which the sum is over all residues in a given segment of secondary structure, H_i is the hydrophobicity of the ith residue, and $\mathbf{s}_i$ is a unit vector pointing from the alpha carbon atom of the *i*th residue to the center of the residue's side chain. The μ_H vector for each segment is displayed as pointing from the center of the segment toward the direction of greater hydrophobicity.

RESULTS

Hydrophobic Moments of Segments of Protein Secondary Structure

In a previous study, we inspected the relative directions and magnitudes of hydrophobic moments of the segments of secondary structure in folded globular proteins [20]. We found that they tend to cancel. The reason for this is that hydrophobic side chains of different segments tend to face each other. For example [see Reference 20, Figure 1] the hydrophobic moments of the four α-helices of the melittin molecules point inwards towards each other and effectively cancel.

A similar pattern is found in the four α-helical structure of hemerytherin from *T. dyscritum* [23] as shown in Figure 1a. This is one of the two 113 residue proteins whose energetics of folding were compared by Novotný et al. [1]. As in the case of melittin, the hydrophobic moments of the four helices tend to oppose those of neighboring helices. This pattern of opposing hydrophobic moments of helices is generally observed in antiparallel α-helical bundles of the type discussed by Weber and Salemme [24].

In two-layer β-sheet structures, as in the α-helical structures, the hydrophobic moments of individual strands tend to point inwards. This is so in the variable domain of the mouse immunoglobulin kappa chain, here termed VL, whose structure was determined by Segal et al. [25] and whose energetics folding were analyzed by Novotný et al. [1]. The nine strands and their associated moments are shown in Figure 2a.

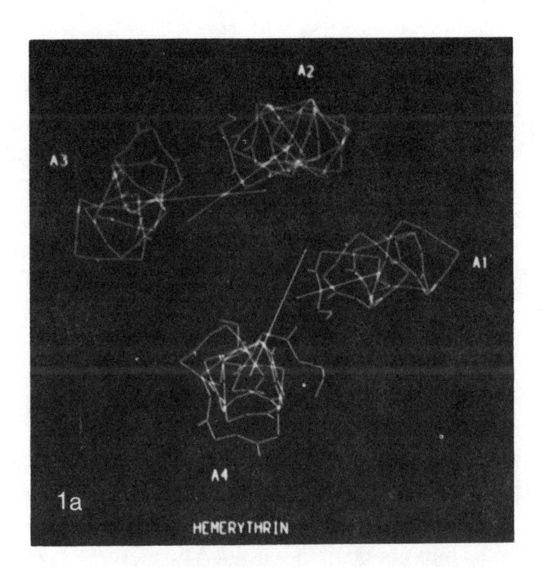

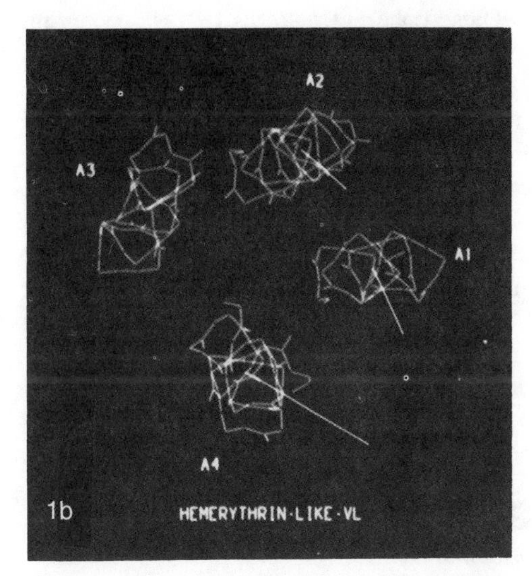

Fig. 1. *Hydrophobic moments of helices in a real and in a misfolded protein. A1 is the first helix in the chain, A2 is the second, and so forth. a) The four α-helical segments from the structure of hemerythrin from* T. dyscritum *[23]. Notice that the hydrophobic moments from the four helices point inwards and oppose each other. Each moment is represented as a line drawn from the center of its segment toward the direction of greater hydrophobicity. b) The incorrectly folded structure of a mouse κ-chain VL domain [25] arranged by Novotńy et al. [1] into the structure of hemerythrin. The hydrophobic moments of the four helices are shown by lines emerging from the centers of the helices. These are scaled by a factor of 2 over those of Figure 1a, for better visibility. (Reproduced in color on p. 342).*

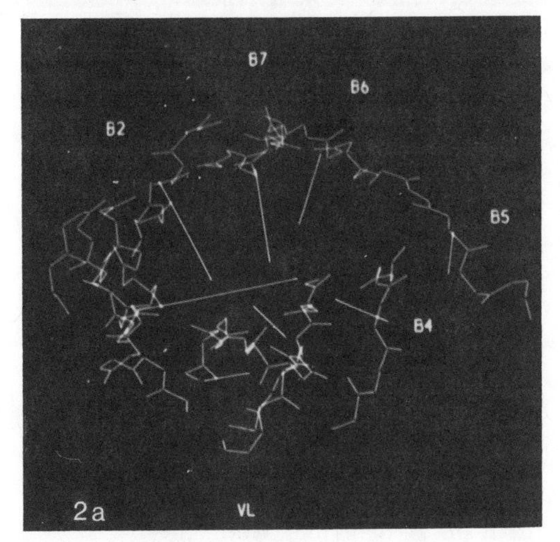

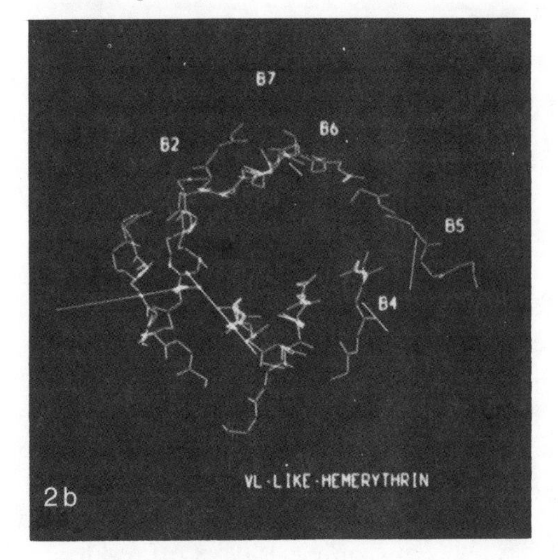

Fig. 2. *Hydrophobic moments of beta sheets in a real and in a misfolded protein. B1 is the first strand of sheet in the polypeptide chain, B2 is the second, and so forth. a. The nine-stranded β-sheet of a mouse κ-chain VL domain [25]. The hydrophobic moment of each strand of the sheet is shown as a line emerging from the center of the strand. The lengths are scaled by a factor of 1.5 over those of Figure 1a. b. The incorrectly folded structures of hemerythrin from* T. dyscritum *[23] as arranged by Novotný et al [1]. The moments are on a scale 2 times those of Figure 1a. (Reproduced in color on p. 342).*

These coherent patterns of residue organization, which give rise to the opposing hydrophobic moments of these actual protein structures, are not present in the incorrectly folded protein molecules devised by Novotný et al [1]. These authors provided useful, improperly folded protein models by replacing the side chains of hemerythrin by those of VL, and vice versa. Then, using the program CHARMM, they made small adjustments in coordinates to reduce the potential energies of the two structures to a local minimum. In the resulting incorrect fold of VL arranged as hemerythrin (Fig. 1b), all of the moments point out towards aqueous solution. The pattern is unlike the one observed in real helical proteins.

A similar lack of coherence is seen in the incorrectly folded hemerythrin molecule, of Novotný et al. arranged in the pattern of the VL domain (Fig. 2b). The hydrophobic moments of the segments of secondary structure do not point generally inwards, as they do in the VL structure; all are small, or point outwards.

CALCULATION OF HYDROPHOBIC ENERGIES

A highly approximate estimate of the hydrophobic component of the free energy of folding can be made from Equation 2, and from a knowledge of the accessible surface areas of residues in a protein.

For the two incorrectly folded models, and their correct analogs, Novotný et al. have tabulated the fractional accessibility of each type of residue in the structure. These accessibilities can be combined with the number and hydrophobicity of each type of residue in Equation 2 to calculate a crude hydrophobic free energy. For these four structures we have done this, and the energies are given in Table I. Both acutal structures are more stable than their incorrectly folded analogs by about 9 kcal mol^{-1}.

DISCUSSION

Energetics of Correctly and Incorrectly Folded Structures

The two incorrectly folded structures, hemerythrin and the immunoglobulin variable domain VL, devised by Novotný et al [1], offer a convenient testing ground for quantitative theories of protein folding. Novotný et al [1] note in their own analysis of these structures, that side-chain nonpolar surface area of the incorrectly folded structures is greater than for the correctly folded structures. This finding is reflected in Table I in terms of hydrophobic free energies, computed from Equation 2. Notice that both actual structures are stabilized by their hydrophobic free energies, and that both hypothetical, misfolded structures are destabilized.

The origin of the higher (destabilizing) energy in the misfolded structures is a combination of greater solvent accessibility of hydrophobic residues, and diminished solvent accessibility of hydrophilic residues. In the VL structure, all residue types other than Gly, His, and Arg lower the hydrophobic free energy. However, in hemerythrin-like VL, a positive contribution to the hydrophobic free energy is contributed by the following residues types: Ala, Cys, Gln, Gly, His Lys, Phe, Tyr, and Val.

It should be emphasized that the hydrophobic free energies of Table I rest on many assumptions and are rough estimates at best. Among other factors, they neglect the amphiphilic character of amino acids such as Lys, Arg, Glu, Tyr, and Trp, in which there are both hydrophilic and hydrophobic portions.

Assessing the Designing Structures With the Aid of the Hydrophobic Moment

The examples of Figures 1 and 2, as well as others [20], suggest that actual protein structures are characterized by coherent patterns of opposed hydrophobic moments in their segments of secondary structure. In contrast, the two hypothetical misfolded structures of Fig-

TABLE I. Hydrophobic Contribution to the Free Energy of Protein Folding, G_H*

Protein	Type of structure	GH/kcal mol^{-1}
Hemerythrin	Actual	−6.4
Hemerythrin-like VL domain	Misfolded	+2.8
Net stabilization of actual structure		−9.2
VL domain	Actual	−7.6
VL-like hemerythrin	Misfolded	+1.4
Net stabilization of actual structure		−9.0

*Computed from Equation 2 as described in the text.

ures 1b and 2b lack these coherent patterns. Thus a visual inspection of a protein in terms of the directions of the hydrophobic moments of its segments may be useful for assessment and design of proposed structures. Elsewhere in this symposium we present a design for a new protein based in part on the principle of opposed hydrophobic moments.

ACKNOWLEDGMENTS

Support from NSF (PCM 82-07520) and NIH (GM31299) is gratefully acknowledged, as well as USPHS National Research Service Award GM 07185 to W.W.

REFERENCES

1. Novotný J, Bruccoleri R, Karplus M (1984): J Mol Biol 177:787.
2. Karplus M, McCammon A (1983): Annu Rev Biochem 52:263.
3. Némethy G, Scheraga HA (1977): Rev Biophys 10:239.
4. Levitt M, Warshel A (1975): Nature 253:692.
5. Kauzmann W (1959): Adv Pro Chem 14:1
6. Némethy G, Scheraga HA (1962): J Phys Chem 66:1773
7. Klapper MH (1973): Kaiser ET, Kezdy FJ., (eds): "Progress in Bioorganic Chemistry," Vol 2, p 55.
8. Pratt L, Chandler (1977): J Chem Physiol 67:3683.
9. Pratt L, Chandler D (1980): J Chem Physiol 73:3434.
10. Tanford C (1979): Proc Natl Acad Sci USA 76:4175.
11. Tanford C (1980): "The Hydrophobic Effect" 2nd ed, New York: John Wiley & Sons.
12. Lee B, Richards RM (1971): M Mol Biol 55:379.
13. Shrake A, Rupley JA (1973): J Mol Biol 79:351.
14. Richmond TJ (1984): J Mol Biol 178:63.
15. Richards FM (1977): Annu Rev Biophys Bioeng 6:151.
16. Nozaki Y, Tanford C (1971): J Biol Chem 246:2211.
17. Wolfenden R, Anderson L, Cullis PM, Southgate, CCB (1981): Biochemistry 20:849.
18. Flauchere J-L, Pliska V (1983): Eur J Med Chem—Chem Ther 18:369.
19. Janin J (1979) Nature 277:491
20. Eisenberg D, Weiss RM, Terwilliger TC (1982): Wilcox W. Faraday Symp Chem Soc 17:109.
21. Rose GD, Geselowitz AR, Lesser GJ, (1985): Lee RH Zehfus, M.H. Science 229:834.
22. Eisenberg D, McLachlin AD (1986) Nature 319:199.
23. Stenkamp RE, Sieker LC, Jensen LH (1978): Acta Cryst Sect B 38:784.
24. Weber P, Salemme FR (1980): Nature 287:82–84.
25. Segal DM, Padlan EA, Cohen GH, Rudikoff, S, Potter, M, Davies, DR (1974): Proc Natl Acad Sci USA 71:4298.

Protein Engineering, pages 187–192

16

The Stabilities of Globular Proteins

Ken A. Dill

Departments of Pharmaceutical Chemistry and Pharmacy, University of California, San Francisco, California 94143

INTRODUCTION

In this chapter, we address two questions: What forces are responsible for the stabilities of the native structures of globular proteins? How accessible are the alternative conformations? Alternative conformations are important, for they specify the thermodynamic reference state. "Stability" can only be defined relative to the reference state; it is the difference in free energy between the native state and the alternative conformations.

THE CONFORMATIONAL FREE ENERGIES OF GLOBULAR PROTEINS

Virtually all types of molecular interaction contribute to the balance of forces responsible for the native structures of proteins. This balance is not yet understood in detail. Our current knowledge of it derives from thermodynamic experiments and theoretical models of folding. For example, it is observed that most globular proteins are marginally stable [1–6]. They melt when the incident thermal energy is only slightly greater than ambient (at temperatures only a few tens of kelvin above room temperature). Free energies of denaturation of small globular proteins are only about 5–15 kcal/mol protein (for temperatures in the range of 10–50°C [1–3], roughly the free energy equivalent of one or two hydrogen bonds. Thus, even the smallest of interactions can contribute significantly to stabilization or destabilization of a globular molecule.

Several observations, to be described below, lead to the view that there are two principal opposing forces responsible for the folding or unfolding (denaturation) of globular proteins: 1) the hydrophobic effect drives the molecule to condense, so as to decrease the unfavorable contacts of the hydrophobic residues with water, whereas, 2) unfolding is strongly favored entropically by the increased conformational freedom of the chain. Support for this view comes from the following observations, explained below:

1) Near room temperature, both the enthalpy and entropy of unfolding are small.

2) At higher temperatures, in the range of 50–100°C, proteins denature, and typical enthalpies and entropies of unfolding are large and positive. This is partly a consequence of large positive heat capacity changes upon unfolding.

3) The heat capacity changes are nearly independent of temperature over the range of 0–100°C [1].

These data may be interpreted as follows. The large positive entropy of unfolding is consistent with an increased number of conformations in the denatured state. On the other hand, the hydrophobic effect should contribute a large negative entropy of unfolding. This follows from transfer experiments [7,8], in which it is observed that dissolving small hydrophobic molecules in water leads to a decreased entropy due to the ordering of the water at the surface of the solute [9]. Therefore, unfolding of the protein should lead to increased exposure of hydrophobic residues to water, and to a decrease in this "hydrophobic" entropy. This hydrophobic entropy should be highly temperature dependent, decreasing in magnitude with increasing temperature, since the ordering of water diminishes at higher temperature. Thus if the balance of forces is primarily due to the hydrophobic effect and the conformational entropy, then the total entropy of unfolding (the sum of the conformational and hydrophobic entropies) should have its maximum positive value at the boiling temperature of the solution, and should decrease with decreasing temperature, as is observed [1]. These two contributions are of opposite sign and have nearly equal magnitude near room temperature. Changes in heat capacity should be little affected by the freedom of the rotational isomeric states of the chain, which are approximately isoenergetic, and should then just reflect effects due to the hydrophobic interactions. Indeed, they are of the same sign and magnitude and show the same temperature independence as heat capacity changes for transfer of small hydrophobic molecules to water.

The temperature dependence of the stability (free energy of unfolding) is not described completely by the hydrophobic effect [8,10]. For the protein, unfolding is favored by increasing temperature, whereas for hydrophobic solutes in water, aggregation, rather than dispersal, is favored by increasing temperature. This is readily understood in terms of the large positive conformational entropy of unfolding of the protein. The conformational entropy contributes to destabilization of the native structure with increasing temperature, but obviously does not play a role for processes involving the transfer of small molecule hydrophobic solutes.

The observations above rule out certain other possible driving forces. For example, condensation could not be driven principally by the burial of electrostatic charges in the protein core, for then the change in heat capacity would be negative [6,11]. Other evidence is also consistent with this view: 1) hydrophobic residues are largely buried in globular proteins, (whereas charged groups are not, and thus their environment should not change much in the unfolding process), and 2) organic solvents and surfactants are denaturants, whereas electrolytes often stabilize the native state. Electrolytes should denature proteins if charge burial were the principal condensation force [4].

In recent years, theoretical methods have emerged with which specific interactions can be studied in proteins. Potential energy parameters have been derived from studies of model systems to provide force fields which are used in molecular mechanics or molecular dynamics simulations [12–14]. The large forces described above, however, do not readily lend themselves to such simulation, for they are largely entropic, and involve many alternative conformations of the solvent and of the backbone and side chains of the protein. Likewise other contributions to the free energy which may be important, such as counterion release and changes in internal modes of vibration, are difficult to simulate for the same reason [although see Refs. 15 and 16]. Moreover, inasmuch as molecular mechanics treats only internal energies, and not free energies, it does not predict temperature dependences. Nevertheless, while detailed simulations currently are not comprehensive in these regards, important deductions on the relative importance of the various energetic contributions are emerging. Regarding the stability of the globular state, one recent effort is noteworthy. Novotny et al. [17] have performed a molecular mechanics simulation to which they added the contribution of the hydrophobic effect through a simple thermodynamic interfacial free-energy term. They purposely imposed an incorrect sequence on the backbone coordinates of a different molecule, then found the local energy minimum in order to study features of models of incorrectly folded proteins. One important conclusion is that without treatment of the hydrophobic effect, even the sign of the free energy of folding can be incorrect, but with even a simple treatment of the hydrophobic effect, the sign of the free energy of folding is predicted correctly.

The role of hydrogen bonds in protein stability, currently a matter of some dispute, may be resolvable by simulation methods. Experiments on urea in water as a model system suggest that the contribution of hydrogen bonds to stability is likely to be small for globular molecules [1,4,18], even though they provide the principal driving force for helix and secondary structure formation in proteins and nucleic acids [19]. Current evidence suggests that there is little change in the average number of hydrogen bonds in globular and unfolded states; intramolecular bonding occurs in the globular state, whereas bonding to water occurs in the unfolded states. Thus it is the average relative bond strength which is at issue. However, even if these and other energetic interactions are small, they are nevertheless likely to be important for the stability of any given protein, since the typical total free energy of folding is so small.

An alternative theoretical approach to the study of stability derives from chain statistical mechanics [20]. It is possible to explore the consequences of the propositions that: 1) the hydrophobic effect is the principal driving force for condensation; 2) the limits of condensation are determined by the steric repulsive forces, which prevent more than one chain segment from occupying the same volume of space; and 3) subject to these constraints, the system will otherwise tend to maximize its entropy, in accord with the second law of thermodynamics. This approach permits us to study the consequences of this simple balance of forces, irrespective of sequence or other specific interactions.

The principal predictions are in qualitative agreement with observations of globular protein structure, as follows [20]. While hydrophobic residues should prefer to be buried in the protein core, some will be exposed at the surface, the consequence of an entropic tendency to distribute randomly between the core and the surface. The fraction buried should increase with molecular weight of the protein. Unfolded (denatured) states of the protein are not near the state of the random coil, which is defined as a chain that obeys random flight statistics [21]; unfolded molecules should be much more condensed (although still largely solvent exposed) than random coil molecules. Short chain molecules (shorter than a few tens of residues) should not have a stable globular state, for if they condensed into a globule, they would have too little interior volume for hydrophobic residues to be suitably protected from the solvent; the hydrophobic advantage would be too small to compensate for the loss of entropy upon folding. It is predicted that proteins should fold into thermodynamically stable domains under certain conditions, for it is not the surface/volume ratio which tends to be minimized through condensation; rather, the chain should tend to configure itself into a geometry for which its surface/volume ratio approximately equals its ratio of hydrophilic/hydrophobic residues. These are the principal examples of the general features of the stability of globular proteins that follow from the simple view that folding is driven largely by the hydrophobic effect but is opposed by the chain conformational entropy.

PROTEIN FOLDING: THERMODYNAMIC OR KINETIC?

The tertiary structure of a protein is determined by the relative free energies of all its possible chain conformations. These free energies are governed by the primary amino acid sequence, solvent character, and external thermodynamic conditions. In principle, the observed tertiary structure can also depend on the pathway of folding, that is, on the time-dependent sequence of configurations that the molecule explores. Either the native structure is thermodynamically stable (i.e., it adopts the lowest possible free energy state), or it is metastable: stable for minutes, hours, or years before it turns over in the organism or the test tube, but unstable on a longer time scale because of the existence of other conformations with even lower free energies. Metastability would imply that there are relatively severe constraints, i.e., that there are kinetic barriers significantly higher than RT, the energy of the incident thermal motions, which prohibit certain pathways of configurational changes. The consequences of metastability are: 1) the rate of conformational change to the state of lowest free energy would be slow on the experimental time scale, thus intermediate states should be detectable, 2) in addition to the free energies of the various chain conformations, the native structure would also depend on the initial configuration (for example, the conformation of the protein as it emerges from the ribosome); and 3) the observed native structure would not necessarily correspond to the state of the global free-energy minimum.

There has been some dispute as to whether native structures of proteins are stable or metastable. The principal observation supporting the hypothesis of thermodynamic stability was due to Anfinsen [22]; that ribonuclease and other proteins can be denatured and renatured to full activity. Denaturation of the large number of native protein molecules in solution results in unfolding to a very large number of different conformations of nearly equal free energy. Thus refolding must begin from this enormous number of initial conformations of the different molecules in solution. It follows that there are an enormous number of refolding pathways. Yet the end result of refolding is convergence to the same native structure. This represents the state of lowest free energy.

The principal motivation for the alternative view that protein structure is determined by kinetic limitations has come from the following argument [Refs. 23, 24, and 20, and the references therein; but see also the discussions in Refs. 25 and 26]. If we let z [$\cong$ 4 (20)] represent the number of rotational isomers of a single-peptide bond, and if rotations are approximately independent, and if n is the number of such bonds, then the number of rotational isomers of a molecule is $N = z^n$. For $n = 100$, $N = 10^{60}$. This number is so large that even at the maximum bond rotation rate, it would take many lifetimes of the universe for a molecule to find one specific conformation, the native structure, by random search. Therefore, the argument goes, there must be relatively specific narrow pathways of folding, so that only a small fraction of all the possible states are explored on the way to the native structure. It is concluded from this view that native structures of proteins are metastable.

There are two flaws in this argument, however. First, it neglects the excluded volume of the chain. When chain conformations can be described by a random flight in space (as for an unfolded molecule in a theta solvent [21]), then for very few of the molecular conformations does the molecule intersect itself, and the above argument provides a good first estimate for the number of chain conformations. But for a globular condensed molecule, most of the random flights are not viable molecular conformations, for they represent configurations in which the chain would pass through itself. The number of viable conformations of globular molecules is thus very much smaller than the number of random coil conformations; it is of the order of

$$N = (z/e)^n$$

where $e = 2.718\ldots$ [20]. Thus the number of conformations which are physically accessible in the globular state for $z = 4$ and $n = 100$ is about 44 orders of magnitude smaller than the number of conformations physically accessible in the random coil state.

The second flaw in the argument above is the assumption that conformations will be explored randomly as the molecule tends toward the native structure. The search would be random only if all the conformations were of equal energy, as would be approximately the case for an unfolded homopolymer molecule condensed chain are highly dependent upon in a theta solvent. But free energies of a single chain conformation, for the residues are sufficiently closely packed in the globular state that their interactions are strong, and the free energies of globular and unfolded molecules will differ significantly. Inasmuch as there are significant free-energy differences, there is a driving force toward structures of lower free energy. This driving force need not be large to be of sufficient importance to render the kinetic argument above invalid. Consider a simple example which closely parallels the problem at hand. Reduction in temperature of a gas in a container, to below its boiling temperature, can result in rapid condensation, just as temperature reduction will cause unfolded proteins teins to fold, often rapidly. But gas condensation requires the exploration of a far greater number of degrees of freedom than protein folding requires. There are of the order of Avogadro's number of molecules in the gas, and three times that number of translational degrees of freedom; there are only of the order of hundreds or thousands of degrees of freedom for the folding of a single protein molecule. It is not the number of degrees of freedom which is important; it is how randomly they are searched. For condensed chain molecules as for gases, the free-energy landscape is not flat; and even small differences in free energy will direct the flow, often rapid, of configurational changes to states of lowest free energy.

The conclusion is that neither kinetic limitations nor metastability need to be assumed to account for the native structures of proteins. Nor is the existence of domains in proteins necessarily due to kinetic necessity. Nor is there currently any evidence to reject the thermodynamic hypothesis. The folding of any given protein can involve slow steps (proline isomerization is a well-known example [26]), but there is no reason to expect that proteins should have any difficulty reaching at least nearly native structures in the vicinity of the global free-energy minimum. We should expect pathways of folding to more closely resemble funnels than tunnels in configuration space.

REFERENCES

1. Privalov PL (1979): Stability of proteins. Small globular proteins. Adv Prot Chem 33:167–241.
2. Tanford C (1968): Protein denaturation. Adv Prot Chem 23:121–282.
3. Pace CN (1975): The stability of globular proteins. Crit Rev Biochem 3:1–43.
4. Kauzmann W (1959): Some Factors in the interpretation of protein denaturation. Adv Prot Chem 14:1–63.
5. Edsall JT, McKenzie HA (1978): Water and proteins. I. The significance and structure of water: its interaction with electrolytes and non-electrolytes. Adv Biophys 10:137–207.
6. Edsall JT, McKenzie HA (1983): Water and proteins. II. The location and dynamics of water in protein systems and its relation to their stability and properties. Adv Biophys 16:53–183.
7. Tanford C (1980): "The Hydrophobic Effect, 2nd ed." New York: John Wiley & Sons.
8. Ben-Naim A (1980) "Hydrophobic Interactions." New York: Plenum Press.
9. Stillinger FH (1980): Water revisited. Science 209:451–456.
10. Creighton T (1985): The problem of how and why proteins adopt folded conformations. J Phys Chem 89:2452–2459.
11. Sturtevant JM (1977): Heat capacity and entropy changes in processes involving proteins. Proc Natl Acad Sci 74:2236–2240.
12. Karplus M, McCammon JA (1983): Dynamics of proteins: Elements and function. Ann Rev Biochem 52:263–300.
13. Zimmerman SS, et al (1977): Conformational analysis of the 20 naturally occurring amino acid residues using ECEPP. Macromolecules 10:1–9.
14. Weiner SJ, et al (1984): A new force field for molecular mechanical simulation of nucleic acids and proteins. J Am Chem Soc 106:765–784.
15. Irakura KK, et al (1985): Transition from B to Z DNA: contribution of internal fluctuations to the configurational entropy difference. Science 229:571–572.
16. Karplus M, Kushick JN (1981): Method for estimating the configurational entropy of macromolecules. Macromolecules 14:325–332.
17. Novotny J, Bruccoleri R, Karplus M (1984): An analysis of incorrectly folded protein models. J Mol Biol 177:787–818.
18. Schellman JA (1955): The thermodynamics of urea solutions and the heat of formation of the peptide hydrogen bond. Compt Rend Trav Lab Carlsberg

Ser Chim 29:223–229, 230–259.
19. Zimm BH, Bragg JK (1959): Theory for the phase transition between helix and random coil in polypeptide chains. J Chem Phys 31:526–535.
20. Dill KA (1985): Theory for the folding and stability of globular proteins. Biochemistry 24:1501–1509.
21. Flory PJ (1953): "Principles of Polymer Chemistry." Ithaca: Cornell.
22. Anfinsen CB (1973): Principles that govern the folding of protein chains. Science 181:223–230.
23. Levinthal C (1968): Are there pathways for protein folding? J Chim Phys 65:44–45.
24. Wetlaufer DB (1973): Nucleation, rapid folding, and globular intrachain regions in proteins. Proc Natl Acad Sci 70:697–701.
25. Harrison SC, Durbin R (1985): Is there a single pathway for the folding of a polypeptide chain? Proc Natl Acad Sci 82:4028–4030.
26. Kim PS, Baldwin RL (1982): Specific Intermediates in the folding reactions of small proteins and the mechanism of protein folding. Ann Rev Biochem 51:459–489.

Protein Engineering, pages 193–199

17

Design of Amphiphilic Peptides

Emil Thomas Kaiser

The Rockefeller University, New York, New York 10021

INTRODUCTION

An era in which it will be possible to design and build biologically active proteins including enzymes from the constitutent amino acids is rapidly approaching [1]. Advances in recombinant DNA technology [2] have made the possibility of constructing proteins a reality. Because of the development of the solid-phase approach [3], the synthesis of peptides which are 40 and even 50 amino acids in length is no longer a daunting undertaking. With improvements in stepwise solid-phase synthesis and in the use of fragment synthesis-condensation [4], it seems likely that the organic synthetic approaches can be extended to polypeptides which are as much as 100 amino acids in length, including some of the small enzymes. The major barrier to the preparation of new proteins and polypeptides with novel biological activities is the inability to predict with confidence the way in which a primary amino acid sequence will fold to give a three-dimensional tertiary structure. While the ground rules for the design of tertiary structure remain to be established, a solid foundation has already been laid for the prediction of amino acid sequences that form certain types of secondary structures. We have found that amphiphilic secondary structures play a very significant role in the biological activity and physical properties of many peptides and proteins which bind to biological interfaces such as membranes [5,6]. In these secondary structures such as the amphiphilic α-helix, one face of the structural region is hydrophobic while the other is hydrophilic. For surface-active peptide and protein molecules, ranging from apolipoproteins through toxins and chemotactic peptides to peptide hormones, amphiphilic secondary structural regions appear to be important to the biological and physical properties. To a first approximation, the possible formation of tertiary structures can be neglected in designing models for these molecules. As a result, as will be discussed below, we have found it possible to develop a rational approach to the design of new amino acid sequences which mimic the important amphiphilic secondary structural regions of the natural systems. This allows us to make model peptides having similar if not enhanced activity relative to the corresponding naturally occurring peptides and proteins.

RECOGNITION OF AMPHIPHILIC SECONDARY STRUCTURES AND PRINCIPLES OF THE DESIGN OF MODELS

While several useful computer programs for the recognition of amphiphilic secondary structures have been developed [7,8], the search for possible amphiphilic secondary structures in peptides can be carried out by very simple visual or graphical methods [6]. For example, the possibility that a peptide or protein may have a region in which an amphiphilic α-helix might be induced in an appropriate amphiphilic environment can be assessed by the use of a helical wheel [9] of the appropriate pitch. A typical helical wheel drawing is given in Figure 1 for the region comprising residues 148–164 of apolipoprotein A-1, the principal protein constituent in high-density lipoprotein. In Figure 1 we show an axial projection of the potential α-helical conformation of these residues, and the amino acids shown in the white areas are hydrophilic while those in the darkened areas are lipophilic. Amphiphilic β-sheets are particularly easily recognized, since they consist of alternating hydrophilic and lipophilic residues.

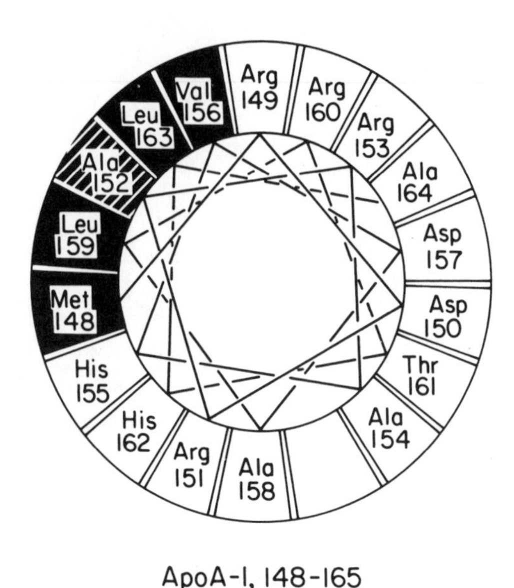

Fig. 1. *Helical wheel projection of the amino acid sequence of residues 148–165 of apolipoprotein A-1. The residues in the white areas are hydrophilic. Those in the dark areas are lipophilic.*

When potential amphiphilic secondary structures are discerned in peptides or proteins, several experimental approaches must be employed to establish the existence of these conformations. Among the indirect methods which can be employed to test the conformational predictions are studies of the binding of the peptides and proteins to amphiphilic surfaces such as those encountered in the cases of phospholipid vesicles or polystyrene beads coated with phospholipid. Additionally, the ability of the peptides and proteins to form stable monomolecular layers at the air-water interface can be examined. We have found that the most useful and convincing approach to confirming the importance of amphiphilic secondary structures in biologically active peptides and proteins comes from the use of peptide synthesis [5,6].

Our approach to peptide design is based on the following hypothesis: If the biological and physical properties of a peptide or protein are dependent on the secondary structural features of a particular region, then in building a model for the region in question, we should be able to design a system with minimum homology to the natural system but with the potential to form a very closely related secondary structure [10]. Thus, if we believe that for a particular region of a peptide or protein a specific amino acid sequence is not crucial but rather that a certain type of secondary structural feature is necessary, then we whould be able to reproduce that secondary structure using an amino acid sequence which has relatively little homology to the natural sequence. In most of our work which has involved the design of peptides which mimic surface-active peptides and proteins we have employed in our solid-phase synthesis naturally occurring amino acids rather than unnatural amino acids in building sequences since we would like to be able to extend the design principles we are developing to use also in the construction of proteins with recombinant DNA methodology.

AMPHIPHILIC HELICAL STRUCTURES AND THEIR DESIGN

Apolipoproteins and Peptide Toxins: Design, Synthesis, and Characterization of Models

We began our research on models for surface-active peptides and proteins by studying the apolipoprotein A-1 system [11–13]. From examination of molecular models of this polypeptide and on the basis of physical measurements, the proposal was made that there might be amphiphilic α-helical regions in this molecule [14]. Subsequently, it was suggested independently by Fitch [15] and McLachlan [16] that these amphiphilic α-helical regions had a repeating character and might be on the order of 22 amino acids in length, punctuated by helix breakers such as a glycine or proline. To test this structural proposal, we constructed a 22-amino-acid peptide which had the potential to form an amphiphilic α-helix corresponding to the prototypic amphiphilic structural unit which appears to be present in apo A-1. The model peptide shown in Figure 2 consisted primarily of three types of amino acids: Leu as the hydrophobic aliphatic residue, Glu as the hydrophilic negatively charged residue, and Lys as the positively charged hydrophilic residue (assuming a pH near neutrality). Our 22-amino-acid model peptide was designed to have the potential to form an amphiphilic α-helix, with a hydrophobic face covering about one third of the surface and the hydrophilic face covering the remaining two thirds. It should also be stressed that the model peptide was designed to have minimal homology to any of the amphiphilic α-helical regions of the naturally occurring proteins. Among the fundamental properties of the 243 amino acid polypeptide apo A-1, which the 22-amino-acid model peptide succeeded in mimicking, were its ability to bind to phospholipid surfaces, its α-helical character in water and water/trifluoroethanol solutions, and its aggregation behavior [11,12]. Additionally, the model showed effective biological activity, behaving as an activator of the enzyme lecithin:cholesterol acyl transferase, an important function of apo A-1 [12].

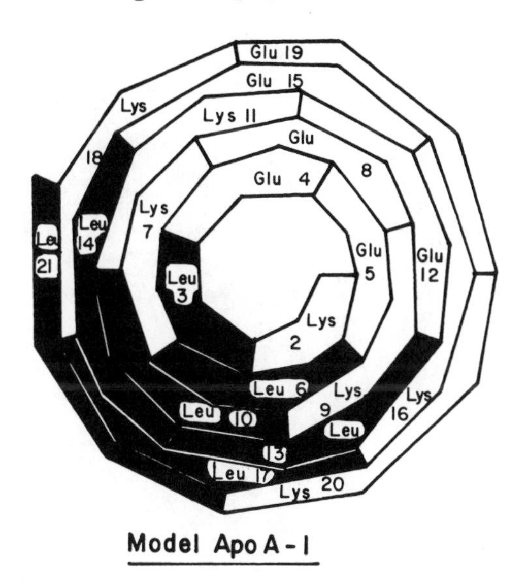

Fig. 2 *Axial projetion of the α-helical conformation of the 22-amino-acid peptide model for apolipoprotein A-1.*

The design of amphiphilic secondary structures was next extended to a system where in addition to the secondary structural region there was a simple "active site" [17,18]. The peptide toxin, melittin, which causes the lysis of erythrocytes, is in its principal form 26 amino acids in length [19]. A segment containing the N-terminal 20 amino acids has the potential to form an amphiphilic α-helix when bound to an appropriate surface. The X-ray structure of the tetrameric form of melittin shows such an amphiphilic α-helix with a kink partway through the sequence due to a proline residue [20]. In addition to the amphiphilic α-helix, there is a cluster of positively charged residues in the C-terminal hexapeptide region. Removal of the C-terminal hexapeptide results in a 20-amino-acid fragment which still binds to the red cell membranes but does not lyse them [21]. As illustrated in Figure 3, we have designed and constructed a model peptide containing an amphiphilic α-helix possessing a very hydrophobic face and in which the Pro residue which causes the kink in the helix of the natural melittin has been replaced with a Ser residue which should not [17]. This model accentuates the amphiphilic α-helical properties of the 1–20 fragment. Maintaining the same C-terminal hexapeptide "active site" region as is found in melittin, the model peptide constructed has an even greater lytic activity than that of the naturally occurring peptide [17,18].

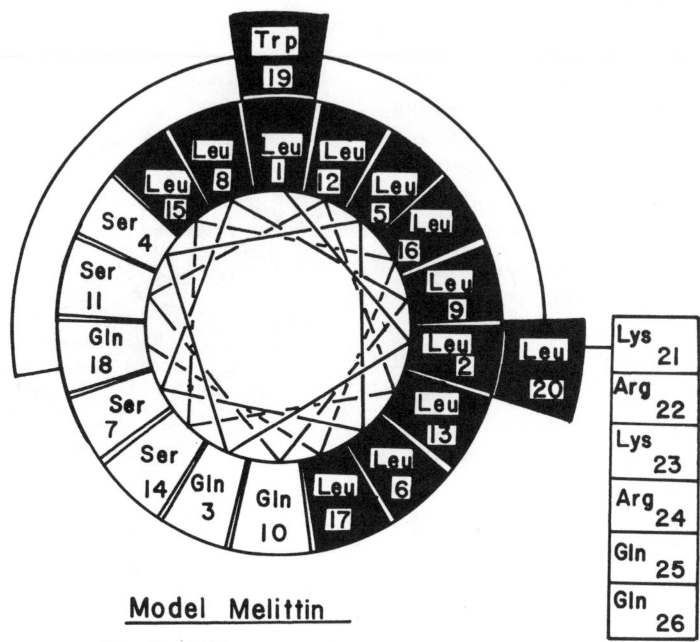

Fig. 3. *Axial projection of the α-helical conformation of a peptide designed to be a melittin model.*

Model Peptide Hormones

Through model building we have demonstrated that amphiphilic secondary features play a major role in hormone structure [6]. There appear to be three general categories of hormones. In the first category are found hormones such as the opioid peptide [Met^5]-enkephalin where there is only a relatively short sequence of amino acids, a "specific recognition site" containing specific amino acids required for eliciting biological activity. Another category of hormones consists of peptides where structures are rather complex and where multiple disulfide bonds may be important, with the possibility that these molecules can form well-defined tertiary structures. In the third category, the group to which our design approach is best addressed, the peptides either have no disulfide bridge or possibly one such bridge, and quite frequently they have a length in the vicinity of 10 to 50 amino acids.

There appear to be at least three advantages to the existence of amphiphilic secondary structures in hormones. The diffusion of the peptide hormone to its receptor may be facilitated by the induction of such regions in the peptide structure [22,23]. Thus, in encountering an appropriate surface the amphiphilic structural region of the hormone which is induced may aid diffusion of the peptide along the surface. As a result, instead of a three-dimensional search there will be a two-dimensional one in the diffusion of the hormone to the receptor. A second point to be considered is that such amphiphilic secondary structural regions may function to hold the "specific recognition site" of the peptide hormone in the precise geometry needed to produce a suitable response from the receptor. Not to be neglected is a third possibility: that peptide hormones are likely to utilize amphiphilic secondary structures in intramolecular interactions with other regions of the peptide molecule, which can lead to stabilization of hormone to enzymatic degradation.

Several peptide hormones have been studied by us along the lines of our structural hypothesis and design concepts. In order to keep this article brief, we will focus on a discussion of

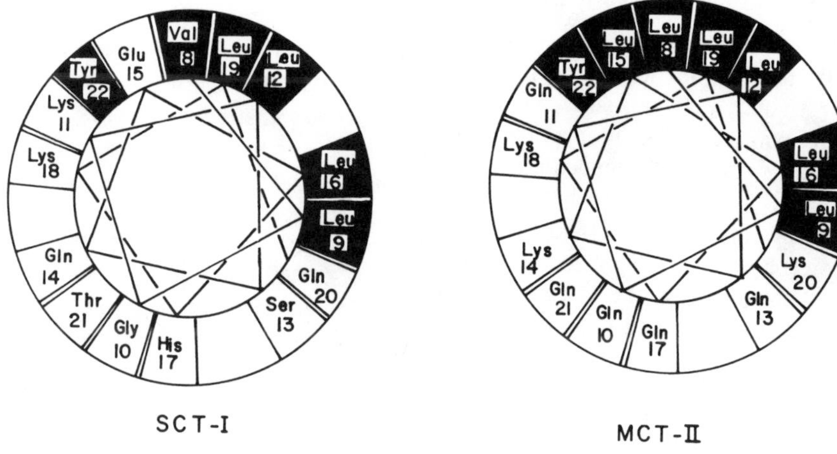

Fig. 4. *Axial projection of the potential α-helical conformation of residues 8–22 in salmon calcitonin I (SCT-I) and a model for calcitonin (MCT-II).*

models for only one such hormone which will illustrate our approach. We have proposed a structural model for the 32-amino-acid calcium-regulating hormone, calcitonin, in which there are three structural regions important to calcitonin's biological activity [24,25]. These regions include a seven-residue cyclic segment at the N-terminus which contains a disulfide bridge between Cys-1 and Cys-7, an amphiphilic α-helix in the region from residue 8–22, and then a hydrophilic spacer unit which includes the region from residue 23 to the C-terminus which is a proline amide. We have described two peptide models for the hormone which we have designed [24,25]. The amphiphilic α-helical structure in the region from residues 8–22 is optimized in both of these models, while the sequence homology to the corresponding region of any natural calcitonin is reduced substantially. Specifically, the sequence homology with natural calcitonins in the hydrophilic face of the amphiphilic α-helix is minimal for the models. In the case of the less effective of the model peptides we designed, there are several changes in the hydrophobic face of the 8–22 amphiphilic α-helix from the sequence of the more active naturally occurring calcitonins like the salmon species. For the more effective model there are relatively few such changes in the hydrophobic face. One significant change in the hydrophobic face of the 8–22 amphiphilic α-helical region for both models is that we utilized a Leu residue rather than a Glu residue at position 15, which is located in what appeared to be otherwise a hydrophobic face. For the more effective calcitonin, while there is only 40% sequence homology to salmon calcitonin in the 8–22 region, the biological activity is comparable. The helical wheel projections of the amphiphilic helices for salmon calcitonin and for this model, MCT-II, are shown in Figure 4. We have found that this model binds with tightness similar to that of the natural system to brain and kidney membranes, has approximately the same behavior in activating adenylate cyclase, and possesses nearly the same hypocalcemic potency in vivo. These results provide strong evidence that the amphiphilic α-helical structure in the region from residues 8–22 is important for binding to calcitonin receptors. An added feature of our results is that our findings with the models show that, although a hydrophilic residue (Asp or Glu) generally occurs on the hydrophobic face at position 15 of the amphiphilic α-helix in the natural calcitonins, it is not required for high biological activity.

Amphiphilic β-Strands and Their Design

We have recently extended our studies of amphiphilic peptides to include peptides designed to form amphiphilic β-strands [26, 27]. In a β-sheet conformation, the peptide backbone is almost completely extended [28], with hydrogen bonds formed between amide groups of neighboring chains. The amino acid side chains project alternately above and below the plane of the peptide backbone. A peptide with alternating hydrophobic and hydrophilic residues will therefore be amphiphilic when in this conformation, having one face hydrophobic and the other hydrophilic. Peptides of this type could form a stable β-sheet structure in aqueous solution through the interaction of the hydrophobic faces.

Physical evidence and some amino acid sequence data suggest that apolipoprotein B (apo B), the major protein component of plasma low-density lipoprotein, consists at least in part of amphiphilic β-strands induced by binding at the surface of the lipoprotein. In order to explore the general properties of amphiphilic β-strands and to have at our disposal apo B models, we have designed and synthesized penta-, nona-, and tridecapeptides with sequences of repeating units of alternating hydrophobic and hydrophilic residues. Acidic and basic residues were also alternated on the hydrophilic face to minimize charge repulsion. The number of repeat units was varied from one to three, enabling us to examine the minimum chain length necessary to form an ordered structure. The tridecapeptide which we have studied most extensively has the amino acid sequence:

NH_2-Val-Glu-Val-Orn-Val-Glu-Val-Orn-Val-Glu-Val-Orn-Val-CO_2H

We expected that this tridecapeptide would be long enough to assume a stable secondary structure when undergoing self-association in aqueous solution and yet short enough to be water soluble. We have shown that our tridecapeptide mimics all the salient physical properties of apo B: namely, formation of a stable monolayer at the air-water interface, preferential stabilization of lipoprotein particles of about 200 Å° diameter, spectra consistent with β-strand structure both in solution and at interfaces, a high tendency to self-associate in aqueous media, and a very high affinity for lipid surfaces. We are currently using the tridecapeptide as well as the related nonapeptide $(Val\text{-}Gly\text{-}Val\text{-}Orn)_2$-Val as tools for the investigation of the specific lipid-protein interactions that occur in low-density lipoproteins.

CONCLUSION

The design of a wide range of peptides that possess a desired secondary structure has been possible through the use of amphiphilicity. In the next stage of protein engineering by this approach, peptides and proteins possessing predicted tertiary structures will be designed through the combination of α-helical and β-strand structural units.

ACKNOWLEDGMENTS

The partial financial support of this research by Public Health Service Progam Project HL-18577 and by the Dow Chemical Company Foundation is gratefully acknowledged.

REFERENCES

1. Kaiser ET (1985): Nature 313:630.
2. Fersht AP, Shi J-P, Wilkinson AJ, Blow DM, Carter P, Waye MMY, Winter GP (1984): Angew Chem (Int Ed) 23:467.
3. Barany G, Merrifield RB (1980): In Gross E, Meienhofer J (eds): "The Peptides, vol 2." New York: Academic Press, p. 3.
4. Nakagawa SH, Kaiser ET (1983): J Org Chem 48:678.
5. Kaiser ET, Kezdy FJ (1983) Proc Natl Acad Sci USA 80:1137.
6. Kaiser ET, Kezdy FJ (1984): Science 223:249.
7. Pongor S, Ulrich PC, Bencsath A, Cerami A (1984): Proc Natl Acad Sci USA 81:2684.
8. Eisenberg D (1985): Ann Rev Biochem 53:595.
9. Schiffer M, Edmundson AB (1967): Biophys J 7:121.

10. Fukushima D, Kaiser ET, Kezdy FJ, Kroon DJ, Kupferberg JP, Yokoyama S (1980): Ann NY Acad Sci 348:365.
11. Fukushima D, Kupferberg JP, Yokoyama S, Kroon DJ, Kaiser ET, Kezdy FJ (1979): J Am Chem Soc 101:3703.
12. Yokoyama S, Fukushima D, Kezdy FJ, Kaiser ET (1980): J Biol Chem 255:7333.
13. Kroon DJ, Kupferberg JP, Kaiser ET, Kezdy FJ (1978): J Am Chem Soc. 100:5975.
14. Segrest JP, Jackson RL, Morrisett JD, Gotto AM (1974): FEBS Lett 38:247.
15. Fitch WM (1977): Genetics 86:623.
16. McLachlan AD (1977): Nature (London) 267:465.
17. DeGrado WF, Kezdy FJ, Kaiser ET (1981): J Am Chem Soc 103:679.
18. DeGrado WF, Musso GF, Lieber M, Kaiser ET, Kezdy FJ (1982): Biophys J 37:329.
19. Habermann E (1972): Science 177:314.
20. Terwilleger TC, Weissman L, Eisenberg D (1982): Biophys J 37:353.
21. Schroder E, Lubke K, Lehmann M, Beitz I (1971): Experentia 27:764.
22. Katchalski-Katzir E, Rishpan J, Sahar E, Lamed R, Hennis YI (1985): Biopolymers 24:257.
23. Kaiser ET (1985) Ann New York Acad Sci 434:321.
24. Moe GR, Miller RJ, Kaiser ET (1983): J Am Chem Soc 105:4100.
25. Moe GR, Kaiser ET (1985): Biochemistry 24:1971.
26. Osterman DG, Mora R, Kezdy FJ, Kasier ET, Meredith SC (1984): J Am Chem Soc 106:6845.
27. Osterman DG, Kaiser ET (1985) J Cell Biochem 29:57.
28. Pauling L, Corey RB (1951): Proc Natl Acad Sci USA 37:729.

Protein Engineering, pages 201–211

18

Theoretical and Experimental Approaches to the Design of Calmodulin-Binding Peptides: A Model System for Studying Peptide/Protein Interactions

Susan Erickson-Viitanen, Karyn T. O'Neil, and William F. DeGrado

E.I. du Pont de Nemours and Company, Central Research and Development Department, Experimental Station, Wilmington, Delaware 19898

INTRODUCTION

Calmodulin is a ubiquitous, small, acidic polypeptide, first described as the heat-stable factor required for calcium-dependent activation of cyclic nucleotide phosphodiesterase [1]. The list of calmodulin-dependent enzymes has expanded rapidly since the initial observations to include certain adenylate cyclases, ATPases, myosin light chain kinase, membrane kinases, and phosphorylase b kinase. Thus, calmodulin appears to play a central role in cellular regulation, motility, and metabolism through modulation of calcium requiring enzyme function [1]. Of special interest in the last few years is the structural and conformational basis of interaction of calmodulin with target enzymes. This review will focus on theoretical and experimental approaches to this problem.

The elucidation of the structural features responsible for calmodulin's ability to modulate the activity of its target enzymes has been hampered by the large size and unknown three-dimensional structure of target proteins. Therefore, there has been considerable effort devoted to the discovery and development of compounds which compete with target enzymes for binding to calmodulin at specific target enzyme-binding sites. Aside from their importance as mechanistic tools for investigating the calmodulin dependence of potential target enzymes, such inhibitors provide an excellent model system for studying calmodulin–protein interactions. Ideally, a calmodulin inhibitor should exhibit the same binding characteristics for calmodulin as target enzymes:

1) The interaction should be calcium dependent; in the absence of calcium the binding should be weak to nonexistent.
2) The inhibitor–calmodulin stoichiometry should be 1:1.
3) The dissociation constant for complex formation should be on the order of 10^{-9} to 10^{-10} M.

Members of the phenothiazine family (e.g., trifluoperazine) were the first reported inhibitors of calmodulin, binding with dissociation constants in the low micromolar range with a stoichiometry of 2:1 [2]. Unfortunately, further studies showed that they also bound to a number of nonspecific sites on calmodulin, limiting their usefulness as structural probes [3]. Nevertheless, studies with these compounds provided the first clues as to the nature of the types of interactions involved in calmodulin's recognition of target enzymes. Using a series of phenothiazine analogs, it could be shown that the primary driving force for drug binding was hydrophobic in nature [4]. Since calmodulin showed greatly reduced affinity for the drugs in the absence of calcium, it appears likely that when calmodulin binds calcium, it exposes a hydrophobic patch, and that it is to this region that inhibitors and target enzymes bind. A second feature important for binding is basicity in target molecules; all inhibitors with reasonable affinities contain one or more positively charged functional groups.

More recently, a number of peptides have been found to bind calmodulin, many with nanomolar dissociation constants [5]. The peptides reported to form tight complexes represent several distinct functional classes including cytotoxic peptides such as the mastoparans [6] and melittin [5], members of the glucagon family [6], and opiate peptides including dynorphin [6] and β-endorphin [7]. Complex formation is calcium-dependent, and for many of these peptides there exists only a single high-affinity binding site. Yet, these peptides vary considerably, not only in their apparent function but also in chain length and amino acid sequence. How can peptides which differ greatly in their primary structure share the common property of binding calmodulin with high affinity? A reasonable hypothesis is that calmodulin recognizes some general structural feature(s) of these peptides rather than a specific amino acid sequence.

As was the case for drug–calmodulin interactions, peptide–calmodulin interactions also appear to involve attractive electrostatic

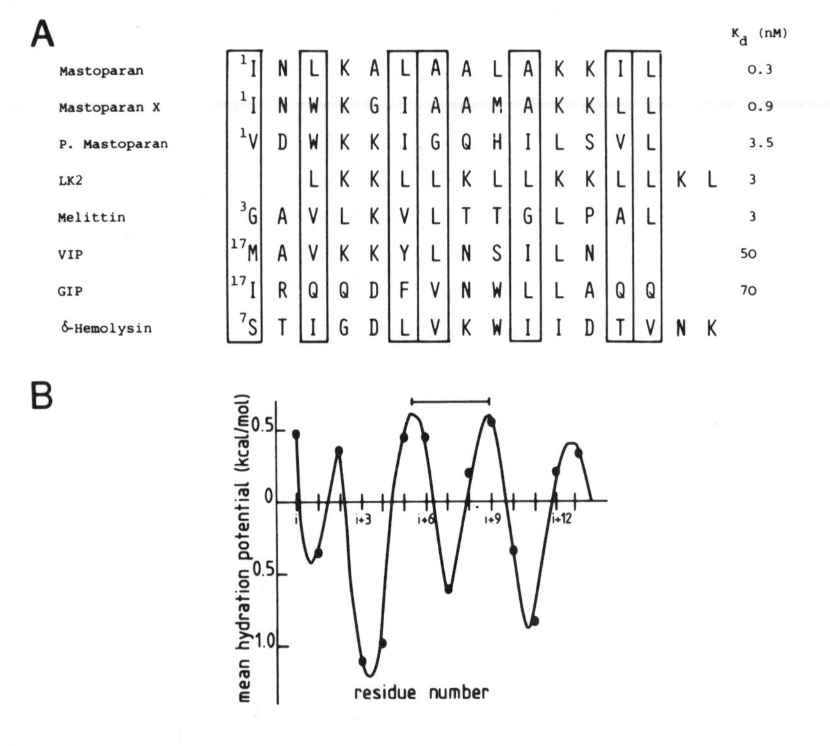

Fig. 1. *A) Aligned partial sequences of calmodulin-binding peptides. The boxes indicate the positions of the invariant hydrophobic residues. B) The mean hydrophobicities for the residue at a given position in the aligned sequences plotted vs. their position in the sequence. The horizontal bar corresponds to the repeat of an α-helix.*

forces. All of the tight binding peptides contain clusters of positively charged residues which are capable of forming multiple electrostatic interactions with acidic regions on calmodulin [8]. This feature of the binding interaction is supported by a detailed study of the influence of ionic strength of the aqueous medium on dynorphin binding; increasing the ionic strength resulted in a decrease in the affinity of the peptide for calmodulin [6]. However, polylysine fails to bind appreciably to calmodulin [8], precluding the possibility that electrostatic interactions alone are responsible for calmodulin binding.

A consideration of the possible secondary structures of the peptides suggests a conformational requirement for calmodulin binding. While the peptides share little exact sequence homology with one another, when properly aligned, hydrophobic residues occupy variant positions along their sequences (Fig. 1A). If the hydrophobicities of the amino acid residues are considered as a function of their positions along the peptide chain, a sine wave with a repeat distance of approximately 3.6 residues is observed (Fig. 1B). This periodicity matches that of an α-helix. As a result, if the peptides adopt helical conformations, the resulting helices are amphiphilic, a conformation in which the hydrophobic and hydrophilic residues segregate nicely on opposite faces of the cylindrical helix. The uninterrupted hydrophobic surface present on the amphiphilic helices might be complementary to the hydrophobic patch on calmodulin which is exposed on calcium binding.

Experimental evidence to support the notion of an amphiphilic α-helix as a key feature common to calmodulin-binding peptides was initially indirect; sequences capable of forming an amphiphilic α-helix were contained within the primary structures of all of the peptide inhibitors with dissociation constants less than 10^{-6} M. Also, circular dichroism spectroscopy of calmodulin–peptide complexes indicated that several inhibitor-peptides, including β-endorphin [9], melittin [5], and mastoparans [10], had random conformations in aqueous solution but adopted helical conformations upon binding to calmodulin. Further studies with deletion peptides of β-endorphin revealed that the calmodulin-binding region of the parent polypeptide chain was localized within a sequence of 12 amino acid residues [7], which had been independently shown to form an amphiphilic α-helix [11].

DESIGN OF CALMODULIN-BINDING PEPTIDES

While these studies were in progress, we were involved in a synthetic approach to demonstrate the structural features responsible for calmodulin–peptide interactions [5]. Perhaps the most direct and satisfying way to prove a structural hypothesis is to synthesize a peptide which embodies the quintessential elements of the hypothetical structure thought to be requisite for a given property. The similarities of the properties of the model peptide to its natural counterpart are an indication of the correctness of the guiding hypothesis. In the case of calmodulin-binding peptides, three peptides were synthesized which epitomize the features of acidic and basic amphiphilic α-helical structures (Fig. 2). Approximately half of the residues in these peptides were leucines, chosen as a residue which was both hydrophobic and favorably disposed to forming α-helices. The remaining residues in peptides 1 and 2 were lysine, a positively charged residue that favors helix formation, or, in peptide 3, glutamic acid, a helix-stabilizing acidic residue. The hydrophobic and hydrophilic residues repeat every 3.5 residues on the average in peptides 1–3, so they will form amphiphilic structures when they adopt helical conformations. The 14-residue peptide 2 is composed of two sequences repeated in tandem of the seven-residue peptide 1. With just seven residues in its sequence, peptide 1 is too short to form a stable α-helix, while the 14-residue peptide 2 should form a marginally stable α-helix due to its increased chain length. If helix formation is obligatory for calmodulin binding, this should translate to higher affinity of peptide 2 versus peptide 1 for calmodulin, as

A

Peptide 1 R-(Leu-Lys-Lys-Leu-Leu-Lys-Leu)$_1$

Peptide 2 R-(Leu-Lys-Lys-Leu-Leu-Lys-Leu)$_2$ **R** =

Peptide 3 R-(Leu-Glu-Glu-Leu-Leu-Glu-Leu)$_2$

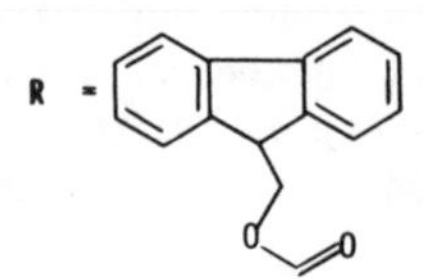

B

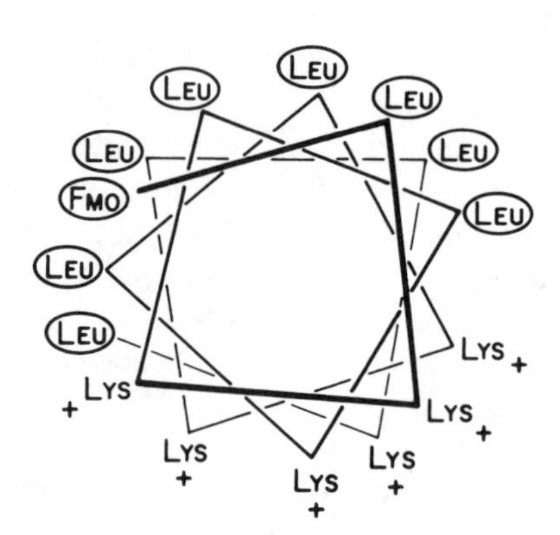

Fig. 2. *A) The amino acid sequences of peptides 1–3. B) An axial helical projection of peptide 2. Note that all the hydrophobic leucyl residues (circled) are segregated from the hydrophilic lysyl residues on opposite sides of the wheel.*

it would be energetically less costly to fold peptide 2 than peptide 1 into a helical conformation. Thus, less binding energy would be spent in folding peptide 2 resulting in enhanced binding of this peptide. Indeed, both peptides 1 and 2 were experimentally shown to bind calmodulin with dissociation constants of 200×10^{-9} and 3×10^{-9} M, respectively. In contrast, the negatively charged analog of peptide 2, peptide 3, failed to bind calmodulin at all, even at micromolar concentrations, illustrating the role of charge in complex formation. Calmodulin had only one high-affinity site for binding peptides 1 and 2, although there were several weaker nonspecific peptide-binding sites. Formation of the high-affinity complex was calcium-dependent and competitive with respect to target-enzyme binding. Finally, it could be shown by circular dichroism spectroscopy that peptide 2 bound to calmodulin in an α-helical conformation. In conjunction with the above mentioned studies with natural calmodulin-binding peptides, these studies with idealized peptides placed on firm experimental ground the hypothesis that a basic amphiphilic α-helix is a structural feature underlying the calmodulin-binding properties of a variety of peptides.

PREDICTING THE STRUCTURE OF CALMODULIN–PEPTIDE COMPLEXES

The above strategy for the design of calmodulin-binding peptides was conceptual in nature, based on the premise that a basic amphiphilic α-helix was a structural feature important for the binding of peptide to calmodulin. It would be desirable to create even more potent and specific inhibitors and increase the generality of the approach so that it might be globally applicable to other peptide/protein complexes. This requires an expansion of the design principles to include not only optimization of amphiphilicity but also optimization of the geometric and electrostatic complementarity of the peptide and protein. This necessitates a knowledge of the structure of the target protein, calmodulin. We therefore set out to predict the structure of calmodulin based on a consideration of the known crystal structures of two proteins with sequence homology to calmodulin [12]. Previously, methodology of predicting the structure of a protein based on sequence homology with another protein of known structure had been established [13]. The modeling procedure involves: 1) aligning the sequences of the proteins of known and unknown tertiary structure to maximize homology, and 2) building computer or physical models in which the most homologous regions occupy geometrically equivalent positions. Regions which display low levels of sequence homology including insertions and deletions are assigned conformations based on intuition, calculation of secondary structures based on prediction schemes, and model building. The geometry of the resulting structure can then be further optimized by energy minimization.

There were two possible calcium-binding proteins with known crystal structure upon which a model for calmodulin could be patterned: carp parvalbumin (CPV) and intestinal calcium-binding protein (ICB). Calmodulin provides an excellent system for testing this modeling approach, since its structure was unknown while the modeling was in progress but crystal structures were under investigation in several laboratories; a low-resolution structure has just been published [14]. Also, while calmodulin shares the function of binding calcium with the two model structure proteins, calmodulin is unique among this class of calcium-binding proteins in having the additional function of binding peptides, hydrophobic drugs, and target enzymes. If the resultant model structure revealed accessible regions of hydrophobicity flanked by acidic residues, i.e., calmodulin inhibitor–binding sites, it would not only validate the predictive model approach to structure elucidation but would provide additional clues as to the rational design of future calmodulin inhibitors with even greater affinities.

The structural organization of calcium-binding proteins was first elucidated by Kretsinger [15], who predicted that the folding motif found in parvalbumin would be general to a number of proteins including calmodulin and troponin C. Two supersecondary structural units repeated in tandem, designated EF hands by Kretezinger, comprise the calcium-binding portion of these proteins. Each EF hand contains a 12-residue loop which positions main-chain and side-chain oxygen atoms in an octahedral geometry appropriate for binding a single calcium ion. The loop is flanked on either side by sequences capable of forming α-helices (Fig. 3). Two E–F hands which are

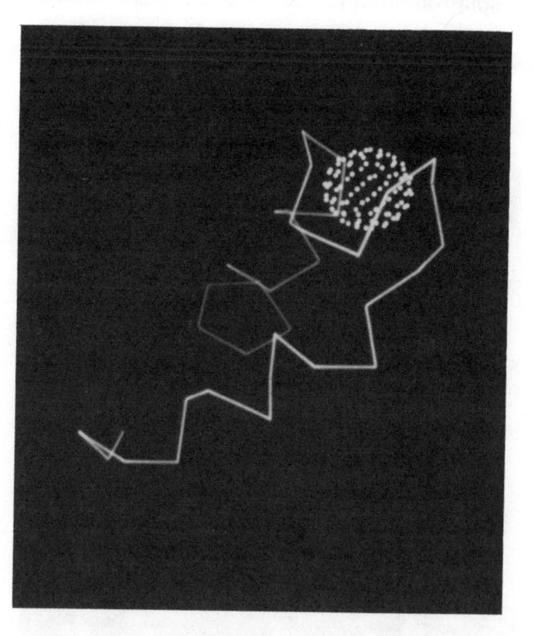

Fig. 3. *The structure of an EF hand. The positions of the Cα atoms are connected by lines. Calcium ions are represented by dots depicting a sphere with a radius 1.8 times that of the van der Waals radius of* Ca^{2+}. *(Reproduced in color on p. 343.)*

TABLE I. Amino Acid Sequences of Calmodulin, Parvalbumin, and Intestinal Calcium-Binding Protein

		Helix E	Loop		Helix F
			* * * # *		
Calmodulin (bovine brain)	ADQLTEEQIAE	FKEAFSLF	DKDGNGTITTKE	LGTVMRSL	GQNPTEA$_{46}$
	$_{47}$E	LQDMINEV	DADGNGTIDFPE	FLTMMARK	MKDTDSEE$_{83}$
	$_{84}$E	IREAFRVF	DKDGNGYISAAE	LRHVMTNL	GEKLTDE$_{119}$
	$_{120}$E	VDEMIREA	NIDGDGEVNYEE	FVQMMTAK	148
Parvalbumin	$_{40}$ADD	VKKAFAII	DQDKSGFIEEDE	LKLFLQNF	KADARALTDG$_{80}$
	$_{81}$E	TKTFLKAG	DSDGDGKIGVDE	FTALVKA	108
Intestinal calcium-binding protein	$_{1}$KSPEE	LKGIFEKY	AKEGLPQLSKEE	LKLLLQTE	FPSLLKGPS$_{44}$
	$_{45}$T	LDELFEEL	DKNGDGEVSPEE	FQVLVKKI	SQ$_{75}$

The amino acid sequences are organized into regions first hypothesized by Kretsinger to form the helices and calcium-binding loops that collectively form an E–F hand. The residues aligned with the *'s indicate the oxygen-containing side chains that are ligands for the Ca^{2+} ion; the residues aligned with the # indicate residues that contribute a backbone carbonyl for complexation with calcium.

positioned adjacently in a protein sequence pack together in a three-dimensionally "good fit" to form a compact globular protein domain (Fig. 4). Calmodulin contains two internally homologous domains, which in turn display approximately 20% sequence homology to either ICB or CPV (Table I). Thus, calmodulin should contain two "E–F dyad" domains, similar to the prototypical example in Figure 4, each of which should bind two calcium ions.

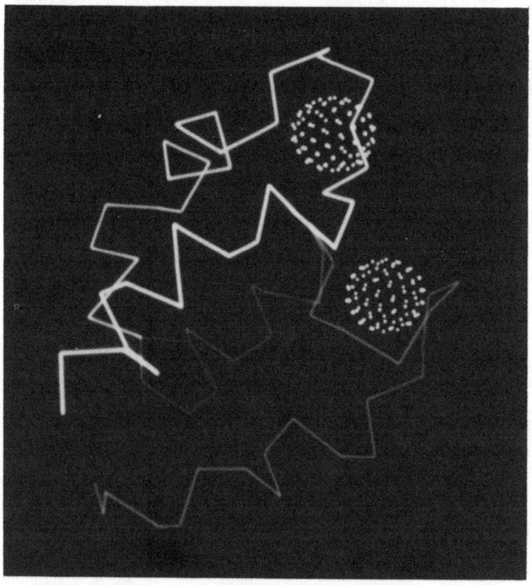

Fig. 4. *Two E-F hands docked together to form an E-F dyad. (Reproduced in color on p. 343)*

Computer models for the individual domains of calmodulin were constructed using interactive computer graphics to replace the side chains of ICB or CPV with the appropriate side chains of calmodulin, while maintaining the positions of the main chain atoms invariant. Since this procedure introduced numerous bad contacts between side chains, the geometry of the resulting structure was optimized by energy minimization. Prediction of the linking region which connects the two domains added a second layer of complexity to the modeling. Fortunately, known physical properties of calmodulin suggested a possible conformational solution to this problem. Calmodulin elutes from size exclusion columns with an anomalously high Stokes radius, suggestive of a nonspherical, bilobed structure. Additional support for a bilobed configuration in which the individual domains are held at some distance from one another came from NMR studies [16] on intact calmodulin and its isolated domains obtained by limited proteolysis. A simple method for connecting the two domains in such a manner is by assigning a helical conformation to the linker region. In fact, the corresponding calmodulin sequence has a high potential to form an α-

helix [17] (Chou Fasman $\langle P_{\alpha} \rangle$ = 1.16), and contains no helix-breaking residues. In addition, the linker sequence's position between two helices obviates the need for separate helix initiation, which is energetically the most costly step in helix formation. Thus, it appeared most likely that this sequence adopted a helical conformation, and the ICB- and CPV-based models for calmodulin were obtained by joining the individual domains with a helical linker (Figs. 5A, B).

The availability of coordinates for both ICB and CPV provided two independent starting points for prediction of the structure of calmodulin-dependent's two domains. If the energy minimization proceeded to a global minimum, then the same structure should result, irrespective of the starting structure. Unfortunately, the energy minimization program appeared to find only local minima in the potential energy surface, as the ICB- and CPV-based structures were nonidentical and bore

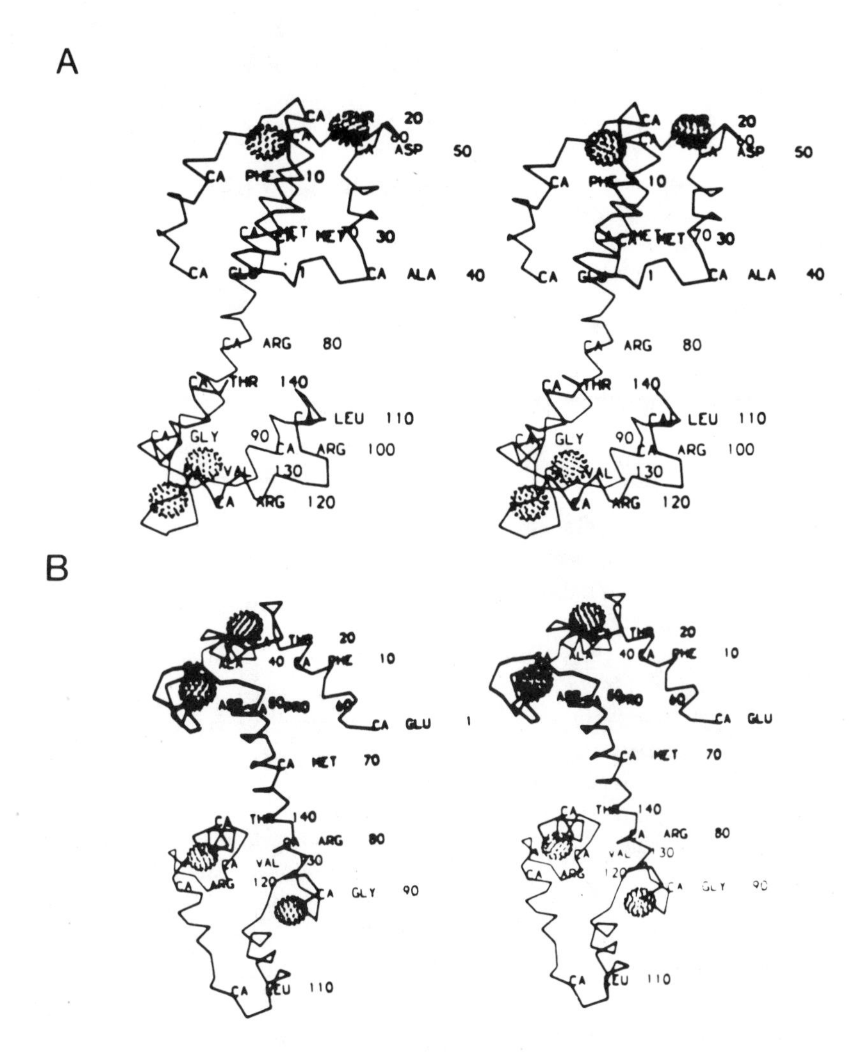

Fig. 5. *Stereo drawings of the Cα backbone for A) the ICB-based model and B) the CPV-based model for calmodulin. The numbers of residues used in these figures correlates with the actual number in calmodulin minus 6.*

closer resemblances to their parent proteins than to one another. The structures shown in Figures 5a, b should therefore be viewed as working models; their degrees of similarity (approximately 3 Å) place limits on the accuracy of the predictions and any conclusions derived therefrom.

Even with these limitations, we can begin to predict the structural features present in the working model of calmodulin that are responsible for the tight binding of basic, amphiphilic helical peptide inhibitors. Again, the known physical and chemical properties of calmodulin were used to guide the modeling. Previous studies comparing the binding of peptide inhibitors to intact calmodulin and the two EF-dyads produced by limited proteolysis had shown that the C-terminal E–F dyad had a greater affinity for mastoparan X than the N-terminal counterpart [6]. In addition, the C-terminal dyad could effectively activate the target enzyme, phosphorylase kinase [18]; therefore, attention was focused on this domain. As discussed in detail above, tight binding of peptides requires multiple positive charges in their structures, so the electrostatic surface of the models for the second (C-terminal) domains of calmodulin were calculated and examined for highly negatively charged regions which might strongly attract such peptides (Fig. 6). One region along the third helix of this domain had a particularly high concentration of negatively charged residues and provided an excellent candidate for the peptide-binding site. Since peptide binding requires hydrophobic as well as electrostatic interactions, it was particularly interesting to note that Val[121] and Ile[125], as well as several residues in the adjacent C-terminal helix (Phe[141]–Val–Gln–Met–Met–Thr–Ala–Lys[148]), are partially exposed to solvent. Reasonable hydrophobic and electrostatic interactions between this site and the basic, amphiphilic α-helical peptides appeared possible, provided the helix bound with its helical axis inclined by approximately 20° with respect to the negatively charged third helix of the second domain (Figure 7).

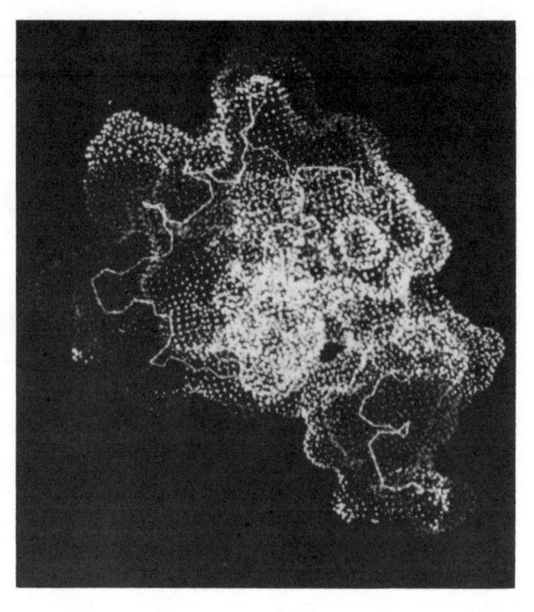

Fig. 6. *Electrostatic potential surface for the second domain of the ICB-based model for calmodulin. Contour levels: red, $V < -10$ kcal/mol; orange, -10 kcal/mol $< V < -3$ kcal/mol; green, -3 kcal/mol $< V < +3$ kcal/mol; blue-green, $+3$ kcal/mol $< V < +10$ kcal/mol; blue, $V > +10$ kcal/mol. The backbone bonds of the structure are shown in white. The second F helix of this domain is in the center with its axis vertically oriented, the second calcium-binding loop is in the upper left, and the second E helix runs along the far left of the diagram. Note the very negative electric potential in the region surrounding the calcium binding loop as well as along the second E helix. (Reproduced in color on p. 343.)*

TESTING THE VALIDITY OF THE MODELS

The motivation for generating a computer model of calmodulin was not solely to reproduce its structure exactly, but rather to obtain an understanding of the principles underlying calmodulin's interactions with peptides. Nevertheless, it is of interest to compare the predicted structure with the very recently reported crystal structure [14]. The most striking feature of the experimentally determined structure is that it is bilobed, with an α-helix connecting the two E–F dyads. The interheli-

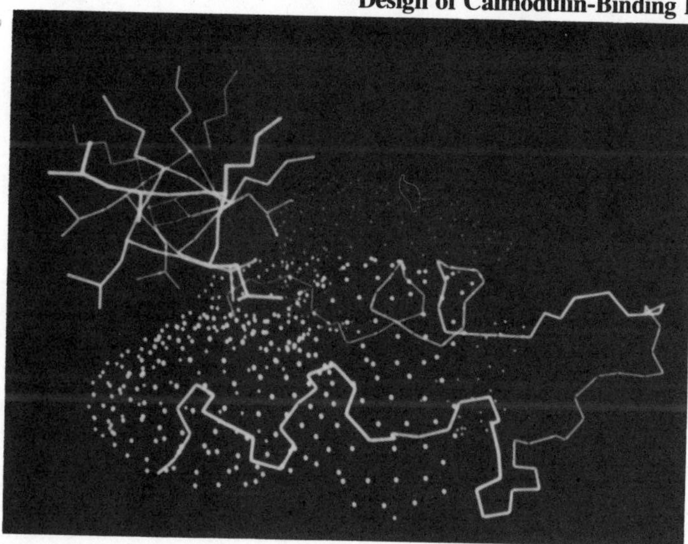

Fig. 7. *Proposed docking orientation for the basic amphiphilic peptide (Leu–Lys–Lys–Leu–Leu–Lys–Leu)$_2$ to the second domain of the ICB-based model for calmodulin. Hydrophobic residues have been colored green, basic residues blue, and acidic residues red for both the peptide and the surface. The surface shown is a solvent accessible surface generated only for those protein residues in the binding site that are capable of contributing to stabilization of the peptide/protein complex. Note that in this orientation there is the potential for both hydrophobic (green–green) and ionic (blue–red) interactions between the peptide and calmodulin. (Reproduced in color on p. 343.)*

cal packing angles found in the calmodulin structure were approximately 95°, close to the value of 100° for the CPV-based model, but somewhat further from the value of 120° observed in the ICB-based model. When the observed structure has been refined to a higher resolution, and coordinates become available, it should be possible to more closely analyze and ultimately resolve the differences between the ICB- and CPV-based models and the actual structure of calmodulin.

In the last year, experimental evidence has accumulated which supports the conformational features proposed by the synthetic- and computer-modeling approaches. These studies represent two lines of research: utilization of affinity labels aimed at modifying specific residues of calmodulin, and synthesis of additional high-affinity peptides based on the predicted model of calmodulin. Jackson and Puett [19] have recently reported specific acylation of a single lysine residue located in each of the two E–F dyads. Lys^{148} was the most reactive residue to labeling by a fluorenyl-based tricyclic compound, and labeling was inhibited by trifluoperazine. It was postulated that this lysine residue may be adjacent to the hydrophobic calmodulin patch involved in hydrophobic drug binding. Indeed, this lysyl residue is in proximity with the postulated peptide-binding site in the models of calmodulin's second dyad.

Examination of the structural models for calmodulin has made it possible to synthesize peptide inhibitors with even greater affinities for calmodulin than peptide 2 described above [20]. Scrutiny of the model in Figure 7 suggested that the hydrophobic patch on the C-terminal dyad was shorter than the helical length of peptide 2. Immediately adjacent to this patch were several acidic residues which were proximal to leucyl side chains near the

TABLE II. Calmodulin-Binding Peptides

Peptide	K_{diss} (nm)	Relative affinity
F*LKKLLKL	200.0	0.015
F*LKKLLKLLKKLLKL	3.0	1.0
KLWKKLLKLLKKLLKLG	0.4	7.5
LKWKKLLKLLKKLLKLG	0.2	15.0

F* = FMOC, N^{α}-9-fluorenylmethoxy-carboxyl.

N-terminus of the peptide. It is likely, therefore, that if a lysine were introduced onto the otherwise uninterrupted hydrophobic face of the peptide helix, enhanced affinity of peptide for protein should result. As a consequence of this and additional considerations, we have recently synthesized two peptides that bind calmodulin with dissociation constants that are subnanomolar (Table II). The systematic variation of the charge of other residues in the amphiphilic peptides should allow the precise definition of the number and types of electrostatic and hydrophobic binding interactions involved.

THE FUTURE

The intent of this review was to highlight some theoretical and experimental approaches to the study of the interactions of an effector molecule with its target protein. Clearly, the premise for studying the interactions of natural and synthetic amphiphilic peptides with calmodulin is that these interactions are mimics of the interaction of calmodulin with target enzymes. Very recent evidence has given this premise credence: A 27-residue peptide has been isolated from a cyanogen bromide digest of myosin light chain kinase that appears to comprise the calmodulin-binding domain of this target enzyme [21]. The peptide binds calmodulin tightly, contains approximately 30% basic residues, and can be shown to be capable of forming a 15-residue amphiphilic α-helix using the Edmondson helical wheel projection. Further studies will reveal whether the basic amphiphilic α-helix is common to the calmodulin-binding domain of other target enzymes. If so, this common feature invites speculation as to the molecular evolution of proteins with disparate functions but common effector modulators.

The second future direction is perhaps even more exciting in that calmodulin binding peptides provide an exquisite model system for the study of how other amphiphilic peptides bind to their receptors. Just as the ability of widely different peptides to bind calmodulin was shown to be related to their ability to adopt amphiphilic helical conformations, it is becoming increasingly evident that this type of conformation is an important feature required for the binding of certain peptide hormones to their biological receptors [22]. Hormone receptors, which contain lipid and carbohydrate components in addition to the protein core, are notoriously difficult to purify; therefore, experimental difficulties limit the kinds of structural questions that can be answered about their interaction with hormone ligand. On the other hand, the known crystal structure of calmodulin, the availability of cloned calmodulin genes [23], and the growing base of information concerning important structural features of ligands and calmodulin itself provide the model necessary to define general rules governing protein/protein interactions. The synthetic peptide approach such as described here coupled with production of altered calmodulins through site-directed mutagenesis can serve as tools by which such a model will be constructed.

REFERENCES

1. Klee CB, Vanaman TC (1982): Adv Prot Chem 35:213–321.
2. Weiss B, Prozialeck WC, Sellinger-Barnette M, Winkler JD, and Schechter L (1985) In Ho Hidaka, Hartshore DJ (eds): "Calmodulin Antagonists and Cellular Physiology." New York: Academic Press.
3. Marshak DR, Lukas TJ, Watterson DM (1985): Biochem 24:144–150.
4. Norman JA, Drummond AH (1980): Mol Pharmacol 16:1089–1094.

5. Cox JA, Comte M, Fitton JE, DeGrado WF (1985): J Biol Chem 260:2527–2534.
6. Malencik DA, Anderson SR (1984): Biochemistry 23:2420–2428.
7. Giedroc DP, Keravis TM, Staros JV, et al (1985): Biochemistry 24:1203–1211.
8. Malencik DA, Anderson SR (1982): Biochemistry 21:3480–3486.
9. Giedroc DP, Ling N, Puett D (1983) Biochemistry 22:5584–5591.
10. McDowell L, Sanyal G, Prendergast FG (1985): Biochemistry 24:2979–2984.
11. Taylor JW, Miller RJ, Kaiser ET (1983): J Biol Chem 258:4464–4471.
12. O'Neil KT, DeGrado WF (1985): Proc Natl Acad Sci (USA) 82:4954–4958.
13. Greer J (1981): J Mol Biol 153:1027–1042.
14. Babu YS, Sack JS, Greenhough TJ, Bugg CE, Means AR, Cook WJ (1985): Nature 315:37–40.
15. Kretsinger RC (1976): Annu Rev Biochem 45:239–266.
16. Aulabaugh A, Niemczura WP, Blundell TL, et al (1984): Eur J Biochem 143:409–418.
17. Chou PY, Fasman GD (1978): Adv Enzymol 47:45–149.
18. Newton DL, Oldewurtel MD, Krinks MH, Shiloach J, Klee, CB (1984): J Biol Chem 259:4419–4426.
19. Jackson AE, Puett D (1984): J Biol Chem 23: 14985–14992.
20. DeGrado WF, Prendergast FG, Wolfe HR Jr, Cox JA (1985): J Cell Biochem 29:83–93.
21. Blumenthal DK, Takio K, Edelman AM, Charbonneau H, Titani K, Walsh KA and Krebs EG (1985): Proc Natl Acad Sci USA 82:3187–3191.
22. Kaiser ET, and Kézdy FJ (1984): Science 223:249–254.
23. Putkey JA, Slaughter GR, Means AR (1985): J Biol Chem 260:4704–4712.

Protein Engineering, pages 213–218

19

Thermal Stability of Proteins

Alexander M. Klibanov and Tim J. Ahern

Department of Applied Biological Sciences, Massachusetts Institute of Technology, Cambridge, Massachusetts 02139

INTRODUCTION

Conditions resulting in protein inactivation are many and all too familiar to those preoccupied with the task of maintaining the biological activity of proteins. The list includes extremes of pH, changes in solvent, the presence of proteases or strong denaturants (such as guanidine hydrochloride and urea), and high temperatures. The mechanisms of thermoinactivation of enzymes are of particular interest because of the benefits of using enzymes as practical catalysts at elevated temperatures, e.g., reduced contamination and viscosity combined with enhanced rates of reaction. The food industry now employs enzymes in large-scale processes at temperatures of the order of 55–65°C (glucoamylase and glucose isomerase) and even as high as 85–110°C (α-amylases). New discoveries such as the remarkable thermostability and high activity of lipase in organic solvents at 100°C [1] in principle extend the feasibility of enzyme catalysis at high temperatures to other industries (e.g., chemical and petrochemical) as well. These applications and others like them should be incentive enough for biotechnologists to familiarize themselves with the phenomenon of heat inactivation of enzymes and its prevention.

EFFECTS OF HEAT ON PROTEIN

Proteins are constantly in motion (Fig. 1). Even at subzero temperatures and in the crystalline state, their constituent atoms undergo vibrations, rotations, and even small translations of the order of 0.2–0.5 Å [2]. At higher temperatures, within the range at which most enzymes exhibit their optimal activity (0–60°C), reversible displacements of whole segments of protein structure, or "breathing," are observed in aqueous solutions. Also, concerted motions required for substrate binding, catalysis and product release are normal phenomena for an active enzyme [3]. Although the protein has some freedom of movement under these conditions, the predominant conformations are dictated by a complex balance of intramolecular, noncovalent interactions: e.g., hydrogen bonds, hydrophobic, ionic, and Van der Waals interactions. At elevated temperatures (45–85°C), radical cooperative intramolecular motions take place: At the temperature at which the noncovalent forces maintaining the native structure of a protein can no longer prevail against the increase in entropy, the protein loses most of its ordered secondary and tertiary structure and is said to have denatured [4–8]. Being less hampered by

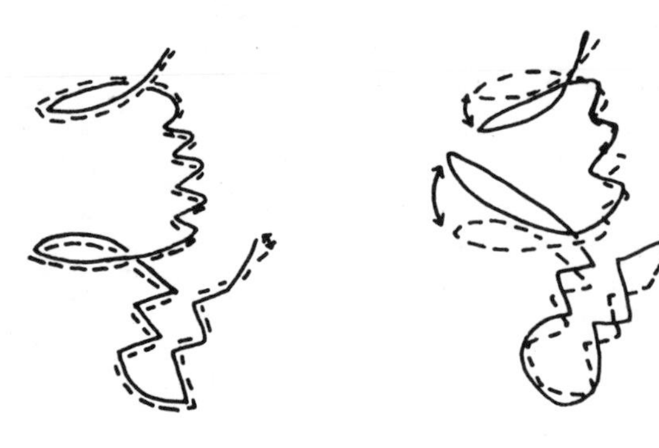
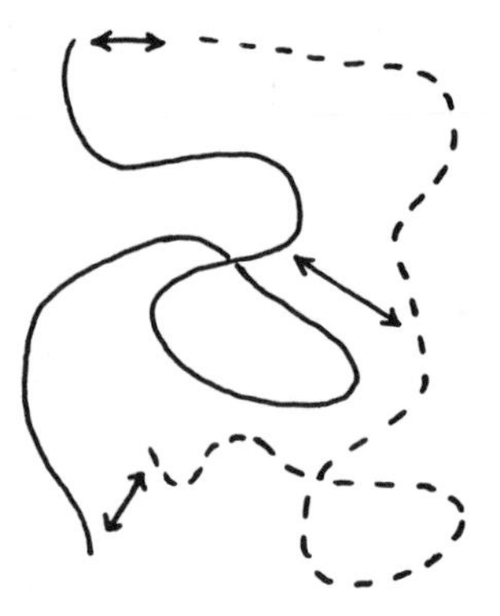

Fig. 1. *The effect of temperature on the degree of motion and order in proteins.*

spatial constraints, the amino acid residues move more freely and acquire more of the characteristics of free amino acids in solution, although they are still held together in a polypeptide chain. In the case of enzymes, denaturation inevitably results in the dispersal of the residues comprising the active site and, as a consequence, catalytic activity is lost.

Although an enzyme above its midpoint temperature (T_m) for thermal denaturation is no longer active, the transition is usually reversible for globular proteins: If an enzyme is subsequently cooled below its T_m, it refolds to the native structure and regains catalytic activity. If heating is prolonged, however, not all of the initial activity returns upon cooling, and eventually the entire enzymatic activity is lost. Thermal inactivation, therefore, consists of two parts: *reversible denaturation* (partial unfolding) and subsequent *irreversible thermoinactivation*. An understanding of these processes is an essential prerequisite for a rational strategy of enzyme stabilization at elevated temperature and should facilitate progress in the field of protein engineering [9,10], in particular, in the design of "superstable" enzymes. Recent studies involving the monomeric, single-chained enzymes bovine pancreatic ribonuclease A [11] and hen egg white lysozyme [12] have established the nature of irreversible thermoinactivation of enzymes at 90–100°C within the pH range of relevance to enzymatic catalysis (pH 4–8).

PROCESSES RESPONSIBLE FOR IRREVERSIBLE ENZYME THERMOINACTIVATION

In principle, the processes resulting in the irreversibility of inactivation can be either conformational or covalent. The mechanisms resulting in the irreversible loss of enzyme activity at high temperature are illustrated in Figure 2.

Conformational

Thermally unfolded molecules can undergo two types of noncovalent transformations: poly-

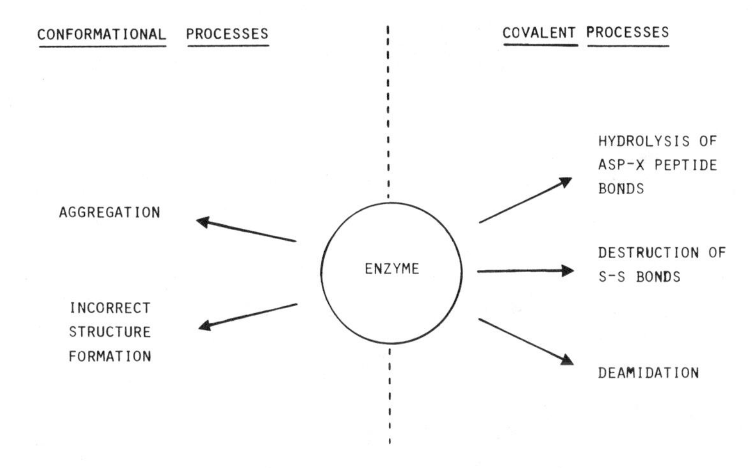

Fig. 2. *The general mechanisms of irreversible thermoinactivation of enzymes.*

molecular (aggregation) and monomolecular (formation of incorrect structures). Upon unfolding, hydrophobic regions of the protein molecule which heretofore were predominantly located in the interior become exposed to the solvent. Such exposure is thermodynamically unfavorable; one of nature's ways to reduce the system's free energy is for the unfolded molecules to interact with one another and thereby rebury the hydrophobic groups. Such an interaction will result in protein aggregation; this phenomenon is well known to anyone who has worked with proteins (or boiled an egg, for that matter).

Concentrated solutions of most enzymes aggregate upon heating. However, in very dilute solutions of enzymes, aggregation is insignificant; yet thermal inactivation nevertheless takes place, and some of the loss of activity is still due to conformational processes [11,12]. This can be explained by the fact that there is more than one way to fold a protein: Upon denaturation the tendency to bury hydrophobic residues—combined with the protein's freedom to sample many conformational states—results in new, kinetically or thermodynamically stable structures which are catalytically inactive. Even after cooling, these "incorrect, scrambled" structures remain because a high kinetic barrier prevents spontaneous refolding to the native conformation. Proteins more readily undergo such transformations if the pH of solution is near their isoelectric point: The reduced net charge on the protein eliminates repulsive interactions that would otherwise hinder incorrect folding of residues upon each other. Incorrect structure formation is arguably "pseudo-irreversible" since the activity lost should return after cooling if one is prepared to wait long enough, but to paraphrase Ephraim Racker, when a process shows no sign of reversal by the time bacteria start growing in the experimental samples, we should feel free to call the reaction irreversible.

Covalent

Conformational processes alone cannot account for the irreversible loss of activity of enzymes, no matter at what pH they are studied. Hence covalent mechanisms are also at work. The integrity of the polypeptide chain is disrupted even under mildly acidic pH: Peptide bond hydrolysis adjacent to aspartic acid residues occurs at rates comparable to those of irreversible thermoinactivation in lysozyme and ribonuclease at pH 4. Measurements of the appearance of new amino and carboxyl termini in the enzymes revealed that hydrolysis of the peptide bonds occurs predominantly at the carboxyl termini of aspartic acid residues. Since aspartic acid has a high frequency in proteins, scissions of the polypeptide chain adjacent to it clearly pose a threat to the activity of enzymes in general.

Another fundamental structural feature shared by many proteins, i.e., disulfide bonds, have rates of destruction under mildly alkaline conditions that may be comparable to the observed rates of thermoinactivation. The OH^--catalyzed β-elimination of cystine that is observed results in the disruption of intramolecular covalent cross-links: These normally play a significant role in maintaining a protein's conformation.

The final process leading to irreversible thermoinactivation of enzymes involves the deamidation of asparagine and possibly glutamine residues. The specific activity of lysozyme decreases with successive deamidations; this process, which occurs at each pH, together with hydrolysis of peptide bonds at pH 4 and disulfide destruction and conformational processes at pH 8, accounts for the total loss of activity of that enzyme at 100°C. In the case of ribonuclease, irreversible inactivation at 90°C is caused at pH 4 by hydrolysis and at pH 8 by a combination of formation of incorrect structures and destruction of S-S bonds. Preliminary results indicate that deamidation of ribonuclease is an important mechanism of its thermoinactivation at pH 6; the same is also true of the dimeric enzyme yeast triosephosphate isomerase.

Since the processes outlined above are of a general nature, they demarcate an upper limit of thermal stability of enzymes, although the contribution of each will vary with the nature of the enzyme and the reactions it catalyzes:

1) Conformational mechanisms of inactivation most likely predominate near the enzyme's isoelectric point.

2) Disulfide bonds undergo destruction even at mildly alkaline pH.

3) Enzymes containing a high percentage of asparagine residues and having electrostatic affinity for negatively charged substrates (e.g., lysozyme) will be adversely affected by deamidation, whereas enzymes whose activity is unaffected by increasing surface electronegativity or whose composition is low in asparagine should be relatively more stable at all pHs.

4) Within the lower pH range, the rate of irreversible thermoinactivation should increase with the frequency of aspartic acid residues.

These findings first point out the inappropriateness of certain strategies for prolonging the activity of active enzymes at high temperatures. Any enzyme engineering that results in a shift of the pH at which an enzyme is to be used nearer its isoelectric point, or an increase in disulfide bonds, asparagines, or aspartic acids, can now be seen to have predictable drawbacks.

As a result of our findings, several approaches to enzyme thermostabilization are now apparent:

Elimination of water from the system. Since water is a reactant in peptide hydrolysis and deamidation, the mediator of alkaline β-elimination of cystine, and the solvent facilitating the mobility leading to incorrect structure formation, it is logical that substitution of another medium for water could stabilize an enzyme with respect to these processes. As indicated by the half-lives of irreversible thermoinactivation of lysozyme at 100°C in various media depcited in Table I, the enzyme is stabilized by several orders of magnitude when incubated as a dry powder or in organic solvents such as cyclohexane, hexadecane, and 1-heptanol. This rationale also explains why the stability of lipase at 100°C in organic media decreases with increasing water content [1]. Since not only lipases [13] but a number of other enzymes including tyrosinase, peroxidase, xanthine oxidase, and cholesterol oxidase also exhibit activity in nearly anhydrous organic media, the elimination of water may prove to be a practical strategy of thermostabilization.

Protein engineering. Since many of the elucidated mechanisms involve specific amino acid residues (Asp, Asn, Cys, and to lesser

TABLE I. The Effect of Solvent on the Stability of Lysozyme Against Irreversible Thermoinactivation at 100°C

Solvent[a]	$t_{1/2}$ (h)
Water	
pH 4[b]	1.42
pH 6[c]	0.17
pH 8[d]	0.01
1-heptanol	100
Hexadecane	120
Cyclohexane	140
As a dry powder	200

[a]Sealed vials containing 1 mg lysozyme (0.94% water) plus solvent (1 ml) were placed in a 100°C glycerol bath. Periodically aliquots were withdrawn, cooled, and assayed for the lysozyme activity toward *M. lysodeikticus* cells at pH 6.0 and 25°C. The water content in the organic solvents was less than 0.007%.
[b]Na acetate, 0.1 M
[c]Na cacodylate, 0.01 M
[d]Na phosphate, 0.1 M

extent, Gln), their replacement should greatly reduce the rate of irreversible thermoinactivation. For example, replacement of Asn, the predominant site of deamidation, with residues approximating its properties and size (e.g., Gln, Ile, or Thr) should stabilize enzymes whose activity is lost due to deamidation (like lysozyme, ribonuclease, or triosephosphate isomerase at pH 6). Similar strategies can be devised for enzymes whose inactivation is due primarily to hydrolysis of aspartic acid residues (like ribonuclease at pH 4) or destruction of disulfide bonds. The practical goal of catalysis by stabilized enzymes at elevated temperatures in aqueous media should be realized when such modifications are combined with established methods hindering *reversible* unfolding (e.g., intramolecular cross-linking or immobilization [14].

The irreversible processes responsible for thermoinactivation of purified enzymes may have significance for investigations of living organisms as well. Argos et al. [15] reported that enhanced thermostabilization of enzymes from thermophiles is the result of a great number of small stabilizing structural increments, one of which is substitution of Asp by Glu. Such a modification would be expected to decrease the rate of irreversible thermoinactivation in the lower range of physiological pH. In addition, deamidation of asparagines has been observed in vivo in over 50 enzymes [16] and has been cited as a symptom of protein aging [17] and a step in the process of protein turnover [18]. Gracy and co-workers [19] have shown that deamidation of the asparagines at the subunit interface of triosephosphate isomerase from mesophilic organisms results in increased susceptibility of the dimer to dissociation. By balancing requirements for protein turnover, thermostability and sites for post-translational modification (such as glycosylation at Asn), evolutionary selection may have optimized amino acid compositions to accommodate the various physiological and environmental constraints facing organisms.

Until now, this paper has concentrated on enzyme stabilization. In some commercially important instances, however, enzymes are already "too stable." For example, persistence of the proteolytic activity of microbial rennet renders it undesirable for cheese-making. Therefore, the ability to more rapidly thermoinactivate microbial rennet would facilitate its acceptance in such an application. The strategies described above need only be applied in reverse to bring about the desired thermo*de*stabilization [20].

SUMMARY

In this work we have discussed the general mechanisms of irreversible thermoinactivation of enzymes in the pH range of physiological and biotechnological interest. The processes elucidated are deamidation of Asn residues, hydrolysis of peptide bonds adjacent to Asp, formation of incorrect (scrambled) structures, and destruction of cystine residues. The contribution of each to overall irreversible thermoinactivation will depend on the solvent, the pH of incubation, and the nature of a given enzyme—e.g, its amino acid composition, isoelectric point, or electrostatic affinity for its substrate. However, due to the general nature of the mechanisms, they will take place in all proteins at high temperature and thus demarcate an upper limit of thermostability of proteins—a half-life of no more than a few hours at 100°C in water in the pH range of 4–8.

These findings provide a basis for rational strategies of enzyme thermostabilization: 1) substitution of organic media for water, and 2) site-specific mutagenesis aimed at replacing the "weak links," i.e., the amino acid residues implicated in irreversible thermoinactivation (Asp, Asn, Cys, and to a lesser extent, Gln). These approaches can be expected to result in decreased rates of irreversible thermoinactivation and thus more thermostable enzymes.

REFERENCES

1. Zaks A, Klibanov AM (1984): Enzymatic catalysis in organic media at 100°C. Science 224:1249–1251.
2. Frauenfelder H, Petsko GA, Tsernoglou D (1974): Temperature-dependent X-ray diffraction as a probe of protein structural dynamics. Nature 280:558–563.
3. Karplus M, McCammon JA (1981): The internal dynamics of globular proteins. CRC Crit Rev Biochem 9:293–349.
4. Kauzmann W (1959): Some factors in the interpretation of protein denaturation. Advan Protein Chem 14:1–63.
5. Tanford C (1968): Protein denaturation. Advan Protein Chem 23:121–282 and (1970) 24:1–95.
6. Lapanje S: "Physicochemical Aspects of Protein Denaturation." New York: John Wiley and Sons, 1978.
7. Privalov PL (1979): Stability of proteins. Advan Protein Chem 33:167–241.
8. Pfeil W (1981): The problem of stability of globular proteins. Molec Cell Biochem 40:3–28.
9. Ulmer KM (1983): Protein engineering. Science 219:666–671.
10. Perry LJ, Wetzel R (1984): Disulfide bond engineered into T4 lysozyme: stabilization of the protein toward thermal inactivation. Science 226:555–557.
11. Zale SE, Klibanov AM (1984): Mechanisms of irreversible thermoinactivation of enzymes. Ann NY Acad Sci 434:20–26.
12. Ahern TJ, Klibanov AM (1985): The mechanisms of irreversible enzyme thermoinactivation at 100°C. Science 228:1280–1284.
13. Zaks A, Klibanov AM (1985): Enzyme catalyzed processes in organic solvents. Proc Nat Acad Sci USA 82:3192–3196.
14. Klibanov AM (1983): Stabilization of enzymes against thermal inactivation. Adv Appl Microbiol 29:1–28.
15. Argos P, Rossman MG, Grau UM, Zuber H, Frank KG, Tratschin JD (1979): Thermal stability and protein structure. Biochem 18:5698–5703.
16. Robinson AB, Rudd CJ (1974): Deamidation of glutaminyl and asparaginyl residues in peptides and proteins. In Horecker BL, Stadtman ER (eds): "Current Topics in Cellular Regulation, vol 8" New York: Academic Press, pp 247–295.
17. Dreyfus JC, Kahn A, Schapira F (1978): Posttranslational modifications of proteins. ibid. vol 14, pp 243–297.
18. Robinson AB, McKerrow JH, Cary P (1970): Controlled deamidation of peptides and proteins: an experimental hazard and a possible biological timer. Proc Nat Acad Sci USA 66:753–757.
19. Yuan PM, Talent JM, Gracy RW (1981): Molecular basis for the accumulation of acidic isozymes of triosephosphate isomerase on aging. Mechanisms of Ageing and Development 17:151–162.
20. Hubble J and Mann P (1984): Destabilization of microbial rennet. Biotechnol Lett 6:341–344.

IV
PURPOSELY MODIFIED PROTEINS AND THEIR PROPERTIES

Protein Engineering, pages 221–224

Overview: Kinetic Aspects of Purposely Modified Proteins

Alan R. Fersht

One of the dreams of the classical enzymologist was to be able to alter the amino acid sequence of an enzyme at will. And now, through the advent of protein engineering, this dream has become reality. Given the gene of an enzyme and a means for its expression, oligodeoxynucleotide-directed mutagenesis may be used to modify the structure of the enzyme with ease. Further, given the three-dimensional structure of the enzyme, the mutations may be rationally engineered and analyzed. The ultimate goal is to use the methods to design and construct enzymes of novel properties to catalyze novel reactions or to improve existing enzymes. To be able to do this requires understanding the relationship between structure and activity of enzymes and the relationship between structure, folding, and stability. Protein engineering is at present being used, on the one hand, for studying the rules of folding of proteins and, on the other, the basis of enzyme catalysis to lay the ground rules for future enzyme design.

REQUIREMENTS OF SYSTEMS FOR STUDYING FOLDING OR CATALYSIS

The requirements of systems for studying folding or catalysis overlap to some extent; e.g., both require high-resolution crystal structures of the proteins and ready expression of their genes. But there are clear priorities for each. For protein folding, the basic requirement is that the protein have a *reversible* transition between folded and unfolded states so that thermodynamic measurements may be made. A small protein ($M_r < 15{,}000$) is probably essential because these are amenable to NMR studies and very high-resolution protein crystallography. This latter characteristic is also useful for the study of enzyme catalysis, but there are other features that may dominate the choice. For example, many of the most interesting examples of enzyme catalysis involve larger enzymes that catalyze complex multisubstrate reactions. Further, many enzymes are oligomers which have interesting subunit interactions, and oligomeric enzymes are of necessity larger than monomeric enzymes.

The overriding factor in studying enzyme catalysis by comparing engineered enzymes is the ability to measure accurately *absolute* rate constants; i.e., rate constants which are independent of enzyme concentration. In conventional steady-state enzyme kinetics, the quantity that is usually measured is the initial rate of reaction, v. This is proportional to $[E]_0$, the concentration of active enzyme in solution. $[E]_0$ is not usually known with accuracy, since enzymes are rarely isolated with the all the protein present being active enzyme. Because of this, accurate steady-state kinetics are normalized by comparison with control reactions under standard conditions. Normalization cannot be done when comparing a mutant enzyme with a wild-type enzyme, because two different enzymes are being compared and not just different samples of the same enzyme. The technique of active-site titration must instead be used. In this, the concentration of active enzyme is determined directly by measuring the amount of a covalently bound or tightly bound intermediate in the reaction [1]. An alternative procedure is to measure the rate constants by pre-steady-state kinetics, as these involve exponential changes that can be designed to be independent of enzyme concentration. Both active-site titration and the application of pre-steady kinetics require that an intermediate accumulates during the reaction. Examples may be found in this volume in the reactions of the tyrosyl-tRNA synthetase [2] and trypsin [3]. The techniques can be extrapolated to studying enzymes, such as dihydrofolate reductase, where the enzyme can be titrated with a tightly bound inhibitor (methotrexate) and the kinetics of binding of NADPH may be monitored [4].

Analysis of catalysis also requires a detailed knowledge of the interactions between the enzyme and substrate, preferably from the crystal structure of an enzyme–substrate complex from X-ray diffraction. One particularly favorable case is that of the tyrosyl-tRNA synthetase, where the enzyme-bound tyrosyl adenylate complex is stable in the absence of tRNA. This has afforded the direct determination of the structure of the complex by X-ray diffraction.

IMMEDIATE GOALS AND STRATEGIES

The long-term goal of designing new proteins de novo is not yet achievable because the laws governing the folding of proteins are largely unknown. But one of the first aims of protein engineering is to investigate protein folding experimentally by preparing mutant enzymes of different thermodynamic and kinetic properties of folding. A promising system is staphylococcal nuclease, described by Shortle in this volume.

What we can do at present is to weave new themes on existing protein structures by making small changes in proteins of known three-dimensional structure. Fortunately, this approach is very productive because it allows the answering of many basic questions in enzymology concerning structure–activity relationships that previously were inaccessible to experiment. At the minimum level of structural change, we can mutate a single side chain in the enzyme. Then, providing nothing radical is done, such as packing a too-large side chain in a confined area or modifying a crucial structural residue, it is likely that the protein structure will be perturbed only slightly. Thus, the catalytic importance of individual residues may be probed. At the basic level, groups that have been postulated to be catalytically important may be removed and the radical effect on catalysis measured. A good example is the modification of Asp-102 in trypsin. At a more subtle level, amino acid side chains that are not so obviously involved in catalysis but just appear to be involved in binding the substrates may be modified to give enzymes of slightly changed activities. Measurements on the modified enzymes give the interaction energies of the side chains and how they are used in catalysis. This approach is exemplified by the studies on the tyrosyl-tRNA synthetase.

Further, many of the goals of industrial protein engineering involve the modification of the properties of existing enzymes, and this may be possible by small modifications of the parent structure. For example, it could well be desirable to alter the kinetic properties k_{cat} and K_M; the specificity; the pH optimum; the temperature stability and optimum; the stability in the presence of chemical reagents; and the isoelectric point of an enzyme. It appears likely some of these goals could be achieved just by simple substitution of amino acid residues.

SOME ACHIEVEMENTS

Industrial Protein Engineering

Experiments on subtilisin by Wells and colleagues reported in this section [5] have already demonstrated the importance of protein engineering in improving the properties of enzymes widely used in industry. For example, they have shown how the stability of the enzyme towards oxidation can be radically improved by the substitution of the oxidation-sensitive residue, methionine. The temperature stability of T4 lysozyme has been increased by engineering a disulphide bond to form an intrachain link between two cysteine residues [6]. Such bonds have been rationally engineered into subtilisin but, as reported by the group from Genex at the meeting, none of these increase the thermostability, despite the favorable geometries. The Imperial College group have recently shown how the pH dependence of subtilisin may be altered by changing surface charges [7]. This procedure should be quite general for perturbing the pK_a-values of catalytic groups in enzymes, and also allows the calculation of dielectric constants of regions of proteins.

Wells and colleagues have also shown how the specificity of subtilisin may be changed in a rational manner.

Basic Structure–Activity Studies

Two famous residues in the postcrystallographic history of enzymology are Tyr-248 of carboxypeptidase and Asp-102 of the serine proteases. Their catalytic roles have been the subjects of much speculation. But the definitive experiments of mutating Tyr-248→Phe and Asp-102→Asn have been documented here by Rutter and colleagues [3]. Note that, as reported by Rutter at the meeting, the mutation Asp→Asn can lead to inversion of the position of protonation of a histidine imidazole ring to which the carboxylate of an aspartate or carboxyamide of asparagine is hydrogen bonded; e.g.,

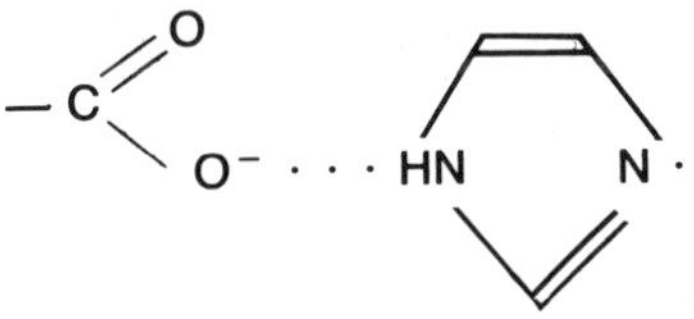

Asp

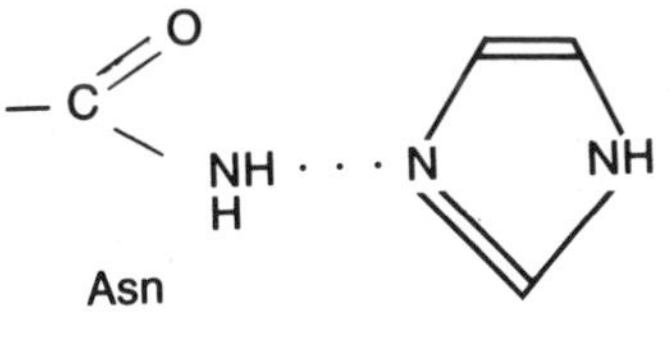

Asn

Thus, the quantitative importance of Asp-102 may not be apparent from the comparison of trypsin and trypsin(Asp→Asn-102). These experiments undoubtedly will be considered in the future as the classic examples of how protein engineering can be used to investigate the role of essential catalytic residues.

The above experiments are a logical extension of the techniques of chemical modification using conventional protein chemistry. Instead of nitrating tyrosine, for example, and increasing its bulk, a more subtle change, Tyr→Phe, can now be made. In general, it is better to make changes that decrease rather than increase steric bulk, since forcing a too-large mass into a protein is bound to distort its structure, whereas a small cavity can be tolerated in proteins. Indeed, small cavities are frequently found.

Experiments on the tyrosyl-tRNA synthetase reported here by Fersht and colleagues are representative of the radically new approach to structure activity relationships offered by protein engineering. The side chains of the active site that form hydrogen bonds with the substrate have been systematically removed and the kinetics of the mutants carefully analyzed. This has afforded the direct experimental determination of the strength of the hydrogen bond in enzyme–substrate interactions and how this binding energy is used in catalysis. The previously unknown mechanism of the enzyme has appeared naturally from the studies: There is direct evidence for a transition state-stabilization mechanism.

REFERENCES

1. Fersht AR (1985): "Enzyme Structure and Mechanism, 2nd edition." New York: WH Freeman & Co, Ch. 4, pp 143–147.
2. Fersht AR, Leatherbarrow RJ (1986): Structure and activity of the tyrosyl-tRNA synthetase. This volume, chapter 24.
3. Rutter WJ, Gardell SJ, Roczniak S, Hilvert D, Sprang S, Fletterick RJ, Craik CS (1986): Redesigning proteins via protein engineering. This volume, chapter 23.
4. Howell EE, Mayer RJ, Warren MS, VillaFranca JE, Kraut J, Benkovic SJ (1986): Active site mutations in dihydrofolate reductase. This volume, chapter 22.
5. Wells JA, Powers DB, Bott RR, Katz BA, Ultsch MH, Kossiakoff AA, Power DS, Adams RM, Heyneker HH, Kunningham BC, Miller JV, Graycar TP, Estell DA (1986): Protein engineering of subtilisin. This volume, chapter 25.
6. Perry LJ, Wetzel R (1984): Disulphide bound engineered into T4 lysozyme: stabilization of the protein toward thermal inactivation. Science 226:555–557.
7. Thomas PG, Russell AJ, Fersht AR (1985): Tailoring the pH dependence of enzyme catalysis using protein engineering. Nature 318:375–376.

Protein Engineering, pages 225–226

Overview: Structural Aspects of Purposely Modified Proteins

Brian W. Matthews

There is no question that oligonucleotide-directed mutagenesis is an incredibly powerful new tool to probe macromolecular structure and function. On the other hand, there remain fundamental limitations that will still make it difficult to obtain definitive answers to the central questions. Modification of "critical" residues in the active site of an enzyme can rule out a proposed mechanism of action (e.g., see the example of carboxypeptidase A [1], but it does not prove that an alternative mechanism of action is correct. Similarly, the substitution of one amino acid by another might increase the stability of a protein, but the question remains whether the effect is due to stabilization of the folded form or destabilization of the unfolded form (or some combination of both).

Another question concerns the effects of single amino acid substitutions on the structure of a protein. Is a given amino acid substitution likely to substantially alter the structure of a protein, or, in other words, to what extent can one "engineer" a protein without destroying it? In this case there is already evidence from classical mutagenesis, in particular for hemoglobin [2] and phage lysozyme [3], that single amino acid substitutions usually cause very little change in the overall structure of the protein. There is also increasing evidence, much of which is summarized in the following chapters, that amino acid substitutions can be made relatively freely without preventing the folding or significantly destabilizing protein structures. A particularly striking example is provided by phage T4 lysozyme, where a temperature-sensitive mutant (Thr 157→ Ile) was selected by classical methods and site-directed mutagenesis subsequently was used to make many different substitutions at the same site [3]. Of this series of lysozymes, the original temperature-sensitive mutant was the least stable.

These results are encouraging in the sense that they suggest that protein structures have evolved to allow, and to compensate for, different mutations (including site-directed ones). On the other hand, it is dangerous to assume that a mutant protein structure can be inferred from the known structure of its parent. Experience with the multiple series of substitutions in T4 lysozyme [3] shows that similar amino acids at a given site can adopt quite different configurations, and, in addition, bound water molecules may substitute for "missing" side-chain atoms. For these reasons, it is essential to determine the structures of mutant proteins directly.

To date, the number of X-ray structures of proteins with oligonucleotide-directed substitutions is limited [1,3–5]. This is in part because of the need to crystallize the mutant proteins and in part because X-ray crystallography is time consuming. Crystallization is something that cannot be predicted, but there has been reasonable success in using conditions similar to those for the parent protein to obtain crystals of mutants [1–5]. Also, high-speed X-ray data collection systems make it feasible to collect multiple data sets rapidly, even for large structures [6]. With the increased availability of this new technology it is likely that we will see a flood of new structural data on mutant proteins within the next few years.

REFERENCES

1. Rutter WJ, Gardell SJ, Roczniak S, Hilvert D, Sprang S, Fletterick RJ, Craik CS (1986): Redesigning proteins via protein engineering. This volume, chapter 23.
2. Fermi G, Perutz MF (1981): In Phillips DC, Richards FM (eds) "Atlas of Molecular Structures in Biology." Vol. 2. Oxford: Clarendon Press.
3. Alber T, Matthews BW (1986): The use of X-ray crystallography to determine the relationship between the structure and stability of mutants of phage T4 lysozyme. This volume, chapter 26.
4. Howell EE, Mayer RJ, Warren MS, Villafranca JE, Kraut J, Benkovic SJ (1986): Active site mutations in dihydrofolate reductase. This volume, chapter 22.
5. Wells JA, Powers DB, Bolt RR, Katz BA, Ultsch MH, Kossiakoff AA, Power SD, Adams RM, Heyneker HH, Cunningham BC, Miller JV, Graycar TP, Estell DA (1986): Protein engineering of subtilisin. This volume, chapter 25.
6. Steitz TA, Joyce CM (1986): Exploring DNA polymerase I of *E. coli* using genetics and X-ray crystallography. This volume, chapter 20.

Protein Engineering, pages 227–235

20

Exploring DNA Polymerase I of *E. coli* Using Genetics and X-Ray Crystallography

Thomas A. Steitz and Catherine M. Joyce

Department of Molecular Biophysics and Biochemistry, Yale University, New Haven, Connecticut 06511

For more than 25 years a central question in molecular biology has concerned the mechanism of DNA replication by DNA polymerases. Among the many issues of interest to us are: 1) the role that the enzyme plays in assuring the fidelity of template-directed DNA synthesis; 2) the mechanism of the enzyme's processivity (i.e., the successive incorporation of nucleotides without enzyme dissociation); and 3) the relationship between polymerization and editing activities. DNA polymerase I (Pol I) of *E. coli* is the first template-directed DNA-synthesizing enzyme for which high-resolution structural information has been obtained [1]; as such, it constitutes an excellent model for investigating the molecular details of replication. Pol I is not itself a replication enzyme (in the sense of replicating the entire chromosome); its major roles in vivo appear to be the repair of damaged duplex DNA and the processing of Okazaki fragments into high-molecular-weight DNA [2]. However, we believe the information we obtain with Pol I will be generally applicable to other replication systems, since the underlying chemical principles are likely to be the same.

A major advantage of Pol I as an experimental system is its simplicity; unlike many other replication enzymes, it is active as a single subunit. The molecule has a molecular weight of 103,000 daltons and has three enzymatic activities: DNA polymerase, a 3′–5′ exonuclease thought to edit out mismatched terminal nucleotides, and a 5′–3′ exonuclease that removes DNA ahead of the growing point of a DNA chain. Pol I has separate binding sites for deoxynucleoside monophosphate and deoxynucleoside triphosphate; the binding of one does not affect the binding of the other. Limited proteolysis removes the 35,000-dalton N-terminal domain that contains the 5′–3′ exonuclease activity [3,4]. The remaining 68,000-dalton fragment (Klenow fragment) has the polymerization and editing activities that are of interest to us and is a suitable size for structural studies.

High-resolution structural analyses of polymerases have not been possible until recently, due to the difficulty of obtaining adequate quantities of material. For Pol I, this problem has been solved by cloning the portion of the structural gene that codes for the Klenow fragment into an expression vector [5]. Overproduction of the fragment has enabled us to obtain large quantities of protein rapidly using few purification steps and has eliminated the possibility of heterogeneity introduced by protease treatment of the intact molecule. The availability of large quantities of homogeneous protein has facilitated the crystallization and determination of the structure of Klenow fragment. Now that polymerases from a number of sources (e.g., alpha subunit of DNA polymerase III of *E. coli*, gene 43 of phage T4, gene 5 of phage T7, DNA polymerase alpha of yeast, herpes simplex virus DNA polymerase) have been cloned, some in high-expression vectors, we can look forward to comparing the crystal structures of several different DNA polymerases.

A combination of structural, biochemical, and genetic studies has led us to conclude that Pol I has three domains, each responsible for a separate enzymatic activity (Fig. 1). As described below, the Klenow fragment structure has two domains, which we believe correspond to distinct polymerization and editing functions. This arrangement is analogous to the situation in DNA polymerase III of *E. coli*, where the polymerization and 3′–5′ exonuclease activities reside on separate subunits (alpha and epsilon, respectively) of the core enzyme [6].

STRUCTURE OF THE KLENOW FRAGMENT

Determination of the Klenow fragment structure was aided by a remarkable development in the technology of X-ray diffraction data collection, the multiwire area detector, which allowed us to measure 2.8 Å resolution data from seven crystals in 14 days. A model of the protein was built by fitting the amino acid sequence [7] into a 3.3 Å resolution electron density map using an Evans and Sutherland PS300 color graphics unit and the computer program FRODO [8]. The structure (Fig. 2) shows that the 605 amino acid polypeptide is folded into two distinct structural domains of approximately 200 and 400 amino acids.

The N-terminal one third of the Klenow fragment (residues 324–517 in the Pol I sequence [8]) forms the smaller of the two domains. This domain has a central core of β-pleated sheet (mostly parallel) with α-helices on both sides. In the crystal, this domain can bind one molecule of dTMP, as well as a divalent metal ion that interacts both with the protein and with the 5′ phosphate of dTMP. The metal is bound to the protein by the carboxyate groups of Asp 355, Glu 357, and Asp 501, with the 5′ phosphate providing a fourth ligand (Fig. 3).

The 400 amino acids of the larger domain

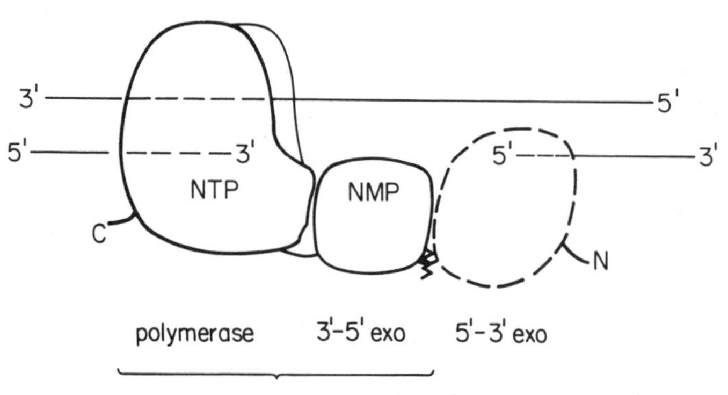

Fig. 1. *Schematic drawing showing the apparent domain structure of Pol I and the enzymatic activities associated with each domain. The solid lines show domains seen in the Klenow fragment structure which has binding sites for deoxynucleoside monophosphate (dNMP) and deoxynucleoside triphosphate (dNTP). The small fragment produced by proteolytic cleavage of Pol I is shown in dashed lines and consists of at least one domain. The guessed domain locations of the 3′ and 5′ termini of a nicked DNA substrate are indicated, but their relative separation is schematic.*

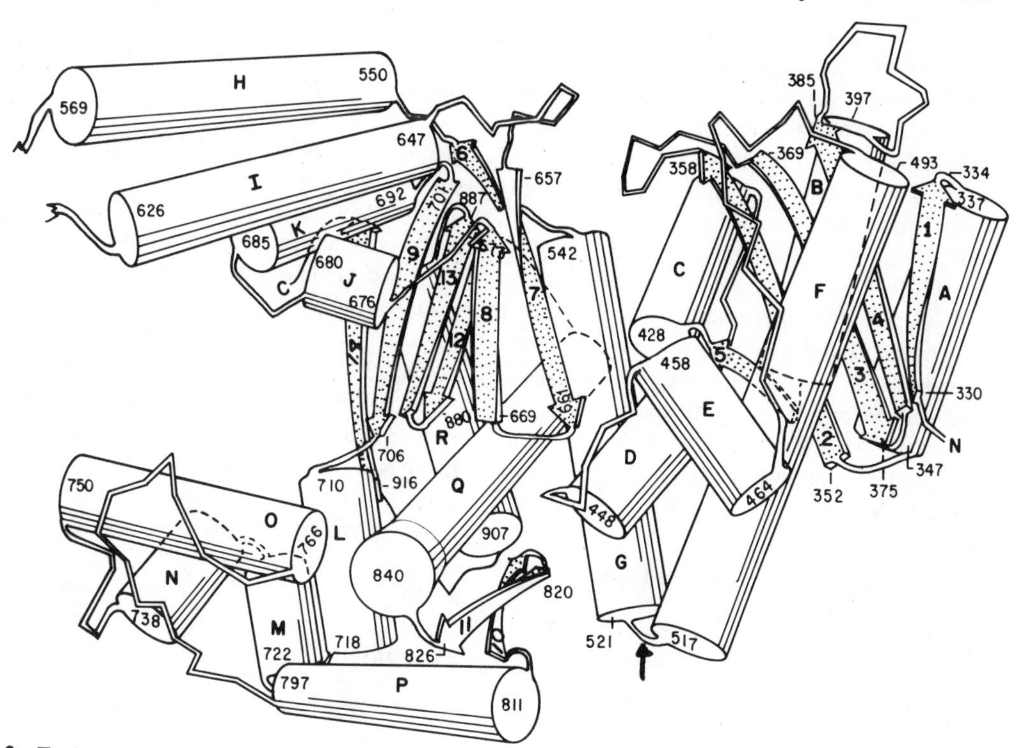

Fig. 2. *Tertiary structure of the Klenow fragment. α-helices are represented by tubes (lettered) and β-sheet by arrows (numbered). The broken lines on the strands between helix H and helix I shows the position of the approximately 50-residue disordered subdomain. The large and small domains are separated by the loop between helices F (residue 517) and G (residue 521).*

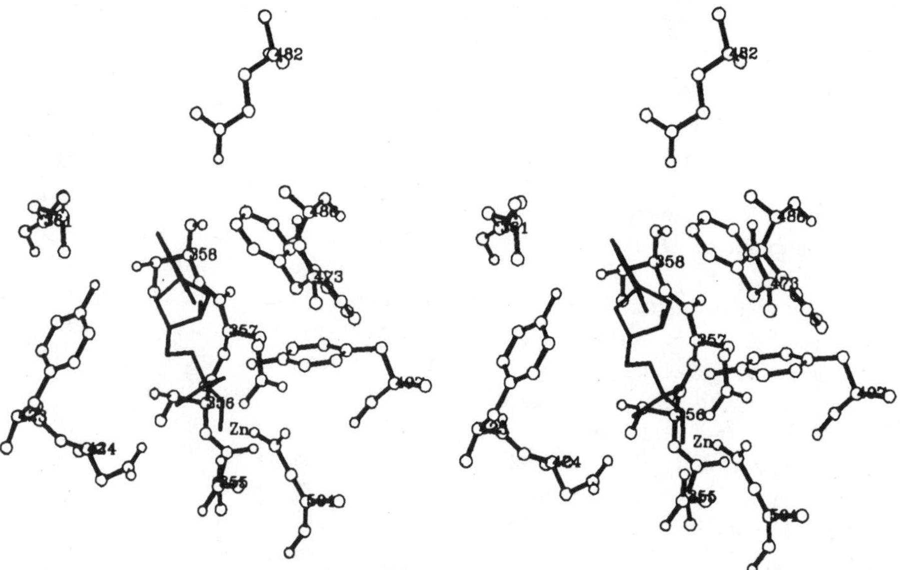

Fig. 3. *Stereo drawing of the binding site for dTMP in the small domain, which we believe to be the active site for the 3′–5′ exonuclease activity. A small molecule (perhaps water or a metal ion) is bound between the carboxyl of Asp 424 and the 5′ phosphate of dTMP but is not shown.*

(521–928) form a structure that contains a very deep cleft, about 20–24 Å wide and 25–35 Å deep. A six-stranded antiparallel β-sheet forms the bottom of the cleft and large protrusions of α-helix form its sides. The structure is similar in shape to a right hand grasping a rod. Thus, one side of the cleft forms a wall, 50 Å long, that can be compared with the curled-over fingers of a right hand. The other side of the cleft is formed primarily by two long α-helices, I and H, projecting from the protein like a thumb. At the tip of the thumb-like protrusion and hanging over the crevice are 50 amino acid residues (not shown in Fig. 2) that appear to be partially disordered in the crystal.

FUNCTION OF THE SMALL DOMAIN

Our main reason for suggesting that the small domain of Klenow fragment has the 3′–5′ exonuclease activity is the location of the dNMP binding site on this domain [1]. dNMP inhibits [9] the 3′–5′ exonuclease reaction (presumably by product inhibition) and therefore probably marks the position of the DNA 3′ terminus in the exonuclease active site. dNMP fails to inhibit polymerization, supporting the idea that the polymerase- and exonuclease-active sites are distinct from one another. Further circumstantial evidence for this location of the exonuclease active site is provided by the protein sequence homology between the small domain and epsilon, the 3′–5′ exonuclease subunit of DNA polymerase III [6]. The homologies, though weak, involve regions of the protein that surround the dNMP site, and residues that interact with the dNMP tend to be conserved [10]. In equating the dNMP binding site with the 3′–5′ exonuclease-active site, we are assuming that the dNMP site observed crystallographically is the identical site responsible for the inhibition observed in vitro. The interactions between the dNMP molecule and the protein support this assumption. In particular, efficient inhibition of the exonuclease reaction requires a free 3′ hydroxyl group [9]; the structure shows that this hydroxyl interacts with the protein, forming a hydrogen bond to the backbone amide of Thr 358. This interaction also makes biological sense if we assume that the dNMP molecule marks the position of the DNA 3′ terminus: A buried 3′ hydroxyl group is compatible with the exonuclease reaction but would be inappropriate for a DNA 3′ terminus at the polymerase active site.

To test the hypothesis that the small domain contains the 3′–5′ exonuclease-active site, but not the polymerase-active site, the DNA encoding the small domain has been cloned in a high-expression vector, allowing overproduction of the protein for study in vitro. We have used site-directed mutagenesis to alter residues involved in the dNMP binding site. Characterization of the resulting mutant proteins has first established that our assignment of the exonuclease active site is correct, and secondly, allows us to identify important catalytic groups.

FUNCTION OF THE LARGE DOMAIN

Model building suggests that the cleft in the large domain can bind double-stranded B-DNA (Fig. 4). The J and K α-helices are placed partially into a major groove and may function to fix the translational position of the DNA on the protein, so that movement of the polymerase would require the protein to spiral along the DNA. The location of the small, disordered, 50-residue subdomain above the cleft suggests that it might bind to the DNA in the complex allowing the protein to completely surround the DNA substrate. This model for the location of the DNA-binding site is strongly supported by theoretical calculations of the electrostatic surface potential [11]. Virtually all of the positive electrostatic charge potential lies within the cleft and describes an approximately spiral path with a pitch of about 34 Å. The rest of the protein surface has negative electrostatic charge potential, as would be expected for an acidic protein. Additional evidence is provided by genetics. Two mutations, *polA5* and *polA6*, which affect the enzyme's interaction with DNA, are both in amino acids within the proposed binding cleft [12].

We have used DNA footprinting to measure the length of DNA covered by Klenow fragment bound at a primer terminus [10]. Combining this length measurement with the model building described above has allowed us to deduce a probable location for the primer terminus and thus the polymerase-active site (Fig. 5). An important conclusion from this work is that the polymerase-active site likely to be located on the large domain; there is no way the primer terminus can reach the small domain unless the protein undergoes a large conformational change on binding DNA. To confirm that the large domain contains the polymerase-active site, we have cloned the DNA encoding the large domain in a high-expression vector and have purified the protein product. The purified large domain has DNA polymerase activity (although at about 50-fold lower specific activity than the intact Klenow fragment) but does not show any measurable 3′–5′ exonuclease activity [20].

A more precise indication of the polymerase active site location is provided by photoaffinity labeling. Rush and Konigsberg have crosslinked the dNTP analogue, 8-azido-dATP, to Klenow fragment and have isolated and sequenced the labeled peptide [10]. The site of crosslinking has been tentatively identified as Tyr 776, located at the end of helix O, pointing towards the proposed DNA binding cleft (Fig. 5). Allowing for some uncertainties in the interpretation of the data, both the photoaffinity labeling and the footprinting data suggest that the polymerase-active site lies somewhere on the arc formed by the N-terminal part of helix Q and the floor of the binding cleft (β-sheets 7, 8, 12, and 13). We have started to probe this region further by in vitro mutagenesis (both localized and site-specific) to identify important active site residues.

As we implied earlier, limitations to our model building (and thus to our interpretation of biochemical or genetic data) are the possibility that the protein may undergo a conformational change upon binding DNA, the likelihood that the 50-residue disordered subdomain will bind DNA, and the chance that the DNA substrate might be distorted. Therefore, much of our future work will be directed towards determining the structure of a complex between Klenow fragment and a DNA substrate. In the presence of EDTA (to prevent 3′–5′ exonuclease activity), the polymerase forms a stable complex with a model substrate consisting of eight base pairs of duplex DNA with a three-nucleotide 5′ extension. Attempts to crystallize this complex are in progress. In the future we hope to be able to use mutants defective in 3′–5′ exonuclease activity in order to study the structure of the ternary complex (enzyme: DNA:dNTP) in the presence of magnesium ion.

RELATION TO OTHER POLYMERASES

We have argued above that Pol I constitutes a good model system for polymerases in general. In this case, one might expect to see some sequence or structural relatedness between Pol I and other polymerases whose primary responsibility is the replication of genomic DNA. In particular, functionally important residues might be conserved. We have found extensive amino acid sequence homologies between the large domain of Klenow fragment and phage T7 DNA polymerase [13]. The strongest homologies involve residues that form the putative DNA-binding cleft (Fig. 6). This is consistent with the notion that both Pol I and T7 DNA polymerase evolved from a common ancestor and that the cleft shares a common and important function, presumably DNA binding and polymerization activity. We have been unable to detect any sequence homologies between Pol I or T7 DNA polymerase and the DNA polymerases from adenovirus, Epstein-Barr virus, and phage T4.

PROCESSIVITY

Pol I, like other polymerases, is a processive enzyme; that is, it incorporates about 20 nucleotides into a growing DNA chain before dissociating [14]. The structure of the Klenow fragment suggests a mechanism for this pro-

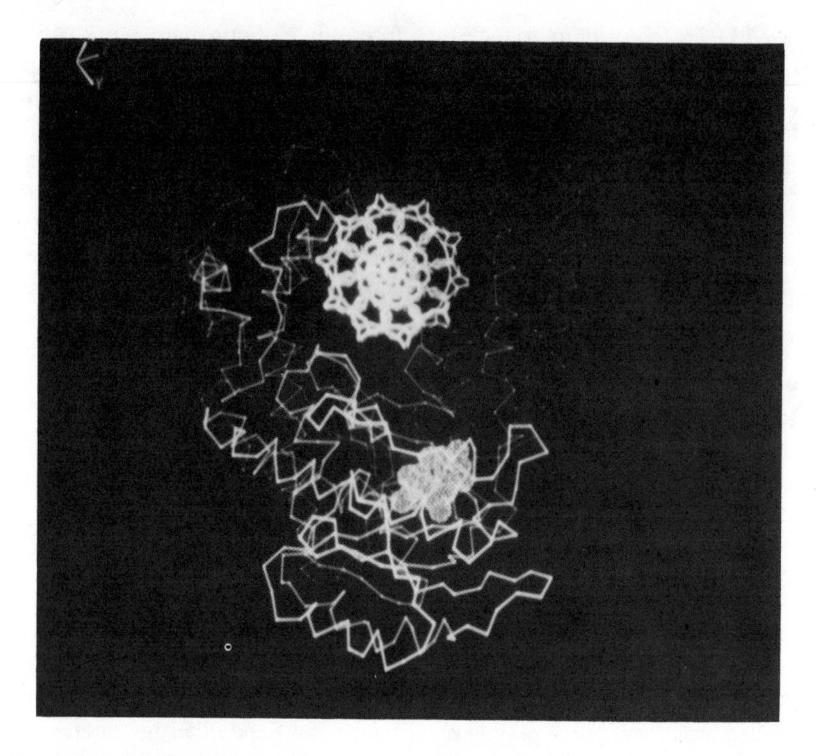

Fig. 4 *Relative orientation of DNA in the cleft and the bound deoxynucleoside monophosphate (in blue). (Reproduced in color on p. 344.)*

cessivity. The 50-residue subdomain that is flexibly attached to the tip of the I and H helices could close off the cleft after DNA binds. If the protein does envelop the DNA substrate, the rate of dissociation of the DNA product may be significantly slower than the rate of polymerization, and therefore could result in processive polymerization. Instead of dissociating from the product, the enzyme would simply slide to the new 3′ terminus between polymerization steps.

FIDELITY OF DNA SYNTHESIS

Clearly, it is of utmost importance that DNA polymerases make as few errors as possible in copying DNA. In vitro experiments indicate that Pol I has an error rate of around 1 in 10^6 when copying a natural DNA template [15]. How is high fidelity achieved? It appears that the 3′–5′ exonuclease activity functions to excise incorrectly incorporated bases [2]. However, for editing to take place, the formation of a mismatched base pair must first be detected at the polymerase active site. How is this done? A detailed answer to this question must await the determination of the structure of the Klenow fragment complexed to DNA containing a mismatched base pair at the primer terminus; for the present, the structure of the native enzyme provides a clue as to how this might be achieved (Steitz, unpublished observations).

As we have described above, the binding cleft, together with the flexible subdomain, could allow the enzyme to surround its DNA substrate, forming a tight orifice through which the duplex product of DNA synthesis must pass. Thus, the enzyme could, in principle, detect the mismatched base pair by scanning either the sugar–phosphate backbone, the major groove, or the minor groove.

We propose that the enzyme contains a "reading head" which can detect mismatched

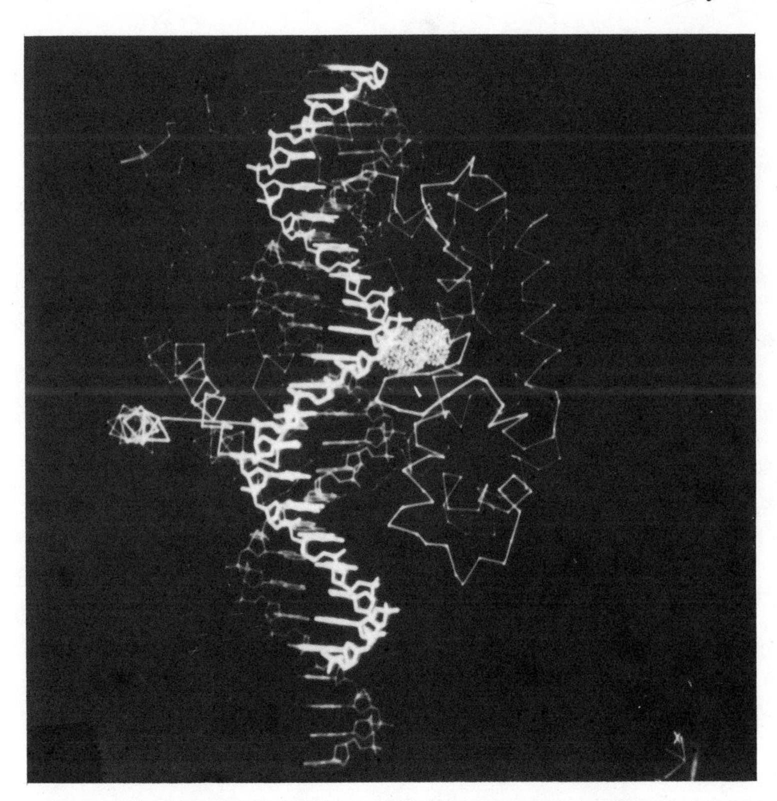

Fig. 5. *Color graphic representation of the Klenow fragment bound to a nicked DNA substrate. The protein α-carbon backbone is shown in purple. The green surface represents the side chain of Tyr 776 that is crosslinked to 8-azido-dATP. The length of DNA shown is protected against DNAse I digestion by binding of the Klenow fragment. The template strand is red-orange, the primer strand is white, and the nick lies between the white and yellow strands of DNA. The protected region upstream of the primer terminus appears to extend some distance beyond the boundary of the protein (especially on the template strand). We believe that this represents steric interference between Klenow fragment and DNAse I. Footprinting with the small molecule methidiumpropyl-EDTA indicates that Klenow fragment protects about eight base pairs upstream of the primer terminus, consistent with the model presented here. (Reproduced in color on p. 345.)*

base pairs in the minor groove. Pol I cannot be scanning the major groove, since substitutions at the 5 position of pyrimidines do not affect polymerization [16]. It is not scanning the DNA backbone structure, since the crystal structures of two mismatched base pairs in a B-DNA duplex, T.G [17] and A.G [18], show little deviation in the sugar–phosphate backbone conformation; however, there are substantial changes, for example, in the locations of the guanine N3 and thymine O2 atoms in the minor groove [17,18]. Since all four arrangements of the Watson–Crick base pairs present the same pattern of hydrogen bond acceptors in the minor groove [19], scanning in the minor groove could be sequence independent; only the mismatched base pair would be different. Thus, an enzyme "reading head" scanning the minor groove could detect the mismatched base pair, stop translocation, and trigger the appropriate changes that move the DNA into the 3′–5′ exonuclease active site. Given the 20–30 Å separation between the proposed polymerase- and exonuclease-active sites in the native structure, editing of a mismatched base pair would require substantial movement of the DNA to the 3′–5′ exonuclease site or of the editing site to the polymerase active site or both. We hope that future biochemical and structural studies of appropri-

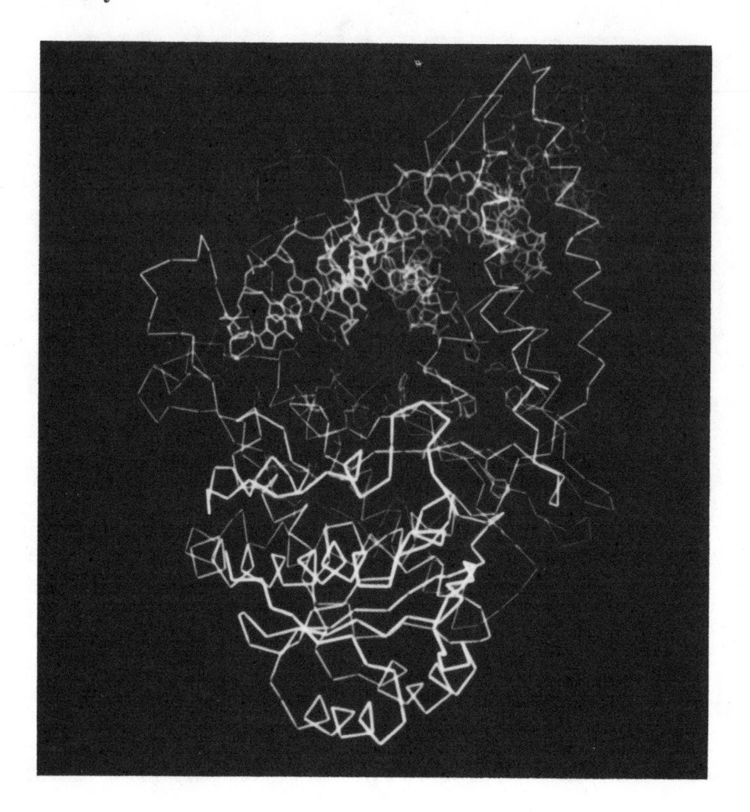

Fig. 6. *Residues that are identical in T7 DNA polymerase and the Klenow fragment shown in orange on a yellow α-carbon backbone. The template strand of the DNA substrate is shown in green, while the primer stand is in blue. There appears to be extensive conservation of side chains pointing into the cleft and in the vicinity of the presumed dNTP binding site labeled by 8-azido-dTP. (Reproduced in color on p. 345.)*

ate enzyme–DNA complexes, together with the detailed analysis of mutants that affect the enzyme's accuracy, will enable us to dissect this complex and intriguing problem.

ACKNOWLEDGMENTS

We wish to thank and acknowledge our collaborators on the research reviewed here: Peter Brick, Paul Freemont, Nigel Grindley, Connie Kline, David Ollis, John Rush, and Jim Warwicker. Research was supported by American Cancer Grant number NP-421 (to Thomas A. Steitz) and USPHS grants GM-28550 (to N.D.F. Grindley), and GM-22778 (to Thomas A. Steitz).

REFERENCES

1. Ollis DL, Brick P, Hamlin R, Xuong, NG, Steitz TA (1985): Structure of large fragment of *E. coli* DNA polymerase I complexed with dTMP. Nature 313:762–766.
2. Kornberg A (1980): "DNA Replication." San Francisco: Freeman.
3. Klenow H, Henningsen I (1970): Selective elimination of the exonuclease activity of the DNA polymerase from *E. coli* B by a limited proteolysis. Proc Natl Acad Sci USA 65:168–175.
4. Brutlag D, Atkinson MR, Setlow P, Kornberg A (1980): An active fragment of DNA polymerase produced by proteolytic cleavage. Biochem Biophys Res Commun 37:982–989.
5. Joyce CM, Grindley NDF (1983): Construction of a plasmid that overproduces the large proteolytic fragment (Klenow fragment) of DNA polymerase I of *E. coli*. Proc Natl Acad Sci USA 80:1830–1834.
6. Scheuermann RH, Echols H (1984): A separate editing exonuclease for DNA replication: The epsilon subunit of *E. coli* DNA polymerase III holoenzyme Proc Natl Acad Sci USA 81:7747–7751.
7. Jones AT (1978): A graphics model building and refinement system for macromolecules. J Appl Cryst 11:268–272.
8. Joyce CM, Kelley WS, Grindley NDF (1982): Nucleotide sequence of *E. coli polA* gene and primary

structure of DNA polymerase I. J Biol Chem 257:1958–1964.

9. Que BG, Downey KM, So A (1978): Mechanism of selective inhibition of 3′ to 5′ exonuclease activity of *E. coli* DNA polymerase I by nucleoside 5′-monophosphates. Biochemistry 17:1603–1606.
10. Joyce CM, Ollis DL, Rush J, Steitz TA, Konigsberg WH, Grindley NDF (1986): Relating structure to function for DNA polymerase I of *E. coli*. In Oxender D (ed): "Protein Structure, Folding and Design," UCLA Symposia on Molecular and Cellular Biology, Vol. 6 New York: Alan R. Liss, Inc., pp 197–205.
11. Warwicker J, Ollis DL, Richards FM, Steitz TA (1985): The electrostatic field of the large fragment of *E. coli* DNA polymerase I. J Mol Biol, 186:645–649.
12. Joyce CM, Fujii DN, Laks HF, Hughes CM, Grindley, NDF (1985): Genetic mapping and DNA sequence analysis of mutations in the *polA* gene of *E. coli*. J Mol Biol, 186:283–293.
13. Ollis DL, Kline C, Steitz TA (1985): Domain of *E. coli* DNA polymerase I showing sequence homology to T7 DNA polymerase. Nature 313:818–819.
14. Bambara RA, Uyemura D, Choi T (1978): On the processive mechanism of *E. coli* DNA polymerase I. Quantitative assessment of processivity. J Biol Chem 253:413–423.
15. Kunkel TA, Loeb LA (1980): On the fidelity of DNA replication. The accuracy of *Escherichia coli* DNA polymerase I in copying natural DNA *in vitro*. J Biol Chem 255:9961–9966.
16. Dale RMK, Ward DC (1975): Mercurated polynucleotides: New probes for hybridization and selective polymer fractionation. Biochemistry 14:2458–2469.
17. Brown T, Kennard O, Kneale G, Rabinovich D (1985): High resolution structure of a DNA helix containing mismatched base-pairs. Nature 315:604–606.
18. Kennard O (1985): Structural studies of DNA fragments: The G.T wobble base pair in A, B and Z DNA; The G.A base pair in B-DNA. J Biomolec Struct Dynam 3:205–226.
19. Seeman NC, Rosenberg JM, Rich A (1976): Sequence-specific recognition of double helical nucleic acids by proteins. Proc Natl Acad Sci 73:804–808.
20. Freemont PS, Ollis DL, Steitz TA, Joyce CM (1986): A domain of the Klenow fragment of *E. Coli* DNA Polymerase I has polymerase but no exonuclease activity. Proteins 1:66–73.

Protein Engineering, pages 237–250

21

Development of a Protein Design Strategy for *Eco*RI Endonuclease

John M. Rosenberg, Bi-Cheng Wang, Christin A. Frederick, Norbert Reich, Patricia Greene, John Grable, and Judith McClarin

Department of Biological Sciences, University of Pittsburgh, Pittsburgh, Pennsylvania 15260 (J.M.R., J.G., J.M.); Biocrystallography Laboratory, VA Medical Center, Pittsburgh, Pennsylvania 15260 (B.-C.W.); Department of Biology, Massachusetts Institute of Technology, Cambridge, Massachusetts 02139 (C.A.F.); Department of Biochemistry and Biophysics, University of California, San Francisco, California 94143 (P.G.)

INTRODUCTION

One of the most intriguing questions in molecular biology today is whether the details of individual DNA recognition systems will form a small number of simple patterns which would lead to the development of a general DNA recognition code. Examples of these systems include the regulation of gene expression by repressors and activators, site-specific genetic recombination, and host-dependent restriction and modification of DNA. Some of these systems, including the *Eco*RI system discussed here, lend themselves especially well to a combination of structural and protein design techniques.

*Eco*RI endonuclease recognizes the (double-stranded) sequence d(GAATTC), uniquely discriminating between it and all other possible hexanucleotides. The enzyme is a small protein of molecular weight 32,000 (276 amino acids) with an accessible gene of known sequence [1,2]. It hydrolyzes the phosphodiester bond between the guanylic and adenylic acid residues resulting in a 5′-phosphate. Although *Eco*RI endonuclease requires Mg^{2+} for phosphodiester bond hydrolysis, tight DNA binding and recognition specificity is retained in the absence of this ion [3–6]. In the absence of Mg^{2+}, the protein acts like a simple binding protein. The association constant for DNA containing its cognate hexanucleotide (GAATTC) is on the order of 10^{11}M, under optimal conditions. This binding constant is independent of DNA length within the range of 12 nucleotide pairs to plasmid DNA. The ratio of the binding free energy at its cognate hexanucleotide to that at noncognate sites is a factor of at least 10^5. This binding specificity indubitably plays a central role in determining the specificity of cleavage sites; however, it is worth noting that the latter may involve additional factors. The ability to bind without cleavage has enabled the growth of cocrystals of *Eco*RI endonuclease and the trideca-nucleotide TCGC*GAATTC*GCG (*Eco*RI site underlined) [7], for which we have recently determined the structure [8–11]. This structure provides opportunities for protein design experiments; however, further discussion of recognition issues is required before they can be described.

Possible Modes of Sequence-Dependent DNA–Protein Interaction

The essence of the recognition process is energy differences between states of the macromolecular system. For a simple DNA binding protein, such as a repressor, this would be the thermodynamic free energy which partitions the system into three classes of states: 1) Cognate binding between the protein and the target (operator) site on the DNA; 2) nonspecific binding of the protein at noncognate sites; and 3) states in which the protein and DNA are physically separate. (Most DNA binding proteins exhibit fairly strong nonspecific binding with the result that the cognate and nonspecifically DNA-bound states predominate under physiological conditions). The in vivo determinant of functionality in these systems is the fractional occupancy of the target sites on the DNA.

The situation is slightly more complicated for an enzyme which catalyzes a chemical reaction at a specific site on DNA. An enzyme functions by lowering the activation barrier separating the reactant and product states. Hence, a sequence-specific enzyme must selectively lower the activation energy only at cognate DNA sequences. The functional measure of specificity here is the ratio of activation energies at the cognate site(s) to the activation energies at all the noncognate sites (which comprise the bulk of the DNA). *Eco*RI endonuclease possesses both types of specificity (binding in the absence of Mg^{2+} and catalytic in its presence) although the two specificities appear very similar. It should be noted that the mechanistic details of the two cases may not be precisely identical. Regarding either the binding or catalytic specificity, the basic question is: What is the source of this energy difference?

The energy differences could, in principle, come from two types of interaction between the protein and the cognate bases: direct and indirect. Hydrogen bonds between the protein and the cognate bases are clear examples of direct determinants of specificity which undoubtedly are central to most, if not all, DNA recognition systems. However, other types of direct physical interaction are also probably significant, including Van der Waals interactions between nonpolar moieties on the protein and bases (e.g., and interaction with a thymine methyl group. See the results reported by Youdarian, et al., for a strongly suggestive example of this type of interaction [12]). Intercalation of aromatic amino acid side chains has been proposed as a mechanism of specificity [13], as has cobinding of divalent cations between the protein and DNA [13–15]. All of these interactions have as a common element the direct interaction between bases on DNA and moieties on the recognition protein.

In indirect recognition, the protein is presumed to interact only with the backbone of the DNA and sequence specificity could be generated in one of two ways: First, Dickerson et al. have noted that the precise conformation of the DNA backbone varies in a sequence dependent way [16–18], leading to the suggestion that proteins could have evolved to recognize these differences. Second, we have noted significant distortions of the DNA within the DNA-*Eco*RI endonuclease complex [8–10] (see also below). We suggested that different sequences of DNA might require differential inputs of energy in order to adopt the altered conformations. In the former, the protein would fit (or not fit) intrinsic sequence dependent differences in the DNA backbone, whereas in the latter, both protein and DNA would be driven towards a common (sequence independent) structure via a sequence-dependent input of energy. In reality, these both represent poles of what could be a continuous spectrum.

Role of Structural Studies in Recognition Problems

Our understanding of the three-dimensional structure of macromolecules is largely based on the results of X-ray crystallographic structure determinations. These structures play important roles in understanding complex macromolecular systems because they provide "snapshots" of the molecules at key functional stages of the system. In DNA-protein interactions, the most important of these stages is the recognition complex. Here, the interactions responsible for the recognition are occuring and can be identified by analyzing the structure of the complex. Thus, the structure itself answers many important questions regarding the recognition mechanism. However, it does not directly yield answers to issues of energetics; these require additional experimental and theoretical analysis. For example, we shall see that certain hydrogen bonds between the protein and the DNA are clearly implicated by the *Eco*RI structure. How much energy do they contribute to the interaction? How are the individual interactions coupled? These and many other questions can only be asked intelligently once the structure is available. In other words, the structure answers some questions and allows others to be framed in such a way that meaningful experiments can be done to answer them.

Role of DNA and Protein Design in Evaluating Recognition Mechanisms

Molecular design is a powerful tool which can be used, in conjunction with other tools, to answer the questions posed above. For example, once an amino acid residue has been implicated in the recognition process, it becomes a candidate for site-directed mutagenisis. The modified enzyme would then be characterized, e.g., through measurements of binding affinity and specificty as well as cleavage rates and specificity. The critical issue is the framing of experimental questions so that one can interpret the consequences of site-directed changes.

DNA–*Eco*RI ENDONUCLEASE STRUCTURE

General Features of the Complex

The cocrystalline DNA–protein complex possess two-fold symmetry. The protein-protein intersubunit diad (two-fold axis of symmetry), the principal diad of the symmetric DNA double helix and the crystallographic two-fold axis all coincide [7]. The molecular boundary, as seen in the Pt-ISIR electron density map, clearly encloses this complex in a well defined globular structure, 50 Å across [8–11] (see Figs. 1 and 2).

Structural Organization of the Protein Subunit

Each *Eco*RI endonuclease subunit is organized into a domain consisting of a five stranded β-sheet surrounded on both sides by α-helices (see Figure 3). This is the well-

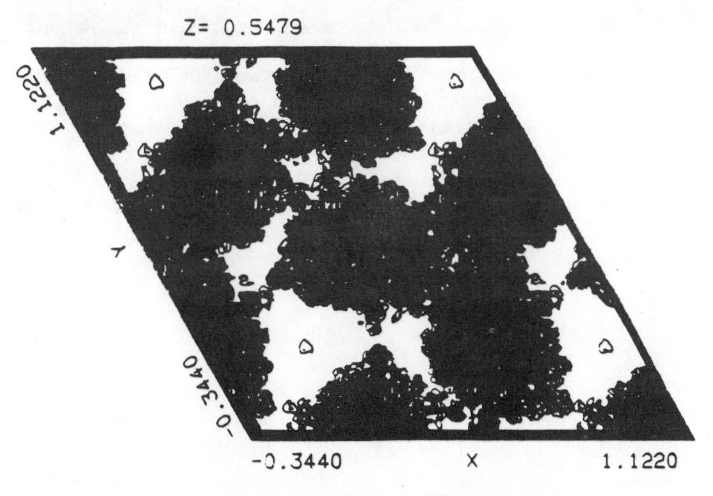

Fig. 1. *A projection down the c-axis of the Pt-ISIR electron density map of the DNA-*Eco*RI endonuclease complex.*

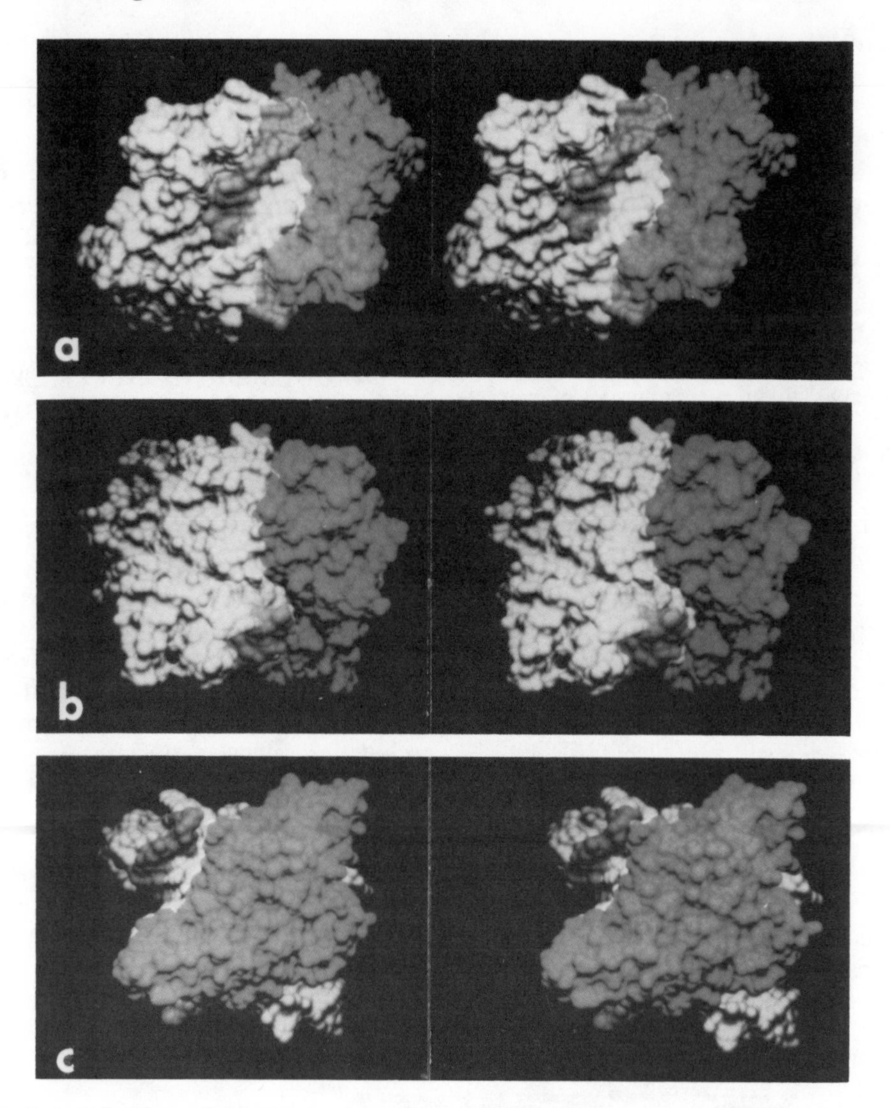

Fig. 2. *Stereo drawings of the solvent accessible surface of the* EcoR*I endonuclease-DNA complex. The subunits of the dimer are colored red and orange and the DNA is colored green and blue. a). The "front" view of the complex, which is a projection down a crystallographic twofold axis. b). The "top" view of the complex, rotated 90° from that in a) so as to view the structure down the c-axis. This results in a view looking approximately down the average DNA helical axis. c). The "side" view of the complex rotated 90° from a) around the c-axis.*

These images were calculated with the programs AMS and RAMS developed by Michael Connolly, modified for use with an Evans and Sutherland PS340 raster graphics system. (Reproduced in color on p. 346.)

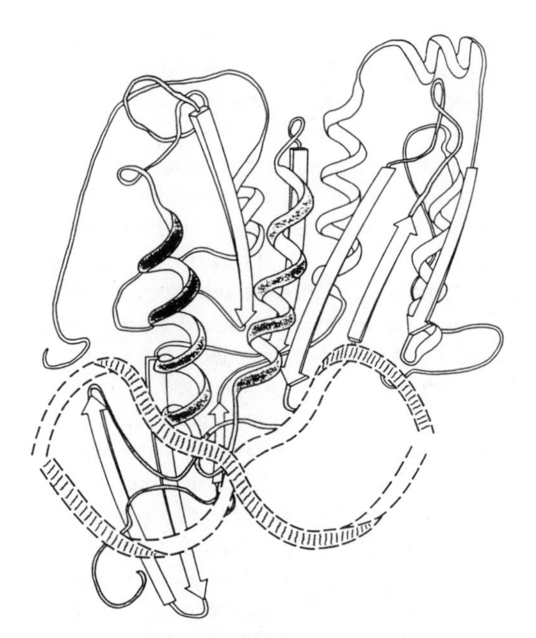

Fig. 3. *Schematic backbone drawing of one subunit of (dimeric)* Eco*RI endonuclease and both strands of the DNA in the complex. The arrows represent β-strands, the coils represent α-helices and the ribbons represent the DNA backbone. The helices in the foreground of the diagram connect the third β-strand to the fourth and the fourth to the fifth. They also interface with the other subunit. The amino-terminus of the polypeptide chain is in the arm near the DNA.*

known α/β architecture [19]. Four of the five strands in the β-sheet are parallel; however, the location of the single antiparallel strand allows us to divide the sheet into parallel and antiparallel three-stranded segments (see Fig. 4). Each of these segments forms a sizable structural unit constructed on a simple three-dimensional pattern in which the physically adjacent elements of secondary structure are essentially contiguous within the primary sequence; i.e., a motif. It is interesting that the parallel motif is the locale for the direct contacts between the protein and DNA bases as well as subunit-subunit interaction, while the antiparallel motif contains the site of DNA strand scission. We also note that the parallel motif is topologically very similar to one-half of the well-known nucleotide binding domain [20]. (The nucleotide binding domain is a six-stranded parallel β-sheet, constructed out of two three-stranded parallel motifs that are very similar to each other).

The principal features of the structure can be noted while following the course of the polypeptide chain. The amino terminus of the polypeptide chain forms part of an extension of the principal α/β domain of the protein, refered to as the "arm," which wraps around the DNA. The polypeptide chain then forms a long α-helix on the surface of the molecule which is followed by a loop into the first strand of the β-sheet. This β-sheet is formed sequentially starting from the outside of the antiparallel motif (see Fig. 3). The next loop, which connects the first and second β-strands, contains another α-helix situated on the surface of the molecule. The loop between the

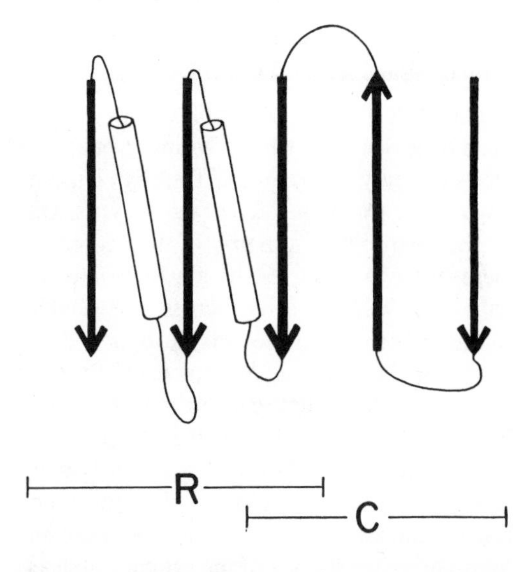

Fig. 4. *Topology diagram of the major α/β domain of* Eco*RI endonuclease. The β-sheet is divided into two overlapping topological segments, the parallel and antiparallel motifs which corresponds to the functional division of the β-sheet into a motif primarily responsible for recognition of the specific DNA sequence, R, and a motif primarily involved in catalytic activity. C.*

second and third (antiparallel) β-strands projects somewhat into the solvent. The third β-strand is a common element of both the antiparallel and parallel motifs. The fact that there is this overlap between the two motifs means that there is an effective means of structural interaction between the regions of the enzyme which are responsible for DNA recognition and the those which are primarily responsible for DNA strand scission activity. The parallel motif is formed sequentially from the middle of the β-sheet to the fifth strand at the edge of the sheet. The α-helices found at the intersubunit interface are the crossover helices [19] of the parallel motif. The α-helix connecting the third β-strand to the fourth is called α-helix 3–4, and the α-helix connecting the fourth β-strand to the fifth is called α-helix 4–5. After exiting the fifth β-strand, the polypeptide chain loops around the surface of the complex, placing the carboxy terminus in proximity with the DNA backbone.

*Eco*RI Endonuclease Possesses an Arm Which Wraps Around the DNA

This "arm" is an extension of the α/β domain (see Fig. 5) which wraps around the DNA partially encircling it, thereby clamping it into place on the surface of the enzyme. Due to the two-fold symmetry of the complex, there are two arms, each of which interacts with the DNA directly across the double-stranded helix from the scissile bond. These arms can be seen as prominent features in Figure 2, where they can be seen to extend from the central portion of the globular protein to envelop the DNA. Each arm is composed of the amino terminus of the protein and a β-hairpin sequentially located between the fourth and fifth strands of the primary β-sheet. (A β-hairpin is a structure consisting of two antiparallel strands of β-sheet connected by a short turn.) Part of the amino terminal portion of the polypeptide chain adds a third β-strand to the β-hairpin, thereby forming a three-stranded antiparallel β-sheet which is the structural foundation of the arm. Thus, there are two β-sheets in each *Eco*RI endonuclease subunit: the primary five-stranded sheet described above and the subsidiary three-stranded sheet described here.

The first 14 amino acid residues of the polypeptide chain form an irregular structure which is sandwiched in between the subsidiary β-sheet and the DNA. The arm mediates several nonspecific DNA-protein contacts which originate from this sandwiched region. Additional DNA backbone contacts are located in the short segment of polypeptide chain which connected the β-hairpin with α-helix 4–5, which follows it in the primary sequence.

Jen-Jacobson et al. have demonstrated by selective proteolysis that the nonspecific contacts between the DNA backbone and some of the amino terminal residues are required for catalytic activity (Jen-Jacobson et al., manuscript in preparation). Many of the resulting proteolytic derivatives retain sequence specific DNA binding but lack strand scission activity.

*Eco*RI Endonuclease Contains Two Catalytic Clefts

The carboxy edge of the antiparallel segment of the β-sheet forms the base of a cleft which binds the third, fourth, and fifth phosphates (counting from the left) of the oligonucleotide TpCpGpCpGpApApTpTpCpGpCpGP. (The scissile bond is at the fifth phosphate). One side of this catalytic cleft is formed by the loops which interconnect the β-strands in the antiparallel motif and which connect β-strand 3 to α-helix 3–4. The scissile bond is facing this side of the cleft. The other side of the cleft is formed by α-helix 3–4 and α-helix 4– 5 from the other subunit. The cleft surface contains many basic amino acid residues which interact electrostatically with the DNA phosphates, contributing to the binding energy.

The catalytic cleft is not fully assembled in our structure. There is a solvent channel, with DNA backbone on one side and protein on the other, which ends at the scissile bond. It is

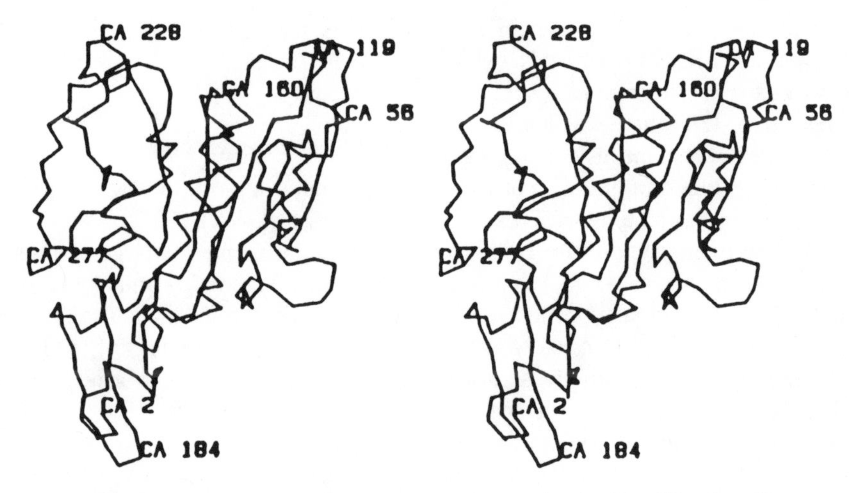

Fig. 5. *A stereo drawing of α-carbon trace of one subunit of* EcoRI *endonuclease.*

through this solvent channel that magnesium probably enters the active site. Once this cation is present, the full site assembles. It is probable that this temporal order is important in the function of the endonuclease (see below). We have recently demonstrated that Mg^{2+} can be perfused into the cocrystals and hydrolytic reaction carried out in situ (Picone et al., manuscript in preparation). The enzyme-product cocrystals still diffract X-rays, and their structure analysis is in progress.

The two symmetrically related clefts, one in each subunit, are approximately 3.5 Å farther apart than the normal separation between the phosphate backbone across the major groove of B-DNA. This increased separation, coupled with the basic residues within the clefts, is probably a major factor promoting the formation of the neokinks noted in the DNA [8].

Structural Features of the DNA

We reported [8] that the DNA conformation in the recognition complex departs from the B-motif in a way which suggests that the DNA is now adopting conformations which would be unstable in the absence of protein. The most striking of these departures are kinks which occur every three base pairs, as summarized in Figure 6. The use of the term "kink" follows the definition of Crick and Klug [21] as an abrupt change in parameters that describe a double helix in three-dimensional space.

It is now known that some DNA sequences are intrinsically bent, even in the absence of proteins [22], whereas the kinks we have noted in the DNA-*Eco*RI endonuclease complex

```
      v           v
T C G C G*A A:T T C G C G
  G C G C T T:A A*G C G C T
        ^           ^
```

Fig. 6. *The sequence of the synthetic olignucleotide with the recognition sequence underlined. * denotes the location of phosphodiester bond hydrolysis resulting in a 5′ phosphate. The hydrolysis reaction requires Mg^{2+} as a cofactor. The dotted line denotes the location of the type I neokink, which is coincident with the crystallographic and molecular twofold symmetry axis. The type I neokink unwinds the DNA by 25° and introduces a bend between the two central blocks, GAA and TTC, of 12° toward the minor groove. Λ denotes the location of the asymmetric type II neokink, which separates the terminal blocks from the central blocks of nucleotide pairs. This kink bends the helical axis by 23°.*

only occur in the presence of protein (the oligonucleotide used in our cocrystals is virtually identical to that studied by Dickerson and colleagues [16–17], which was not kinked in their structures). The intrinsic bends and kinks are probably structurally different from those which require the binding of a specific protein, and these two situations are certainly thermodynamically distinct. We feel that it is important that our terminology reflect these differences and refer to the intrinsically stable kinks and bends as such (or simply as kinks and bends). Those which require a specific binding protein are termed neokinks and neobends. Thus, Richmond et al. [23] observed that the DNA in their nucleosome structure contained "sharp bends" and/or possible kinks which would be neobends in our terminology.

We feel that one of the more intriguing observations to emerge from the *Eco*RI endonuclease structure is that the repertoire of conformational states intrinsically accessible to DNA has been expanded to include additional "neoconformations" which are stabilized by the binding of a protein. Specifically, we define a neoconformation as a structural distortion that is imposed on the double helix by a binding protein and that is not seen in the absence of protein.[1] This should not be taken to exclude the possibility that thermally transient fluctuations in DNA structure would include neoconformations in the absence of protein. Indeed, fluctuations of this sort may well be important intermediates in the formation of DNA–protein complexes. However, the bulk of DNA molecules at any instant would not be in a neoconformation, according to our definition, unless they were bound to a protein. We suspect that neoconformations will be a general feature of many nucleic acid–protein interactions. We also suspect that there will be a finite number of well-defined, structurally feasible neoconformations, analogous to the set of tight turns enumerated for protein structure [see, for example, reference 19]. Neoconformations may also have a role in sequence specificity alluded to above, i.e., some sequences may accept the distortion imposed by the protein more readily than other sequences.

DNA–Protein Interactions

The major groove of the recognition hexanucleotide (GAATTC) appears to be filled with protein, forming a large, complementary interface. All the base–amino acid interactions appear to be located in the major groove of the DNA, while the edges of the bases which are exposed in the minor groove are open to the solvent. Thus, the direct interaction of complementary surfaces in the major groove is a major determinant of the recognition specificity.

A crucial component of the DNA–protein interface is formed by the amino terminal ends of the crossover α-helices in the parallel motif. Since the DNA–protein complex possesses twofold symmetry, α-helices from both subunits participate in the formation of a four-helix bundle which inserts into the major groove of the DNA. One of the roles of the neokinks is to make room for this bundle, which would not fit into the major groove of conventional B-DNA.

The α-helix which connects the third and fourth β-strands makes an angle of approximately 60° with the average DNA helix axis, as shown in Figure 7. The polypeptide chain turns sharply at the end of the α-helix so that the amino acid residues at the amino-terminal end of the helix, those in the bend and in the adjacent stretch of chain, are in close proximity to the DNA. This α-helix is also adjacent to the molecular twofold axis, and the amino terminus of the helix is physically adjacent to the amino terminus of the symmetry-related helix from the other subunit. These two α-helices form a symmetric module which is

[1]This is based on the definitions of *neo* as "in a new or different form or manner" and "new chemical compound isomeric with or otherwise related to (such) a compound" [24].

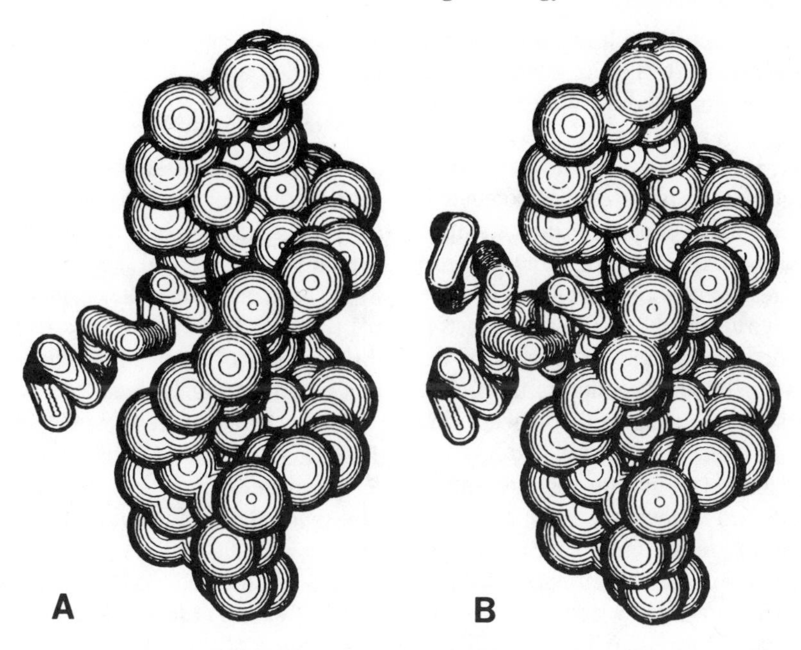

Fig. 7. *A). The double-stranded DNA in the* Eco*RI endonuclease-DNA complex and the α-helix, which connects the third and fourth β-strands. B). The same view as in A) with the symmetry related α-helix from the other subunit included. The two α-helices form the inner module, which recognizes the inner tetranucleotide, AATT.*

responsible for the direct interactions between the endonuclease and the bases of the inner tetranucleotide (AATT). This structural unit will be referred to as the inner recognition module. Two symmetry-related pairs of amino acid side chains make contact with two symmetry-related pairs of sequential adenine residues. Interestingly, each pair of adjacent adenines interacts with one amino acid from each subunit (see Fig. 8). In our current interpretation of the electron density map, these residues are glutamic acid 144 and arginine 145. The position of the arginine side chain in our current model is consistent with a hydro-

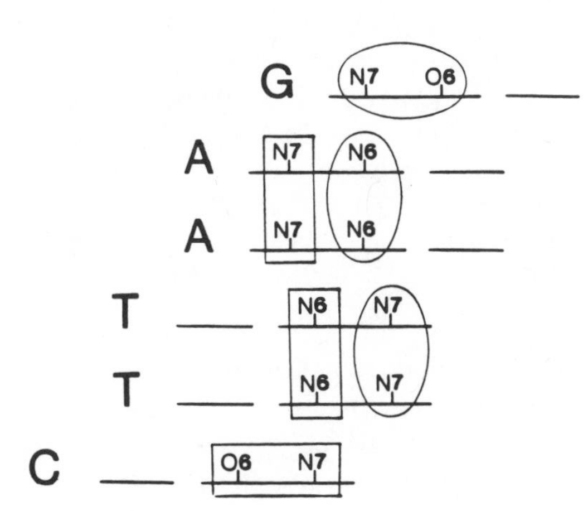

Fig. 8. *A schematic drawing depicting the interactions between base pairs in the* Eco*RI recognition site and amino acid side chains of* Eco*RI endonuclease. Rectangles denote interactions from one subunit and ovals denote interactions from the symmetry related subunit. The proposed interactions are arginine 200 hydrogen bonding to the N7 and O6 of the GC base pairs, arginine 145 hydrogen bonding to the two N7 moieties of adjacent adenines and glutamic acid 144 hydrogen bonding to the N6 groups of adjacent adenines.*

gen-bonding interaction with the N7 moieties of adjacent adenines, while the glutamic acid is probably hydrogen bonding to the exocyclic N6 groups of both adenines.

Separate α-helical modules are responsible for the direct contacts between the protein and the two outermost G–C pairs of the canonical hexanucleotide, GAATTC. These modules are called the outer recognition modules. They are identical by virtue of the twofold symmetry, and each independent module consists of the cross-over α-helix which connects the fourth and fifth strands of the principal β-sheet. This helix has many interactions with both α-helices of the inner recognition module and is thereby positioned so that it projects its amino terminus into the major groove of the DNA. At this stage of our analysis, it appears likely that arginine 200 interacts with the guanine in a manner predicted by Seeman, et al. [25].

PRECISE AND DEGENERATE RECOGNITION OF DNA

A key test of our understanding of sequence specific recognition will be passed when we can alter (via site-directed mutagenesis) the sequence recognized by *Eco*RI endonuclease to other hexanucleotides in a predetermined way. This is an ambitious goal, and it is reasonable to expect that initial attempts will be less than totally successful. One mode of partial success would be to convert the enzyme from one which recognizes a unique hexanucleotide (GAATTC) into one with reduced specificity. This is likely, because alteration of buffer composition has already been shown to have this effect (see below) in what historically has been referred to as *Eco*RI* specificity [26]. Careful analysis of the patterns of reduced specificity has already provided valuable information on the recognition mechanism. For example, we were able to correctly predict the hydrogen-bonding sites on the DNA which we subsequently observed in the structure by analyzing the pattern of *Eco*RI* specificity [27]. These considerations are directly relevant to protein design experiments; therefore that argument is reproduced below, in modified form.

*Eco*RI* Sites Form Hierarchies

*Eco*RI endonuclease loses its normally tight specificity under a variety of buffer conditions and recognizes many nucleotide sequences similar to the canonical site, GAATTC [26,28–31]. These buffer conditions include: elevated pH (8–9.5), Mn^{2+}, low ionic strength and by the addition of organic compounds such as glycerol or ethylene glycol.

When Polisky et al. first observed the *Eco*RI* phenomenon, they noted that some RI* sequences are hydrolyzed much more rapidly than others [26]. In particular, these authors noted that the canonical *Eco*RI site is always hydrolyzed much more rapidly than any RI* site. These results were extended by Goodman et al., who noted that there is a clear hierarchical order in the hydrolysis rates at the leftmost base (G in the canonical sequence). Data collected by Gardner et al. [32] and Rosenberg and Greene [27] showed that cleavage rates at all *Eco*RI* sites can be represented by hierarchies. The hierarchy at the leftmost base is: G >> A > T >> C [33]. That is, GAATTC is hydrolyzed much more rapidly than AAATTC, etc. The hydrolysis rates for A and T are much closer than any others in the hierarchy. The hierarchies at the next two positions (both adenine in the canonical hexanucleotide) are identical: A >> [G, C] >> T [27]. (GAATTC >> GAGTTC = GACTTC >> GATTTC, etc.). The hierarchies at the last three positions are complementary to those at the first three, due to the intrinsic symmetry of the *Eco*RI site. T >> [G, C] >> A at both positions where thymine is found in the canonical sequence and C >> T > A >> G at the right-hand end of the hexanucleotide.

The RI* Hierarchies Can Be Correlated With Hydrogen Bonds

The theoretical foundation for this analysis is the work of Seeman et al., who showed that if a recognition protein utilized hydrogen bonds for specificity, two hydrogen bonds per base pair would be required to uniquely discriminate between all the bases [25]. A single hydrogen bond to a base would result in degeneracies in that two different bases would be recognized. The particular pair of degenerate bases depends on the particular hydrogen bond, as summarized in Table I. These ideas have been applied to restriction enzymes which contain degeneracies in their recognition sequences [5, 27, 34].

The basic assumptions of the RI* hydrogen bonding model are as follows:

1. There are two hydrogen bonds between *Eco*RI endonuclease and each purine in the hexanucleotide, as indicated by the X-ray structure.

2. Under RI* conditions, one or more of these base–protein hydrogen bonds is randomly replaced by hydrogen bonds to water molecules, reducing the recognition specificity.

3. The rate of hydrolysis is correlated with the total number of hydogen bonds between the enzyme and the bases, i.e., each RI* sequence can be positioned within its apropriate hierarchy by simply counting base–protein hydrogen bonds.

Two protein–guanine hydrogen bonds (see Fig. 8) correlate with the observed hierarchy at the left end of the hexanucleotide: They join arginine 200 to N7 and O6 of the guanine. These allow complete discrimination between guanine and the other bases. However, if the hydrogen bond to O6 is "lost" (replaced by hydrogen bonds to water), the remaining bond to N7 is unable to discriminate between adenine and guanine (because both possess identical N 7 atoms); i.e., the Pu degeneracy occurs. If the bond to N7 is lost while that to O6 is retained, the enzyme is unable to discriminate between guanine and thymine; i.e., the Gt degeneracy occurs. (The O4 of thymine

TABLE I. Single Hydrogen Bond Degeneracies

Bases equal	Symbol[a]	Compliment[b]	Groove	Hydrogen bonds
A and G	Pu	Py	Major	N7 (purine)
T and C	Py	Pu	Major	N7 (purine′)[c]
A and C	Ac	Gt	Major	N6(A) and N4(C) or O4(T′) and O6(G′)
G and T	Gt	Ac	Major	O4(T) and O6(G) or N6(A′) and N4(C′)
A and T	At	At	Minor	N3(A) and O2(T)[d]
G and C	Gc	Gc	Minor	N2(G)

[a]The two bases are indicated with the purine in capitals and the pyrimidine in lower case, (e.g., Ac represents the simultaneous recognition of adenine and cytosine), except for the purine (Pu) and pyrimidine (Py) degeneracies.
[b]Symbol of degeneracy on complementary strand of DNA.
[c]Purine′, C′ etc., refers to the purine. C etc., on the complementary strand. (A degeneracy on one strand of DNA automatically implies another degeneracy on the complementary strand. Hence, degeneracies occur in complementary pairs. This complementarity restores twofold recognition symmetry, which appears to be violated by many of those restriction enzymes that normally recognize degenerate sequences [5].)
[d]An additional contact to exclude N2(G) is necessary.

occupies a three-dimensional position in the DNA very close to that occupied by O6 of guanine, so either could accept this hydrogen bond.) Both hydrogen bonds would have to be lost before a cytosine could be accepted at the left-hand end of the hexanucleotide.

The position of an RI* sequence within the hierarchy is determined simply by the number of hydrogen bonds between the protein and the bases: two with guanine (plus 10 with the other five base pairs, for a total of 12 hydrogen bonds); one with either adenine or thymine (plus the additional 10 for a total of 11 hydrogen bonds); and none between the protein and cytosine (for a total of 10). Thus, our model clearly predicts the observed hierarchy G >> [A, T] >> C. Secondary interactions would be expected to order those bases with an equal number of hydrogen bonds (A and T in this case; although there is a relatively small difference in the rate for adenine vs. that for thymine).

The hierarchies at the second- and third-base positions (which are both adenine in the canonical RI site) follow similarly. Recall that the structure showed hydrogen bonds between the protein and the N7 and N6 moieties of both adenine residues. Loss of a hydrogen bond at N6 leads to Pu (failure to discriminate between purines) as before. Similarly, loss of a hydrogen bond at N7 results in the Ac degeneracy (because N4 of cytosine and N6 of adenine occupy essentially the same positions). Thus A, G, C, and T would have two, one, one, and zero protein-base hydrogen bonds, respectively. We thereby obtain the hierarchy: A > > [G, C] > > T, which is what is actually observed.

Relationship Between Physiological and *Eco*RI* Activities

From a mechanistic viewpoint, the problem is not to "explain" the *Eco*RI* activity; rather, it is to understand its absence under physiological conditions. The hierarchical spectrum of *Eco*RI* sites is just what one should expect from a recognition mechanism based solely on hydrogen bonds (that form more or less independently of each other). Loss of a single hydrogen bond would be expected to reduce the interaction energy by 1 to 4 kcal. The resulting reduction in association constant or catalytic rate constant would be one or two orders of magnitude—just what is observed in the *Eco*RI* hierarchies. However, under physiological conditions, there is no detectable activity at RI* sites. The mystery is compounded by the observation that *Eco*RI endonuclease will bind to plasmid DNA lacking an intact *Eco*RI site, without detectable hydrolysis of the DNA in the presence of $Mg;^{2+}$ [Rosenberg, et al., unpublished]. That is, even though the enzyme does have a higher binding affinity for the canonical *Eco*RI site, the differential specificity cannot be simply explained via simple binding constants. What is the basis of this incremental specificity, and how does the change to *Eco*RI* buffer conditons allow cleavage at additional sites?

Our current working hypothesis is that there are conformational changes in the endonuclease (and DNA) such that functional recognition and cleavage sites are formed in an obligate temporal order. In the structure described above, we noted that although there were many DNA–protein interactions at the recognition interface; the cleavage site was not fully assembled. Furthermore, we hypothesize that there is physical coupling between the individual components of the DNA–protein interaction. As a result, the conformational change between the initial (inactive) and final (active) forms assumes cooperative properties. By this we mean that the enzyme retains the inactive conformation under physiological conditions until virtually all the sequence specific DNA–protein interactions have formed. This allows relatively subtle effects to dramatically alter the relative population of these two states. Furthermore, it is not unreasonable to argue that *Eco*RI* buffer conditions relax the "cooperativity" of the transition and/or alter its point of "onset"—i.e., the number of sequence-specific DNA–protein interactions required for an appreciable fraction of the population to assume the active form.

PERSPECTIVE ON PROTEIN DESIGN

The preceding considerations are critical to the protein design experiments we are planning to use in order to test our ideas about sequence specific recognition. In particular, they emphasize the importance of the characterization procedures employed to test the functional consequences of particular changes. It will be necessary to test the mutant proteins in the binding reaction (without Mg^{2+}) as well as under catalytic conditions, specifically with both "standard" and *Eco*RI* buffers, because different effects might be noted in these assays. For example, the proteolysis results of Jen-Jacobson et al., which we noted previously, demonstrate that it is possible to abolish catalytic activity while retaining binding specificity by modifying the enzyme at a location distal to the actual cleavage site. We suspect that some site-directed changes might have similar effects.

The structure suggests several obvious candidates for site-directed mutagenisis. The amino acid residues which directly hydrogen bond to the bases are clearly of interest—i.e., glu 144, arg 145, and arg 200. For example, reducing the hydrogen-bonding potential at the recognition site should lead to recognition of degenerate sequences. If it were possible to remove arg 145 without perturbing the rest of the interface, we would expect that there would no longer be hydrogen bonds to the N7 atoms of the adenines. This would lead to recognition of G–Ac–Ac–Gt–Gt–C in the notation of Table I (GAATTC + GCATTC + GACTTC + GCCTTC + GAAGTC + . . .).

Subtler changes could lead to a different type of degeneracy. Consider replacing glutamic acid 144 by glutamine. This would substitute two hydrogen bond donors for two acceptors. In the current interpretation of the electron density map, the hydrogen bonds from both adenine N6 groups are accepted by one of the glu 144 oxygen atoms. (We cannot completely exclude the possibility that both oxygen atoms are involved until the crystallographic refinement process is complete.) This leads to several predictions for the glutamine replacement. If there are hydrogen bonds to the other 144 side-chain oxygen atoms (e.g., from another amino acid side chain) the glutamine side chain's nitrogen atom would be forced towards the DNA. This would replace pairs of hydrogen bond acceptors with pairs of donors, leading to the recognition site GGGCCC. On the other hand, if there were no restriction on the orientation of the glutamine side chain's oxygen and nitrogen atoms, then either one could hydrogen bond to the DNA, leading to recognition of four sites: GAATTC + GGGTTC + GAACCC + GGGCCC. All of these predictions require the assumption that the mutation does not disrupt the entire interface or the conformational change discussed above.

The preceding discussion has provided examples of our current thinking regarding the use of protein design as a tool for analyzing cause-and-effect relationships in the *Eco*RI endonuclease problem. Specifically, we are designing mutations to test our working hypotheses. We are excited about the probability that the combination of structure, protein design, thermodynamics, and kinetics will enable us to formulate a detailed, quantitative model that matches the observed data. When this is done, we will be able to claim that we really understand how this enzyme recognizes its target DNA sequence.

REFERENCES

1. Greene PJ, Gupta M, Boyer HW, Brown, WE and Rosenberg JM (1981): Sequence analysis of the DNA encoding the *Eco*RI endonuclease and methylase. J Biol Chem Vol. 256:2143–2153.
2. Newman AK, Rubin RA, Kim S-H, Modrich P (1981): DNA sequences of structural genes for *Eco*RI DNA restriction of modification enzymes. J Biol Chem 256:2131–2139.
3. Modrich P (1979): Structures and mechanisms of DNA restriction and modification enzymes. Q Rev Biophys 12:315–369.
4. Halford SE, Johnson NP (1980): The *Eco*RI restriction endonuclease with bacteriophage λ DNA Biochem J 191:593–604.
5. Rosenberg JM, Boyer HW, Greene PJ (1981): The structure and function of the *Eco*RI restriction en-

donuclease, In Chirikjian JG (ed): "Gene Amplification and Analysis:Volume 1: Restriction Endonucleases." Elsevier/North-Holland, pp 131–164.

6. Jack WE, Rubin RA, Newman A, Modrich P (1981): Structures and mechanisms of *Eco*RI DNA restriction and modification enzymes, In Chirikjian JG (ed): "Gene Amplification and Analysis Volume I; Restriction Endonucleases." Elsevier/North-Holland, pp 165–179.
7. Grable J, Frederick CA, Samudzi C, Jen-Jacobson L, Lesser D, Greene PJ, Boyer HW, Itakura K, Rosenberg JM (1984) Two-fold symmetry of crystalline DNA-*Eco*RI endonuclease recognition complexes. J Biomolec Struct Dynam 1:1149–1160.
8. Frederick CA, Grable J, Melia M, Samudzi C, Jen-Jacobson L, Wang B-C, Greene PJ, Boyer HW, Rosenberg JM (1984): Kinked DNA in crystalline complex with *Eco*RI endonuclease. Nature 309:327–331.
9. McClarin JA, Frederick CA, Grable J, Samudzi CT, Wang B-C, Greene P, Boyer HW, and Rosenberg JM "Structural Studies on a DNA-*Eco*RI Endonuclease Recognition Complex." Proceedings from the Conference: The Fourth Conversation in Biomolecular Stereodynamics.
10. Rosenberg JM, McClarin JA, Frederick CA, Wang B-C, Boyer HB, Greene P "The 3 Å Structure of a DNA-*Eco*RI Endonuclease Recognition Complex." Proceedings of the Conference: The Molecular Evolution of Life. Sponsored by the Nobel Institute for Chemistry and the Royal Swedish Academy of Sciences.
11. McClarin JA, Frederick CA, Wang B-C, Greene P, Boyer HW, Grable J, Rosenberg JM (1986): 3.0 Å structure of *Eco*RI endonuclease. Submitted to Science.
12. Youderian P, Vershon A, Bouvier S, Sauer RT, Susskind MM (1983): Changing the DNA-binding specificity of a repressor. Cell 35:779–783.
13. Rich A, Seeman NC, Rosenberg JM (1977): Protein recognition of base pairs in a double helix. In Vogel HJ (ed): "*Nucleic Acid-Protein Recognition.*" New York: Academic Prss, pp 361–374.
14. Rosenberg JM, Seeman NC, Kim JP, Suddath FL, Nicholas HB, Rich A (1973): Double helix at atomic resolution. Nature 243:150–154.
15. Seeman NC, Rosenberg JM, Suddath FL, Kim JP, and Rich A (1976): RNA double-helical fragments at atomic resolution: I. The crystal and molecular structure of sodium adenylyl-3′,5′-uridine hexahydrate. J Mol Biol 104:109–144.
16. Dickerson RE, Drew HR (1981): Structure of a B-DNA dodecamer: II. Influence of base sequence on helix structure. J Mol Biol 149:761–786.
17. Dickerson RE, Drew HR (1981): Kinematic model for B-DNA. Proc Natl Acad Sci USA 78:7318–7322.
18. Dickerson RE (1983): Base sequence and helix structure variation in B- and A-DNA. J Mol Biol 166:419–441.
19. Richardson JS (1981): The anatomy and taxonomy of protein structure. Adv Protein Chem 34:167–339.
20. Rossmann MG, Liljas A, Branden CI, Banaszak LJ (1975): Evolutionary and structural relationships among dehydrogenases. In Boyer P (ed): "The Enzymes," Vol 11, pp 61–102.
21. Crick FHC, Klug A (1975): Kinky helix. Nature 255:530–533.
22. Wu H, Crothers DM (1984): The locus of sequence-directed and protein-induced DNA bending. Nature 308:509.
23. Richmond TJ, Finch JT, Rushton B, Rhodes D, Klug A (1985): Structure of the nucleosome core particle at 7 Å resolution. Nature 311:532–537.
24. Gove PB (1963): "Webster's Seventh New Collegiate Dictionary." G & C. Merrian Co.
25. Seeman NC, Rosenberg JM, Rich A (1976): Sequence-specific recognition of double helical nucleic acids by proteins. Proc Natl Acad Sci USA 73:804–808.
26. Polisky B, Greene P, Garfin DE, McCarthy BJ, Goodman HM, and Boyer HW (1975): Specificity of substrate recognition by the *Eco*RI restriction endonuclease. Proc Natl Acad Sci USA 72:3310–3314.
27. Rosenberg JM, Greene P (1982): *Eco*RI* specificity and hydrogen bonding DNA 1:117–124.
28. Hsu M, Berg P (1978): Altering the specificity of restriction endonuclease: Effect of replacing Mg^{2+} with Mn^{2+}. Biochemistry 17:131–138.
29. Woodbury CP Jr, Downey RL, von Hippel PH (1980): DNA site recognition and overmethylation by the *Eco*RI methylase. J Biol Chem 255:11526–11533.
30. Malyguine E, Vannier P, Yot P (1980): Alteration of the specificity of restriction endonucleases in the presence of organic solvents. Gene 8:163–177.
31. Woodhead JL, Bhave N, Malcolm ADB (1981): Cation dependence of restriction endonuclease *Eco*RI activity. Eur J Biochem 115:293–296.
32. Gardner RC, Howarth AJ, Messing J, Shepherd RJ (1982): Cloning and sequencing of restriction fragments generated by *Eco*RI*. DNA 1:109–115.
33. Goodman HM, Greene PJ, Garfin DE, Boyer HW (1977): DNA site recognition by the *Eco*RI restriction endonuclease and modification methylase. In Vogel HJ (ed): Nucleic Acid–Protein Recognition. New York: Academic Press. pp 239–259.
34. Smith HO (1979): Nucleotide sequence specificity of restriction endonuclease. Science 205:455–462.

Protein Engineering, pages 251–256

22

Active Site Mutations in Dihydrofolate Reductase

Elizabeth E. Howell, Ruth J. Mayer, Mark S. Warren, Jesus E. Villafranca, Joseph Kraut, and Stephen J. Benkovic

The Agouron Institute La Jolla, California 92037 (E.E.H., J.E.V.) Department of Chemistry, The Pennsylvania State University, University Park, Pennsylvania 16902 (R.J.M., S.J.B.); Department of Chemistry, University of California/San Diego, La Jolla, California 92093 (M.S.W., J.K.)

INTRODUCTION

Dihydrofolate reductase (DHFR, EC 1.5.1.3) catalyzes the NADPH-linked reduction of dihydrofolate to tetrahydrofolate, a vital step in the metabolic pathway for DNA synthesis. A reaction mechanism for this enzyme which incorporates all the available data from X-ray crystallography, NMR, and kinetic experiments has recently been proposed [1,2]. In this mechanism, the pteridine ring of bound dihydrofolate is activated at C6 for hydride transfer by protonation of N5. The path the proton follows has not been defined and the active site aspartic acid 27 may protonate N5 directly or indirectly (via a fixed water molecule, #403). In the indirect mechanism, Asp27 may stabilize the less predominant enol tautomer of dihydrofolate, allowing protonation of N5 from the 4-OH group. Hydride transfer then occurs from C4 of the A (re) face of the nicotinamide ring of NADPH to the si face of the protonated pteridine ring at C6. Carbonium ion character at C4 of the nicotinamide ring in the transition state is stabilized by interactions with three nearby oxygen atoms from the Ile13, Thr45 and Ile94 residues [3].

In the current model for productive substrate binding [1], Asp27 forms hydrogen bonds with Thr113 and with both N3 and the 2-amino group of dihydrofolate. Thr113 hydrogen bonds indirectly to the 2-amino group of dihydrofolate through a fixed water molecule (#405). Similar interactions have also been seen in the X-ray structure of the chicken DHFR–biopterin–$NADP^+$ ternary complex [4]. However, in this model the Asp27 is more than 5 Å from N-5 of dihydrofolate, so the pathway for proton donation most likely involves an indirect transfer via water #403.

With the above hypotheses in mind, we seek to construct mutant enzymes with designed substitutions in the active site that will allow us to assess the role of any particular amino acid in the catalytic mechanism. In particular, conserved residues proposed to interact with the substrate were chosen for mutational analysis. The specific mutations we have constructed by oligonucleotide-directed mutagenesis of the cloned *Escherichia coli* DHFR gene include Asp27 (GAT codon) → asn (AAC) [5], and Thr 113 (AGC codon) → val (GTG) [6]. A Ser27 (AGC codon) mutant gene was obtained as a primary site revertant of the Asn27 DHFR gene [7]. The availability of specifically designed mutant enzymes will now allow us to dissect the respective contributions of residues 27 and 113 toward substrate binding and catalysis as well as to evaluate directly the role of Asp27 in the protonation mechanism.

CONFORMATIONAL EFFECTS OF MUTATIONS

Prior to extensive characterization of the mutant enzymes, several criteria were evaluated to establish that the conformation of the mutant proteins had not been dramatically altered by the amino acid substitution. These criteria included: a) the equilibrium distribution between two predominant enzyme conformations as described by Cayley et al. [8]; b) any effects on binding of the substrate remote from the mutational site (NADPH for the above mutants); and c) any alterations in the binary X-ray structure of the mutant protein–MTX complex as obtained by difference Fourier mapping techniques.

The Thr113 → val mutant protein has been evaluated by techniques a and b, and the Asp27 → asn or ser DHFRs by b and c. All the results obtained so far indicate minimal conformational alterations and are summarized below.

Thr113 → val

Binding of NADPH to the wild-type (wt) enzyme is biphasic when monitored by fluorescence quenching using a stopped-flow spectrometer. Cayley et al. [8] have described the data as conforming to the following scheme

$$\begin{array}{l} E_1 + NADPH \underset{k_{off}}{\overset{k_{on}}{\rightleftharpoons}} E_1 \cdot NADPH \\ \downarrow\uparrow k_2 \\ E_2 \end{array}$$

(Scheme 1)

where k_2 is the rate of conversion of E_2 to E_1 and k_{on} and k_{off} are rate constants describing the binding interactions of NADPH with E_1. The equilibrium distribution between E_1 and E_2 is described by the relative amplitudes of the two phases. The values for k_2, k_{on} and the relative amplitudes for wt and Val113 DHFRs can be seen in Table I [6]. Except for a twofold increase in the relative concentration of E_2 for the Val113 mutant, the values are quite comparable.

Asp27 → Asn or Ser

The Asn27 and Ser27 DHFRs were crystallized as binary complexes with the inhibitor methotrexate (MTX) [7]. The crystals were isomorphous (within 0.5%) with those of the MB1428 DHFR–MTX complex whose structure has been previously determined [1,3,9]. This DHFR is isolated from a trimethoprim resistant strain of *E. coli* and is found to have a Lys154 → glu substitution [10]. Diffraction data to 1.9 Å resolution were collected and difference Fourier maps were initially calculated using the previously determined phases for the MB1428 structure. Later, wild-type DHFR phases were calculated and used as a starting point for further refinement.

Refinement of the Asn27 DHFR–MTX structure indicates it is almost identical with the wt DHFR–MTX structure. No significant movements in the protein backbone chain or in the bound ligand are observed. The minor perturbations observed in the active site involve movement of two fixed water molecules (#403 and 567) by 0.9 and 1.4 Å. These movements are probably due to alterations in the hydrogen bonding networks which arise from the asparagine substitution.

The refined structure obtained for the Ser27 DHFR–MTX complex also indicates only minor alterations in the active site area. The gamma oxygen (OG) of serine now occupies the space just below the position where oxygen OD1 of Asp27 in the wt DHFR structure existed and a new water molecule is inserted in the space previously occupied by oxygen OD2 of Asp27. Additionally, a small van der Waals gap in the structure is observed where the OG of Ser27 is 4 Å from the 2-amino

TABLE I. Effects on Binding of NADPH by the Val113 Mutant DHFR

	k_{on} ($M^{-1}sec^{-1}$)	k_2 (sec^{-1})	E_1:E_2 (relative amplitudes)
wt (Thr113)	1.6×10^7	0.030	1.2:1
Val113	2×10^7	0.035	1:2.5

The relative amplitudes and k_2 and k_{on} values were determined according to Scheme 1. The conditions were 25 mM Tris, 50 mM MES, 25 mM ethanolamine, 0.1 M NaCl pH 7.0 buffer, 25°C, 0.46–2 μM DHFR, 10–100 μM NADPH, 340 nM.

group of MTX—too great a distance to represent a hydrogen bond.

KINETICS

Asp27 → Asn or Ser

The expression of a cloned *E. coli* DHFR gene in *E. coli* yields a protein preparation containing some contaminating wt DHFR due to expression of the chromosomal DHFR gene. Also, for the Asn27 protein, deamidation of asparagine is a potential concern. This contaminating activity can be removed in many cases by differential elution of affinity columns and/or by isoelectric focusing. The isoelectric points of the wild-type and Asn27 mutant enzymes are 4.5 and 4.8, allowing the mutant enzyme to be purified by isoelectric focusing [5]. The Ser27 mutant DHFR was purified by elution from a MTX affinity column using 50 mM KH_2PO_4 pH 8 buffer containing 0.5 M KCl. The wt enzyme, in contrast, requires pH 9.0 and 1 M KCl washes prior to elution with 2 mM folate [11].

The pH profiles obtained for the kinetic parameters of the wt, Asn27 and Ser27 enzymes are illustrated in Figure 1 [7]. The wt profile corresponds well with those previously published [12]. The pH profiles for the two mutant enzymes are quite different, indicating an increased activity at lower pH values. In fact, if it were possible to assay the activity accurately at pH values lower than 4.5, it appears the mutant Ser27 catalytic rate would equal or surpass the wt rate. Unfortunately, dihydrofolate insolubility and decomposition of NADPH make it very difficult to measure the activity at low pH values.

The increasing catalysis at low pH values for both mutant enzymes may reflect protonation of dihydrofolate at N-5 in solution (N-5 pK_a of free DHF = 3.8 [13]), with protonated dihydrofolate then acting as a substrate for the mutant enzymes. In effect these mutations have probably eliminated the proton donation step in the wt enzyme's catalytic mechanism, instead utilizing protonated dihydrofolate as the substrate. Additionally, binding of nonprotonated dihydrofolate must occur in the mutant enzymes, for example, at pH 7.0, and result predominately in nonproductive complex formation.

The K_M values for dihydrofolate are fairly constant as a function of pH (at least above pH 5) for the wt, Asn27, and Ser27 enzymes. These values are obtained at saturating concentrations of NADPH and are approximately 1.2 μM, 44 μM, and 135 μM, respectively. The difference in binding affinity is in accord with our structural findings and suggests loss of hydrogen-bonding interaction(s).

The K_M values for NADPH are quite similar for the three enzymes and are 0.94 μM for wt, 1.47 μM for the Asn27 DHFR, and 1.72 μM for the Ser27 DHFR at pH 7.0. This observation suggests that the primary effect of the mutation is associated with the substrate binding site, not the cofactor site.

The above results strongly support the mechanism proposing that Asp27 is of central importance in transfer of a proton to substrate. Additionally, in the wild-type mechanism the ionized form of Asp27 does not appear to stabilize transition-state binding of the positively charged dihydropteridine ring relative to ground-state binding as the k_{cat} values of the mutant enzymes approach the wt rate at low pH.

Thr113 → val

A typical set of data and the method of graphical analysis used to obtain k_{cat} and

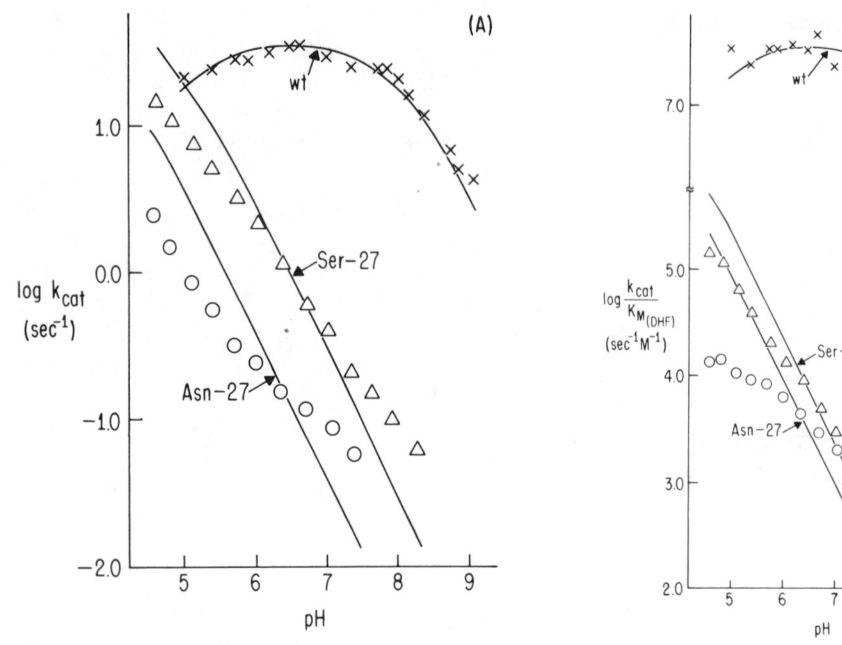

Fig. 1. *pH profiles of (A) log* k_{cat} *and (B) log* $k_{cat}/K_{M(dihydrofolate)}$ *for wt DHFR (x) Asn27 mutant DHFR (o) and Ser27 mutant DHFR (delta). Assays were performed at 30°C in 0.033 M succinic acid + 0.044 M imidazole + 0.044 M diethanolamine + 10 mM β-mercaptoethanol buffer [15]. Ranges of enzyme concentrations were 0.069–6.2 nM for wt, 80–600 nM for Asn27 DHFR and 19–1,900 nM for Ser27 DHFR. The dihydrofolate concentration was varied and the NADPH concentration was saturating at all pH values. Curves giving the best fit to the data were generated by using Eq. 5, reference 12, for the wt or by Eq. 1 and 2, in reference 7, for the Asn27 and Ser27 mutants.*

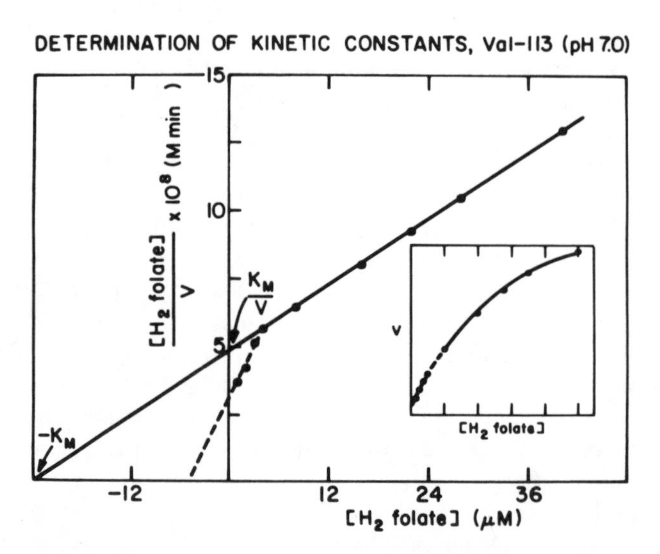

Fig. 2. *An example of the Michaelis-Menton plot (inset) and the graphical analysis used to obtain Michaelis-Menton parameters for the Val113 mutant DHFR, pH 7.0. The conditions are as indicated in Table II.*

TABLE II. Kinetic Constants for the Val-113 Mutant DHFR

	k_{cat} (sec^{-1})	k_{cat}/K_M (DHF) ($M^{-1}sec^{-1}$)	K_M (DHF) (μM)
wt (Thr113)	17.0	2.0×10^7	1.0
Val113	14.0	5.6×10^5	25

These kinetic parameters were determined using 25 mM Tris, 50 mM MES, 25 mM ethanolamine, 0.1 M NaCl buffer pH 7.0, 25°C, 9–30 nM DHFR, 100 μM NADPH and 1–50 μM dihydrofolate.

$K_{M(dihydrofolate)}$ values for the Val113 mutant DHFR are illustrated in Figure 2 [6]. The non-zero intercept in the Michaelis-Menten plot and the biphasic behavior in the S/V versus S plot are a result of contaminating wild type enzyme, presumably from the *E. coli* chromosomal gene. The 2% wt contamination, calculated from the reciprocal plot by the method of Spears et al. [14], is consistent with this contaminating enzyme being a product of the chromosomal gene. For these relative velocities and K_M values, the the graphical analysis provides results accurate to 5% for the high K_M enzyme.

The values at pH 7 for k_{cat} and $K_{M(dihydrofolate)}$ for the Val113 mutant are summarized in Table II. The k_{cat} for the mutant reaction is virtually identical and the K_M for dihydrofolate is increased 25-fold as compared with wt DHFR values. This increase in K_M for the mutant DHFR implies that the hydrogen bond between the 2-amino group of dihydrofolate and water #405 is lost due to the valine substitution.

The pH dependence of both k_{cat} and k_{cat}/K_M from pH 5.0–9.5 is shown in Figure 3. Stone and Morrison have suggested the basic pK_a (>pH 8) observed for the wt enzyme corresponds to a perturbed value for Asp27 [12]. The pK_a values observed for the mutant Val113 enzyme are 0.4–0.5 pH units lower than those obtained for the native enzyme. One simple interpretation of this pK_a decrease is that elimination of the hydrogen bonding interaction between Thr113 and Asp27 destabilizes the acid form of Asp27.

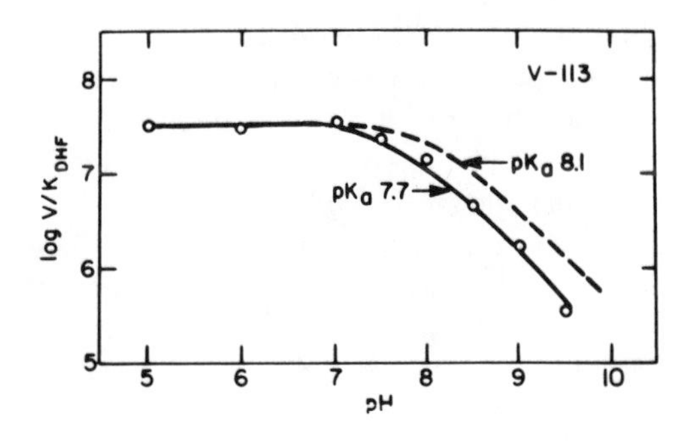

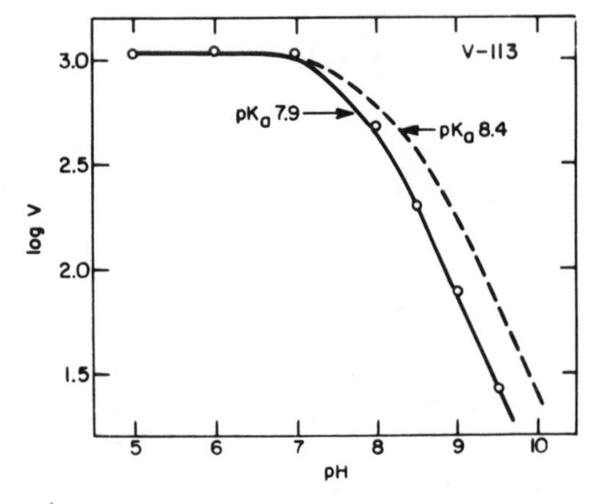

Fig. 3. *The pH dependence of* k_{cat} *and* $k_{cat}/K_{M(dihydrofolate)}$ *for the Val113 mutant DHFR where the solid lines are the theoretical curves fitting the data, determined by using Eq. 5 in reference 12. The dashed lines are the theoretical curves expected if the* pK_a *were the same as that of the wild-type enzyme [12] and were calculated using the pH independent values* ($k_{cat}, k_{cat}/K_M$) *obtained for the Val113 mutant DHFR. The conditions are as indicated in Table II. The units are* min^{-1} *for* k_{cat} *and* M^{-1} min^{-1} *for* k_{cat}/K_M.

The results from the Val113 mutant enzyme indicate that loss of binding energy (a 25-fold increase in K_M for dihydrofolate) has not been translated into a loss of catalytic activity. Thus the strictly conserved Thr113 residue may be required for maintaining a low K_M value. In short, this is an active site residue whose interaction with dihydrofolate is limited solely to binding and is not involved in the catalytic processing reaction; that is, the interaction of Thr113 with substrate does not result in any extra stabilization of the transition state to facilitate catalysis. Additionally, these results imply that release of tetrahydrofolate is probably not the rate determining step in the native enzyme. These deductions assume that tetrahydrofolate binding is affected by the mutation in the same manner as dihydrofolate, and also that no changes in rate determining step occur due to the mutation. Previously, isotope effects utilizing NADPD have shown hydride transfer is not the rate-limiting step in DHFR catalysis [12].

CONCLUSION

Our results from this first round of mutagenesis in dihydrofolate reductase reveal a mechanism of enzyme action that is surprisingly malleable. We find binding of substrate can be substantially impaired without a concomitant loss in catalytic activity. Also elimination of the proposed proton donor in DHFR leaves an active enzyme which can still utilize protonated substrate at low pH values. Clearly this ability to design specific mutations, which has become available with the advent of recombinant DNA technology, will facilitate the study of structure–function relationships in proteins.

ACKNOWLEDGMENTS

This work was supported by the Office of Naval Research Contract N00014-85-K-0663 (to J.K.), Public Health Fellowship F32 GM09375 (to E.E.H.), and NIH Grant GM10928 (to J.K.). We also thank Stuart Oatley for refining the crystal structures of the wt, Asn27 and Ser27 DHFRs in complex with methotrexate.

REFERENCES

1. Bolin JT, Filman DJ, Matthews DA, Hamlin RC, Kraut J (1982): Crystal structures of *Escherichia coli* and *Lactobacillus casei* dihydrofolate reductase refined at 1.7 Å resolution. J Biol Chem 257:13650–13662.
2. Kraut J, Matthews DA (1986): Dihydrofolate Reductase. In Jurnak F, McPherson A (eds): "Biological Macromolecules and Assemblies, Volume 3." New York: John Wiley and Sons (in press).
3. Filman DJ, Bolin JT, Matthews DA, Kraut J (1982): Crystal structures of *Escherichia coli* and *Lactobacillus casei* dihydrofolate reductase refined at 1.7 Å Resolution. J Biol Chem 257:13663–13672.
4. Personal communication, David Matthews and David Filman.
5. Villafranca JE, Howell EE, Voet DH, Strobel MS, Ogden RC, Abelson JN, Kraut J (1983): Directed mutagenesis of dihydrofolate reductase. Science 222:782–788.
6. Chen JT, Mayer RJ, Fierke CA, and Benkovic SJ (1986): Site specific mutagenesis of dihydrofolate reductase from *E. coli.* J Cellular Biol 29:73–82.
7. Howell EE, Villafranca JE, Warren MS, Oatley SJ (1986): The functional role of aspartic acid 27 in dihydrofolate reductase revealed by mutagenesis. Science 231:1123–1128.
8. Cayley PJ, Dunn SMJ, King RW (1981): Kinetics of substrate, coenzyme and inhibitor binding to *Escherichia coli* dihydrofolate reductase. Biochemistry 20:874–879.
9. Matthews DA, Alden RA, Bolin JT, Freer ST, Hamlin R, Xuong N, Kraut J, Poe M, Williams M, Hoogstein K (1977): Dihydrofolate reductase: X-ray structure of the binary complex with methotrexate. Science 197:452–455.
10. Bennett CD, Rodkey JA, Sondey JM, Hirschmann R (1978): Dihydrofolate reductase: The amino acid sequence of the enzyme from a methotrexate resistant mutant of *Escherichia coli.* Biochemistry 17:1328–1337.
11. Baccanari DP, Stone D, Kuyper L (1981): Effect of a single amino acid substitution on *Escherichia coli* dihydrofolate catalysis and ligand binding. J Biol Chem 256:1738–1747.
12. Stone SR, Morrison JR (1984): Catalytic mechanism of the dihydrofolate reductase reaction as determined by pH studies. Biochemistry 33:2753–2758.
13. Poe M (1977): Acidic dissociation constants of folic acid, dihydrofolic acid and methotrexate. J Biol Chem 252:3724–3728.
14. Spears G, Sneyd JGT, Loten EG (1971): A method for deriving kinetic constants for two enzymes acting on the same substrate. Biochem J 125:1149–1151.
15. Ellis KJ, Morrison JF (1982): Buffers of constant ionic strength for studying pH-dependent processes. Methods Enzymol 87:405–426.

Protein Engineering, pages 257–267

23

Redesigning Proteins via Genetic Engineering

William J. Rutter, Stephen J. Gardell, Steven Roczniak, Donald Hilvert, Stephen Sprang, Robert J. Fletterick, and Charles S. Craik

Hormone Research Institute and Department of Biochemistry and Biophysics, University of California, San Francisco, California (W.J.R., S.J.G., S.R., S.S., R.J.F., C.S.C.) and Laboratory of Bio-organic Chemistry and Biochemistry, Rockefeller University, New York, New York 10021 (D.H.)

INTRODUCTION

Typically form does not follow function with regard to protein structure. Current understanding of the relationship between the three-dimensional structure of a protein and its physiological function is at best limited. In particular, substrate specificity and the mechanism of catalysis at the atomic level remain hypothetical, even when the three-dimensional structure of the enzyme has been determined, the amino acid residues in the active site are defined, the kinetic features of the catalysis are elucidated, and enzyme substrate adducts or enzyme modifications have been produced. At the present state of our knowledge, we think it prudent to study proteins for which there is substantial structural and functional information to lay the groundwork for more difficult experiments that create completely novel structures. Two prototype hydrolytic enzymes were selected for study: carboxypeptidase A and trypsin. These two enzymes have been among the most thoroughly studied at the level of both structure and mechanism. The genes encoding these proteins [1,2] as well as several other related genes have been isolated and fully characterized in our laboratory over the last several years. These genes have been employed as subjects for site-directed mutagenesis in order to test particular structure/function hypotheses. In each case unexpected and revealing information has been obtained.

PROBING THE CATALYTIC MECHANISM OF CARBOXYPEPTIDASE A

Pancreatic carboxypeptidase A (CPA) catalyzes the hydrolysis of aromatic and branched-chain aliphatic amino acids from the carboxy terminus of peptides and proteins. The enzyme also hydrolyzes the corresponding ester derivatives. The structure of bovine CPA has been investigated extensively. These studies have resulted in the elucidation of its amino acid sequence [3] as well as the determination of its three-dimensional structure in the native state [4,5] and when complexed with a variety of ligands [6]. The deduced structure of the enzyme substrate complex suggests that only two amino acids, Glu270 and Tyr248, as well as zinc and water, are close enough to the scissile peptide bond to be directly involved in the catalytic process (Fig. 1). Various mechanisms involving these moieties have been proposed to account for catalytic activity [7–12].

Cleavage of peptide and ester substrates may involve direct nucleophilic attack of Glu270 thus resulting in the formation of a mixed-anhydride intermediate (Fig. 2), or alternatively, water may be the nucleophile assisted by Glu270 acting as a general base catalyst. In either case, Tyr248 previously has been presumed to be a general acid catalyst that contributes a proton to the incipient amine anion generated during the cleavage of peptide substrates. This proposal was structurally based on X-ray diffraction studies which showed that the Tyr248 hydroxyl is ordinarily at the surface of the enzyme but moves within hydrogen bonding distance of the cleaved peptide bond as a result of substrate binding. This postulated role for the phenolic hydroxyl was also consistent with the decreased peptidase and increased esterase activities which typically accompanied chemical modification of Tyr248 [14–16]. Ostensibly the Tyr248 hydroxyl was presumed not required for the cleavage of ester substrates because of the greater stability of the alkoxide anion formed during ester hydrolysis.

Replacement of Tyr248 with phenylalanine by site-directed mutagenesis probes the contribution of the hydroxyl moiety to catalysis [17]. If Tyr248 were acting as a general-acid catalyst, then the mutant protein should be inactive on peptide substrates yet still retain activity towards ester substrates. The experiments were carried out with rat CPA cDNA, which was obtained previously from a rat pancreatic cDNA library [1]. The deduced amino acid sequence of the rat CPA cDNA is 78% homologous with the bovine enzyme; further, the residues implicated in catalysis or ligand binding are conserved between the two species. Site-directed mutagenesis was accomplished with synthetic oligomers and a single-stranded DNA template.

Expression of the wild-type enzyme (CPA-WT) or the engineered variant (CPA-Phe248) was carried out in yeast using the α factor pheromone system [18,19]. The expression vectors introduced into yeast utilize the α factor promoter to direct the synthesis of an mRNA transcript coding for a chimeric protein which contains the leader segment of the α factor precursor [20] fused to proCPA-WT or proCPA-Phe248. Analysis of RNA from the yeast transformants revealed the presence of an RNA species of the appropriate size that cross-hybridized with the rat CPA cDNA. The hydbrid proteins are expected to be secreted from the yeast by virtue of sorting information contributed by the leader segment of the α factor precursor. Furthermore, during transit along the secretory pathway, the fusion proteins are processed by endogenous yeast proteases which recognize the canonical spacer peptide, LysArgGluAlaGluAla, at the junction of the α factor leader and proCPA sequences [21,22]. Conditioned media from yeast transformants contained rat CPA-immunoreactive proteins that comigrated with rat proCPA and CPA during SDS–PAGE. Thus, yeast is capable of processing the fusion proteins to yield proCPA and, to a limited extent, of converting proCPA to CPA. The remainder of the proCPA can be converted completely to CPA by treatment with trypsin.

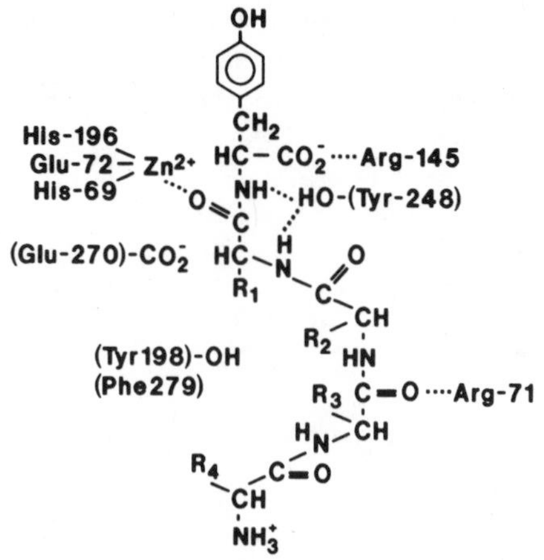

Fig. 1. *The postulated productively bound complex of carboxypeptidase A and a polypeptide substrate. (Adapted from ref. 4).*

CPA produced in yeast and CPA isolated from the rat pancreas exhibit identical cata-

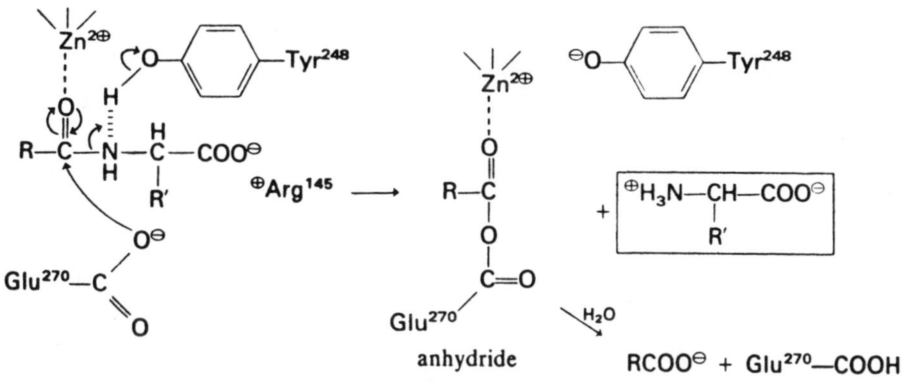

Fig. 2. *Postulated mechanism of action of carboxypeptidase A in which Glu270 acts as a nucleophilic catalyst. (From ref. 13.)*

lytic characteristics (kcat and Km values) towards a variety of substrates. Unexpectedly, CPA–WT and CPA–Phe248 also display similar kcat values when assayed with peptide and ester substrates (Table I). These data indicate that the Tyr248 phenolic hydroxyl is not required to mediate general acid catalysis during the hydrolysis of peptide substrates. If the Tyr248 hydroxyl acts as a proton donor at all, it can be replaced by an alternative group in the active site with virtual impunity.

Although Tyr248 is somewhat distant from the site of peptide bond scission [4] (Fig. 1), the selective inhibition of peptidase activity which accompanies its derivatization [23] suggests that Tyr198 may be the general acid catalyst. Therefore, a mutant in which Tyr198 is replaced by phenylalanine and a double mutant containing Phe198 and Phe248 were constructed. These variants exhibit peptidase activity which is also similar to CPA-WT (Table I), thus eliminating Tyr198 from consideration as the proton donor. The small decreases in kcat that are observed following removal of Tyr198 or Tyr248 hydroxyls may reflect a role of these moieties in transition-state stabilization [24].

Another amino acid residue which might fulfill the role of general-acid catalyst is Glu270 [25,26]. It is conceivable that the proton abstracted from H_2O by Glu270 acting as a general base may be subsequently transferred to the leaving group during the breakdown of the tetrahedral intermediate (Fig. 3).

TABLE I. Comparison of the Michaelis-Menten Parameters Exhibited by the CPA Variants

Variant	Substrate	k_{cat}	K_m
CPA-Phe248	Peptide	0.5	7
	Ester	NC	1.5
CPA-Phe198	Peptide	0.7	NC
	Ester	NC	NC
CPA-Phe198Phe248	Peptide	0.25	8
	Ester	NC	1.5

The phenolic hydroxyls of Tyr248 or Tyr198 are not required for catalytic activity; however, the Tyr248 OH appears to be involved in substrate binding. The peptide and ester substrates used in this study were carbobenzoxy-glycyl-glycyl-phenylalanine and benzoyl-glycyl-O-phenyllactate, respectively. Data is presented as the ratio of the kinetic parameters of the CPA variant and the wild-type enzyme.
NC, no change.

The present findings indicate that the decreased peptidase activity accompanying chemical modification of Tyr248 cannot be due to a deleterious effect on the ability of the phenolic hydroxyl to participate in general acid catalysis; rather it seems to reflect steric or electrostatic factors which accompany derivatization of the phenolic ring. Therefore, to account for the differential effect of Tyr248 modification on peptidase and esterase activities, one must invoke mechanistic differences

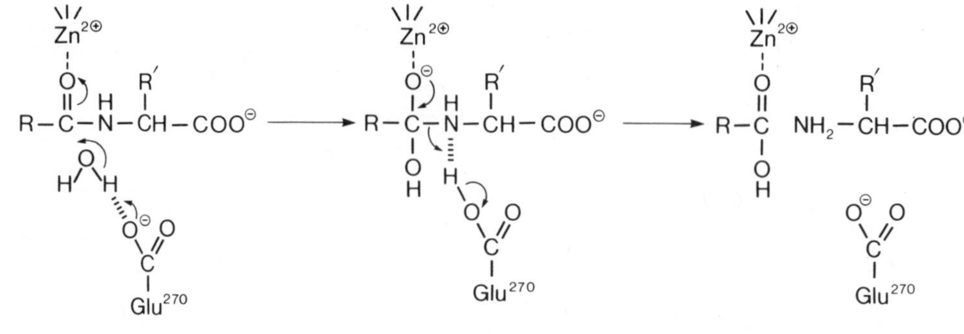

Enzyme Substrate Complex — Tetrahedral Intermediate — Enzyme Product Complex

Fig. 3. *Mechanistic scheme for carboxypeptidase A in which Glu270 acts as both a general-base catalyst and proton donor to the leaving group.*

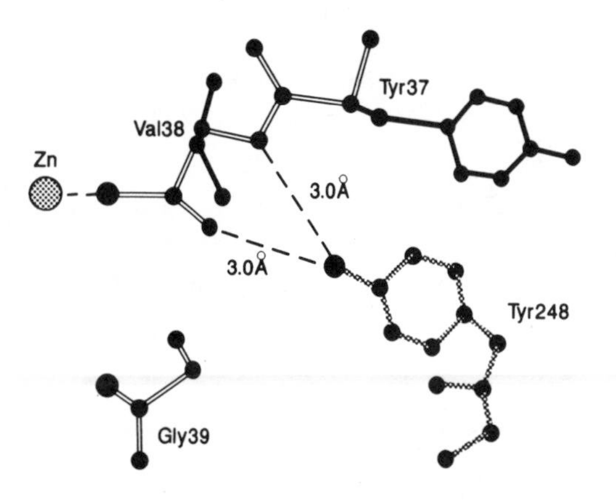

Fig. 4. *Hydrogen-bonding interactions of the Tyr248 phenolic hydroxyl in the bovine CPA-PCI complex. Note that Gly39 is hydrolyzed from the remainder of the inhibitor but remains trapped within the binding pocket.*

in the manner in which these different classes of substrates are handled at the active site [8].

Although the kcat values are not dramatically changed by removal of the Tyr248 phenolic hydroxyl, there is a significant increase in the Km value of the peptide substrate (Table I). The contribution of the phenolic hydroxyl to ligand binding was conclusively demonstrated by a 70-fold increase in the Ki of the potato carboxypeptidase inhibitor (PCI) [27] due to the Tyr248-to-phenylalanine replacement. This result is consistent with the X-ray structure of the bovine CPA-PCI complex which indicates that the phenolic hydroxyl of Tyr248 forms hydrogen bonds to both the amide proton and the carboxylate anion of Va138 of the inhibitor (Fig. 4) [25]. The apparent decrease in binding affinity (1.6–2.5 kcal/mol) exhibited by CPA-Phe248 is consistent with estimates for the predicted loss of two intermolecular hydrogen bonds [28].

This study establishes that the Tyr248 hydroxyl is involved in ligand binding but does not make a crucial contribution to the catalytic process. Additional alterations of amino acid residues within the active site may help to further elucidate the catalytic mechanism of CPA.

REDESIGNING TRYPSIN: ALTERATION OF SUBSTRATE SPECIFICITY

Trypsin is a pancreatic protease that belongs to a large family of homologous serine proteases. These enzymes utilize the amino acid, serine, at their active site and appear to have the same catalytic mechanism [29,30]. This biologically important class of proteins includes enzymes involved in protein degradation (trypsin, chymotrypsin, elastase, α-lytic protease, *S. griseus* protease A, *S. griseus* protease B), blood coagulation (thrombin), clot dissolution (plasmin), complement fixation (C1 protease), pain sensing (kallikrein), and fertilization (acrosomal enzyme). The diverse biological functions of the structurally homologous serine proteases are presumably the result of the different constellations of amino acids utilized by each enzyme for substrate binding.

To investigate the role of specific amino acids in the substrate specificity of trypsin, we isolated and sequenced the rat pancreatic trypsinogen gene [2] and cDNA [31]. A full-length copy of the trypsinogen coding sequence (including the signal peptide) was constructed that was uninterrupted by intervening sequences. A heterologous expression system for the controlled biosynthesis of wild-type and substrate-binding pocket mutant trypsinogens was then established using mammalian cells [32].

Although the three-dimensional structure for rat pancreatic trypsin is not known, the primary structure has 74% identity with bovine trypsin, whose three-dimensional crystal structure is known [33,34]. The active site residues His57, Asp102, and Ser195 are present in both enzymes and are located in regions with sequence identity. Similarly, Asp189 at the base of the substrate-binding pocket, which presumably confers the substrate specificity for arginyl and lysyl substrates, is present in the rat sequence. The glycine residues at positions 216 and 226, which ostensibly allow entry of large amino acid side chains into the pocket, are also conserved. Indeed, when the modeled structure of rat trypsin is compared with the actual structure of bovine trypsin complexed with pancreatic trypsin inhibitor [35] (with Lys15 of the inhibitor in the substrate binding pocket), there are no substitutions within 7.6 Å of the ligand side chain. We therefore used the bovine trypsin structures complexed with either the pancreatic trypsin inhibitor or benzamidine (an arginine analog) [36,37] as a model for the rat trypsin complexed with lysine and arginine substrates, respectively.

Comparison of the related tertiary structures of trypsin and elastase [38] suggests that glycine residues 216 and 226 are appropriate initial targets for probing the structural basis of the substrate specificity of trypsin. The analogs of Gly216 and Gly226 in elastase are Val216 and Thr226 [39], which restrict the substrate specificity of elastase to small hydrophobic amino acids. Therefore, we expected that the substrate-binding properties of trypsin would be altered if these residues were modified.

Assuming that the structures of rat and bovine trypsin are identical, the three-dimensional coordinates of any amino acid at positions 216 and 226 can be modeled to determine the distances between the amino acid replacements at these sites and the rest of the protein–ligand complex. Although the same binding pocket is shared by cationic substrates, the amino acid residues comprising the pocket are employed differently for lysyl substrate binding than for arginyl substrate binding. Both arginyl and lysyl substrates form hydrogen bonds with a water molecule which is bound in the specificity pocket by hydrogen bonds to the backbone carbonyl oxygens of Trp215 and Val227. The lysine side chain is bound in the specificity pocket by direct hydrogen bonds with Ser190 and by indirect hydrogen bonds (mediated through the water molecule) with Asp189.

In contrast to lysine, the longer arginine side chain extends deeper into the binding pocket than the lysine side chain, displacing the water molecule and forming a cyclic network of direct hydrogen bonds with Asp189. The guanidinium group fills the base of the substrate-binding pocket. Small changes in the tertiary structure of the trypsin substrate complex will result from adding methyl groups at the interface between the ligand and the specificity pocket. Since catalytic activity is dependent on the substrate alignment, which is determined in part by the specificity pocket, these changes may affect the kinetic constants Km, which relates to binding affinity, and kcat, which is a measure of catalytic activity. From modeling studies, the Trypsin-216Ala mutant is predicted to show relatively better activity for arginine as compared to lysine because the water molecule that is presumably displaced by Ala216 does not take part directly in binding of arginine. On the other hand, the Trypsin-226Ala mutant should show relatively enhanced lysine activity relative to arginine, because there is more space to accommodate the steric conflicts of Asp189, the methyl group at position 226, and the substrate at the base of the pocket (see Fig. 5).

To test these hypotheses, glycine residues at 216 and 226 were replaced by alanine residues by site-directed mutagenesis, resulting in three trypsin mutants (Trypsin-216Ala, Trypsin-226Ala, and Trypsin-216Ala,226Ala). Both the wild-type and the mutant enzymes were expressed in an SV40 expression system in COS cells (monkey kidney). A mammalian expression system was chosen to maximize the probability that native disulfide bond formation and secretion of the zymogen would occur. The wild type and a revertant enzyme in which the alanine at position 226 was site-specifically reverted to a glycine exhibited identical catalytic parameters. Results presented in Table II show that addition of a methyl group (alanine vs. glycine) in either position 216 or 226 compromises the ability of the pocket to accept either lysine or arginine substrates (Km's are higher). However, the catalytic activity can be altered in a discriminatory manner. The Trypsin-216Ala mutant does indeed operate more selectively on arginine substrates than wild type, and the Trypsin-226Ala mutant operates more selectively on lysine substrates. The double mutant Trypsin-216Ala,226Ala has a very low catalytic activity (1000-fold lower than normal) due to the greatly constricted binding pocket, showing a modest preference for lysine substrates. In addition, unpredictably, the Trypsin-226Ala and Tryspin-216Ala,226Ala mutants exhibit an altered conformation which resembles trypsinogen as detected by native gel electrophoresis [32]. A similar analysis shows that this conformation is converted to a trypsin-like conformation upon binding of a substrate analog. Thus, the two different conformational states of trypsin and trypsinogen are not determined by the zymogen peptide alone, but may also be produced by subtle changes in the interior of the molecule, namely the addition of a methyl group at amino acid position 226.

PROBING THE CATALYTIC MECHANISM OF TRYPSIN

The catalytic triad–Ser195, His57, and Asp102—is present in the catalytic sites of all serine proteases, from bacteria to humans. A serine residue was originally shown to be required for catalysis on the basis of its unique reactivity with organophosphates such as diisopropylfluorophosphate (DIFP) [40]. A histidine residue in trypsin was implicated in catalysis through its specific alkylation by the affinity label tosyl–lysine chloromethylketone (TLCK) [41]. The high-resolution crystal structure of trypsin revealed that His57 and

TABLE II. Modification of Arginine/Lysine Specificity of Trypsin*

	Arg		Lys		
	k_{cat}	K_m	k_{cat}	K_m	Arg/Lys[a]
Trypsin-216Ala	1.0	30	0.3	30	3.0
Trypsin-226Ala	0.01	40	0.1	25	0.05
Trypsin-216Ala, 226Ala	0.001	15	0.0005	2	0.3

*The data depicted are ratios of substrate-binding pocket mutant and wild-type trypsin kinetic parameters. The absolute values obtained are approximated for purposes of clarity.

[a]Ratio of kcat/Km values.

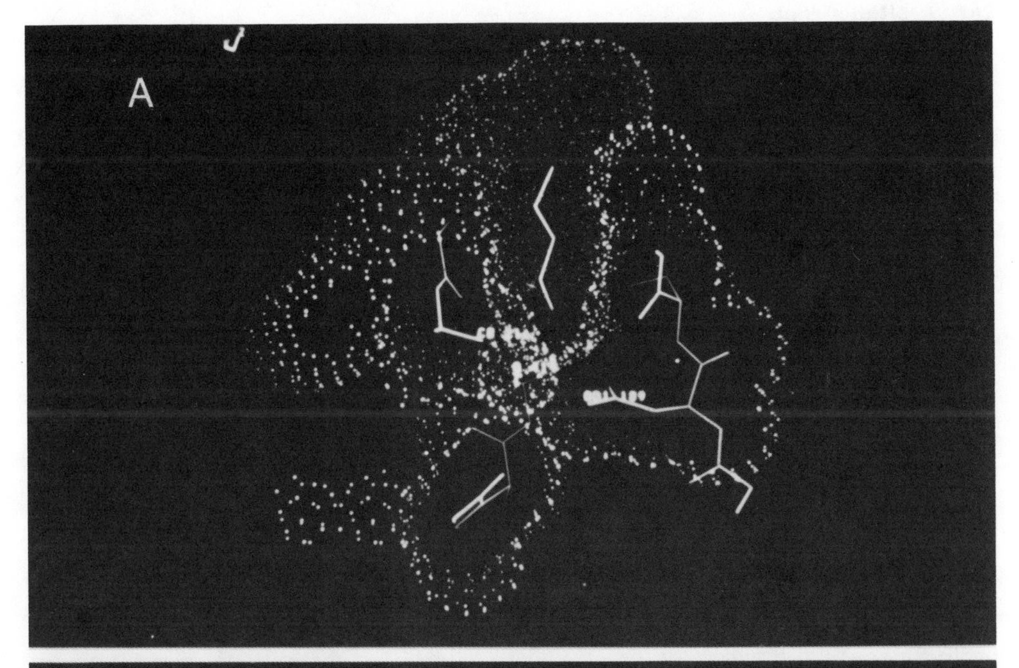

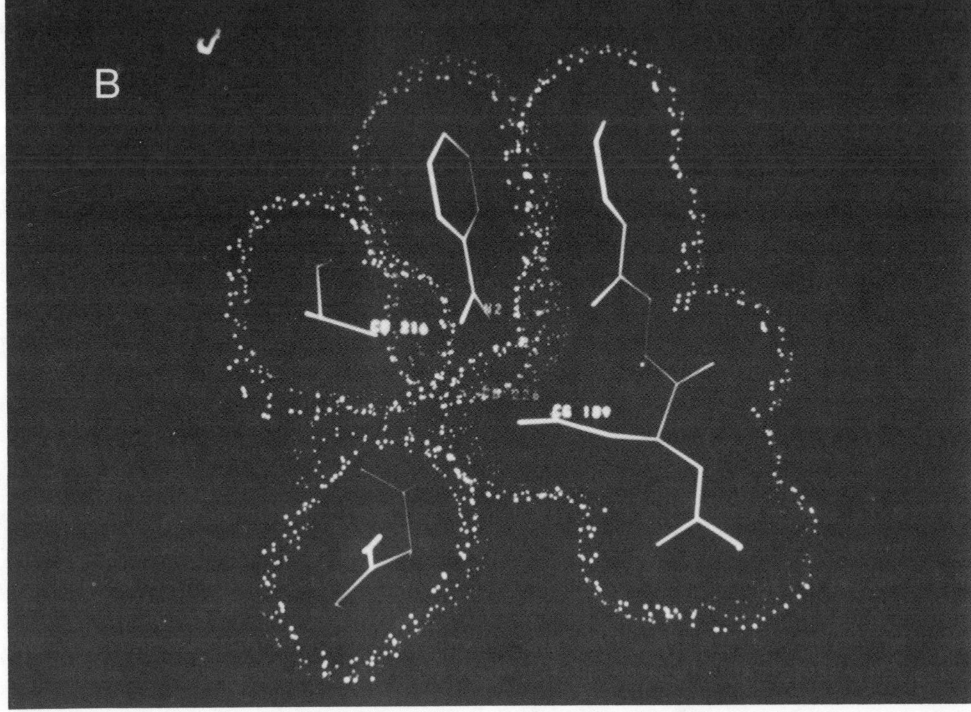

Fig. 5. *Space-filling representations of the substrate-binding pocket of mutant trypsins. The polypeptide backbone is shown for the chain segments, including amino acids 189 to 191, 214–217, and 224–227. The atoms are shown with their van der Waals contact surfaces. The enzyme atomic surface is blue and the ligand atomic surface is red. Atomic coordinates for the drawing were obtained from the Brookhaven Protein Data Bank (entry set 3PTB). (A) The Lys15 of pancreatic trypsin inhibitor bound in the trypsin specificity pocket. The van der Waals surface of the beta carbon of Ala216 overlaps that of the water molecule 414 and of Lys15. (B) Benzamidine bound in the trypsin specificity pocket. The van der Waals surface of the beta carbon of Ala226 overlaps that of the guanidinium group of benzamidine. The van der Waals overlap of the gamma carbon of Asp189 with the ligand is appropriate because a hydrogen bond is formed at this contact. (Reproduced in color on p. 347).*

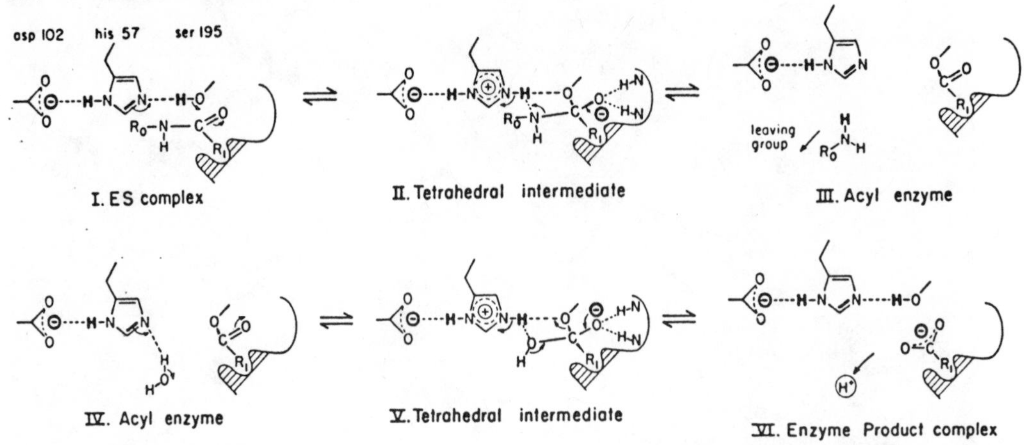

Fig. 6. *Proposed mechanism for serine protease hydrolysis of peptides. In this representation, the proton shuttle is concerted. (Adopted from Stroud, et al. [42] with permission.)*

the β-hydroxyl group of Ser195 are within hydrogen-bonding distance of one another [33,34]. These and other data are consistent with the catalytic mechanism in which His57 abstracts a proton from Ser195 and donates a proton to the leaving group (Fig. 6). Surprisingly, the three-dimensional structure also revealed that the carboxyl group of the buried Asp102 was within 2.8Å of the imidizole ring of His57, suggesting that the three residues, Asp102, His57, and Ser195, act as a "proton relay system." Thus, the aspartic acid, through its effect on His57 and Ser195, may aid in catalytic activity. Consistent with this theory, the Asp102 residue is invariant in the crystal structures of all studied serine proteases. However, its role in catalysis has never been verified, since selective chemical modification of the aspartic acid residue has not been possible. Evidence for its participation has relied on indirect methods such as nuclear magnetic resonance studies of the imidazole protons [43], neutron diffraction studies on deuterated trypsin [44], and quantum mechanical studies of the "proton relay system" [45]. These studies suggest that Asp102 shares the proton with the imidazole of His57, thereby 1) increasing the basicity of His57, 2) orienting the imidazole ring into a proper position for interacting with the substrate, and 3) increasing the nucleophilicity of Ser195.

We have directly examined the role of Asp102 in the catalytic mechanism of the serine proteases by replacing this residue in trypsin with an asparagine, an isosteric amino acid, by site-directed mutagenesis [46]. The mutant enzyme Trypsin-102Asn was expressed to high levels (10 mg/liter) by establishing a stable eucaryotic cell line that secreted trypsinogen into the culture medium. Chinese hamster ovary cells were cotransfected with a plasmid containing either the wild-type or the mutant trypsinogen gene under transcriptional control of the early promoter from Simian Virus 40 and a plasmid that encoded the bacterial phosphotransferase gene (neo). The neo gene confers resistance to the aminoglycoside antibiotic G418 [47], and permits the phenotypic selection of a cell line that expresses the gene. Cells that coexpressed the trypsinogen and neo genes were screened for overexpressors, and the cell line that produced the highest level of trypsinogen was expanded into mass culture (10 liters). Wild-type trypsin and Trypsin-102Asn were purified to homogeneity by a combination of ion-exchange and affinity chromatography techniques. Wild-type trypsin isolated from this expression system showed identical physical and catalytic properties as the natural rat enzyme. Trypsin-102Asn, however, exhibited remarkably different catalytic properties

TABLE III. Relative Activity of Trypsin-102Asn to Wild-Type Trypsin at Neutral and Alkaline pH*

	k_{cat}	K_m
Neutral pH (7.0)	.0002	2
Alkaline pH (10.5)	.15	10

*Trypsin-102Asn mutants have dramatically lower catalytic activity at neutral pH but approach wild-type activity at alkaline pH. The values shown are ratios of Trypsin-102Asn and wild-type trypsin.

TABLE IV. Relative Reactivity of Trypsin-102Asn to Wild-Type Trypsin with Active Site Titrants*

	Trypsin	Trypsin-102Asn
DIFP[a]	1	.0001
MUGB[b]	1	<.002
TLCK[c]	1	.2

*Trypsin-102Asn shows dramatically lowered reactivity with Ser195-directed active site reagents but not with a histidine-specific reagent. Values shown are relative to wild-type trypsin.
[a]DIFP, diisopropylfluorophosphate, phosphorylates Ser195.
[b]MUGB, 4-methylumbelliferyl p-guanidinobenzoate, monitors the acylation reaction.
[c]TLCK, tosyl L-lysine chloromethylketone, specifically alkylates histidine 57.

(Table III). Kinetic analysis below pH 7 indicated that the acylation reaction follows a titration curve similar to that found with the native enzyme, except that the pKa of histidine is lowered about 1.5 pH units to 5.3, and the maximal rate constant is at least 5000-fold lower than that for wild-type trypsin. Thus, Asp102 is clearly important but not absolutely required for catalytic activity. The acylation of Trypsin-102Asn still occurs at a rate 400-fold greater than alkaline hydrolysis of the same ester substrae. Above pH 7, the rate of the reaction increases in direct proportion to the solvent hydroxide ion concentration, such that at pH 10.5 the kcat value of the Trypsin-102Asn catalyzed hydrolysis of esters is about 15% that of the wild-type enzyme (Table III).

To determine the chemical reactivity of the active site residues of Trypsin-102Asn, we employed the active site reagents DIFP, 4-methylumbelliferyl p-guanidinobenzoate (MUGB), and TLCK (Table IV). The reactivity of Trypsin-102Asn with the Ser195-specific reagent DIFP is approximately 10,000-fold less than that observed for wild-type trypsin. In addition, the mutant enzyme is at least 500-fold less reactive than wild-type trypsin with the acylating reagent, MUGB. Thus, the nucleophilicity of Ser195 is greatly compromised in Trypsin-102Asn. On the other hand, the reactivity of Trypsin-102Asn with TLCK (a His57-specific affinity label agent) is only decreased 5-fold. This suggests that the histidine can be properly positioned in the active site of the mutant enzyme for reaction with the affinity label. The primary effect on catalysis of the aspartic acid to asparagine replacement at position 102 therefore appears to be on Ser195, implying that it is the network of hydrogen bonds involving Asp102, His57, and Ser195 that leads to the strong nucleophilic character of Ser195 which ultimately accounts for its role in catalysis. This supports the importance of the catalytic triad: all the residues are required for optimal function.

Surprisingly, Trypsin-102Asn displays a unique set of catalytic characteristics that qualitatively differentiate it from the parent enzyme. Trypsin-102Asn is a good catalyst at basic pH because of its ability to utilize solvent hydroxyl in the catalytic reaction. It is not yet known whether the hydroxyl interacts directly with the histidine residue, with the serine residue, or facilitates the reaction in some other manner. This is the first example of the production via genetic engineering of an enzyme with a qualitatively different reaction mechanism. It illustrates the possibility of producing new structures that show reduced activity under physiological conditions but are efficient catalysts under other circumstances.

Although the solution studies described above provide substantial information on the functional consequences of the site-specifically mutated enzymes, a three-dimensional structural analysis is required to fully understand the modified functional state. Therefore, we crystallized Trypsin-102Asn by vapor diffusion against polyethylene glycol at pH 6.0 in the presence of benzamidine. The crystals attained 0.5–1.0 mm dimensions and are suit-

Fig. 7. *Crystals of trypsin 102 Asn. The crystals range in size from 0.5–1 mm in length and were grown in polyethyene glycol at pH 6.0 in the presence of benzamidine using vapor diffusion and the hanging drop method.*

able for study by X-ray crystallography (Fig. 7). The packing of the molecules in the rat trypsin crystal differs from that of the bovine crystal. This difference is probably not a consequence of the engineered mutation at position 102. The three-dimensional structure of the rat structure has been determined by molecular replacement methods using the atomic coordinates of the bovine enzyme as a starting model [48].

The three-dimensional structure of rat Trypsin-102Asn is currently determined at 2.3 Å resolution. This permits the visualization of side-chain positions and hydrogen-bonding patterns. From the similarity in primary structure between the rat and bovine enzymes and the overall tertiary structure similarity between the serine proteases with known structures, we assumed that rat trypsin would be structurally similar to bovine trypsin. The X-ray structure results confirm that there is nearly exact structural identity between the rat and bovine enzymes. There is 0.5 Å deviation between similar atoms of the polypeptide backbone in the two structures. Although subtle differences were found between the two structures at the active site region, the asparagine residue in Tryspin-102Asn occupies the same site as Asp102 in the bovine structure. An analysis of the conformational differences of these and other residues of the enzyme structure is presently in progress.

These studies involving genetics, recombinant DNA technology, enzymology, protein chemistry, and crystallography reveal the importance of a coordinated multidisciplinary approach to structure/function studies. Results from these studies will provide a deeper understanding of the mechanism of action of naturally occurring proteins and provide a practical framework for designing new proteins with different structures and functions for a variety of purposes.

ACKNOWLEDGMENTS

We thank Ms. L. Spector for preparing the manuscript and gratefully acknowledge grant support from NSF PCM830610 to W.J.R. and DMB8608086 to C.S.C.

REFERENCES

1. Quinto C, Quiroga M, Swain WF, Nikovits WC Jr, Standring DN, Pictet RL, Valenzuela P, Rutter WJ (1982): Proc Nat Acad Sci USA 799:31–35.
2. Craik CS, Choo Q-L, Swift GH, Quinto C, MacDonald RJ, Rutter WJ (1984): J Biol Chem 259:14255–14264.
3. Bradshaw RA, Ericsson LH, Walsh KA, Neurath H (1969): Proc Nat Acad Sci USA 63:1389–1392.
4. Quiocho FA, Lipscomb WN (1971): Adv Prot Chem 25:1–78.
5. Rees DC, Lewis M, Lipscomb WN (1983): J Mol Biol 168:367–387.
6. Rees DC, Lipscomb WN (1981): Proc Nat Acad Sci USA 78:5455–5459.
7. Lipscomb WN (1980): Proc Nat Acad Sci USA 77:3875–3878.
8. Vallee BL, Galdes A, Auld DS, Riordan JF (1983): In "Sprio TG (ed): "Metal Ions in Biology," vol 5. New York: John Wiley and Sons, pp 26–75.
9. Kaiser ET, Kaiser BL (1972): Acta Chem Res 5:219–224.

10. Mock WL, Chen JT (1980): Arch Biochem Biophys 203:542–552.
11. Breslow R, Wernick DL (1977): Proc Nat Acad Sci USA 74:1303–1307.
12. Makinen MW, Wells GB, Kang S (1985): In Eichhorn GL, Marzilli LG (eds): "Advances in Inorganic Biochemistry," vol 6. Amsterdam: Elsevier, pp 1–69.
13. Walsh C (1979): In "Enzymatic Reaction Mechanisms." San Francisco: WH Freeman & Co, p 106.
14. Simpson RT, Riordan JF, Vallee BL (1963): Biochemistry 3:616–622.
15. Riordan JF, Sokolovsky M, Vallee BL (1967): Biochemistry 6:3609–3617.
16. Urdea MS, Legg JI (1979): J Biol Chem 254:11868–11874.
17. Gardell SJ, Craik CS, Hilvert D, Urdea M, Rutter WJ (1985): Nature 317:551–555.
18. Emr SD, Schekman R, Flessel MC, Thorner J (1983): Proc Nat Acad Sci USA 80:7080–7084.
19. Brake AJ, Merryweather JP, Coit DG, Heberlein VA, Masiarz FR, Mullenbach GT, Urdea MS, Valenzuela P, Barr PJ (1984): Proc Nat Acad Sci USA 81:4642–4646.
20. Kurjan J, Herskowitz I (1982): Cell 30:933–943.
21. Julius D, Schekman R, Thorner J (1984): Cell 36:309–318.
22. Julius D, Brake A, Blau L, Kunisawa R, Thorner J (1984): Cell 37:1075–1089.
23. Cueni L, Riordan JF (1978): Biochemistry 17:1834–1842.
24. Jencks W (1975): Adv Enzymol 43:219–405.
25. Rees DC, Lipscomb WN (1982): J Mol Biol 160:475–498.
26. Monzingo AF, Matthews BW (1984): Biochemistry 23:5724–5729.
27. Hass GM, Ryan CA (1981): In Lorand L (ed): "Methods in Enzymology," vol 80. New York: Academic Press, pp 778–791.
28. Fersht AR, Shi J, Knell-Jones J, Lowe DM, Wilkinson AJ, Blow DM, Brick P, Carter P, Waye MMY, Winter G (1985): Nature 314:235–238.
29. Neurath H, Walsh KA, Winter WP (1967): Science 158:1638–1644.
30. Kraut J (1977): Ann Rev Biochem 46:331–358.
31. MacDonald RJ, Stary SJ, Swift GH (1982): J Biol Chem 257:9724–9732.
32. Craik CS, Largman C, Fletcher T, Roczniak S, Barr PJ, Fletterick R, Rutter WJ (1985): Science 228:291–297.
33. Stroud RM, Kay LM, Dickerson RE (1974): J Mol Biol 83:185–208.
34. Huber R, Kukla D, Bode W, Schwager P, Bartel K, Diesenhofer J, Steigemann W (1974): J Mol Biol 89:73–101.
35. Bode W, Schwager P, Huber R (1976): "Miami Winter Symposium," vol 11. New York: Academic Press, pp 43–76.
36. Bode W, Walter J, Huber R, Wenzel HR, Tschesche H (1984): Eur J Biochem 144:185–190.
37. Bode W, Schwager P (1975): J Mol Biol 98:693–717.
38. Sawyer L, Shotton DM, Campbell JW, Wendel PL, Muirhead H, Watson HC, Diamond R, Ladner RC (1978): J Mol Biol 118:137–208.
39. Shotton DM, Watson HC (1970): Nature 225:811–816.
40. Dixon GH, Go S, Neurath H (1956): Biochim Biophys Acta 19:193–195.
41. Shaw E, Mares-Guia M, Cohen W (1965): Biochemistry 4:2219–2224.
42. Stroud RM, Krieger M, Koeppe RE II, Kossiakoff AA, Chambers JL (1975): "Proteases and Biological Control," Cold Spring Harbor, Cold Spring Harbor Laboratory, pp 13–32.
43. Robillard G, Shulman RG (1974): J Mol Biol 86:541–558.
44. Kossiakoff AA, Spencer SA (1980): Nature 228:414–416.
45. Umeyama H, Hirono S, Nakagawa S (1984): Proc Nat Acad Sci USA 81:6266–6270.
46. Craik CS, Roczniak S, Largman C, Rutter WJ (1986): Science, submitted.
47. Davies J, Smith DI (1978): Ann Rev Microbiol 32:469–518.
48. Sprang S, Standing T, Fletterick R, Finer-Moore J, Stroud R, Xuong N-H, Hamlin R, Rutter WJ, Craik CS (1986): Science, submitted.

Protein Engineering, pages 269–278

24

Structure and Activity of the Tyrosyl-tRNA Synthetase

Alan R. Fersht and Robin J. Leatherbarrow

Department of Chemistry, Imperial College of Science and Technology, London SW7 2AY, United Kingdom

INTRODUCTION

The first enzyme of known three-dimensional structure to be studied by protein engineering is the tyrosyl-tRNA synthetase from *Bacillus stearothermophilus* [1,2]. Its properties illustrate well what is required of an enzyme-system for a rapid kinetic analysis of the relationship between structure and activity. The tyrosyl-tRNA synthetase is almost an ideal system from the point of view of kinetics. It catalyzes the aminoacylation of tRNA in a two step reaction, activation (eq. 1) followed by transfer (eq. 2). In the absence of added

$$E + Tyr + ATP \rightleftharpoons E \cdot Tyr\text{-}AMP + PP_i \quad (1)$$

$$E \cdot Tyr\text{-}AMP + tRNA^{Tyr} \rightarrow Tyr\text{-}tRNA^{Tyr} + AMP + PP_i \quad (2)$$

pyrophosphate or tRNA, the E.Tyr–AMP complex is stable and may be assayed or handled in solution. This leads to several important consequences. The amount of active enzyme may be accurately titrated by the formation of E·[^{14}C]Tyr–AMP which may be trapped on a nitrocellulose filter disk. The rate constants for both activation and transfer are nicely in the time range of stopped-flow and quenched-flow studies (t 1/2 ~ 15 ms) and can be measured on very small amounts of material using the inherent fluorescence of the enzyme or readily available radio-labelled derivatives. Further, a stable crystalline E·Tyr–AMP complex is formed so that the structure of the E·Tyr–AMP complex has been solved [3] in addition to that of the free enzyme [4]. This is a rare possibility since enzyme-intermediate complexes usually decompose on a time scale of milliseconds while the gathering of X-ray data usually takes hours.

The enzyme is a symmetrical dimer of M_r 2 × 47,500. Kinetic studies indicate interesting subunit interactions since the enzyme in solution binds tightly only 1 mol of tyrosine or 1 mol of tRNA per mol of dimer and forms rapidly only 1 mol of tyrosyl adenylate [5].

The gene of the enzyme has been cloned into M13 and is expressed in large quantities from *Escherichia coli* that has been infected with the phage [6]. This is convenient for three reasons. First, oligodeoxynucleotide-directed mutagenesis is conducted very conveniently in M13. Second, Sanger's dideoxy sequencing [7] is conducted in M13, and so we routinely sequence the complete mutant enzyme genes to search for the desired and adventitious mutations. (This takes just one day per mutant using a set of synthetic primers spaced along the sequence.) Third, the tyrosyl-tRNA synthetase form *B. stearothermophilus* is thermostable, like many enzymes from thermophiles, while the enzyme from *E. coli* is unstable at elevated temperatures. The relatively low amounts of *E. coli* tyrosyl-tRNA synthetase (and many other host enzymes) in preparations of the *B. stearothermophilus* enzyme produced from the cloned gene are precipitated completely by incubation for 30 min at 56°C; at this temperature, the enzyme from *B. stearothermophilus* is stable. This is important when performing steady-state kinetic measurements on mutants of low activity since their activities would otherwise be obscured by host enzyme. (Note also that genetic recombination between the cloned gene and the chromsomal gene is minimized when the two are from different species.)

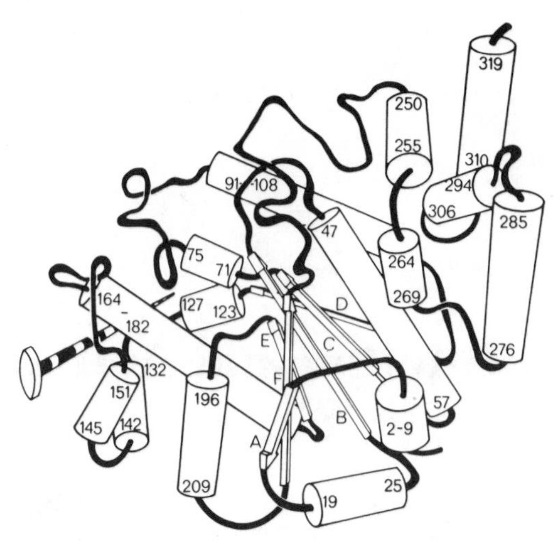

Fig. 1. *Structure of the tyrosyl adenylate binding domain of the tyrosyl–tRNA synthetase. (From Blow and Brick [4].)*

STRUCTURE OF TYROSYL-tRNA SYNTHETASE AND ITS COMPLEX WITH TYROSYL ADENYLATE

Although there are 419 amino acid residues in each subunit, only the first 319 are seen in the crystal structure, (Figure 1) [6]. The 100 amino acid residues of the C-terminus are too mobile, either temporally or spatially, to be resolved. A truncated enzyme consisting of residues 1–319 was constructed by deletion of a segment of the gene and found to be identical to the full-length enzyme in the activation reaction [12]. It neither aminoacylates or binds tRNA. The structural organization of the enzyme parallels the functional, there being separate domains for the separate activities.

The enzyme-bound tyrosyl adenylate complex (Figure 2) is a nice example of complementary interactions between an enzyme and substrate. Sketched in Figure 2 are the hydrogen bonds that are made. Note in particular the binding site for the hydroxyl of tyrosine. This makes two hydrogen bonds with the enzyme: one to the carboxylate of Asp176 and the other to the hydroxyl of Tyr34. These interactions are particularly important because it is they that distinguish between the binding of phenylalanine and tyrosine to the enzyme and are thus responsible for the fidelity of protein biosynthesis. This is done with such precision by these interactions in the tyrosyl-tRNA synthetase—an accuracy of one part in 10^5—that the enzyme does not require an editing mechanism [9].

STRATEGY FOR STUDYING STRUCTURE–ACTIVITY RELATIONSHIPS

A characteristic of enzyme catalysis is that the binding energy between the enzyme and substrate is utilized in catalysis and specificity. There is a simple equation relating the rate constant k_{cat}/K_M (the "specificity constant") of the Michaelis-Menten equation to the binding energy of the enzyme and substrate [10,11] (eq. 3, where ΔG_s is the

$$RT\ell n(k_{cat}/K_M) = RT\ell n(kT/h) - \Delta G^{\ddagger} - \Delta G_s \quad (3)$$

binding energy between the enzyme and substrate, $\Delta G^{\ddagger}$ is the energy of activation of the chemical steps, and k and h are the constants of Boltzmann and Planck, respectively.)

It can be seen from eq. 3 that, since ΔG_s is algebraically negative for favorable interactions, binding energy is used directly to increase the value of the specificity constant. Using site-directed mutagenesis, we are able to alter the binding energy term ΔG_s by changing the side chains of the amino acid residues

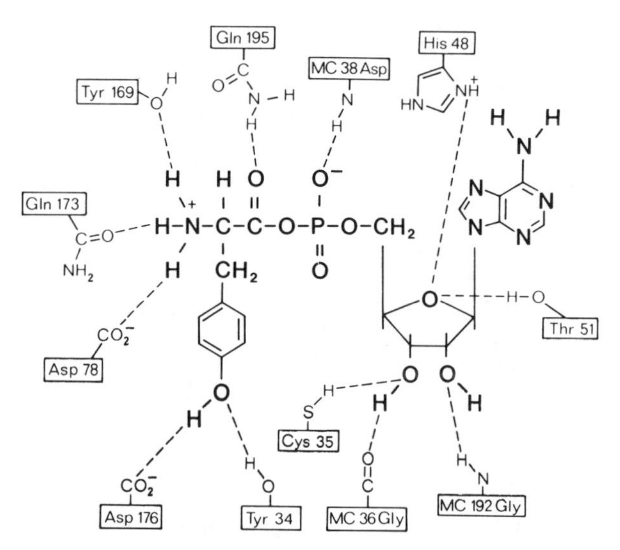

Fig. 2. *Sketch of the active site for the binding of tyrosyl adenylate and the possible hydrogen bond interactions. All the interactions from side chains have now been confirmed by directed mutagenesis.*

that interact with the substrate. The changes of binding energy may be calculated from measured values of k_{cat}/K_M for mutant and wild-type enzymes using eq. 4 (where

$$\Delta G_i = RT\ell n(k_{cat}/K_M)_{mut}/(k_{cat}/K_M)_{wt} \quad (4)$$

ΔG_i is the change of interaction energy on mutating the wild-type enzyme (subscript wt) to mutant (subscript mut). (Mutation of side chains that do not take part directly in catalysis does not alter $\Delta G^{\ddagger}$.)

Equation 3 was orginally formulated for the variation of a series of substrates reacting with the same enzyme. Specifically, it applied to the comparison of a specific substrate with isosteric or smaller substrates: steric repulsion on attempting to cram a too-large substrate into an active site clearly obscures any attempt to measure binding energies but will measure repulsion energies. When applied to enzymes, the same caveat applies: Mutation must be made to smaller or isosteric side chains. Further, it must always be borne in mind that mutation of a side chain could cause a conformational change in the enzyme or induce other artifacts. Ways of testing for this are discussed later. Equation 4 thus indicates a strategy for determining the importance of specific interactions between an enzyme and substrate in catalysis: the group that is being investigated is replaced by a smaller (or isosteric) group and the values of k_{cat}/K_M are measured. We now describe two sets of results that shed general light on the nature of enzyme catalysis and specificity.

HYDROGEN BONDING IN BINDING AND SPECIFICITY

The hydrogen bond is a ubiquitous feature of biological interactions: It has an essential role in determining the structure of proteins and nucleic acids; it is a major determinant of specificity in enzyme catalysis and in biological information transfer; and it can directly influence the rate of enzymic reactions by stabilizing ionic charges that are formed in the transition state. Hydrogen bonding in macromolecules and their complexes in aqueous solution is a complex phenomenon because water competes for the hydrogen bonding sites

[12,13]. The calculation of the overall energetics is consequently difficult, and there is little knowledge of the energies involved. Examination of the structure of the enzyme-bound tyrosyl adenylate complex (Figure 2) shows a large number and variety of hydrogen bonds that may be readily mutated. Such mutational experiments should provide an experimental means of analyzing the energetics and role of hydrogen bond. Accordingly, the initial strategy of our experiments was to alter residues that form hydrogen bonds with the substrates.

The data from the mutation of a large number of hydrogen bonds may be summarized thus [14]: Deletion of a side chain between enzyme and substrate to leave an unpaired, uncharged, hydrogen bond donor or acceptor weakens binding energy by only 0.5–1.5 kcal/mol. But, the presence of an unpaired and charged donor or acceptor weakens binding by a further 3 or so kcal/mol. These values are much lower than the absolute strengths of hydrogen bonds in vacuo and are the consequence of hydrogen bonding in aqueous solution being an *exchange process:* Hydrogen-bonding groups on enzyme and substrate in water are hydrogen bonded with water and so on, forming the enzyme-substrate complex. The enzyme and substrate just exchange their hydrogen bonds with water to form bonds with each other (eq. 5, where –H = hydrogen-bonding group and –B =

$$E\text{–}H\cdots OH_2 + HOH\cdots B\text{–}S = [E\text{–}H\cdots B\text{–}S] + HOH\cdots OH_2 \quad (5)$$

hydrogen bond accepting group.) The process is thus very nearly isoenthalpic [14]. On deletion of one of the hydrogen bonding groups, as in eq. 6, a hydrogen bond inventory, i.e., counting the number

$$E \quad OH_2 + HOH\cdots B\text{–}S = [E \quad B\text{–}S] + HOH\cdots OH_2 \quad (6)$$

of hydrogen bonds on each side of the equation, shows that the process does not lead to the loss of the full strength of a hydrogen bond. The loss of the bond in the enzyme–substrate complex on the right in eq. 6 is partly compensated by the loss of the a hydrogen bond between the enzyme and water on the left. The loss of a hydrogen bond to a charged group in the complex is much more important. For example, on mutation of Tyr-Phe164, the positively charged $-NH_3^+$ group of the substrate is left partly unsolvated. A hydrogen-bond inventory of the mutation (eq. 7) shows

$$E \quad H_2O + H_2O\cdots H_3N^+\text{–}S = [E \quad H_3N^+\text{–}S] + HOH\cdots OH_2 \quad (7)$$

that a strong hydrogen bond between water and the charged group is lost on forming the enzyme–substrate complex. Accordingly, we find that the presence of an unpaired charged hydrogen-bonding group weakens binding by some 3.5–4.5 kcal/mol. Detailed discussion and further examples are given in reference 14.

HYDROGEN BONDING IN CATALYSIS

How does the enzyme catalyze the activation of tyrosine? The chemical mechanisms of the reaction, illustrated in Figure 3, involves the simple attack of the carboxylate ion of tyrosine on the alpha-phosphoryl group of ATP to generate a pentacoordinate intermediate, which eliminates magnesium pyrophosphate [15]. The classical enzymatic processes of acid–base or covalent catalysis would not seem to be of any use here, as the reaction consists of the attack of a fully ionized good nucleophile on an activated compound with a good leaving group (Mg-pyrophosphate from ATP). Protein engineering has in fact revealed the mechanism of the reaction and has shed light on a general mechanism of enzymatic catalysis. There are two famous publications in the history of enzymology which propose that enzymes can utilize the binding energy with their substrates to increase the rate of chemical catalysis. Haldane [16] suggested in

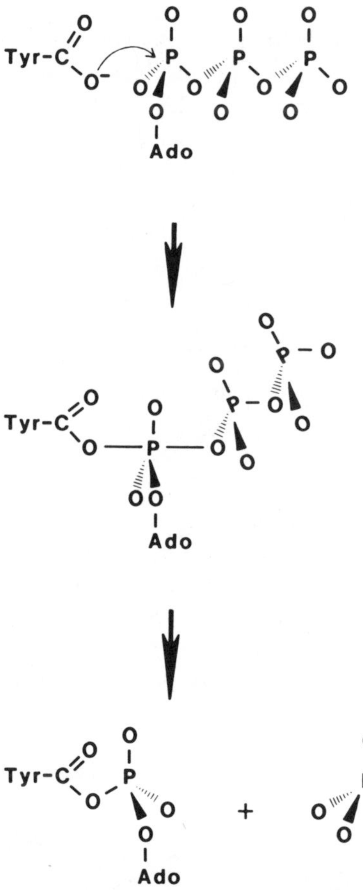

Fig. 3. *Chemical mechanism for the formation of tyrosyl adenylate. The transition state involves a pentacoordinate phosphorus. (From Leatherbarrow et al. [18].)*

1930 that an enzyme could catalyse a cleavage reaction by the binding sites on an enzyme helping pull the substrate apart and, conversely, in the reverse reaction, helping push the reagents together. Pauling [17] put this in more mechanistic terms by stating that the structure of an enzyme should be complementary to the structure of the transition state of its substrate rather than the unreacted substrate so that the substrate is strained towards the unstable transition state— the strain theory of enzyme catalysis. However, theoretical considerations [10,11] show that reaction rate is optimized simply when the binding energies of interactions are realized in the enzyme–transition state complex rather than the enzyme–substrate complex, without the need to invoke substrate distortion–the concept of transition state stabilization (Figure 4). In this way, the improved binding energy on going from substrate to transition state lowers the activation energy of the reaction and so increases the catalytic rate. Further, hydrogen bonds should be particularly important in mediating such differential binding effects, because the strength of hydrogen bonding varies strongly with interatomic distance and so is sensitive to the movement of atoms during the reaction.

The advent of protein engineering allows transition state stabilization to be tested directly. An amino acid side chain which interacts with the substrate can be altered to remove the interaction. If the side chain binds equally

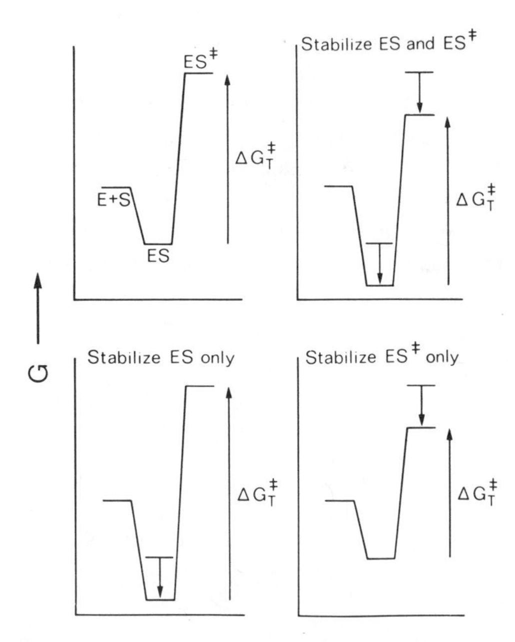

Fig. 4. *Effects of changes in binding energy on the energy levels of enzyme-bound complexes. An interaction which stabilizes the transition state only, and not the ground state, leads to an increased value of* k_{cat}. *(From Fersht [11].)*

well with the substrate in both the ground and transition states then removal should raise K_S and not affect k_{cat}. Conversely, if the side chain binds the substrate only in the transition state, then removal should leave K_s unaltered, and just lower the value of k_{cat} (Figure 4).

Such a mechanism has indeed been found for the activation step of the tyrosyl-tRNA synthetase [18]. There is a binding site for the gamma-phosphoryl group of ATP, consisting of the hydroxyl of Thr 40 and the imidazole of His45 (Figure 5), that provides little binding energy for unreacted ATP but stabilizes the transition state of the reaction. As seen in Table I, mutation of Thr40 and His45 hardly alters the dissociation constant of ATP from the E.Tyr.ATP complex but dramatically affects catalysis. It can do this by taking advantage of the change in geometry of the ATP about the alpha-phosphoryl group of ATP as it goes from 4-coordinate to 5-coordinate. This could enable the gamma-phosphoryl group to swing into its binding site (Figure 6).

Removal of the side chains that bind ATP elsewhere also lowers k_{cat} significantly, but has a much smaller effect on K_S for ATP (Table II) [19]. The preferential binding of ATP in the transition state enhances catalysis by a factor of greater than 10^5. Removal of side chains that bind to tyrosine (Tyr→Phe34, Tyr→Phe169) does not significantly alter k_{cat} but increases the dissociation constant of tyrosine from the E ·Tyr–ATP complex.

Showing that the transition state binds better to the enzyme does not prove which particular mechanism is causing this [20]. Whatever the ambiguities, it is inferred that during the reaction the tyrosine does not move relative to the enzyme, but the ATP and enzyme move to a more optimal conformation as the transition state is reached.

IMPROVING THE ENZYME IN VITRO

One result of particular interest for designing enzymes to catalyze reactions in vitro is that we have been able to improve quite dramatically the affinity of the enzyme for ATP. This was done rationally by considering the hydrogen bonding between the enzyme and tryosyl adenylate. All the residues illustrated in Figure 2 are conserved in the highly homologous tyrosyl-tRNA synthetase from *E. coli,* apart from Thr51. A proline occupies this position, and a proline cannot make the necessary hydrogen bond with ATP. We also suspected from preliminary crystallographic data that the bond between Thr 51 and the ribose ring oxygen is too long to be energetically favorable, at about ~3.5 A instead of the normal -0-0- distance of ~2.8A in an -OH · · · O- hydrogen bond. Accordingly, Thr51 was mutated to Ala51, which resulted in a two-fold improvement in K_M for ATP

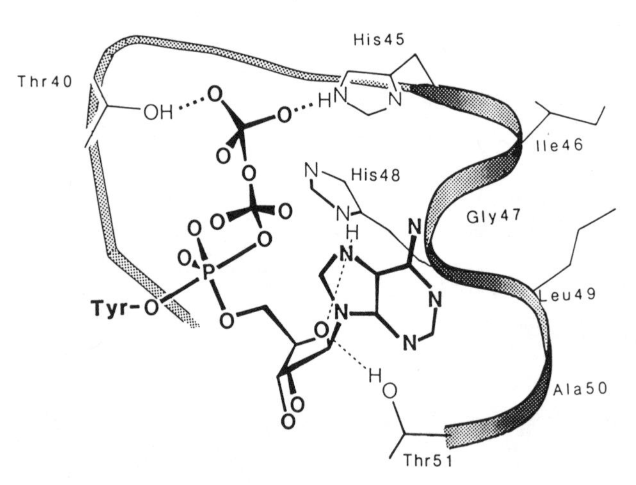

Fig. 5. *Model building the pentacoordinate transition state into the enzyme. (From Leatherbarrow et al. [18].)*

TABLE I. Presteady-State Kinetic Parameters for the Formation of Tyrosyl Adenylate

Enzyme	k_3 (s^{-1})	K_S Tyr (μM)	K_S ATP (mM)
TyrTS	38	12	4.7
TyrTS(His–Gly45)	0.16	10	1.2
TyrTS(Thr–Ala40)	0.0055	8.0	3.8
TyrTS(Thr–Ala;His–Gly45)	0.0012	4.5	1.1

From Leatherbarrow et al. [18]

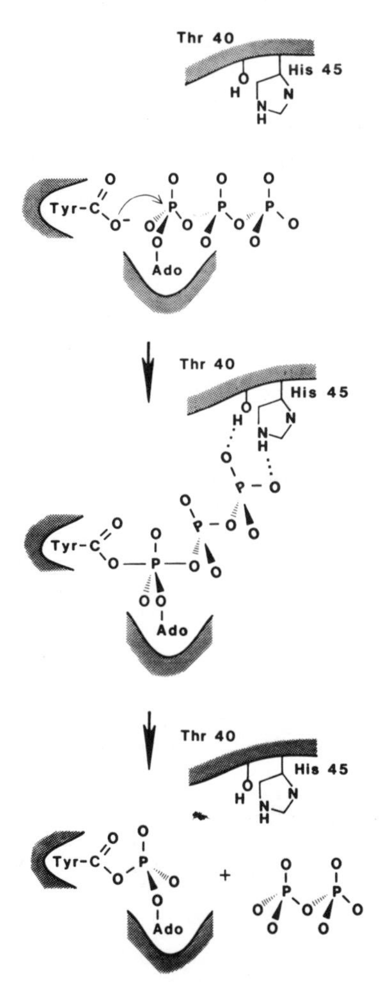

Fig. 6. *Possible mechanism for the transition state stabilization by Thr40 and His45. The binding energy of the two residues is not realized until the transition state is reached. (From Leatherbarrow et al. [18].)*

(Table III), and to Pro51, which resulted in a 100-fold better K_M for ATP in aminoacylation. More recently, we have shown directly that the interatomic distance between Thr51 and the ring ribose is too long. The distance of ~3.5 A is correct for an -SH · · · O- bond (Figure 7), and we find that mutation of Thr→Cys51 leads to an increase in the affinity of the enzyme for ATP [21].

We have thus generated a family of enzymes with different amino acids at position 51 exhibiting increasingly improved values of K_M for ATP (Table III). The improved K_Ms are at the slight expense of k_{cat}. This results in each one of the series being the most active at a different concentration range of ATP (Figure 8). Thus, one can tailor an enzyme for the most appropriate substrate concentration in vitro.

TABLE II. Activation of Tyrosine by Tyrosyl-tRNA Synthetases

Enzyme	k_3 (s^{-1})	K_s ATP (mM)	K_s Tyr (μM)
Wild type	38	4.7	12
(321-419)	34	5.2	12
Tyrosine binding-site mutants			
Tyr→Phe34	35	4.4	29
Tyr→Phe169	35	4.6	1,320
ATP binding-site mutants			
Cys→Ser35	4.7	4.8	8
Cys→Gly35	4.0	4.5	11
His→Gly48	9.9	9.9	23
Thr→Ala51	75	4.7	12

From Wells and Fersht [19]

TABLE III. ATP Dependence of Aminoacylation of tRNA

Enzyme	k_{cat} (s^{-1})	K_M (mM)	k_{cat}/K_M ($s^{-1}M^{-1}$)
TyrTS	4.7	2.5	1860
TyrTS(Ala51)	4.0	1.25	3200
TyrTS(Cys51)	2.9	0.29	8920
TyrTS(Pro51)	1.8	0.019	95800

From Fersht et al. [21]

DETECTION OF STRUCTURAL CHANGES BY THE DOUBLE MUTANT TEST

There is one recurrent problem in mutating residues: Does the alteration of a side chain cause just a localized effect, or does it cause changes that are propagated through the enzyme? Gross changes in structure may be detected by X-ray crystallography (>0.1 Å for structures determined at high resolution), or with perhaps even more sensitivity by NMR. As an alternative to these techniques, we have introduced a simple protein-engineering test to detect the propagation of changes in an enzyme or enzyme–substrate complex — the "double mutant" test [22]. This was inspired by the enormous change in ATP affinity on the mutation of Thr→Pro51, which places a helix-destabilizing residue in a helix (Figure 1). The change in affinity is too large to be accounted for by just the removal of the poor hydrogen bond, and so we reasoned that the proline must be causing a structural change. The residue most likely to be affected is the third residue towards the N-terminal from the proline, i.e., His48. By making the combination of single and double mutants (Thr→Pro51, His→Gly48, and [His→Gly48;Thr→Pro51]), we showed that the effect of mutation of Thr→Pro51 was entirely lost when there was glycine at position 48 rather than histidine (Figure 9) [22]. This procedure may be used as a general method to detect whether the mutation at one position affects binding at another. Figure 9 may be generalized as in Figure 10. Two residues, A and B are separately mutated to generate AB→A′B and AB→AB′, and doubly mutated to give A′B′. If the mutations are entirely independent, i.e., cause no structural changes in the free enzyme

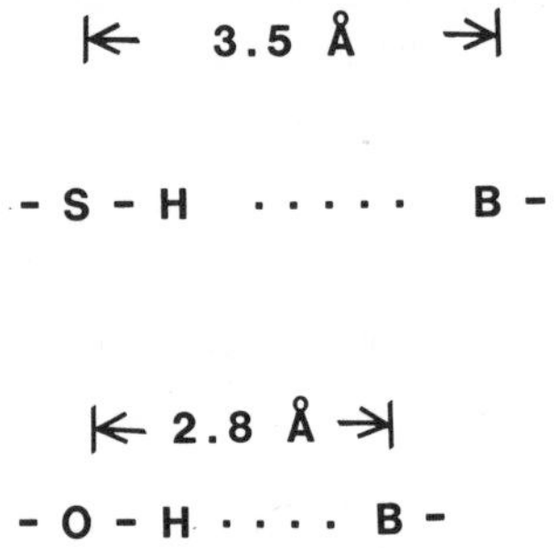

|← 3.5 Å →|

-S-H ····· B-

|← 2.8 Å →|

-O-H ···· B-

Fig. 7. *The hydrogen bond between the thiol group and base is longer that between the hydroxyl group and a base. (From Fersht et al. [21].)*

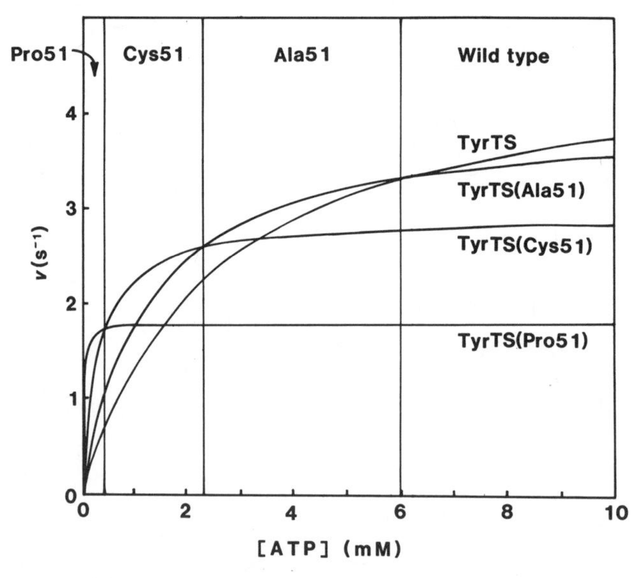

Fig. 8. *The dependence of rates of aminoacylation or tRNA by mutant tyrosyl-tRNA synthetases (at position 51) with varying concentrations of ATP. Each enzyme has a region in which it is the most active. (From Fersht et al. [21].)*

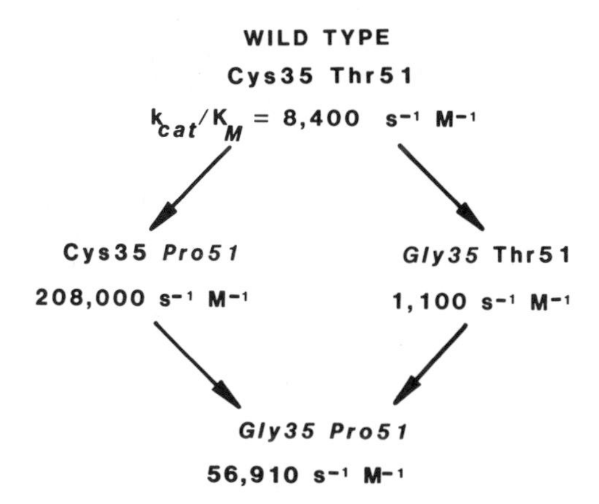

Fig. 9. *Free energy differences between single and double mutants of the tyrosyl-tRNA synthetase (From Carter et al. [22].)*

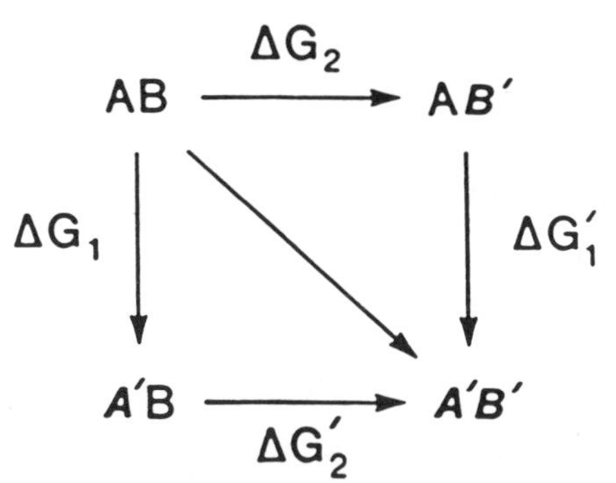

Fig. 10 *General scheme for a double mutant test (From Carter et al. [22].)*

or enzyme–substrate complex that affect one another, then the free energy changes will be independent of each other and so $\Delta G_1 = \Delta G'_1$ and $\Delta G_2 = \Delta G'_2$.

HALF-OF-THE SITES ACTIVITY: REVERSIBLE DISSOCIATION OF THE ENZYME

Why is the enzyme a dimer? Are the monomers unstable or inactive? To begin to answer these questions, we have engineered a mutation at the subunit interfaces which forces the enzyme to dissociate under certain conditions. Residue Phe164 lies on the symmetry axis and the side chain from one subunit is in hydrophobic contact with that of its symmetry-related partner. We therefore mutated Phe→Asp164 in the expectation that at high pH the dimer would be forced apart by the negatively charged carboxylates [23]. This was found. The monomer is stable and reversibly reassociates at high concentrations and low pH. However, it is inactive and binds tyrosine only very weakly, if at all, but it does bind tRNA. We are at present extending this approach to produce heterodimers containing one set of mutations on one subunit and a different set on the other. We hope this will be done by preparing the mutant Phe→Lys164. On mixing monomeric TyrTS(Asp164) with TyrTS (Lys164), the heterodimer TyrTS(Asp164, Lys164) should be preferentially formed because it will be stabilized by a buried salt-bridge.

REFERENCES

1. Winter G, Fersht AR, Wilkinson AJ, Zoller M, Smith M (1982): Nature(Lond) 299:756–758.
2. Fersht A, Shi JP, Wilkinson AJ, Blow DM, Carter P, Waye MMY, Winter GP (1984): Angewandte Chemie 23:467–473.
3. Rubin J, Blow DM (1981): J Mol Biol 145:489–500.
4. Blow DM, Brick P (1985): In Jurnak F and McPherson A (eds): "Biological Macromolecules and Assemblies, Vol 2." New York: John Wiley & Sons, pp 442–469.
5. Jakes R, and Fersht AR (1975): Biochemistry 14:3344–3350.
6. Wilkinson AJ, Fersht AR, Blow DM, Winter G (1983): Biochemistry 22:3581–3586.
7. Sanger F, Nicklen S, Coulsen AR (1977): Proc Natl Acad Sci USA 74:5463–5467.
8. Waye MMY, Winter G, Wilkinson AJ, Fersht AR (1983): EMBO Journal 2:1827–1829.
9. Fersht AR, Shindler JS, Tsui WC (1980): Biochemistry 19:5520–5524.
10. Fersht AR (1974): Proc R Soc Lond B 187:397–407.
11. Fersht AR (1985): "Enzyme Structure and Mechanism, 2nd edition." New York: WH Freeman & Co, pp 311–317.
12. Klotz IM, Franzen JS (1962): J Am Chem Soc 84:3461–3466.
13. Cantor CR, Schimmel PR (1980): "Biophysical Chemistry, Pt 1," New York: WH Freeman & Co, p 277.
14. Fersht AR, Shi JP, Knill-Jones J, Lowe DM, Wilk-

inson AJ, Blow DM, Brick P, Carter P, Waye MMY, Winter G (1985): Nature (Lond) 314:235–238.
15. Lowe G, Tansley G (1984): Tetrahedron 40:113–117.
16. Haldane JBS (1930): "Enzymes." Longmans Green and Co, p 182.
17. Pauling L (1948): Am Sci 36:51–58.
18. Leatherbarrow RJ, Fersht AR, Winter G (1985): Proc Nat Acd Sci USA 82: 7840–7844.
19. Wells TNC, Fersht AR (1985): Nature 316:656–657.
20. Fersht AR (1985): "Enzyme Structure and Mechanism, 2nd edn." New York: WH Freeman & Co, pp 331–343.
21. Fersht AR, Wilkinson AJ, Carter P, Winter G (1985): Biochemistry 24:5858–5861.
22. Carter PJ, Winter G, Wilkinson AJ, Fersht AR(1984): Cell 38:835–840.
23. Jones DH, McMillan AJ, Fersht AR, Winter G (1985): Biochemistry 24:5882–5857.

Protein Engineering, pages 279–287

25

Protein Engineering of Subtilisin

James A. Wells, David B. Powers, Richard R. Bott, Brad A. Katz Mark H. Ultsch, Anthony A. Kossiakoff, Scott D. Power, Robin M. Adams, Herb H. Heyneker, Brian C. Cunningham, Jeff V. Miller, Thomas P. Graycar, and David A. Estell

Genentech, Inc. (J.A.W., D.B.P., R.R.B., B.C.C., M.H.U., A.A.K.), and Genencor, Inc. (S.D.P., R.M.A., H.H.H., B.A.K., J.V.M., T.P.G., D.A.E.), South San Francisco, California 94080

INTRODUCTION

Subtilisin is one of the most highly studied and well-understood enzymes. It is a serine endopeptidase (MW ≅ 27,500) that is secreted in large amounts from a wide variety of *Bacillus* species. The protein sequence of subtilisin is determined from five different species of *Bacillus* [1,9,40] The three-dimensional structure has been determined to 2.5 Å resolution [2,3] and recently to 1.8 Å resolution [4] for the *B. amyloliquefaciens* enzyme. X-ray crystallographic analysis reveals that, although subtilisin is genetically unrelated to the mammalian serine proteases, it has a similar active site structure. The enzyme mechanism and the identification of catalytic and substrate binding residues have been elucidated for subtilisin through extensive kinetic, chemical modification [6], and crytallographic [13,14,27,41] studies.

The large data base for subtilisin makes it an attractive model enzyme to apply protein engineering technology to elucidate structure–function relationships as well as the process of protein secretion. This article summarizes site-directed mutagenesis experiments designed to map the pathway for secretion and processing of subtilisin. Studies are described where active site residues involved in catalysis, and in substrate specificity, have been altered by genetic engineering to understand better their function and to design enzymes with new properties. Finally, this article describes the introduction of disulfide bonds into subtilisin to study their effect on secretion, on protein structure, and on enzyme stability. It is significant that subtilisin represents the largest industrial enzyme market, primarily as an additive in laundry detergents. Thus, in addition to being a good model enzyme for protein engineering, there is much interest in improving the utility of subtilisin for industrial purposes.

SUBTILISIN GENE STRUCTURE AND SECRETION

The subtilisin genes from *Bacillus amyloliquefaciens* [7,8], *B. subtilis* [9], and *B. licheniformis* [42,43] have been cloned, sequenced, and expressed in *B. subtilis*. The sequences of the cloned subtilisin genes suggest that the protein is produced as a larger precursor, designated preprosubtilisin (Fig. 1). The subtilisin precursor contains a signal-like sequence followed by a 75-amino-acid prosequence joined to the mature enzyme (Fig. 1). Interestingly, the neutral protease gene from *B. amyloliquefaciens* [8] and *B. subtilis* [10] also contains a prosequence, but it shares no homology with that of subtilisin.

Cell fractionation and Western blot analysis detected a membrane-associated precursor that was identified as preprosubtilisin [11]. This precursor can be processed in an autoproteolytic fashion, as diagrammed in Figure 2. The evidence for this is that when mutations are introduced that inactivate the *B. amyloliquefaciens* subtilisin enzyme (such as Asp32→Asn or large deletions in essential coding sequence), mutant precursors accumulate in the membrane and further processing is inhibited. Inhibition of processing can be alleviated partially by exogenous addition of active subtilisin, or by expression of inactive *B. amyloliquefaciens* subtilisins in *B. subtilis* hosts containing an active chromosomal subtilisin gene. Thus, subtilisin can be processed autocatalytically like the mammalian pancreatic enzyme, trypsin. In contrast to other secreted proteins [12], the release of mature subtilisin from the cell membrane can proceed by an autoproteolytic mechanism independent of proteolysis by signal peptidase.

These studies illustrate that, unlike postranslational chemical modification of proteins, the production of mutant proteins depends on the biology of the expression system. Although the expression system may not always be able to produce a particular desired mutant enzyme, in some cases (e.g., Asp32→Asn), such mutants can be exploited to better understand the biology of the expression system.

MUTAGENESIS OF THE CATALYTIC SITE

The structure and mechanism of action of serine proteases is perhaps the most extensively studied of any enzyme class [1–3,5,6].

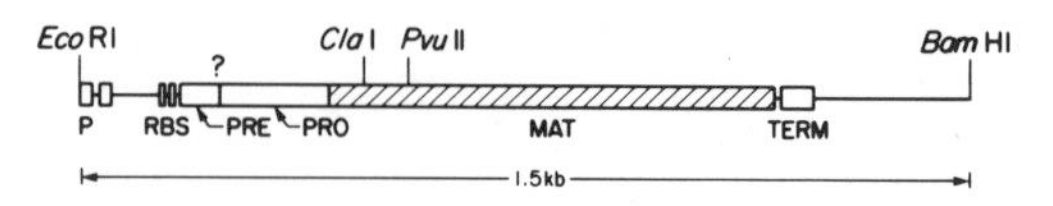

Fig. 1. *Functional sequence diagram of the subtilisin gene. P: region defining potential promoter; RBS: potential ribosome binding site; PRE: signal-like sequence; PRO: putative pro-peptide; MAT: mature enzyme sequence; TERM: transcription terminator sequence; A question mark indicates that the precise junction between the pre and pro sequences is unknown. (From Wells et al. [7].).*

The residues directly involved in catalysis of peptide bond hydrolysis by subtilisin are diagrammed in Figure 3. To test the effect of charge and hydrogen bonding, Asp32 has been substituted with asparagine, serine, and glutamic acid [16]. However, it can only be inferred that such mutants have very low activity because they are incapable of autoproteolytic processing. Further work is necessary to express these mutants in quantities suitable for characterization.

The oxyanion binding site (see Fig. 3) has been altered by replacing Asn155 with Thr155 [17]. The threonine hydroxyl group can be a similar hydrogen bond donor; however, modeling studies showed that it should be beyond direct hydrogen-bonding distance. Although production of mature Thr155 subtilisin was dramatically reduced, sufficient quantities of enzyme were obtained for kinetic characterization. In this mutant the k_{cat} (a measure of $E \cdot S \rightarrow E\text{–}S^{\neq}$) was 2,500-fold reduced while K_m (a measure of E·S dissociation) was essentially unchanged. Although a hydrogen bond between the peptide carbonyl oxygen and Asn155 may be weakly formed in the Michaelis complex (E·S in Figure 3), it is more strongly formed in the transition state because the kinetic effects were observed almost exclusively on k_{cat} and not on K_m. From the ratio of catalytic efficiencies (k_{cat}/K_m) of wild type (Asn155) to mutant (Thr155), a free-energy transition state stabilization difference of −4.7 kcal/mole was calculated for the wild type compared to the mutant. These results are consistent with studies of mutations pro-

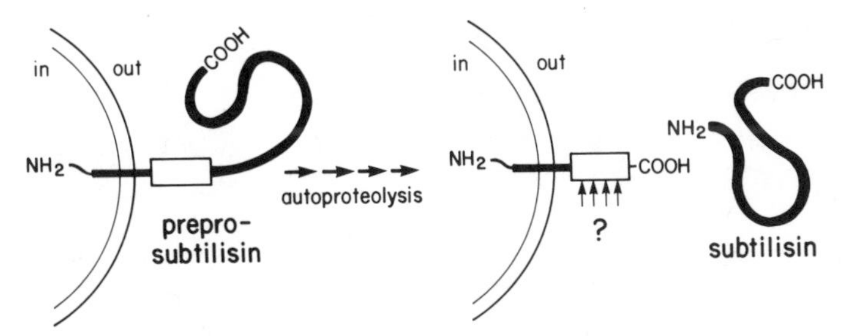

Fig. 2. *Processing model proposed for subtilisin [11]. Subtilisin can be released from the cell membrane in a posttranslational autoproteolytic process.*

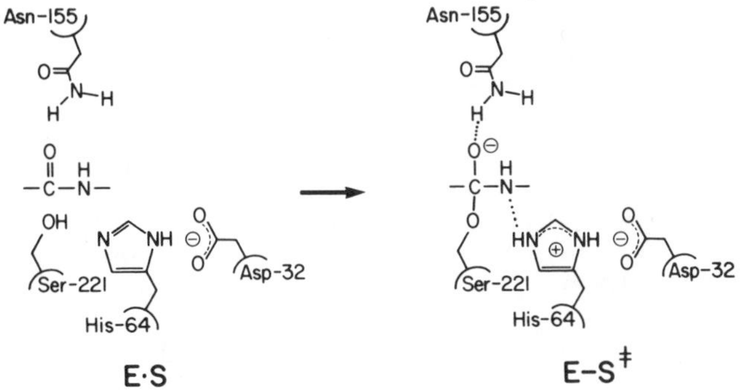

Fig. 3. *Diagram of the catalytic site of the subtilisin showing the catalytic triad (Ser221, His64, and Asp32) and the oxyanion-binding residue, Asn155. The substrate binds in such a way as to juxtapose the carbonyl of the susceptible peptide bond and the catalytic Ser221 (E·S complex). The proton on the Ser221 hydroxyl is accepted by His64 coincident with attack on the carbonyl group in the peptide bond, producing a tetrahedral transition state intermediate (E–S$^{\neq}$). A possible role of Asp32 may be to orient His64 and to stabilize the positively charged His64 produced in the transition state, as proposed for trypsin [15]. The oxyanion is presumed to be stabilized by a hydrogen bond to the amide of Asn155 (indicated by dashed lines) as well as to the peptide amide-NH from Ser221 (not shown) [13,14].*

duced in tyrosyl-tRNA synthetase [18] where it was estimated that hydrogen bonds between charged acceptors and neutral donors (analogous to the oxyanion and Asn155 hydrogen bond) contribute a binding energy of −3.5 to −5 kcal/mol.

CASSETTE MUTAGENESIS

Two examples are discussed below where at a particular site it was difficult to predict which amino acid substitution would be the most desirable. In these cases it was necessary to produce a number of mutants and determine the substitution that would give the desired property. Making multiple mutations over a defined sequence by conventional site-directed mutagenesis in M-13 would be laborious, owing to the low and highly variable efficiency of mutagenesis. A "cassette mutagenesis" strategy was developed [19] that allowed more efficient production of multiple mutations over a defined sequence (see Fig. 4). In this procedure, unique and silent restriction sites are introduced to flank closely the target codon. Duplex synthetic DNA cassettes are inserted between these restriction sites to restore the coding sequence, except over the target codon. Oligonucleotides can be synthesized in pools to reduce the necessity to purify one oligonucleotide per mutant. These pools are grouped so that all codon members can be

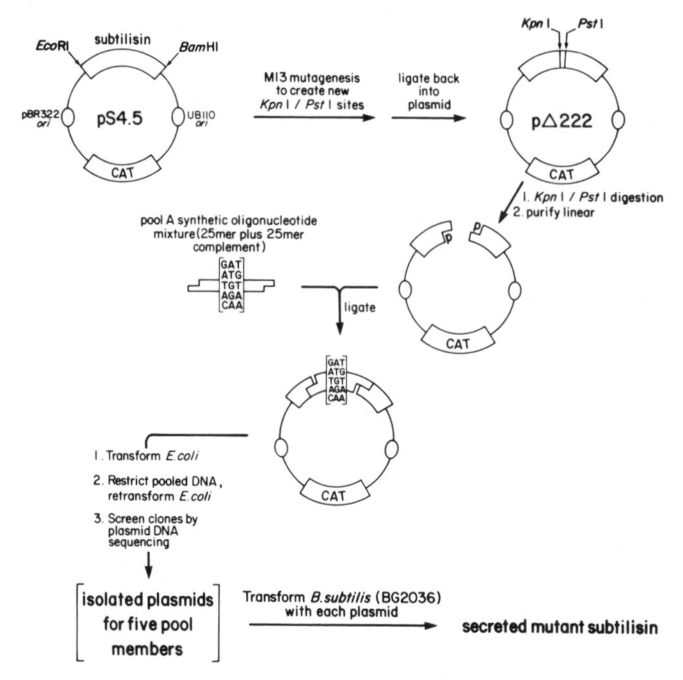

Fig. 4. *The cassette mutagenesis method. The EcoRI-BamHI fragment containing the B. amyloliquefaciens subtilisin gene was cloned into M-13mp9 and mutagenized to create a new KpnI and PstI site. The mutagenized gene was cloned back into the plasmid, pBS42, and digested with KpnI and PstI to create a gapped linear molecule that was purified. This was ligated, in separate reactions with four pools of duplex synthetic oligonucleotide cassettes (25 bp in length) containing five codons each. The cassette restored the coding sequence, introduced an altered codon at position 222, and eliminated both the KpnI and PstI sites. These ligations were transformed into E. coli and total plasmid DNA was prepared. This DNA was restricted with KpnI to eliminate any plasmids not containing a cassette, and the resulting transformants were screened for desired mutations by DNA sequencing. Isolated plasmids were transformed into a protease deficient host, BG2036. (From Wells et al. [19].)*

identified by a single track of DNA sequence. In the pool shown in Figure 4, a single A-track can distinguish all the members. In this way all 19 possible amino acid substitutions have been isolated at the two different positions discussed below. The condition of finding unique and silent restriction sites which closely flank the target codon can be met readily because the cassette need not restore the restriction sites that are initially introduced [19]. Inspection of the subtilisin gene sequence reveals that any codon can be set up for cassette mutagenesis using a synthetic DNA cassette of less than 26 base pairs.

ENGINEERING OXIDATIVE RESISTANCE

Although subtilisin is remarkably stable to inactiviation by denaturants, it is highly susceptible to inactivation by chemical oxidants [6]. Peptide mapping studies have indicated that oxidation of methionine222 to the sulfoxide (next to the catalytic site Ser221) results in nearly 90% loss of enzyme activity [20]. It was anticipated that such a loss in activity would be prevented by replacing the methionine with a residue that is insensitive to oxidation. However, it was impossible to predict

TABLE I. Relative Specific Activities of Codon-222 Mutant Subtilisins[a]

Codon-222	Percent relative specific activity
Cys	138.0
Met	100.0
Ala	53.0
Ser	35.0
Gly	30.0
Thr	28.0
Asn	15.0
Pro	13.0
Leu	12.0
Val	9.3
Gln	7.2
Phe	4.9
Trp	4.8
Asp	4.1
Tyr	4.0
His	4.0
Glu	3.6
Ile	2.2
Arg	0.5
Lys	0.3

[a]Mutant enzymes were purified and assayed as described in reference 21.

the optimal substitution; this methionine is conserved in all subtilisins that have been sequenced, and it occupies a partially buried position. Furthermore, the production of a sulfoxide introduces only a single oxygen into the structure, yet this has dramatic functional consequences.

All 19 amino acid replacmements were produced at codon 222 by the cassette mutagenesis method (Fig. 4). When these mutant plasmids were expressed in *B. subtilis,* all produced processed enzymes. The enzymes were purified and their specific activities were determined (Table I) [21]. The most active mutants were produced by small amino acid substitutions. Large, bulky, or charged amino acids were found to have low specific activities. Amino acids containing a branched β-carbon (i.e., Thr, Val, Ile) were poorer than their nonbranched counterparts (i.e., Ser, Leu, Met). This may be a consequence of the tight packing around the β-carbon [2–4]. The higher specific activity of the cysteine mutant compared to the wild type was attributed to an increased k_{cat} [21]. In fact, none of the other substitutions were more catalytically efficient (i.e., greater k_{cat}/K_m) than the wild type. It was unexpected that the residues considered most homologous to the wild-type methionine (i.e., Leu, Val, Ile) [22] were not the best alternatives. These results emphasize the need to make multiple mutations to find the optimal substitution in the absence of good predictive tools.

It was shown that, although the wild-type and Cys222 enzymes were rapidly inactivated by 1 M H_2O_2, the Ser222, Ala222, and Leu222 enzymes were completely resistant to peroxide over a 1-hour time period [21]. In addition to producing enzymes with greater oxidative stability, these studies confirmed the identification of Met222 as a residue whose modification can produce critical effects on subtilisin activity. In this context, site-specific mutagenesis of α-1 antitrypsin inhibitor has been used to replace the reactive site methionine residue and engineer greater oxidative stability into this protein [23,24].

To characterize the effect of substitutions at position 222 on enzyme function, the three-dimensional X-ray structure was determined for Phe222 at 2.0 Å and refined to an R factor of 0.16 [25]. Interpretation of the difference density map shows a slight main-chain perturbation and that the phenylalanine ring from Phe222 can sterically interfere with substrate binding at the S-1′ site. This may account for the 20-fold reduction in specific activity (Table I).

ALTERATION OF SUBSTRATE SPECIFICITY

Subtilisin, unlike the pancreatic serine proteases, has a very broad substrate specificity [5,26]. Crystallographic studies of subtilisin [13,14,27,41] have shown that subtilisin contains an extended crevice for binding of peptides. By convention, the substrate residue contributing the carbonyl of the scissile peptide bond (Fig. 3) is termed the P-1 residue

[28]. Modeling studies suggested that substitutions at glycine166 could affect substrate binding. Because of uncertainties in predicting the effect of substitutions at position 166, we produced 19 mutants by cassette mutagenesis [26,29].

The effects of mutations at position 166 on substrate specificity were probed by a series of peptide p-nitroanilide substrates differing only in their P-1 residues [26,29]. The kinetic parameters k_{cat} and K_m were determined by progress curve analysis [21]. In cases where substrates showed product inhibition, the kinetic parameters were determined from initial rate measurements.

Some of the results for mutations that alter the charge of the residue at position 166 are shown in Table II. These data show several trends. First, the positively charged substitutions are more efficient catalysts (up to 500-fold for Lys166) than wild type against a substrate containing glutamic acid at the P-1 position. In addition, Asp166 is better against a substrate containing arginine at P-1 than is Asn166 or wild type. Furthermore, Lys166 is poorer against the arginine substrate. The alanine P-1 substrate has similar kinetic parameters among the mutant enzymes. This is expected because the alanine side chain cannot extend far enough into the binding region to interact substantially with the enzyme.

The wild-type enzyme is best against large hydrophobic amino acids (e.g., Phe and Tyr), while Asp166 and Glu166 are poor against these substrates. Although, compared to wild type, Arg166 and Lys166 are greatly improved for hydrolysis of the glutamate substrate, these catalytic efficiencies for the glutamate substrate, are much below the wild-type enzyme or these same mutant enzymes against the phenylalanine substrate. Among many possible explanations for these results is that the binding of glutamate substrate is poor relative to phenylalanine because of electrostatic repulsion with glutamate 156 located near the top of the P-1 binding cleft. Also, the arrangement of the lysine or arginine side chain at position 166 may depend on the bound substrate. Indeed, a recent X-ray crystallographic study of lysine166 in the absence of substrate reveals that this side chain is highly disordered [30].

TABLE II. Effect of Charged Amino Acid Substitutions at Position 166 on P-1 Substrate Specificity[a]

	P-1 Substrate ($k_{cat}/K_m \times 10^{-4}$)			
Position 166	Phe	Ala	Glu	Arg
Gly (wild type)	40	1	0.002	0.4
Asp	0.5	0.4	<0.001	1
Glu	2	0.4	<0.001	ND
Asn	13	1	<0.001	0.1
Gln	60	3	<0.001	0.06
Lys	50	3	1	0.02
Arg	40	5	0.1	ND

[a]Tetrapeptide substrates, succinyl-L-Ala-L-Ala-L-Pro-L-X-p-nitroanilide, were synthesized by Dr. John Burnier at Genentech. X represents the P-1 amino acid. ND = not determined.

The magnitude of the k_{cat}/K_m values determined for the charged substitutions interacting directly with a complementary charged substrate are lower than expected. From estimates of strengths of specific salt bridges in α-chymotrypsin (−2.9 kcal/mol) [31] and in phenylalanyl-tRNA synthetase (−2.7 kcal/mol) [32] we might anticipate catalytic efficiency ratios to wild type of roughly 10^2. Although the values were in this range for the interaction of the glutamate P-1 substrate with the positively charged 166 mutants (i.e., Lys166 and Arg166 are increased 500 and 50, respectively, over wild type), the catalytic efficiency ratios for Asp166 against the arginine P-1 substrate was only 2.5-fold and 10-fold increased over wild type and Asn166, respectively. Crystallographic analysis of these mutants with bound substrate analogs may help to provide an explanation of these observations.

The k_{cat} values were found to vary as dramatically as the K_m values (data not shown). This is perhaps not too surprising, realizing that the geometry of the enzyme-bound substrate in the transition state ($E\text{–}S^{\neq}$) is tetrahedral while that of the ground state (E·S) is trigonal. Crystallographic studies of enzyme-bound product complexes [14], boronic acid peptides [13], and chloromethyl ketone peptide inhibitors [27] suggest that in the tetrahedral intermediate the P-1 side chain moves deeper into the S-1 crevice by as much as 1 Å. The changes in k_{cat} observed for the 166 mutants are consistent with altered binding of the P-1 side chain in the transition state complex. However, the effects on k_{cat} may also result from differences in the orientation of the scissile peptide bond of the substrate and the catalytic site.

There are at least two other reports of altered enzyme specificity by site-directed mutagenesis. A mutation in tyrosyl-tRNA synthetase [18] in the tyrosine binding site (Tyr34→Phe) decreased the ability to distinguish between tyrosine and phenylalanine by 15-fold. Site-specific mutagenesis of rat trypsin [33] in the P-1 binding site (Gly216→Ala, Gly226→Ala) has altered the selectivity of this enzyme for arginine and lysine substrates. These mutations produced changes in k_{cat} as well as K_m as observed for mutations produced in subtilisin. However, the trypsin mutants had largely reduced catalytic efficiencies (40- to 4,000-fold) toward these substrates compared to wild-type trypsin. While the specificity changes produced in subtilisin were as large as those produced in trypsin, the catalytic efficiencies of mutant subtilisins were actually increased relative to wild type toward some substrates (Table II).

ENGINEERING OF DISULFIDE BONDS INTO SUBTILISIN

Subtilisin contains no disulfide bonds or cysteine residues. This makes subtilisin a good model for production of disulfide bonds in secreted proteins in vivo. To test the feasibility of secretion of a subtilisin with an engineered disulfide bond, we chose to introduce a disulfide bond whose geometry would conform well to naturally occurring disulfide bonds [34,35] and would not cause significant main-chain atom rearrangement if the bond formed. This disulfide site (Ser24→Cys, Ser87→Cys) was on the back side of the molecule away from the active site and therefore should not effect activity.

The double-mutant enzyme (Ser24→Cys, Ser87→Cys) was made, was found to be secreted, and had the same specific activity as the wild type [36]. Denaturing gels (SDS–PAGE) in the presence and absence of reducing agent suggested the disulfide bond had formed. The X-ray crystal structure for this mutant [37] confirmed that the disulfide bond was formed and that it conformed to a right-handed disulfide bond geometry as predicted from modeling studies (X_3 is +96°) (Fig. 5). The structure further showed that there was very little alteration in the protein structure except for the cysteine residues 24 and 87 that formed the disulfide bond.

To test the flexibility of the protein structure to accommodate a disulfide bond, a second disulfide mutant was constructed (Tyr21→-

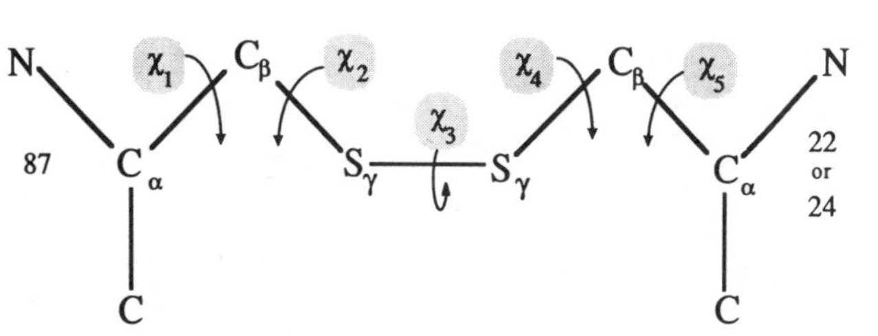

Fig. 5. *Model disulfide bond showing dihedral bond angles X_1 to X_5.*

Ala; Thr22→Cys; Ser87→Cys). Modeling studies in this case predicted this would give a left-handed disulfide bond geometry. (Note that this disulfide utilizes a cysteine at position 87, which is in common with the other disulfide mutant mentioned above). This disulfide mutant was also found to be secreted with an intact disulfide bond as judged by the migration differences in reduced and oxidized denaturing gels [36]. Structural analysis of this disulfide showed it had left-handed disulfide bond geometry (X_3 is $-98°$) [37]. However, the dihedral angles corresponding to X_2 and X_4 were unusual for left-handed disulfide bonds [34].

Studies of the thermostability of these mutants were complicated by autolysis. Neither disulfide mutant was more stable to heat-induced autolytic inactivation; however, both were more stable than their corresponding reduced derivative (i.e., double cysteine) [36]. The reduction potentials were measured and showed that the Cys24/Cys87 disulfide bond was four times stronger than the Cys22/Cys87 disulfide.

It is significant that these disulfide mutants of subtilisin were secreted and oxidized in vivo and that both cysteines in the disulfide were novel to the protein. These studies extend upon previous work in T4 lysozyme [38] and dihydrofolate reductase [39], where single cysteines were introduced near preexisting cysteines and disulfides were oxidized in vitro.

SUMMARY

We have shown examples of the utility of protein engineering of subtilisin to alter substrate specificity, to improve stability to chemical oxidation, to introduce disulfide bonds, and to study the role of particular residues in enzyme catalysis. Furthermore, particular mutant enzymes have proved useful in the study of maturation and of secretion of subtilisin. Although at present the functional consequences of particular amino acid replacements are not very predictable, protein engineering will enlarge the data base and should ultimately permit the deliberate engineering of enzyme properties.

Two functionally important and conserved residues (Met222 and Gly166) have been shown to accommodate all amino acid substitutions yielding mutant enzymes that maintain some activity. The X-ray structural analyses that have been performed so far indicate that substitutions at positions 166 and 222 cause only local perturbations in structure, albeit with dramatic functional consequences. Although subtilisin structure and activity appear to tolerate these substitutions, the enzyme has apparently evolved a catalytically most-optimal residue. At position 222, methionine has been chosen even at the expense of oxidative stability, perhaps because subtilisin is unlikely to be challenged with strong oxidants in nature. This may serve to illustrate that the highest probability of improving the properties of proteins by protein engineering will be to engineer those properties that have not been selected for in nature.

REFERENCES

1. Markland FS, Smith EL (1971): In Boyer PD (ed): "The Enzymes, Vol III, New York: Academic Press, pp 561–608.
2. Wright CS, Alden RA, Kraut J (1969): Nature 221:235–242.
3. Drenth J, Hol WGJ, Jansonius J, Kockoek R (1972): Eur J Biochem 26:177–181.
4. Bott R, Katz B, Ultsch M, Kossiakoff T (manuscript in preparation).
5. Philipp M, Bender ML (1983): Mol Cell Biochem 51:5–32.
6. Svedsen IB (1976): Carlsberg Res Commun 41:237–291.
7. Wells JA, Ferrari E, Henner DJ, Estell DA, Chen EY (1983): Nucleic Acids Res 11:7911–7925.
8. Vasantha N, Thompson LD, Rhodes C, Banner C, Nagle J, Filpula D (1984): J Bacteriol 159:811–819.
9. Stahl ML, Ferrari E (1984): J Bacteriol 158:411–418.
10. Yang MY, Ferrari E, Henner DJ (1984): J Bacteriol 160:15–21.
11. Power SD, Adams RM, Wells JA (1986): Proc Natl Acad Sci USA 83:3096–3100.
12. Michaelis S, Beckwith J (1982): Ann Rev Microbiol 36:435–465.
13. Matthews DA, Alden RA, Birktoft JJ, Freer ST, Kraut J (1975): J Biol Chem 250:7120–7126.
14. Robertus JD, Kraut J, Alden RA, Birktoft JJ (1972): Bicohem 11:4293–4303.
15. Kossiakoff AA, Spencer SA (1981): Biochem 20:6462–6473.

16. J Wells, D Estell, and S Power, unpublished results.
17. Wells JA, Cunningham BC, Graycar TP, Estell DA (1986): Phil Trans Roy Soc Lond A 317:415–423.
18. Fersht AR, Shi JP, Knill-Jones J, Lowe DM, Wilkinson AJ, Blow DM, Brick P, Carter P, Waye MMY, Winter G (1985): Nature 314:235–238.
19. Wells JA, Vasser M, Powers DB (1985): Gene 34:315–323.
20. Stauffer CE, Etson D (1969): J Biol Chem 244:5333–5338.
21. Estell DA, Graycar TP, Wells JA (1985): J Biol Chem 260:6518–6521.
22. Dayhoff MO, Schwartz RM, Orcutt BC (1978): In Dayhoff MO (ed): "Atlas of Protein Sequence and Structure, Vol V, Supplement 3" pp 345–352.
23. Rosenberg S, Barr PJ, Najarian RC, Hallewell RA (1984): Nature 312:77–80.
24. Courtney M, Jallat S, Tessier LH, Benavente A, Crystal RG, Lecocq , JP (1985): Nature 313:149–151.
25. Bott RR, Ultsch M (manuscript in preparation).
26. Estell DA, Graycar TP, Miller JV, Powers DB, Burnier JP, Ng PG, Wells JA (1986): Science 233:659–663.
27. Robertus JD, Alden RA, Birktoft JJ, Kraut J, Powers JC, Wilcox PE (1972): Biochem. 11:2439–2449.
28. Schechter I, Berger A (1967): Biochem Biophys Res Commun 27:157.
29. Wells JA, Powers DB, Bott RR, Graycar TP, Estell DA: (1987) Proc Natl Acad Sci USA Vol. 84 (in press).
31. Fersht AR (1972): J Mol Biol 64:497–509.
32. Mulivor R, Rappaport KP (1973): J Mol Biol 76:123–134.
33. Craik CS, Largman C, Fletcher T, Roczniak S, Barr PJ, Fletterick R, Rutter WJ (1985): Science 228:291–297.
34. Richardson JS (1981): Adv Protein Chem 34:167–339.
35. Thornton JM (1981): J Mol Biol 151:261–287.
36. Wells JA, Powers DB (1986): J Biol Chem 261: 6564–6570.
37. Katz BA, Kossiakoff AA (1985): J Bio Chem 261: 15480–15485.
38. Perry LJ, Wetzel R (1984): Science 226:555–557.
39. Villafrance JE, Howell EE, Voet DH, Strobel MS, Ogden RC, Abelson JN, Kraut J (1983): Science 222:782–788.
40. Nedkov P, Oberthür W, Braunitzer G (1983): Hoppe-Seyler's Z Physiol Chem 364:1537–1540.
41. Poulos TL, Alden RA, Freer ST, Birktoft JJ, Kraut J (1976): J Biol Chem 251:1097–1103.
42. Powers DB, Wells JA (manuscript in preparation).
43. Jacobs M, Eliasson M, Uhlen M, Flock J (1985): Nucleic Acids Res 13:8913–8927.

Protein Engineering, pages 289–297

26

The Use of X-Ray Crystallography to Determine the Relationship Between the Structure and Stability of Mutants of Phage T4 Lysozyme

Tom Alber and Brian W. Matthews

Institute of Molecular Biology and Department of Physics, University of Oregon, Eugene, Oregon 97403

INTRODUCTION

The finding that synthetic ribonuclease spontaneously adopts an enzymatically active conformation in water showed that the amino acid sequence of a protein can contain all the information required to establish and maintain its native structure [1]. Beyond this, knowledge of the physical basis of protein structure is painfully sketchy. How does a particular sequence of amino acids specify a particular structure and disfavor all alternative structures (including random disordered forms)? What are the contributions of individual amino acids to the stability of a protein? What are the consequences of single amino acid changes? These are basic questions without ready answers.

The forces that contribute to protein stability—van der Waals forces, hydrogen bonding, electrostatic forces, hydrophobic interactions, chain entropy and so on—were identified through thermodynamic studies of protein denaturation. The results of these studies, however, cannot be factored into the contributions of individual amino acids in the sequence. In addition, the relative magnitudes of the different forces stabilizing a protein are not accurately known.

With a detailed knowledge of the structure of a protein and access to a high-speed computer, the interactions thought to be important for stability can be quantitatively modeled. Unfortunately, the interactions within proteins and with the solvent are extremely complex, and simplifying assumptions that compromise the accuracy of the calculations are made. In addition, since protein stability is the small difference in free energy between the native conformation and all other conformations, difficulties in treating of the denatured state make rigorous mathematical modeling problematical. As a result, it is presently not possible to calculate the contribution of a particular amino acid to protein stability from first principles. There is no quantitative theory for distinguishing the amino acids that are essential for stability from those that make no contribution or even destabilize the observed native structure.

In light of the difficulties described above, it is fortunate that temperature-sensitive (ts) mutants can be used experimentally to identify amino acids that are involved in essential interactions. Due to the ease of isolating ts mutants, phage T4 lysozyme provides a good model system for studying the contributions of individual amino acids to protein thermal stability. By comparing the X-ray crystal structures of wild-type and ts mutant lysozymes, specific differences in their three-dimensional structures can be correlated with changes in stability.

In this chapter we describe the structure of phage T4 lysozyme, discuss some of the methods used to generate and study mutants with altered thermostability, and highlight some of the basic insights that have come from comparing wildtype and mutant lysozymes. Many different mutations that destabilize T4 lysozyme have been found. All forces that are potentially important for stability can play a role at each site in the protein. Studies of randomly generated single amino acid substitutions have inspired further investigation of site-directed mutants. These studies have begun to provide information about the relative importance of particular interactions. The side chain of Thr 157, for example, has been found to be part of an important network of hydrogen bonds in the wild-type protein. A new ion pair that forms when Thr 157 is replaced by arginine does not adequately substitute for the wild-type hydrogen-bond network, even though the new ion pair contains two hydrogen bonds of its own. A challenge for the future will be to continue to identify important stabilizing interactions and to find characteristics that distinguish them for structural features that do not enhance stability.

THE X-RAY CRYSTAL STRUCTURE OF PHAGE T4 LYSOZYME

The X-ray crystal structure of phage T4 lysozyme provides a detailed view of the interactions within the protein. The structure was determined by the method of multiple isomorphous replacement [2] and refined to a high degree of accuracy (R = 19.3%) at high resolution (1.7 Å) [3]. The average error in atomic positions is on the order of 0.1–0.2 Å.

Figure 1 shows a drawing of the α-carbon backbone. Even though it is a small protein (164 amino acids), T4 lysozyme contains two distinct domains. The N-terminal domain (residues 1–60) contains all the β-sheet in the structure, as well as two α-helices. The C-terminal domain (residues 80–164) is like a barrel whose bottom and staves are composed of seven α-helices. The domains are joined by a long α-helix (residues 60–80) that traverses the length of the molecule. Amino acids 162–164 are disordered in the crystals.

The active site is in the deep cleft between the domains. Substrate binding requires a breathing motion of the domains. Analysis of the atomic motions in the crystal structure suggests that the "hinge" for this breathing motion is centered around residue 67 in the long connecting α-helix [3].

LYSOZYME MUTANTS

T4 lysozyme is produced late in the phage life cycle and is required for the lysis of the host bacterium. Streisinger and co-workers developed a method of screening for mutants of the enzyme that are mildly temperature-sensitive compared to the wild-type protein [4]. The screen, illustrated in Figure 2, is based on the fact that lysozyme activity can be assayed directly on Petri plates seeded with T4 phage. The ts mutants—induced by "random" mutagenesis—produce less and less lysozyme activity at progressively higher growth temperatures (Fig. 2). Even at the highest growth temperature (43°C), however, the mutants produce enough lysozyme activity to form phage plaques. The variant lysozymes purified from these T4 mutants have all been found to be moderately more sensitive to reversible thermal denaturation than the wild-type protein [5,6].

These mildly temperature-sensitive mutants have two advantages for studies of protein stability. First, since the proteins are stable at low growth temperatures, they can be purified in the large quantities required for structural

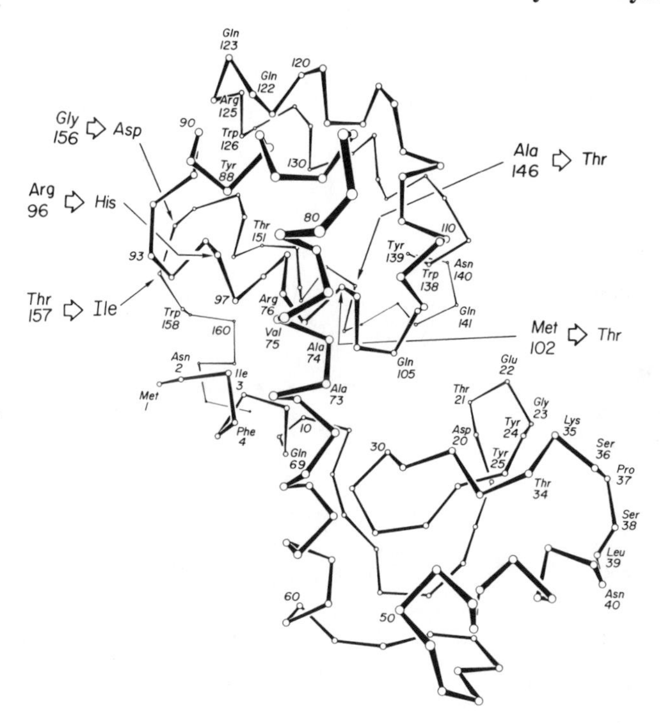

Fig. 1. *Drawing of the α-carbon backbone of phage T4 lysozyme. The N- and C-terminal domains of the protein are clearly visible. The "waist" between the domains includes the long connecting α-helix (residues 60–80) and the chain termini. Five "randomly" induced temperature-sensitive mutants that have been studied crystallographically are indicated. Thermodynamic and X-ray structural analyses of a series of site-directed mutants have been used to probe the relative importance of different interactions involving Thr 157.*

and physical characterization. Second, since the effects on stability are small (changes in ΔG of unfolding are on the order of 2–4 kcal/mole [5,6]), they are more likely to result from changes in single interactions in the protein. In theory, this should allow the changes in stability to be more readily correlated with particular structural features.

Over a dozen ts mutants have been sequenced, but no simple pattern in the nature or location of the mutations was found. Five mutants—Arg 96 → His, Met 102 → Thr, Ala 146 → Thr, Gly 156 → Asp, and Thr 157 → Ile — were subjected to detailed thermodynamic and/or structural analysis [5-9]. The location of these substitutions is indicated in Figure 1. Two of the substitutions (Ala 146 → Thr and Gly 156 → Asp) cause structural changes that propagate through the protein. In contrast, the structural changes associated with the other substitutions are localized to the site of the mutation.

The changes in structure have been analyzed in terms of their effect on solvent accessible surface area, helix-forming potential, and the calculated energy of the native conformation. The first two parameters reveal suggestive trends, but the number of structures investigated is too small to draw firm conclusions [9].

Taken together, these studies support the idea that all forces thought to stabilize proteins—such as van der Waals forces, hydrophobic interactions, hydrogen bonds, electrostatic interactions, and so on—potentially act in concert at each site in the molecule. The

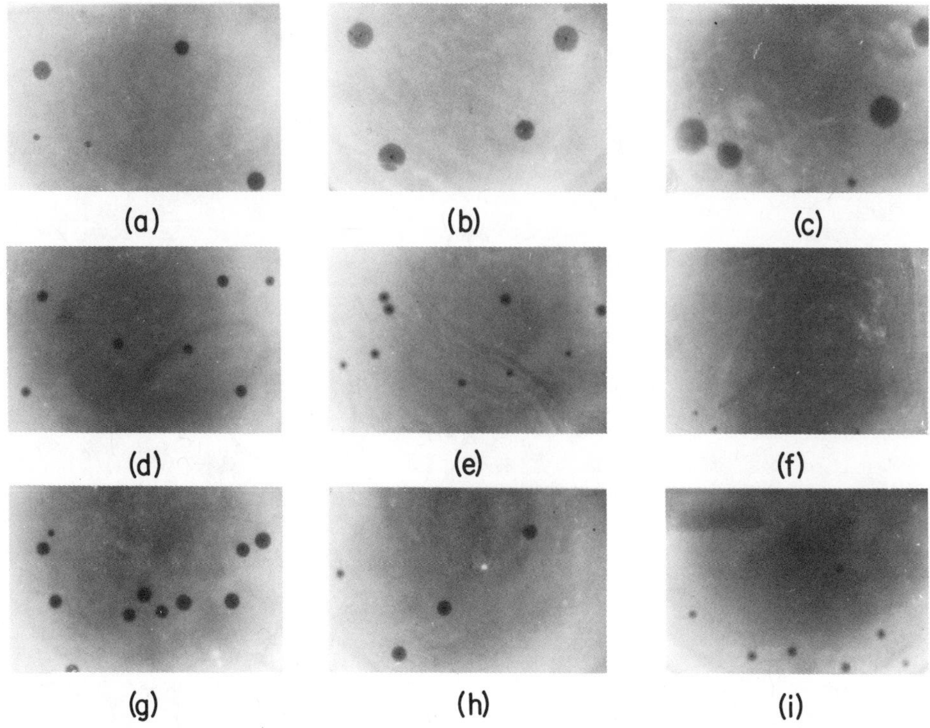

Fig. 2. *Test for T4 phage producing a temperature-sensitive lysozyme. Bacteria mixed with a small number of T4 phage are incubated on Petri plates at an elevated growth temperature for 6 hours and then exposed to chloroform vapors at room temperature for 12–18 hours. The combined action of chloroform and phage lysozyme breaks open the uninfected bacteria surrounding the phage plaques. The size of this "halo" surrounding the dark central plaque is a measure of the amount (and activity) of lysozyme present in the plaque. Panels (a), (b), and (c) show Petri plates seeded with wild-type phage grown at 31°, 37°, and 43°C, respectively. The size of the halo increases with increasing growth temperature. The halos produced by the ts lysozyme mutants Arg 96→ His (d–f) and Thr 157 → Ile (g–i) after growth at 31°, 37°, and 43°C are shown for comparison. While the ts mutants produce enough lysozyme to form plaques even at 43°C, the size of the halo decreases with increasing growth temperature. This correlates with reduced thermal stability of the phage lysozyme.*

observed stability of the protein is the global sum of all the local contributions [9].

SITE-DIRECTED MUTANTS

The best understood ts mutation is Thr 157 → Ile. Thr 157 is located on the surface of the protein in a loop between two α-helices. As shown in Figure 3, the γ-hydroxyl group of the Thr side chain is involved in intramolecular hydrogen bonds, and the γ-methyl group makes van der Waals contacts with neighboring atoms. Hydrogen bonds to the α-methyl group are lost when Thr 157 is replaced by Ile. The mutation also causes a shift in the position of the α-methyl group and a change in the arrangement of solvent molecules at this site. However, structural changes do not propagate through the protein.

Site-directed mutagenesis provides a powerful method for analyzing the relative contributions of hydrogen bonding, van der Waals contacts, and solvent structure at this site, because several of the natural amino acids are structurally related to Thr, the wild-type amino acid. For example, a measure of the importance of hydrogen bonding is provided by Ser, since it contains the hydroxyl group but not

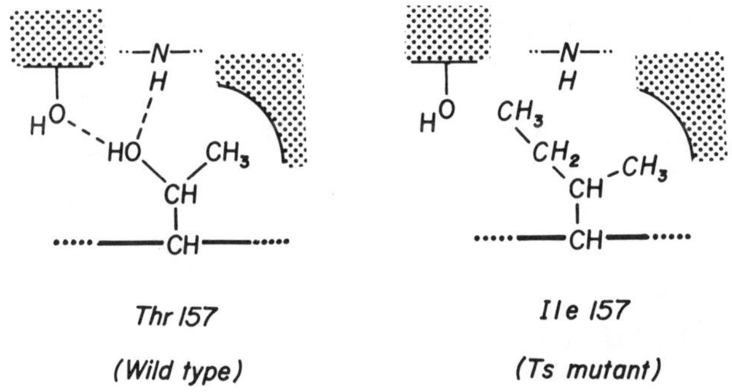

Thr 157 (Wild type)

Ile 157 (Ts mutant)

Fig. 3. *Schematic drawing showing some of the intramolecular interactions of Thr 157. Hydrogen bonds to the side chain hydroxyl group of Thr 155 and the main chain amide of Asp 159 are eliminated in the ts mutant Thr 157 → Ile. The mutation also causes a shift in position of the α-methyl group and changes in the surrounding solvent (not shown).*

the methyl group of Thr. The role of van der Waals forces is tested by Val, since it is similar in shape to Thr but cannot hydrogen bond. Introducing Gly at position 157 may allow water to play the role of the Thr hydroxyl group. In addition, charged amino acids such as Arg and Asp can be introduced to see if electrostatic forces can substitute for the interactions in the wild-type protein.

The two primer method of oligonucleotide-directed mutagenesis [10; Rossi and Zoller, this volume] was used to target mutations to the codon for Thr 157 in the cloned T4 lysozyme gene [11]. Instead of using mutagenic primers with unique sequences, novel mixtures of primers were used to simultaneously create different substitutions. For example, a mixture of mutagenic 18-mers specifying A, G, or C in the first position of the 157 codon, A, G, or T in the second position, and T in the third position was used to rapidly construct lysozymes with Ser, Val, Gly, Leu, Ile, Arg, His, Asp, and Asn at residue 157. Mutants were distinguished from wild-type genes by preferential hybridization to the radioactive primer mixture and then identified by DNA sequencing. The mutant genes were cloned in an expression vector in *Escherichia coli* for purification of large quantities of the encoded mutant proteins.

To date, directed mutant lysozymes with Ser, Gly, Ile, Leu, Asn, Asp, Glu, His, and Arg at position 157 have been isolated. The temperature at the midpoint of the reversible thermal denaturation transition (T_m) of the mutants is compared with wild type (Thr 157) and the original ts lesion (Ile 157) in Table I. A striking finding is that all the mutants are less stable than wild type. This suggests that the methyl and hydroxyl groups of Thr 157 both contribute to stability. Apparently, hydrogen bonds and van der Waals forces both play a role at this site.

In considering these results in detail, it cannot be assumed that the three-dimensional structure of the protein remains invariant when one amino acid is replaced by another. The X-

TABLE I. Relative Stabilities of Lysozymes With Different Substitutions at Position 157.

Amino acid at position 157	Melting temperature (°C ± 1°, measured at pH 2) relative to wild type
Threonine (wild type)	—
Asparagine	−2
Serine	-2.5
Glycine	−4
Aspartic acid	−4
Leucine	−5
Arginine	−5
Glutamic acid	−6
Histidine	−8
Isoleucine (ts mutant)	−11

ray structures of randomly induced ts mutants show that the conformation of the protein can adjust to accommodate a new side chain [7–9]. Consequently, it is necessary to determine the X-ray crystal structure of each mutant to understand its susceptibility to thermal denaturation.

CRYSTALLOGRAPHIC STUDIES OF SITE DIRECTED MUTANTS

Since the X-ray crystal structure of T4 lysozyme is known, it is straightforward to determine the structures of mutants that do not alter the packing of molecules in the crystal. Because water occupies over 50% of the crystal volume, the crystal lattice can often tolerate considerable changes in protein sequence and structure. The mutant lysozymes crystallized to date pack the same way as the wild-type protein, and knowledge of the wild-type lysozyme structure has been used to determine the structures of the mutants. The difference Fourier method allows the differences between the structure of the wild-type protein and an isomorphously crystallized mutant to be displayed directly.

Figure 4 shows one zone of the X-ray diffraction pattern of crystals of wild-type (4a) and Thr 157 → Arg mutant (4b) lysozymes. Small differences in the intensities of analogous reflections can be seen. When the differences in diffracted intensity are used to obtain coefficients for the calculation of an electron density map, the resulting map shows the differences between the mutant and wild-type proteins. Two examples of slices through such difference maps are shown in Figure 5. Changes in electron density caused by the mutations Thr 157 → Arg (Fig. 5a) and Thr

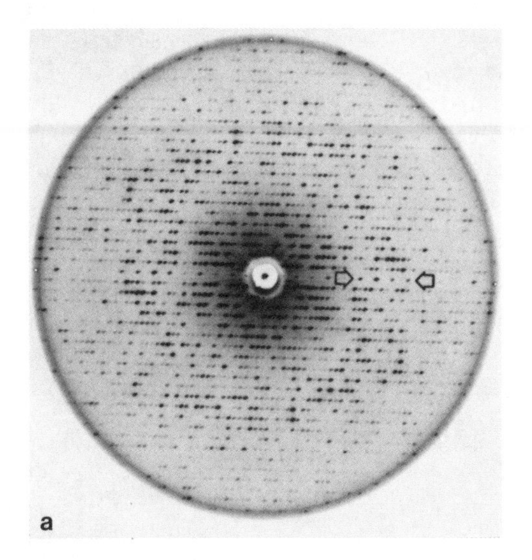

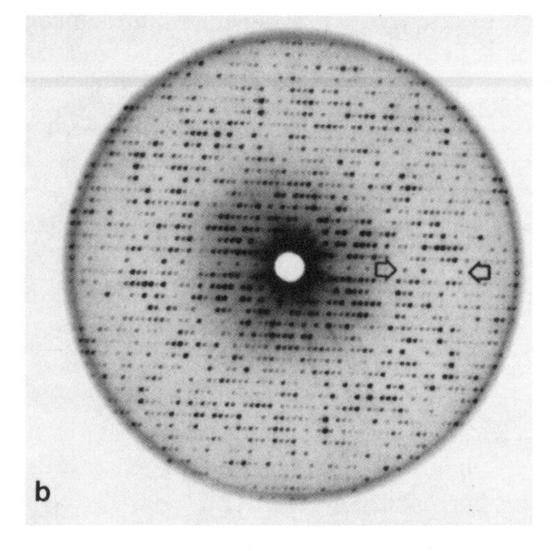

Fig. 4. *X-ray diffraction patterns (h0l zone) of wild-type (a) and Thr 157 → Arg mutant (b) lysozymes. The full set of diffraction data consists of many such zones. The resolution of the spots (or "reflections") increases with distance from the center of the pattern. Film (a) (precession angle = 20°) shows higher resolution data than (b) (precession angle = 15°). Higher resolution data provide a sharper image of the protein electron density and allow atomic positions to be determined more accurately.*

The spacings between the reflections on the two films are determined by the packing of molecules in the crystals. The intensities of the reflections are a measure of the average distribution of atoms in the lysozyme molecules and the surrounding solvent of crystallization. The spacings between the reflections on the two films are essentially identical, indicating that the cyrstals are isomorphous (i.e., the molecular packing is not perturbed by the mutation). Though the intensities of analogous reflections are quite similar, small differences are apparent. The arrowheads bracket four reflections whose relative intensities change, and many other subtle differences can be seen on inspection. These intensity changes are due to differences in the structures of the mutant and wild-type proteins.

157 → Asn (Fig. 5b) are shown superimposed on a schematic drawing of the wild type structure.

In practice, analysis of difference maps is followed by high-resolution crystallographic refinement of the mutant structures (W. Hen-

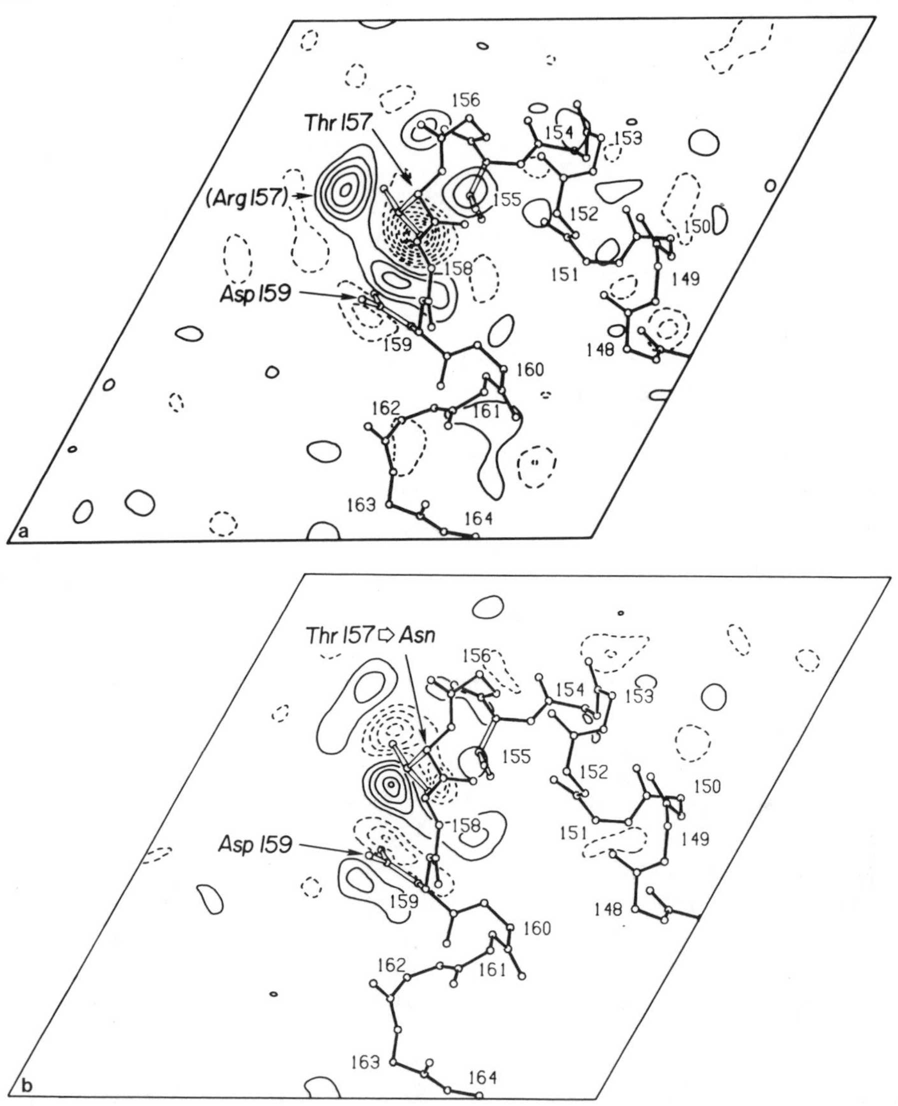

Fig. 5. *Single sections of difference electron density maps showing changes due to the substitutions Thr 157 → Arg (a) and Thr 157 → Asn (b). Amplitudes (F_{mutant}–$F_{wild\text{-}type}$) to 2.0 Å resolution and phases of the refined wild type structure were used to calculate the maps. A drawing of the peptide backbone in this region is superimposed on the difference electron density. For clarity, the side chains of all amino acids except Thr 155, Thr 157, and Asp 159 (open bonds) have been omitted from the drawing.*

Positive electron density due to atoms in the mutant that are not present in the wild-type structure is indicated by solid contours. For example, the solid contours marking the position of additional atoms in the side chain of Arg 157 are indicated in (a). Where the mutant lacks atoms that are present in the wild type protein, the difference map shows negative electron density denoted by broken contours. For exmaple, the loss of the Thr γ-hydroxyl group caused by the Thr 157 → Arg mutation results in striking concentric rings of broken contours surrounding the position of the hydroxyl group (a). When a mutation causes a shift in the position of a group of atoms, the difference map shows paired negative and positive features. For example, Asp 159 moves toward Arg 157 to form an ion pair (a), but in contrast, Asp 159 moves away from Asn 157 (b).

drickson, this volume). In the case of the lysozyme mutants, about 16,000 X-ray reflections can be measured to a resolution of 1.7 Å. The atomic coordinates and a parameter related to the motion of each atom are then adjusted by a least-squares refinement procedure to optimize agreement with the measured reflection intensities. Even well-refined protein structures—of which phage lysozyme is an example—can have average errors in their atomic coordinates of 0.1–0.2 Å. The uncertainty is worse for mobile atoms often found on the molecular surface.

To date, X-ray crystal structures of site-directed mutants with ser, asn, leu, his, and arg at position 157 have been determined at 1.7 Å resolution. The structure of Ser 157 mutant has been refined (R = 17.4%), and all the mutant proteins listed in Table I have been crystallized. Several important results and conclusions come from these studies.

1) Hydrogen bonds contribute more to thermal stability at position 157 than van der Waals contacts. This view is supported by the finding that both Ser and Asn can confer the bulk of the thermal stability that is lost when Thr 157 is replaced by Ile. Crystallographic studies show that both Ser and Asn restore the intramolecular hydrogen bonds that are disrupted by the Ile substitution (Fig. 3). With Asn at position 157, conformational shifts in the polypeptide backbone allow this to occur (Fig. 5b).

Surprisingly, the Thr 157 → Ser mutation results in a large increase in the mobility of the γ-hydroxyl group of the 157 side chain but no change in its average position. The γ-methyl group of Thr 157 apparently restricts the motion of the γ-hydroxyl group. These results are consistent with the view that stronger noncovalent bonds occur between groups that are more constrained by the native structure to occupy positions that favor interaction [12,13].

2) A new surface ion pair involving residue 157 does not adequately substitute for the hydrogen bonds to the Thr 157 hydroxyl group. The crystal structure of the Arg 157 mutant shows that Asp 159 moves toward the site to form a nonnative ion pair. This ion pair restores about half of the free energy of stabilization lost in the Thr 157 → Ile mutation. A similar degree of stabilization has been attributed to the ion pair between the N- and C-termini on the surface of bovine pancreatic trypsin inhibitor [14].

3) Crystallographically observed hydrogen bonds do not stabilize proteins to the same extent. The Arg 157 → Asp 159 salt bridge contains two new hydrogen bonds in addition to the ionic interaction, yet the Arg 157 mutant is temperature sensitive.

4) The wild-type solvent structure is not sufficient to confer wild-type stability. His 157 restores the wild type arrangement of solvent molecules, but this mutant is intermediate in stability between wild-type (Thr) and the original ts lesion (Ile).

5) Bound water may stabilize proteins. This conclusion is supported by the result that T4 lysozyme with Gly at position 157 is almost as stable as the wild-type protein. A bound water molecule may stabilize the protein by restoring the hydrogen bonds involving the side chain hydroxyl group of Thr 157. Crystallographic studies are in progress to test this idea.

These conclusions are stated in terms of effects on the native structures of the mutant and wild type lysozymes. It is also possible that some of the effects we observe arise from changes in the free energies of the denatured forms of the proteins.

It should be noted that the X-ray structure of each of the site-directed mutants contains unexpected features. For example, dispersed positional shifts occur in response to Asn 157. The mobility of the hydroxyl group of Ser 157 is increased compared to Thr 157. Arg 157 forms a new ion pair. The side chain of Leu 157 adopts a conformation that is very different from the similarly branched Asn. The protein structure adjusts even in response to seemingly conservation substitutions. These results emphasize the importance of high-resolution structural studies for the interpretation of the effects of mutations.

It has been encouraging to find that, despite these structural adjustments, isomorphous crystals of many mutants of T4 lysozyme can be obtained. This means that the way is open to detailed structural studies designed to systematically explore current hypotheses concerning the physical basis of protein stability.

ACKNOWLEDGMENTS

These studies are part of a collaborative effort involving many colleagues in the Institute of Molecular Biology. We thank, in particular, W. Baase, E.N. Baker, W. Becktel, J. Bell, B-L. Chen, S. Cook, F.W. Dahlquist, S. Daopin, T.M. Gray, M.G. Grütter, L. McIntosh, D. Muchmore, J. Schellman, D.E. Tronrud, L.H. Weaver, K. Wilson and J. Wozniak.

The work was supported in part by grants from the National Institutes of Health, The National Science Foundation, and the M.J. Murdock Charitable Trust. T.A. was a fellow of the Helen Hay Whitney Foundation.

REFERENCES

1. Gutte B, Merrifield RB (1969): The total synthesis of an enzyme with ribonuclease A activity. J Am Chem Soc 91:501–502.
2. Remington SJ, Anderson WF, Owen J, TenEyck LF, Grainger CT, Matthews BW (1978): Structure of the lysozyme from bacteriophage T4: An electron density map at 2.4 Å resolution. J Mol Biol 118:81–98.
3. Weaver LH, Matthews BW (1985): Structure of bacteriophage T4 lysozyme refined at 1.7 Å resolution. J Mol Biol. (in press)
4. Streisinger G, Mukai F, Dreyer WJ, Miller B, Horiuchi S (1961): Mutations affecting the lysozyme of phage T4. Cold Spring Harbor Symp Quant Biol 26:25–30.
5. Schellman JA, Lindorfer M, Hawkes R, Grütter M (1981): Mutations and protein stability. Biopolymers 20:1989–1999.
6. Hawkes R, Grütter MG, Schellman J (1984): Thermodynamic stability and point mutations of bacteriophage T4 lysozyme. J Mol Biol 175:195–212.
7. Grütter MG, Hawkes RB, Matthews BW (1979): Molecular basis of thermostability in the lysozyme from bacteriophage T4. Nature (London) 277:667–669.
8. Grütter MG, Weaver LH, Gray TM, Matthews BW (1983): Structure, function and evolution of the lysozyme from bacteriophage T4. In Matthews CK, et al (eds): "Bacteriophage T4." Washington, DC: American Society for Microbiology, pp 356–360.
9. Alber T, Grütter MG, Gray TM, Wozniak JA, Weaver LH, Chen B-L, Baker EN, Matthews BW (1986): Structure and stability of mutant lysozymes from bacteriophage T4. UCLA Symposia on Molecular and Cellular Biology, "Protein Structure, Folding, and Design." New York: Alan R Liss, Inc pp 307–318.
10. Zoller MJ, Smith M (1984): Oligonucleotide-directed mutagenesis: A simple method using two oligonucleotide primers and a single-stranded DNA template. DNA 3:479–488.
11. Owen JE, Schultz DW, Taylor A, Smith GR (1983): Nucleotide sequence of the lysozyme gene of bacteriophage T4: Analysis of mutations involving repeated sequences. J Mol Biol 165:229–248.
12. Goldenberg DP (1985): Dissecting the roles of individual interactions in protein stability: Lessons from a circularized protein. J Cell Biochem 29:321–335.
13. Creighton TE (1984): "Proteins: Structures and Molecular Properties." New York: WH Freeman and Co, pp 152–157, 321–328.
14. Brown LR, DeMarco A, Richarz R, Wagner G, Wuthrich et al (1978): The influence of a single salt bridge on static and dynamic features of the globular solution conformation of the basic pancreatic trypsin inhibitor. Eur J Biochem 88:87–95.

Protein Engineering, pages 299–305

27

Cytotoxic Proteins: Photoaffinity Labeling and Site-Directed Mutagenesis of an Active Site Residue in Diphtheria Toxin

R. John Collier

Department of Microbiology and Molecular Genetics, Harvard Medical School and Shipley Institute of Medicine, Boston, Massachusetts 02115

INTRODUCTION

Among the wide range of proteins created in the course of evolution, some have the capacity to kill cells. Well-studied classes of such cytotoxic proteins are: bacterial toxins that act on cells of eukaryotic hosts (e.g., diphtheria toxin); plant toxins, which may be lethal for animals that ingest them (e.g., ricin); and microbicidal toxins, which are produced by certain microbes and act on certain of their ilk (e.g., colicins, yeast-killer toxins).

Not surprisingly, a great many modes of attack have been evolved. One is to form holes in the cell envelope (e.g., colicin E1), and another, to degrade the target cell's DNA (colicin E2). A third, highly favored mode is to inactivate some component of the cell's protein synthesis apparatus. Certain bacteria release bacteriocins to block protein synthesis in other bacteria (e.g., colicin E3), while others release toxins to inhibit protein synthesis in animal hosts (e.g., diphtheria and *Pseudomonas* toxins). Also, several plant toxins act at the level of ribosomes (e.g., ricin, modeccin). This aspect of metabolism therefore seems to be a major area of vulnerability.

Cytotoxic proteins present phenomena of interest to both basic and applied science. From the standpoint of medicine, we would like to understand how and why bacterial toxins kill their hosts; and these and other toxins may well have applications in targeted killing of certain populations of cells in the body—for example, in the treatment of certain types of cancer. The toxins also represent useful probes for the cell biologist and intriguing problems of protein structure and activity to be dissected by biochemists, biophysicists, and molecular biologists.

In this Chapter I focus on diphtheria toxin as a case study of a complex and interesting cytotoxic protein. I summarize current knowledge of its structure and activity, present recent results of photoaffinity labeling, and site-directed mutagenesis of an active site residue, and speculate on the evolutionary origin of this molecule.

ADP-RIBOSYLATION

It has been shown that introduction of one to two molecules of an active fragment of diphtheria toxin (DT; Mr 58,342) into a sensitive animal cell is sufficient to kill the cell [1]. There is also evidence that ricin, colicin E3, and certain other toxic proteins may be similarly efficient. How is such efficient killing accomplished?

The lethality of DT depends crucially on the fact that it is a highly specific enzyme with the capacity to inactivate a single, key component of cellular metabolism. (Other toxins that act at a multiplicity of about one molecule/cell are probably also enzymes.) DT, after appropriate activation steps, catalyzes transfer of the ADP-ribosyl group of NAD onto a single site on elongation factor 2 (EF-2) [2]:

$$\text{NAD} + \text{EF}-2 \rightleftharpoons \text{ADP-ribosyl-EF}-2 + \text{nicotinamide} + \text{H}^+$$

The elongation factor, which is required for growth of nascent polypeptide chains of ribosomes, is thereby inactivated; protein synthesis ceases; and the cell dies. The lethality of DT therefore depends on the fact that one to two molecules are capable of inactivating all of the ca. 2 million EF-2 molecules within the cell over a period of hours to days.

Since 1968, when DT was found to act by ADP-ribosylation of EF-2, several other bacterial toxins have been shown to act by ADP-ribosylation mechanisms. *Pseudomas* toxin (exotoxin A), which has no apparent sequence homology to DT, uses precisely the same reaction to kill cells. Cholera toxin and its close relatives ADP-ribosylate a regulatory subunit of adenylate cyclase and thereby elevate cAMP levels in cells [3]). This does not kill the cell, but in intestinal epithelium it causes ion and fluid fluxes that result in diarrhea. Pertussis toxin ADP-ribosylates a different regulatory subunit of the cyclase and thereby perturbs cAMP metabolism [4]. Finally, there is a recent report that one of the botulinum toxins (C2) has ADP-ribosyl transferase activity, but the target protein is not known [5].

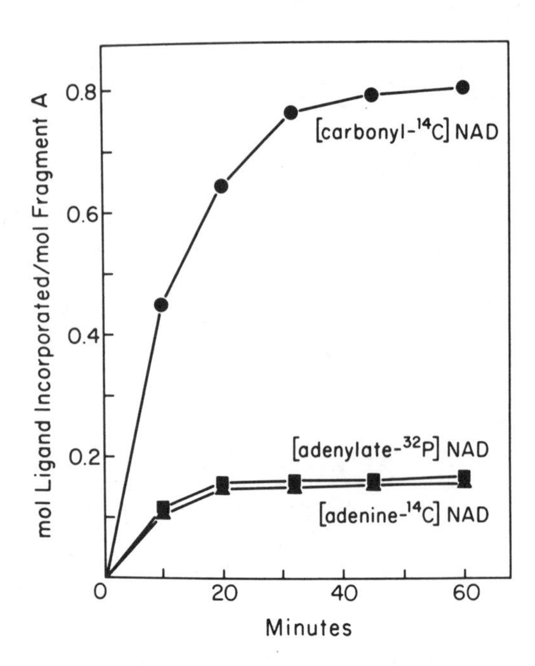

Fig. 1. *Photoaffinity labeling of diphtheria toxin fragment A with 3 native NAD preparations. Reaction mixtures containing 20 μM fragment A and 40 μM NAD radiolabeled in the nicotinamide (●), adenylate phosphate (■), or adenine (▲) moiety were irradiated with 254 nm UV light at 4°C. Aliquots were removed at intervals, and trichloroacetic acid-precipitable label was determined.*

The evolutionary origins of, and relationships among, these toxins form an intriguing puzzle, which is briefly discussed below.

What determines the specificity of DT for EF-2? It has been found that EF-2 is the only protein in eukaryotic cell extracts that is ADP-ribosylated by DT, and the basis of this specificity is curious indeed. The ADP-ribosylation site on EF-2 was shown by Bodley and co-workers to be a hypermodified histidine residue termed diphthamide [6]. There is only one diphthamide residue per EF-2 molecule, and it appears that no other protein in the cell besides EF-2 contains this residue. Diphthamide arises from posttranslational modification of a specific histidine by a sequence of at least three enzymic reactions [7]. The unmodified EF-2 appears to retain its ability to function in protein synthesis, and only after the histidine is modified does EF-2 become a substrate for

DT. Diphthamide certainly did not evolve to make the cell vulnerable to toxin attack! The question of its normal function will be addressed below.

LOCATING THE TOXIN'S ACTIVE SITE

Given that ADP-ribosylation is commonly used by various bacterial toxins to modify host cell proteins, we may ask if the various toxins act by the same catalytic mechanism. To probe this question, we initiated experiments to identify active site residues. Our initial attempts to use photolabeling with DT revealed a novel photochemical reaction, which has led to the identification of what may be a crucial active site residue.

Diphtheria toxin is released from *Corynebacterium diphtheriae* as a single, 535-residue polypeptide chain, which is a proenzyme [2]. For expression of ADP-ribosyl transferase activity, the toxin must be cleaved at an arginine-rich region within the first of two disulfide loops in the toxin. This creates two large fragments, fragment A (N-terminal, ca. 193 residues) and fragment B (C-terminal, ca. 342 residues) linked by a disulfide bridge. Rupture of this disulfide permits separation of fragments A and B and expression of ADP-ribosyl transferase activity by A.

We and others had shown earlier that fragment A contained a single NAD binding site (K_D ca. 8 μM; [8]), and we decided that photoaffinity labeling with NAD might enable us to identify a residue in this site. Instead of employing photolabile derivatives of NAD, we decided to employ the native dinucleotide and rely on the aromatic moieties to promote labeling. This strategy proved fruitful.

Figure 1 shows results obtained when purified fragment A was irradiated at 254 nm in the presence of three individual preparations of NAD, containing ^{14}C in carbonyl group of the nicotinamide moiety, or in the adenine moiety (uniformily labeled), or ^{32}P in the adenylate phosphate [9]. It is readily apparent that label from the nicotinamide-labeled NAD was much more efficiently incorporated than that from the other preparations. In fact, the efficiency of labeling frequently was close to stoichiometric. This, together with the finding that the photoaffinity-labeled fragment A was enzymically inactive, hinted that this reaction should be probed further.

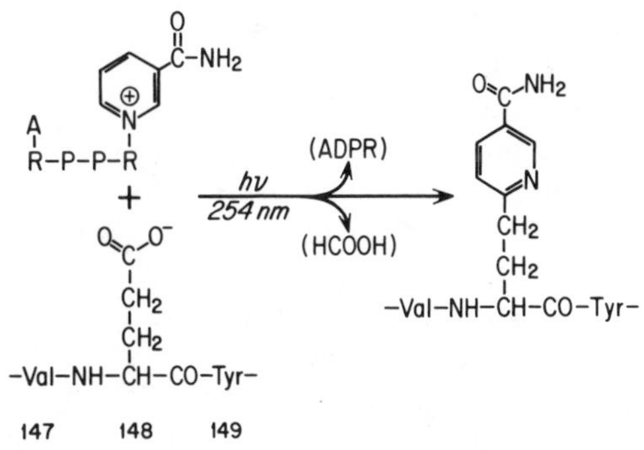

Fig. 2. *UV-induced transfer of the nicotinamide moiety of NAD to Glu-148 of fragment A.*

To make a long story short, we found that the label incorporated from [*carbonyl*$^{-14}$C] NAD was located at a single site in fragment A, position 148, which corresponds to Glu in the native molecule. The next question was the nature of the photoproduct at position 148. To probe this question we isolated the thermolytic tripeptide Val-X-Tyr, corresponding to residues 147–149 of the photolabeled fragment A, and subjected it to a variety of analyses, including nuclear magnetic resonance (through collaboration with Normal Oppenheimer, University of California, San Francisco) and mass spectrometry (through collaboration with James McCloskey, University of Utah). The results [10] gave the unique structure shown in Figure 2. The entire nicotinamide moiety from NAD, but none of the rest of the molecule, had become attached via the number 6 ring carbon, to the γ-methylene group of Glu-148; and the Glu-148 side chain had been decarboxylated in the process. This structure had not been described before, in any system. We have not studied the photochemical reaction mechanism in detail, but we believe that the primary excitation event is photon absorption by the nicotinamide ring and suspect that a free radical mechanism may be involved.

The major implications of the structure of the photoproduct for the present purposes are that Glu-148 may be an active site residue. We infer that, regardless of the photochemistry involved, the γ-methylene group of Glu-148 must be close to, or contiguous with, the number 6 nicotinamide ring carbon of NAD in the dinucleotide–fragment A complex. This would place the Glu-148 side chain carboxyl within a short radius of the number 6 nicotinamide ring carbon, and possibly in contact with the nicotinamide-ribose linkage that is ruptured during ADP-ribosyl transfer. Thus the carboxyl group of Glu-148 might in fact participate directly in catalysis. This intriguing possibility led us to use site-directed mutagenesis to replace this residue.

We and others had already obtained clones of portions of the DT gene (the whole toxin gene can be cloned only under P4 conditions), and we subjected one of these clones (which encoded the entire A moiety plus about half of B) to site-directed mutagenesis to replace Glu-148 with either Asp or Gly [11]. Early results indicated that either substitution gave an enzymically inactive fragment A product, and we then characterized the Asp-148 mutant in greater detail. To date, our results indicate that the Asp-148 form of fragment A has less than 0.6% the specific enzymic activity of the native fragment (Fig. 3), but binds NAD with little or no loss of affinity. Also, the Asp-148 mutant shows the same level of trypsin sensitivity as the wild-type fragment A. Thus, substitution of Asp for Glu at position 148 apparently causes little or no perturbation of folding, but virtually or completely abolishes ADP-ribosyl transferase activity. This implies that the precise spatial position of the side chain carboxyl at position 148 is crucial for this enzymic reaction. This in turn is consistent with the notion that this group may participate directly in catalysis.

Assuming that the photoaffinity labeling reaction described does indeed label an active site residue of DT, we would like to know how generally useful the method may be. Can

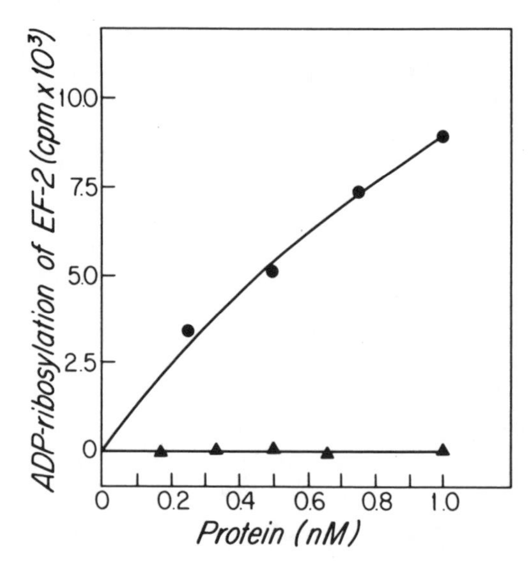

Fig. 3. *ADP-ribosyl transferase activity of trypsin-activated Glu-148 (wild-type) and Asp-148 (mutant) toxin peptides. Periplasmic fractions containing the wild-type or mutant gene fragments were digested with trypsin under mild conditions, to generate fragment A, and then were titrated in the ADP-ribosylation assay. Glu-148, closed circles; Asp-148, closed triangles.*

we use it to label active sites of other ADP-ribosyl transferases? We have good indication that *Pseudomonas* toxin, which catalyzes the same reaction as DT, is also effectively photolabeled at a specific site with [carbonyl-^{14}C]NAD, although the labeling efficiency is somewhat lower than with DT. Cholera toxin and pertussis toxin are currently being tested. What about other NAD-lined enzymes? To date we have tested three dehydrogenases, with essentially negative results in all cases [9]. Thus, the question of the general utility of the method remains open. There are indications it is not unique to DT, but whether it will extend to all of the ADP-ribosyl transferases, or even beyond, remains uncertain.

If the method were to prove generally useful as an active site labeling method for ADP-ribosylating toxins, this would have possible applications with regard to vaccine development, as well as for understanding basic aspects of toxin action. Methods of preparing vaccines against certain ADP-ribosylating toxins are being actively sought (e.g., pertussis toxin, *Pseudomonas* toxin), and ways of making improved vaccines against others (e.g., DT) are always welcome. If one can completely eliminate toxic activity by substituting or deleting a single active site residue, then structural perturbations should be minimal and immunogenic activity retained. Photoaffinity labeling, coupled with site-directed mutagenesis, may therefore lead to a new generation of recombinant DNA vaccines.

THE QUESTION OF ENTRY

In the discussion above, I neglected an important aspect of DT action–namely, how it gains access to its substrates. In the case of DT and certain other cytocidal toxins, it is clear that the toxin acts *directly* to modify a cell constituent within the bounds of plasma membrane, thus necessitating that the toxin's active site traverse a membrane at some level. The general phenomenon of protein insertion into, and passage across, membranes is complex and poorly understood, and this is likewise true of toxin entry mechanisms. There is no single toxin whose entry is understood completely. However, DT and a few other toxins have been studied in some detail, and a model for DT entry has emerged that is at least plausible.

The model may be summarized briefly, as follows. The first step in entry appears to be attachment to cell surface receptors. This is believed to occur via interaction with a receptor-binding site on the C-terminal region of the toxin's B fragment. The nature of the receptor remains uncertain, but Eidels and coworkers [12] have evidence that a high Mr glycoprotein, or a family of such molecules, may constitute the receptor. Next, the toxin-receptor complex is endocytosed into coated vesicles and delivered to endosomes. Within the endosomes, or derivative vesicles, the toxin is exposed to low pH conditions, ca. pH 4–5 (13). This induces exposure of buried hydrophobic regions present in the N-terminal half of B [14]. These hydrophobic regions, some of which are sufficiently long to span a bilayer [15], are believed to promote insertion of the toxin into the endosomal membrane, and this insertion somehow mediates transfer of the A fragment across the membrane and into contact with the cytosol. The toxin has been shown to insert into artificial lipid bilayers under low pH conditions to form ion-conductive channels [16, 17]. It has also been shown that when a receptor-bound DT at the cell surface is exposed to low pH, the A fragment enters the cells directly, without the mediation of endocytosis.

There is good evidence to support certain aspects of this model and weak evidence for other aspects. The receptor-binding step seems to be general for toxins that act on intracellular targets, while the nature of the receptor varies widely among toxins. The endocytosis event also appears to be common. Acid-dependence of entry is found in various animal viruses but is not universal; likewise, it is not universal among toxins, since ricin intoxication is not stimulated by acid treatment.

It is clear that elucidation of the actual entry mechanisms will require all the tools of cell biology, biochemistry, biophysics, and molecular biology that can be amassed. However similar or diverse the mechanisms employed, they will be of interest in understanding more general phenomena of protein–membrane interactions and will probably have several applications related to medicine. For example, to target toxins to specific classes of cells, various workers have attached portions of toxins to antibodies directed against specific cell-surface antigens. Clearer understanding of the entry mechanisms will greatly facilitate the choice of specific portions of toxins to attach and of the modes of attachment.

EVOLUTIONARY CONSIDERATIONS

The fact that the substrate protein for DT was found in animal cells but not in bacteria (except for Archaebacteria) led Uchida and co-workers to propose that DT may have evolved from an ancestral eukaryotic protein [18]. The fact that the structural gene for DT is carried on a phage genome heightened such speculation. Only recently has such speculation received direct support, however, Two reports appeared in 1984 describing the existence of an endogenous enzyme from eukaryotic cell extracts that ADP-ribosylated EF-2 [19, 20]. Although these reports remain to be confirmed, the existence of such an enzyme would at least provide a rationale for the existence of the diphthamide residue on EF-2. Presumably, ADP-ribosylation of EF-2 may regulate protein synthesis, under an as-yet-unknown set of circumstances.

Why was this activity not seen earlier? Lee and Iglewski [19] reported that cell extracts contain a heat-labile inhibitor of the enzyme under normal conditions, and the inhibition must be released for enzyme activity to be expressed. Lee and Iglewski found that the enzyme co-purified with EF-2, and that the inhibitor was removed during this purification [19].

If this general picture is confirmed, it will provide an additional clue to the puzzle of how DT and *Pseudomonas* toxin evolved. Presumably, one or more events of gene capture took place in the course of interactions between prokaryotes and eukaryotes. The events that occurred in transforming the captured gene, or genes, into a toxin are unclear. However, to act on a particular eukaryotic cell, the ADP-ribosyl transferase would have to be resistant to the particular inhibitor of the target cell. This might occur by mutation of the enzyme after bacterial capture. Alternatively, if various families of ADP-ribosyl transferase:inhibitor complexes exist in which cross-inhibition does not occur, then it may be that the endogenous enzyme captured from one eukaryote was used against another eukaryote with an inhibitor of different specificity.

Additional questions arise as to how the functional domains arose for receptor attachment and transmembrane transfer. Did they derive from domains that were once parts of the enodgenous ADP-ribosyl transferase, or were they once parts of other eukaryotic or prokaryotic proteins?

Finally, how do other ADP-ribosylating toxins relate to DT and *Pseudomonas* toxin? Moss and co-workers have identified enzymes in eukaryotic cells that act in somewhat similar fashions to cholera toxin [21], but there is no hard evidence to support the notion that there may be an evolutionary relationship. Such questions provide intriguing material for both those interested in molecular evolution and those focused on molecular pathogenesis.

ACKNOWLEDGMENTS

This work was supported in part by Public Health Service Research Grant AI-22021 from the National Institute of Allergy and Infectious Diseases and Grant CA-39217 from the National Cancer Institute.

REFERENCES

1. Yamaizumi M, Mekada E, Uchida T, Okada Y (1978): One molecule of diphtheria toxin fragment A introduced into a cell can kill the cell. Cell 15:245–250.
2. Collier RJ (1975): Diphtheria toxin: Mode of action and structure. Bact Rev 39:54–85.
3. Gill DM, Meren R (1978): ADP-ribosylation of membrane proteins catalyzed by cholera toxin: Basis of the activation of adenylate cyclase. Proc Natl Acad Sci USA 75:3050–3054.
4. Katada T, Ui M (1982): Direct modification of the membrane adenylate cyclase system by islet-activating protein due to ADP-ribosylation of a membrane protein. Proc Natl Acad Sci USA 79:3129–3133.
5. Simpson LL (1984): Molecular basis for the pharmacological actions of *Clostridium botulinum* type C_2 toxin. J Pharmacol Exp Ther 230:665–669.
6. Van Ness BG, Howard JB, Bodley JW (1980): ADP-ribosylation of elongation factor 2 by diphtheria toxin: NMR spectra and proposed structures of ribosyl-diphthamide and its hydrolysis products. J Biol Chem 255:10710–10716.
7. Moehring TJ, Danley DE, Moehring JM (1984): In vitro biosynthesis of diphthamide, studied with mutant chinese hamster ovary cells resistant to diphtheria toxin. Mol Cell Biol 4:642–650.
8. Kandel J, Collier RJ, Chung DW (1974): Interaction of fragment A from diphtheria toxin with nicotinamide adenine dinucleotide. J Biol Chem 249:2088–2097.
9. Carroll SF and Collier RJ (1984): NAD binding site of diphtheria toxin: Identification of a residue within the nicotinamide subsite by photochemical modification with NAD. Proc Natl Acad Sci USA 81:3307–3311.

10. Carroll SF, McCloskey JA, Crain PF, Oppenheimer NJ, Marschner TM, Collier RJ (1985): Photoaffinity labeling of diphtheria toxin fragment A with NAD: Structure of the photoproduct at position 148. Proc Natl Acad Sci USA 82:7237–7241.
11. Tweten RK, Barbieri JT, Collier RJ (1985): Diphtheria toxin: Effect of substituting aspartic acid for glutamic acid 148 on ADP-ribosyltransferase activity. J Biol Chem 260:10392–10394.
12. Eidels L, Proia RL, Hart DA (1983): Membrane receptors for bacterial toxins. Microbiol Rev 47:596–620.
13. Draper RK, Simon MI (1980): The entry of diphtheria toxin into the mammalian cell cytoplasm: Evidence of lysosomal involvement. J Cell Biol 87:849–854.
14. Collins CM (1984): Diphtheria toxin: Dinucleotide binding and conformation. Doctoral thesis, University of California, Los Angeles.
15. Greenfield L, Bjorn MJ, Horn G, Fong D, Buck GA, Collier RJ, Kaplan DA (1983): Nucleotide sequence of the structural gene for diphtheria toxin carried by corynebacteriophage β. Proc Natl Acad Sci USA 80:6853–6857.
16. Kagan BL, Finkelstein A, Colombini M (1981): Diphtheria toxin fragment forms large pores in phospholipid bilayer membranes. Proc Natl Acad Sci USA 78:4950–4954.
17. Donovan JJ, Simon MI, Draper RK, Montal M (1981): Diphtheria toxin forms transmembrane channels in planar lipid bilayers. Proc Natl Acad Sci USA 78:172–176.
18. Uchida T, Gill DM, Pappenheimer AM Jr. (1971): Mutation in the structural gene for diphtheria toxin carried by temperate phage β. Nat New Biol 233:8–11.
19. Lee H, Iglewski WJ (1984): Cellular ADP-ribosyltransferase with the same mechanism of action as diphtheria toxin and *Pseudomonas* toxin A. Proc Natl Acad Sci USA 81:2703–2707.
20. Sitikov AS, Davydova EK, Ovchinnikov LP (1984): Endogenous ADP-ribosylation of elongation factor 2 in polyribosome fraction of rabbit reticulocytes. FEBS Letters 176:261–263.
21. Moss J, Stanley SJ, Watkins PA (1980): Isolation and properties of an NAD- and guantidine-dependent ADP-ribosyltransferase from turkey erythrocytes. J Biol Chem 255:5838–5840.

Protein Engineering, pages 307–314

28

Biological Function of Modified Yeast and Mammalian *ras* Genes in Their Heterologous Systems

Deborah DeFeo-Jones, Kelly Tatchell, and Edward M. Scolnick

Department of Virus and Cell Biology, Merck Sharp & Dohme Research Laboratories, West Point, Pennsylvania 19486 (D.D.-J., E.M.S.), Department of Biology/G5, University of Pennsylvania, Philadelphia, Pennsylvania 19104 (K.T.)

The *ras* genes represent divergent members of a multigene family which is highly conserved among eucaryotes as evolutionarily distant as man and yeast. Such broad conservation suggests that these genes must have some basic and perhaps essential function in the eucaryotic cell. *Ras* genes, which code for a 21,000 dalton protein (called p21), were originally identified as the viral oncogenes of the transforming retroviruses Harvey (Ha) and Kirsten (Ki) murine sarcoma virus (MUSV). These oncogenes (called *v-Ha-ras* and *v-Ki-ras*, respectively) were subsequently used as molecular probes to explore the origin and distribution of their transforming sequences in normal vertebrate cells—studies that ultimately led to the identification and isolation of their normal cellular counterparts (c-ras-genes) [1].

While the clearly established oncogenic potential of the viral *ras* genes created a great deal of interest in their mechanism of action, it was the discovery of mutated forms of the cellular (c) *ras* genes in a number of human cancers which gave impetus to the study of their normal cellular function. It was discovered that DNA derived from some human tumors could transform an established mouse fibroblast cell line (NIH 3T3 cells). These transformed fibroblasts lost their normal property of contact inhibition, and instead grew into dense foci. The subsequent cloning of the DNA sequences responsible for the transformation event revealed that they were altered cellular *ras* genes. To date, three members of the *ras* gene family have been isolated from vertebrate cells: *c-Ha-ras*, *c-Ki-ras*, and *N-ras* (for the neuroblastoma cell line from which it was isolated). [1–3].

The viral *ras* gene products differ from their cellular proto-oncogenes at two amino acid residues; the glycine at amino acid residue 12 found in all cellular *ras* genes is replaced by either arginine or serine in *v-Ha-ras* and *v-Ki-ras*, respectively, while the alanine at amino acid 59 of the cellular genes is replaced by a threonine in both v-*ras* proteins. The transforming cellular *ras* genes have suffered point mutations which substitute anyone of several amino acids for either Glycine-12 or Gluta mine-61 [3].

RAS proteins are found associated with the inner plasma membrane of cells via a palmitic acid moiety which is covalently linked to a C-terminal amino acid of the protein; this cellular location probably represents the functionally active site since mutant p21 proteins that fail to associate with the plasma membrane are transformation defective. P21 proteins, which contain a threonine, instead of the normal alanine, at amino acid position 59, are capable of autophosphorylating this residue. The p21 protein specifically binds with GDP and GTP and has an intrinsic GTP hydrolysing activity. When compared to the normal p21 protein, this hydrolysis of GTP is significantly impaired in mutated oncogenic *ras* proteins, strongly suggesting that this activity is intimately involved in the normal functioning of the p21 protein [1,4,5].

RAS GENES ARE HIGHLY CONSERVED

Genomic DNA from a wide variety of eucaryotic species hybridizes to viral *ras* DNA. The DNA sequences from these cross-hybridizing species reveal that they code for proteins with considerable homology to mammalian *ras*. The *Drosophila* and slime mold *ras* genes code for proteins with 75% homology to p21 *ras* [6,7]. Such conservation implies that *ras* may have some important, it not essential, function in all eucaryotes. The prediction of this hypothesis is that mutations in *ras* that render p21 nonfunctional will have deleterious effects on the viability or growth of the organism. Unfortunately this hypothesis cannot be tested easily in mammals, but it can be tested in those organisms with well-developed genetics such as Drosophila or yeast. The recent demonstration that the yeast *Saccharomyces cerevisiae* contains two *ras* homologous genes, call *RAS* 1 and *RAS* 2, provides an organism in which detailed genetic analysis of *ras* gene function can be accomplished [8,9,]. Transformation in yeast occurs by homologous recombination, thus allowing the replacement of a normal gene with one altered by in vitro manipulation. Although methods of directing DNA into the germline of many eucaryotes now exists, the transforming DNA does not integrate via homology and normal sequences cannot be removed [10,11]. The unique property of homologous integration in yeast, coupled with observations that specific point mutations drastically affect both the biochemical and biological activity of the *ras* proteins, makes this sytem an extremely exciting one in which to perform protein structure–function studies.

The yeast and mammalilan *ras* proteins share many structural, immunological, and biochemical features. The yeast and mammalian *ras* genes are approximately 65% homologous over the first 165 N-terminal amino acids; conservative amino acid changes in the yeast proteins account for about one half of the divergent amino acids in this region. Comparison of the remaining C-terminal portions of the two genomes shows broad differences (see Figure 1); while both the mammalian and Drosophila *ras* genes encode proteins of 188 or 189 amino acids, the yeast *RAS* 1 and *RAS* 2 genes have primary products of 309 and 322 amino acids, respectively. The larger size of the yeast *RAS* 1 and *RAS* 2 proteins is a consequence of 112 and 125 additional C-terminal amino acids, respectively, as well as seven additional N-terminal amino acids [9,12]. Since all mammalian *ras* genes show poor conservation of their C-terminal sequences, the major comparative difference with the yeast *RAS* genes lies more in the larger size of this nonconserved region. It is thought that this region of divergence among all the *ras* genes may be involved in the specificity of action of the individual genes, though there is no data to indiate this. The extreme C-termini of the yeast *RAS* genes again shows homology to mammalian *ras*; in particular, the cysteine at amino acid position 186 is conserved. This amino acid has been shown to be necessary for the membrane localization of mammalilan *ras* [4]. The yeast *RAS* genes are more homologous to the "normal" mammalian *ras* genes than the mutated ones; that is they encode a glycine, a glutamine and an alanine at positions analogous to amino acids 12, 61, and 59, respectively, of the mammalian *ras* genes [9, 12].

Though the yeast *RAS* 1 and *RAS* 2 proteins are larger (34Kd and 35 Kd, respectively), they still share antigenic determinants with the mammalian *ras* proteins and possess both the GDP/GTP specific binding and GTP hydro-

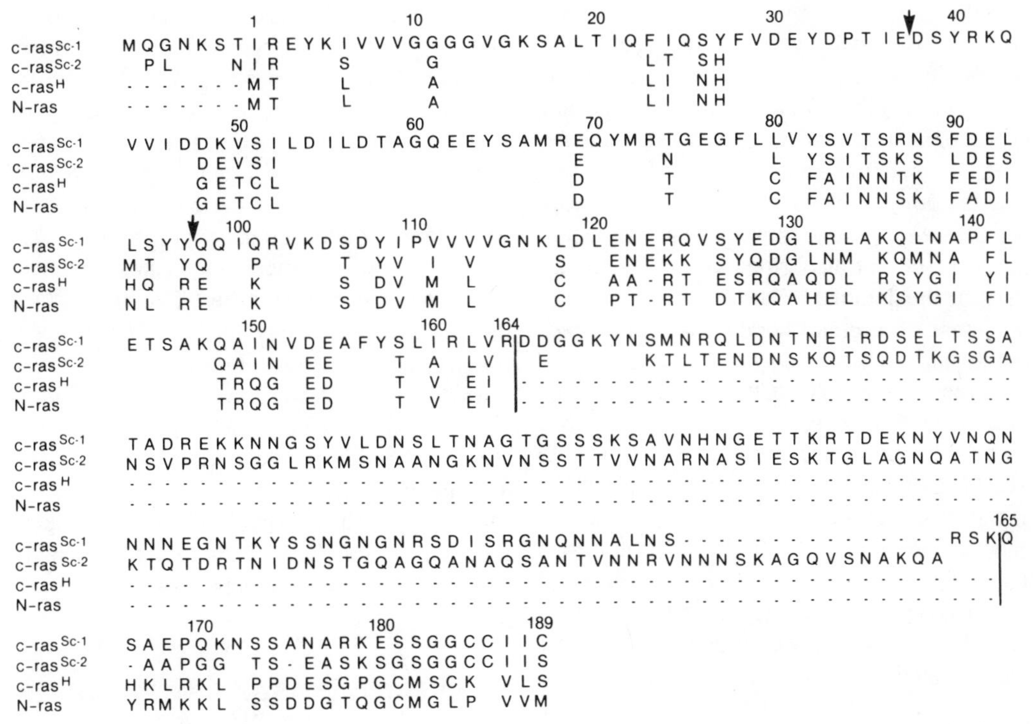

Fig. 1. *Comparison of the amino acid sequence of* RAS *1 and* RAS *2 with c*-Ha-ras *and N*-ras. *The amino acids have been numbered according to the c-*Ha-ras *numbering system. The dashed lines are positions where no amino acids are present. Empty spaces represent where amino acids are identical between all four genes. Between position 164 and 165, amino acids are present only in* RAS *1 and* RAS *2. Arrows indicate positions where introns would occur, if shown, in the nucleotide sequence of c*-Ha-ras *(From Dhar et al. [12].)*

lyzing activity of their mammalian counterpart, p21. In addition, substitution of either the normal glycine or glutamine residue (at position 12 or 61, by mammalian *ras* numbering) of the yeast *RAS* protein with a known "activating" amino acid mutation (i.e., valine at position 12) drastically impaired the GTPase activity of the yeast *RAS* proteins, similar to what is found with mutated mammalian *ras* proteins [13–15].

This data indicated that the yeast and mammalian *ras* proteins had clearly conserved biochemical functions. However, it is difficult to equate the biological function of two proteins based purely on their biochemical activities. The ability of p21 to bind and hydrolyze GTP is a property shared with other know guanine nucleotide-binding proteins such as the G-proteins of the adenylate cyclase system and transducin. In fact, recent sequence data indicates that p21 shares some structural homology with these proteins [5,16,17]. In order to show that yeast *RAS* was not just some distant relative of these other proteins, it was important to establish that the yeast and mammalian *ras* proteins were biologically functional in their heterologous systems. Such cross-species functionality would indicate that at least their immediately associated functions (i.e., interactions with secondary proteins) are conserved, thus making the yeast *RAS* system relevant to the mammalian one. To do such studies, it is necessary to have defined biological assays in which to test the respective *ras* genes.

BIOLOGICAL ASSAYS

Functional assays have been developed that can assess biological activity of a given *ras* gene in mammalian and in yeast cells. In the mammalian system, the NIH 3T3 transfection assay has proven to be a powerful tool in the identification of transforming DNA sequences. This assay was used to identity the oncogenic *ras* sequences present in some human tumors. The procedure involves exposure of a recipient cell monolayer (usually mouse NIH 3T3 cells) to some donor DNA (in this case the donor DNA would be plasmids bearing different *ras* gene constructs) and the scoring for foci of transformed cells. This assay tests for the dominant transforming phenotype of mutant *ras* proteins but does not, of course, test for any possible "normal" function [18].

In the yeast system it has been shown that haploid cells lacking either *RAS* 1 or *RAS* 2 are viable; however, if these cells lack both yeast *RAS* genes they fail to grow, demonstrating that these genes provide some essential function in yeast [19,20]. In addition, some haploid yeast strains lacking *RAS* 2 alone fail to grow on nonfermentable carbon sources such as ethanol or glycerol at 37°C, a defect which can be eliminated by increased gene dosage of *RAS* 1 (21). Therefore, the ability of any *ras* gene to function in yeast could be assayed by two criteria: 1) The ability to restore viability to yeast cells lacking both *RAS* genes and 2) the ability to allow yeast cells with a disruption of *RAS* 2 to grow on ethanol or glycerol.

CONSTRUCTION OF YEAST/MAMMALIAN *ras* HYBRIDS

These biological assays were used to test a series of *ras* genes, some representing the "natural" gene and others representing chimeric constructions between yeast and mammalian *ras* (22). Figure 2 diagrams the latter molecules. The yeast *RAS* 1 gene was chosen for the construction of chimeras because of the fortunate occurrence of restriction sites which break the molecule into regions encompassing what would be approximately the first exon (from the AUG start to the ClaI site), the second and third exons (from the ClaI to AccI site), and the fourth exon (from the AccI site to termination) of the cellular mammalian *ras* genes (see Figure 1). The v-*Ha-ras* genome was used to contribute the mammalian portion of the chimeras because of its lack of introns and the presence of specific mutations which enable it to transform NIH 3T3 cells with a high efficiency. Oligo directed mutagenesis was used to create the ClaI and AccI sites found in the yeast *RAS* 1 gene in the v-*Ha-ras* gene, thus allowing for the even shuffling of exon domains between the two genes.

The resulting chimeric molecules (detailed in Figure 1) were subcloned into either a plasmid (called P14) containing a high-efficiency transcriptional element for mammalian cells (termed the LTR) or a yeast shuttle vector (called AAH5), which is a high copy, 2μ containing plasmid carrying a yeast promoter and terminator element and the auxotrophic marker for leucine (LEU2). This plasmid vector when transformed into yeast cells, allows for high expression of any fragment subcloned into it [22].

MODIFIDED YEAST AND MAMMALIAN *ras* GENE ARE BIOLOGICALLY FUNCTIONAL IN THEIR HETEROLOGOUS SYSTEMS.

Table I is a summary of results obtained with the various *ras* gene molecules in the different biological assay systems. In brief, apparently all of the constructs were biologically functional in the yeast system; most importantly, both the v-*Ha-ras* and c-*Ha-ras* genes were active. However, there is apparent variation in the degree to which the different *ras* genes can function in the yeast system. For example, the ability to grow on glycerol was somewhat dependent on whether or not the gene construct contained the additional yeast C-terminal information (See Figure 2 and Table I). The ability of the various constructions to transform mouse NIH 3T3 cells was somewhat more complicated but highly instructive. While Hybrid A and Hybrid C, when it contained the activating mutation arg-12, could transform mammalian cells, no transformation occurred using a full length *RAS* 1 or *RAS* 2 gene, with or without an activating mutation, nor with Hybrid B and D, both of which contained the activating mutations pres-

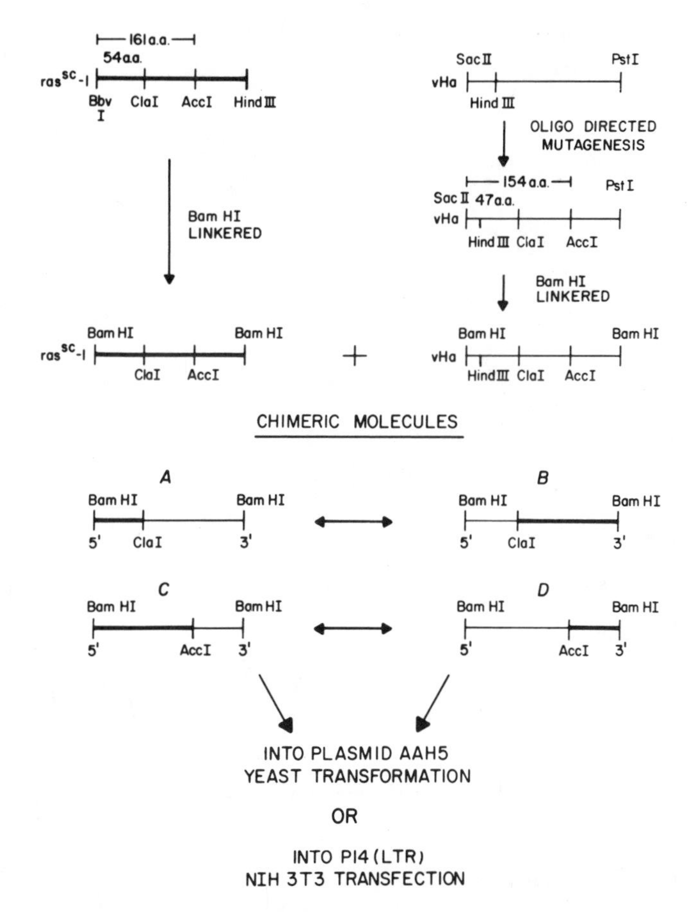

Fig. 2. *Construction of yeast/mammalian hybrid genes. The heavy horizontal lines represent yeast* RAS *1 sequences; the light horizontal lines represent* v-Ha-ras *sequences. The yeast* RAS 1 *gene was digested with the restriction enzyme BbvI in order to remove upstream yeast promotor signals which could interfere with transcription in mammalian cells. It was secondarily digested with the enzyme Hind III, the ends flushed, ligated to Bam HI linkers and the final fragment subcloned into the BamHI site of PBR322.* v-Ha-ras *was subcloned into M13mp8 as a 720bp BamHI fragment and oligonucleotide directed mutagenesis performed as described in Reference 15, introducing a ClaI site and an AccI site at positions in frame and analogous to those sites found in yeast* RAS *1. The resulting DNA molecule was digested with BamHI and subcloned into that site in PBR322. The chimeric molecules A,B,C and D were constructed by switching regions between the yeast* RAS *1 gene and the* v-Ha-ras *gene as diagrammed. The ClaI and AccI sites found in the yeast* RAS *1 gene, and substituted into the* v-Ha-ras *gene fragmented these* ras *genes into regions encompassing approximately the first exon (AUG start codon to ClaI site), the second and third exons (ClaI to AccI sites) and the fourth exon (AccI site to terminator sequence) domains of the mammalian* ras *genes (see Fig. 1). Hybrid C was further modified by introducing potentially activating mutations into this clone, thus creating two variants (Hybrid C^{leu} and Hybrid C^{thr}). All molecules were secondarily subcloned into the BamHI site of the Ha-MUSV-p14 plasmid molecule or flush end ligated into the AAH5 yeast shuttle vector plasmid. (From DeFeo-Jones et al [22].) © 1985 by the American Association for the Advancement of Science).*

ent in v-*Ha-ras* (See Figure 2). Though the transforming activities of the hybrid constructs that did function are lower than those of the mammalian *ras* genes (see footnote to Table I) this data showed that the first two exons of mammalian *ras* could be replaced by the corresponding region of yeast *RAS* and be biologically functional [22].

TABLE I. Biological Activity of Different *ras* Genes and Yeast/Mammalian Hybrid Genes in Both the Yeast and Mammalian Assay Systems

	Yeast assay[b]		Mammalian assay[c]
Gene tested[a]	Growth on glycerol	Viability of *RAS* minus cells	(tranformation of NIH 3T3 cells)
Yeast *RAS* 1	+++	Yes	−
Yeast *RAS* 2	+++	Yes	−
v-*Ha-ras*	+	Yes	+++
c-*Ha-ras*	++	Yes[d]	++
Yeast *RAS* 1^{Leu}	+++	Yes	−
Hybrid A	+	N/D	++
Hybrid B	+++	Yes	−
Hybrid C	+	Yes	−
Hybrid D	+++	Yes	−
Hybrid C^{Leu}	+	Yes	++
Hybrid C^{Thr}	+	Yes	+
del-*RAS* 1^{Leu}	++	Yes	+

[a]Genes with the superscript Leu (leucine) or Thr (threonine) contain the activating mutations leu-61 or thr-59 [22].
[b]In the yeast system, expression of the normal yeast *RAS* 1 and *RAS* 2 genes is representative of a wild-type phenotype.
[c]In the mammalian system, the highly transforming v-*Ha-ras* gene is the basis for scoring the transforming activity of the other genes tested.
[d]Wild-type yeast cells and yeast cells containing the various *ras* constructs have doubling times of 1.5–2 hours. Yeast cells containing c-*Ha-ras* as the only active *ras* gene have a doubling time of 5.5–6 hours [23].

A MODIFIED YEAST RAS GENE CAN TRANSFORM MAMMALIAN CELLS

Considering that while the full length *RAS* 1 gene and Hybrid B did not transform while Hybrid C, with an activating mutation, did lead to the reasoning that perhaps the full length yeast *RAS* gene was unable to transform mammalian cells because of the larger size of the divergent C-terminal region.

It was known from data, both in yeast and bacteria, that this region of the yeast protein was susceptible to proteolytic degradation [15]. If this occurred when the yeast protein was synthesized in the mammalian cell as well, it would prevent proper localization of the protein to the plasma membrane and, hence, transformation of the cell. Taking this into account, a yeast *RAS* 1 gene was designed which lacked most of this nonhomologous region of the yeast genome, retained the very C-terminus of the yeast gene, and also contained a potentially activating mutation (see Figure 3) [22]. When tested, this modified yeast *RAS* 1 gene (called *del-RAS* 1^{leu}), which is composed entirely of yeast coding sequences, transformed the NIH 3T3 cells with an efficiency comparable to Hybrid C^{thr} (see bottom of Table I).

SUMMARY

Yeast *RAS* provides a system in which well defined functional analyses, at both the biochemical and biological level, can be performed on specifically altered *ras* proteins. Given the overall cross-species funtionality of the yeast and mammalian *ras* genes, it is very likely that their immediately associated functions are conserved, thus making the yeast system highly relevant to the mammalian one. Several phenotypic and biochemical changes have already been observed as a result of various modifications made in the yeast *RAS* genes. When a point mutation which is known to increase the transforming activity of a mammalian *ras* gene was introduced into the

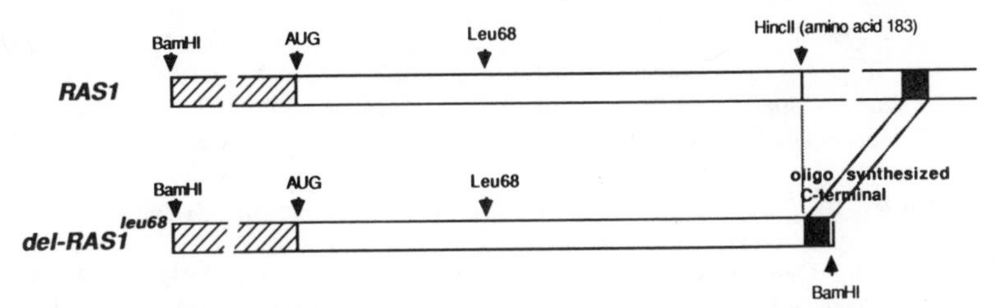

Fig. 3. *Construction of modified yeast* RAS *1 gene (del-*RAS *1^{leu}. This mutant was constructed by isolating an N-terminal BamHI/HincII fragment of a* RAS *1 gene containing the mutation leucine 68 (analogous to leucine 61 in mammalian* ras*) which was intact through the HincII site located at amino acid 183. This fragment was then ligated to a purposely designed synthetic complementary oligonucleotide which regenerated nine c-terminal amino acids of the yeast gene (184, 185 and 303-309), the yeast stop codon (TGA) and a BamHI end. The resultant molecule was subcloned into the BamHI site of the Ha-MUSV-p14 plasmid or flush end ligated into the AAH5 yeast shuttle vector plasmid. From DeFeo-Jones et al. [22].*

corresponding amino acid position of the yeast *RAS* 2 gene, it was found that yeast cells containing this altered gene were unable to sporulate, no longer accumulated glycogen, became nonviable upon nitrogen starvation, and had somewhat elevated intracellular cAMP levels [24]. In vitro assays using membranes prepared from these cells showed increased adenylate cyclase activity; and the altered *RAS* 2 protein exhibited reduced GTPase activity [15,24]. Findings such as these have expanded the parameters within which one can investigate the functional properties of modified *ras* proteins. This, together with the ease of identifying second-site and pseudorevertants in yeast, and the facility with which one can clone the genes responsible for the reversion has opened the way for elucidating both the mechanism of action and the pathway in which *ras* genes normally function.

REFERENCES

1. Ellis RW, Lowry DR, Scolnick EM (1982): The viral and Cellular p21 (ras) gene family. In: "Advances in Viral Oncology (Vol. 1): Cell-derived Oncogenes," Klein G (ed) Raven Press, New York, pp 107–126.
2. Tabin C, Bradely S, Bargmann C, Weinberg R, Papageorge A, Scolnick EM , Dhar R, Lowy DR, Chang E (1982): Mechanism of activation of a human oncogene. Nature 300:143–148.
3. Shih TY, Weeks MO, (1984): Oncogenes and cancer: the p21 *ras* gene. Cancer Investigations 2:109–123.
4. Willumsen BM, Christensen A, Hubbert NL, Papageorge AG, Lowy DR (1984): The p21 *ras* C-terminus is required for transformation and membrane association. Nature 300:583–585.
5. Gibbs JB, Sigal IS, and Scolnick EM (1985): *ras* p21; Biochemical properties of the normal and oncogenic forms. TIBBS 10:350–353.
6. Neumen-Silberberg FS, Schejter E, Hoffmann FM Shito B (1984): The Drosophila *ras* oncogenes: Structure and Nucleotide Sequence. Cell 37:1027–1033.
7. Reymond CD, Gomer RH, Mehdy MC, Firtel RA (1984): Developmental Regulation of a Dictyostelium Gene Encoding a Protein Homologous to Mammalian *ras* Protein. Cell 39:141–148.
8. DeFeo-Jones D, Scolnick EM, Koller R, Dhar R (1983): *ras*-related gene sequences identified and isolated from *Saccharomyces cerevisiae*. Nature 306:707–709
9. Powers S, Kataoka T, Fasano O, Goldfarb M, Strathern J, Broach J, Wigler M (1984): Genes in *S. cerevisiae* encoding proteins with domains homologous to the mammalian *ras* proteins; Cell 36:607–612.
10. Rothstein RJ (1983): One step gene disruption in yeast. In: "Methods in Enzymology, Vol. 101," pp 202-211
11. Gordon JW, Ruddle FH (1985): DNA-mediated genetic transformation of mouse embryos and bone marrow—a review. Gene 33:121–136.
12. Dhar R, Nieto A, Koller, R, Defeo-Jones D, Scolnick EM (1984): Nucleotide sequence of two ras^H-related genes isolated from the yeast *Saccharomyces cerevisiae*. Nucleic Acid Res 12:3611–3618.
13. Papageorge AG, DeFeo-Jones D, Robinson PS, Temeles G, Scolnick EM (1984): *Saccharomyces cervisiae* synthesizes proteins related to the p21 gene product of *ras* genes found in mammals. Mol Cel Biol 4:23–29.
14. Temeles G, DeFeo-Jones D, Tatchell K, Ellinger MS, Scolnick EM (1984): Expression and characterization of ras mRNA's from *Saccharomyces cerevis-*

iae. Mol Cell Biol 4:2298–2305.

15. Temeles G, Gibbs JB, D'Alonzo JS, Sigal IS, Scolnick EM (1985): Yeast and mammalian *ras* proteins have conserved biochemical properties. Nature 313:700–703.
16. Lochrie MA, Hurley JB, Simon MI (1985): Sequence of the alpha subunit of photoreceptor G-protein: Homologies between Transducin, *ras* and Elongation Factors. Science 228:96–98.
17. Tanabe T, Nukada T, Nishikawa Y, Sugimoto K, Suzuki H, Takahashi H, Noda M, Haga T, Ichiyama A, Kangawa K, Minamino N, Matsuo H, Numa S. (1985): Primary structure of the alpha-subunit of transducin and its relationship to *ras* proteins. Nature 315:242–245.
18. Lowy DR, Rands E, Scolnick EM, (1978): Helper-independent transformation by unintegrated Harvey sarcoma virus DNA. J Virol 26:291–298.
19. Tatchell K, Chaleff D, DeFeo-Jones D, Scolnick EM (1984): Requirement of either of a pair of *ras*-related genes of *Saccharomyces cerevisiae* for spore viability. Nature 309:523–527.
20. Kataoka T, Powers S, McGill C, Fasano O, Strathern J, Broach J, wigler M (1984): Genetic analysis of yeast *RAS* 1 and *RAS* 2 genes. Cell 37:437–445.
21. Tatchell K, Robinson LC, Breitenbach M (1985): *RAS* 2 of *Saccharomyces cerevisiae* is required for gluconeogenic growth and proper response to nutrient limitation. PNAS 82:3785–3789.
22. DeFeo-Jonjes D, Tatchell K, Robinson LC, Sigal IS, Vass WC, Lowy DR, and Scolnick EM (1985): Mammalian and yeast *RAS* gene products: Biological Function in their Heterologous Systems. Science 228:179–183.
23. Kataoka T, Powers S, Cameron S, Fasano O, Goldfarb, M, Broach J, and Wigler M (1985): Functional homology of mammalian and yeast *RAS* genes. Cell 40:19–26.
24. Toda T, Uno I, Ishikawa T, Powers S, Kataoka T, Brock D, Cameron S, Broach J, Matsumato K, Wigler M (1985): In Yeast, *RAS* Proteins are Controlling Elements of Adenylate Cyclase. Cell 40: 27–36.

Protein Engineering, pages 315–322

29

Site-Specific Mutants of pp60$^{v\text{-}src}$

Mark A. Snyder and J. Michael Bishop

G.W. Hooper Foundation and Department of Microbiology and Immunology, University of California, San Francisco, California 94143 (M.A.S., J.M.B.), California Biotechnology, Mountain View, California 94043 (M.A.S.)

INTRODUCTION

One area of study in molecular biology concerns the structure–function relationships of proteins. That is, how is a protein's observable activity modulated by specific structures within the protein itself? By way of analogy, this is similar to asking how each part of a car contributes to make a car do what it does.

One approach to this problem is to make changes in the amino acid sequence of the protein; this chapter will focus on two such techniques. The first is to alter previously-identified structural elements in a protein and observe how these changes affect the protein's function(s). In the analogy above, one might alter the part of the automobile known as the "carburetor", learning, of course, that the "carburetor" affects the "engine" function, but not "braking" or "radio."

The second, more general method is to randomly alter the amino acid sequence of a protein in the hopes of destroying, and hence identifying, important functional regions. Analogously, one would fire a pistol at various parts of the automobile; doing so would reveal important regions in the front and bottom of the car, with the middle and top being less interesting.

One class of proteins which has borne much fruit from these techniques are the oncogenic tyrosine kinases, responsible for in vitro transforming activity and in vivo tumorigenicity of a variety of retroviruses. An understanding of how these proteins function would lead not only to insights into the genesis of cancer, but would also provide insight into the regulation of cellular growth and differentiation. This review will focus on what has been learned from v-*src*, the oncogene of Rous sarcoma virus. It should be noted that elegant studies have also been performed on the oncogenes *fps* and *mos*, as well as *ras*, which is not a tyrosine kinase.

TECHNIQUES OF MUTAGENESIS

Three methods have been used in the mutation of v-*src*. The first, called oligonucleotide-directed site-specific mutagenesis [1,2], uses a synthetic oligonucleotide to prime the synthesis in vitro of a mutated, complementary strand of a gene or part of a gene which has

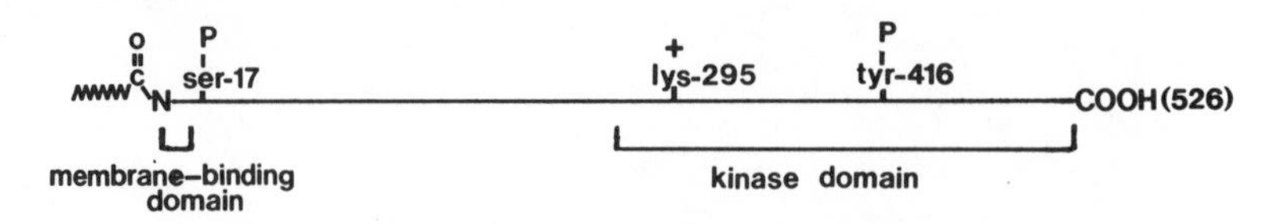

Fig. 1. *The major structural features of pp60*$^{v\text{-}src}$.

been isolated in a single-stranded form. This technique is very powerful, and can change any amino acid into any other.

The second method, called bisulfite mutagenesis [3], is a chemical modification of double-stranded DNA at particular restriction enzyme sites, resulting in the conversion of a cytosine residue to uracil, yielding a G–C to A–T transition. This technique can, in one step, yield a pool of mutagenized genes altered at all of the sites recognized by a particular restriction enzyme.

The third method, linker insertion and linker insertion–deletion [4], also alters the amino acid sequence at or near restriction sites in the DNA. In this technique, 8–12 base pair synthetic fragments of DNA, containing a sequence recognized by a particular restriction enzyme, are inserted in the gene at a site or sites recognized by other restriction enzymes. By judicious choice of sites and length of the synthetic fragment, insertions, substitutions, or deletions of stretches of varying lengths of amino acids are possible. This method produces rather more drastic changes in the primary structure of the protein; like bisulfite mutagenesis, however, it can be used to rapidly generate predictable changes throughout the gene.

THE STRUCTURE OF pp60 $^{v\text{-}src}$, THE PROTEIN PRODUCT OF v-*src*

The cardinal features of the primary structure of pp60$^{v\text{-}src}$ are shown in Figure 1. The viral gene product is 526 amino acids long and contains two major sites of phosphorylation—serine-17 and tyrosine-416. pp60$^{v\text{-}src}$ functions as a protein tyrosine kinase, and in this regard one residue, lysine-295, has been shown to be important in the catalytically obligatory binding of ATP. The kinase domain resides in the carboxy-terminal half of the protein. pp60$^{v\text{-}src}$ is located predominantly at the inner face of the plasma membrane and is known to be myristylated at the amino-terminal glycine. The expression of pp60$^{v\text{-}src}$ in an appropriate cell type in culture is sufficient to transform that cell; the presence of pp60$^{v\text{-}src}$ in a viral genome is sufficient to make the virus tumorigenic. For a more detailed discussion of pp60$^{v\text{-}src}$ and other oncogenic tyrosine kinases, the reader is referred to reference 5.

The elucidation of these structural elements of pp60$^{v\text{-}src}$ led to several questions: 1) What are the functions of the serine and tyrosine phosphorylations on pp60$^{v\text{-}src}$? 2) What sequences control myristylation, and what role does myristylation play in pp60$^{v\text{-}src}$ function? 3) What effect does loss of ATP binding have on the ability of pp60$^{v\text{-}src}$ to transform cells in culture and cause tumors? 4) What other regions can be identified which are important to the functioning of pp60$^{v\text{-}src}$?

MUTATIONS AT TYROSINE-416

One way to examine the function of tyrosine-416 phosphorylation would be to eliminate this modification and then study the properties of the resulting mutant. Any amino acid lacking a hydroxyl group in principle could serve, but converting tyrosine to, say, glycine could have other effects as well. Tyrosine has a large, planar, hydrophobic side chain, while glycine does not. The resulting change in the nature of the side chain could lead to changes in the secondary and tertiary structure of the protein not related to loss of phosphorylation, and hence give misleading results.

The ideal replacement would maintain the side-chain structure of tyrosine and simply lack the phenolic hydroxyl group. Such an amino acid exists, of course, and it is phenylalanine. The codon for tyrosine-416 is TAC, while one codon for phenylalanine is TTC. Such a change can be simply and quickly engineered by oligonucleotide-directed site-specific mutagenesis; this technique was used by Snyder et al. [6,7] to generate a mutant called SF1.

The mutant gene and its wild-type counterpart were placed into a plasmid which would allow their expression in mouse fibroblasts (which are easier to work with than cells from the natural host for Rous sarcoma virus, the chicken). The introduction of either wild-type or mutant gene into the fibroblasts rapidly produced large numbers of foci. Adherent mammalian cells normally grow as a monolayer in tissue culture; cell division ceases when an individual cell is in contact with its neighbors. When cells are transformed, this growth inhibition is lost, and cells pile up over each other. A group of such cells, arising from a single progenitor, is termed a focus, and focus formation is one of the hallmarks of the transformed state.

Thus, tyrosine-416 phosphorylation, by at least one criterion, is dispensible for transformation. A second criterion for transformation is the ability of those cells which normally adhere to the bottom of a tissue culture dish to grow in soft agar. Several foci from wild-type and SF1-pp60$^{v\text{-}src}$ expressing cells were seeded into soft agar, and all formed colonies of cells equally well. By this second parameter also, the loss of tyrosine-416 phosphorylation does not affect transformation.

A final growth change signaling transformation involves cell shape. Normal adherent cells in culture are flat and elongated. Upon transformation, these cells become extremely rounded. While this was the case for cells transformed by wild-type pp60$^{v\text{-}src}$, cells transformed by SF1-pp60$^{v\text{-}src}$ became only somewhat shortened. This was the first indication that tyrosine-416 phosphorylation was involved in some aspect of pp60$^{v\text{-}src}$ function. What is responsible for the difference in cell shape? The rounding up of transformed cells is typically accompanied by the disintegration of actin stress filaments, resulting in a major change in the cytoskeleton. A partial retention of actin cables could explain the intermediate shape of SF1-transformed cells; upon examination, however, these cells were also shown to be totally lacking in actin cables. The change in cell shape remains unexplained.

If tyrosine-416 phosphorylation has no great effect on cell growth, does it affect the tyrosine kinase activity of pp60$^{v\text{-}src}$ itself? One measure of the kinase activity is phosphorylation of the heavy chain of immunoglobulins by pp60$^{v\text{-}src}$ in an immune complex. By this assay, the activity of SF1-pp60$^{v\text{-}src}$ is comparable to wild-type; activity towards the exogenously added substrate enolase appears to be reduced, however, by some two to fourfold.

In an attempt to study the kinase activity of SF1-pp60$^{v\text{-}src}$ in vivo, two cellular parameters were measured: the total cellular phosphotyrosine level, and the phosphorylation of the cellular protein pp36^{cell}. Upon transformation by v-*src*, the level of total phosphotyrosine in cells increases some tenfold, from roughly 0.02% of total cell phosphoamino acids to 0.1%–0.3%, due to the increased phosphorylation of a number of cellular targets. One major target is a protein with molecular weight of 36,000 daltons called pp36^{cell}. Whether overall cellular phosphotyrosine or the specific phosphorylation of pp36^{cell} was measured, SF1-pp60$^{v\text{-}src}$ was equally as active as the wild-type protein. By a variety of measurements, then, phosphorylation of tyrosine-416 does not appear to play a major role in the modulation of kinase activity.

As noted previously, pp60$^{v\text{-}src}$ also has a major phosphorylation site at serine-17. When pp60$^{v\text{-}src}$ labeled in vivo with ^{32}P was analyzed, this phosphorylation was unaffected. However, a small amount of phosphotyrosine was still present in the carboxy-terminal half of the molecule. The significance of this finding remains unexplored.

Could the modification at tyrosine-416 somehow affect noncatalytic interactions of pp60$^{v\text{-}src}$ with cellular proteins? Approximately 5%–10% of pp60$^{v\text{-}src}$ is found associated in a cytoplasmic complex with two cellular proteins of molecular weights 89,000 and 50,000 daltons. This complex is believed to be involved in the transport of pp60$^{v\text{-}src}$ from its site of synthesis in the cytoplasm to the inner face of the plasma membrane. While in the complex, pp60$^{v\text{-}src}$ is not phosphorylated on tyrosine. Could tyrosine-416 phosphorylation be a signal for transfer of pp60$^{v\text{-}src}$ from the complex to the membrane? The answer appears to be no, since only normal amounts of SF1-pp60$^{v\text{-}src}$ are found in the complex, with the rest localized in a membranous fraction.

If tyrosine-416 phosphorylation does not appear to control any of the obvious transforming functions and activities of pp60$^{v\text{-}src}$, does it play a role in the control of tumorigenicity? Here, the answer is a resounding yes. Mouse fibroblasts transformed with wild-type pp60$^{v\text{-}src}$ readily form tumors when as few as 10^5 cells are injected intradermally into syngeneic mice. However, tumors were never observed when as many as 5×10^7 SF1-transformed cells were given. What is the reason? When this experiment was repeated using nude mice, which lack the T-cell mediated immune system, SF1-transformed cells grew as readily as wild-type. The inability of SF1-transformed cells to grow in a syngeneic host appears to derive from the failure of SF1-pp60$^{v\text{-}src}$ to suppress a lethal immune response in the host animal; alternatively, SF1 could be causing an immunogenic effect not produced by the wild-type to produce a lethal immune response in the host animal. The failure to phosphorylate tyrosine-416, therefore, has led to a separation of transformation from tumorigenicity.

The results obtained with the site-specific mutant of pp60$^{v\text{-}src}$ were confirmed and expanded by Cross and Hanafusa [8],who took advantage of appropriate restriction sites near the tyrosine-416 codon to make a linker insertion-deletion mutant. The mutant, called T10-1, has the five-amino-acid sequence glu$_{412}$-asp-asn-glu-tyr$_{416}$ replaced by the tripeptide arg-ser-asp. The T10-1 v-*src* gene was placed into a viral vector and this construct used to infect chickens and also chicken fibroblasts in culture.

The T10-1 virus was found to induce foci in cultured cells, but fewer foci were observed and they appeared several days later than foci induced by wild-type virus; transformed cells were able to grow well in soft agar. The tyrosine kinase activity of T10-1 was measured by the immunoglobulin phosphorylation method, and found to be 50% of the wild-type level. The overall level of cellular phosphotyrosine, however, increased in T10-1 transformed cells to virtually the same extent as cells transformed by wild-type pp60$^{v\text{-}src}$. When T10-1 pp60$^{v\text{-}src}$ itself was examined, phosphorylation at serine-17 was found to be normal, with residual phosphotyrosine in the carboxy-terminal half of the molecule amounting to some 10% of wild-type levels. Tryptic mapping yielded several minor spots from T10-1, one or more of which may result from minor tyrosine phosphorylation(s). Additionally, the T10-1 mutation did not affect either the rate of synthesis or the half-life of the mutant pp60$^{v\text{-}src}$ in cells.

The T10-1 virus was also used to study tumorigenicity of the mutant in chickens. The virus induced small tumors which grew slowly. However, given a long incubation, tumors started to grow rapidly, reaching a size comparable to that of wild-type induced tumors. Both SF1 and T10-1, then, are greatly reduced in their tumorigenic potential; the apparent in vivo differences (SF1-transformed cells do not grow in an immune-competent mouse, whereas T10-1 virus yields tumors in a normal chicken) may reflect differences in the mutants themselves, or in intrinsic properties of the two systems.

Thus, the role of tyrosine-416 phosphorylation in pp60$^{v\text{-}src}$ function is subtle, yet dramatic. It appears to control the ability of pp60$^{v\text{-}src}$ to make some change, presumably at the cell surface, which allows a transformed cell to escape immune surveillance. The phosphorylation does not appear to affect, in any dramatic way, the ability of pp60$^{v\text{-}src}$ to transform cells in culture. Whether the change in tumorigenicity is due to some subtle alteration in kinase activity remains to be determined. These site-specific mutants of tyrosine-416 offer a unique tool for the explanation of the transforming and tumorigenic functions of pp60$^{v\text{-}src}$.

MUTATIONS AT SERINE-17

The other major site of phosphorylation, serine-17, has been modified by Cross and Hanafusa [8] by linker insertion and insertion–deletion. Two changes have been made at arginine-15, part of the recognition sequence for cAMP-dependent protein kinase, the cellular enzyme presumed responsible for phosphorylation of serine-17. One mutant (NY11-1) has a tetrapeptide in place of arginine, and another mutant (NY11-4) a pentapeptide. A deletion (NY11-7) was also made from residues 15 through 27 which eliminates the phosphorylation site entirely. All mutants have been expressed by placing them back into Rous sarcoma virus vectors.

The deletion mutant behaves, in all parameters studied, like the wild type. The *src* gene carrying the NY11-7 mutation induced focus formation to the same degree as wild-type and with the same latency. Chicken fibroblasts expressing NY11-7 pp60$^{v\text{-}src}$ were morphologically transformed and grew in soft agar; additionally, the mutant virus induces tumors in chickens with virtually the same efficiency as wild-type. Phosphorylation of serine-17 does not appear to be needed for the transforming or tumorigenic functions of pp60$^{v\text{-}src}$.

Does serine-17 phosphorylation control kinase activity? The answer here seems to be no. NY11-7 pp60$^{v\text{-}src}$ has normal levels of activity towards immunoglobulin heavy chain, and the transformed cells have as much total phosphotyrosine as cells transformed by wild-type pp60$^{v\text{-}src}$. Additionally, neither the rate of synthesis nor the half-life of NY11-7 is greatly affected by the mutation, indicating that serine-17 phosphorylation does not control pp60$^{v\text{-}src}$ expression or stability.

The two mutations at arginine-15 greatly reduce, but do not abolish, phosphorylation on serine-17. As with the deletion mutant, NY11-1 and NY11-4 induce foci, morphologically transform cells, and possess normal kinase activity towards immunoglobulin heavy chain. The deletion mutant still has phosphoserine in the amino-terminal half of the protein; tryptic mapping indicated that this was due to one or more minor phosphoserines distinct from serine-17. These minor phosphoserines were also present in the arginine-15 mutants, where they comprised the majority of the residual total phosphoserine in the protein. Finally, it was noted that the level of phosphotyrosine in these mutants was normal; combined with the results mentioned above, the data indicate that phosphorylations of serine-17 and tyrosine-416 are independent of each other. Unlike tyrosine-416, the role of phosphorylation of serine-17 remains a mystery.

MODIFICATIONS AT LYSINE-295

Over the years, a number of experiments have suggested that pp60$^{v\text{-}src}$ may possess nonkinase functions which are required for complete transformation of cells in culture and for oncogenesis in animals (see reference 5 and references therein). One way to test this would be to eliminate the kinase activity without altering any of the remaining properties of the protein. Lysine-295 has been recently identified as important in ATP binding; alteration of this amino acid, if the change were subtle enough, might abolish kinase activity without affecting any other putative functions.

To accomplish this, Snyder et al. [9] used oligonucleotide-directed site-specific mutagenesis to convert this lysine to a methionine. The choice of methionine is perhaps not obvious, but was made primarily because the side-chain backbone structures of methionine and lysine are virtually isostructural: the substitution of a sulfur atom for carbon does not greatly alter bond distances or angles. The major effect of the change is the replacement of a positively charged amino group by a hydrogen atom, eliminating the interaction with the negatively-charged ATP molecule necessary for catalysis.

The resulting mutant, called SF2, was expressed in rat fibroblasts. Cellular parameters of transformation detailed above were examined, and in every case the cells were found to be untransformed; additionally, cells expressing SF2-pp60$^{v\text{-}src}$ were unable to form tumors in nude mice. As expected, the protein encoded by SF2-pp60$^{v\text{-}src}$ was found to be totally inactive as a kinase. These data suggest that kinase activity is an absolute requirement for at least most of the commonly-observed features of transformation.

The one surprising feature of this mutant was discovered when the phosphorylation of the SF2-pp60$^{v\text{-}src}$ protein itself was examined. As mentioned above, serine-17 is believed to be phosphorylated by a cellular cAMP-dependent protein kinase. When SF2-pp60$^{v\text{-}src}$ was isolated from cells labeled in vivo with phosphate, no phosphorylation of any kind was detected. Similarly, serine-17 of wild-type but not SF2-pp60$^{v\text{-}src}$ could be phosphorylated in vitro by the catalytic subunit of cAMP-dependent protein kinase. While lack of tyrosine-416 modification could be explained by lack of an autophosphorylating activity, the absence of serine-17 phosphorylation was puzzling.

One precedent for such a loss of phosphorylation exists. The conversion of bacterial phosphorylase *a* to phosphorylase *b* results from dephosphorylation of serine-12 of this enzyme. The reaction is dependent on the presence of ATP at the active site of phosphorylase *a*; in the absence of bound ATP, the serine residue is forced into a pocket which is inaccessible to the converting phosphatase. While this process is a dephosphorylation rather than a phosphorylation, as in pp60$^{v\text{-}src}$, the principle is the same. The phosphorylase data suggests that, although the amino-terminal and kinase domains of pp60$^{v\text{-}src}$ were thought to be independent of each other, the phosphorylation of serine-12 may be an ordered event and dependent upon prior binding of ATP and the active site. The significance of this finding is not yet known.

MUTATIONS AT THE MYRISTYLATION SITE

As mentioned above, pp60$^{v\text{-}src}$ is myristylated at glycine-2 after the initiating methionine is cleaved off. One obvious function for this modification is to aid in or direct the binding of pp60$^{v\text{-}src}$ to the plasma membrane. This observation led to several questions: 1) What sequences are responsible for directing the myristylation event? 2) What effect, if any, does myristylation have on kinase activity? 3) Is myristylation necessary for transformation and/or tumorigenicity?

As a first step, Cross et al. [10] used linker insertion–deletion to make several deletions within the amino-terminal 9 kilodaltons of pp60$^{v\text{-}src}$. Substitution of a tri- or tetrapeptide for amino acids from position 15 to positions 27, 49, or 81 had little effect on kinase activity or transforming capacity. Like wild-type pp60$^{v\text{-}src}$, these three mutants associated with the plasma membrane and were myristylated. In contrast, mutants with di- or tri- peptides substituted for amino acids 2–81 or 2–15 were transformation defective, and the mutant proteins neither associated with the membrane nor were myristylated. However, both mutants possessed between 50% and 100% of normal kinase activity, and the overall increase in total cellular phosphotyrosine was comparable to that observed with wild-type.

These studies led to several conclusions. First, sequences from residues 2–15 are required for membrane association, myristylation, and cell transformation. Indeed, since even larger deletions beginning at position 15 were still myristylated, the region from position 2–15 appears to be the *only* region necessary to direct this modification. In a subsequent study [11], this region was further narrowed to the first 7–10 amino acids of pp60$^{v\text{-}src}$. Second, myristylation appears to have very little effect on kinase function; indeed, other workers have found that the first 25% of the entire protein can be deleted without significant loss of activity. This result extends the observations described above which indicated that the area around serine-17 was dispensible for transformation and tumorigenesis.

Third, myristylation is a requirement for the membrane association of pp60$^{v\text{-}src}$: those mutants which failed to be myristylated were not able to bind to the plasma membrane. This result also provides insight into the role of membrane binding in pp60$^{v\text{-}src}$ function, since the nonmyristylated mutants, although possessing nearly normal levels of kinase activity, were nontransforming. This suggests that some critical target for pp60$^{v\text{-}src}$ tyrosine kinase activity is located at or near the cell membrane, and that membrane association is required for access to this target.

The questions concerning the role of myristylation and membrane binding have also been addressed by Kamps et al. [12] using oligonucleotide-directed site-specific mutagenesis. In this study, glycine-2 was replaced by either alanine or glutamic acid. Neither of these mutants was myristylated, which suggests that the enzyme which carries out this reaction has a pronounced specificity for the acceptor amino acid. As expected, neither mutant was able to induce morphological transformation. Both possessed normal levels of protein kinase activity, and cells expressing these mutants showed the expected increase in total cellular phosphotyrosine.

Besides defining the roles of myristylation and membrane binding in relation to pp60$^{v\text{-}src}$ function, these mutants also provide a way to dissect some of the processes involved in morphological transformation. Since all of the nontransforming myristylation mutants still phosphorylate, or cause to be phosphorylated, many cellular targets in vivo, many of these targets may simply be adventitious substrates. By analyzing those proteins which are phosphorylated only by wild-type pp60$^{v\text{-}src}$, it should be possible to determine which cellular proteins must be modified by pp60$^{v\text{-}src}$ in order to cause transformation.

OTHER IMPORTANT REGIONS IN pp60$^{v\text{-}src}$

The studies described so far have focused on readily-identifiable structural features of pp60$^{v\text{-}src}$. In an attempt to identify other regions which are involved in pp60$^{v\text{-}src}$ function, Bryant and Parsons [13,14] have used bisulfite mutagenesis to construct mutants at a *Bgl* I site in the *src* gene. Four mutants were obtained, one each at amino acids 430 through 433. All of these mutants had greatly diminished or absent kinase activity in vitro, and a similar loss of activity towards the cellular target pp36^{cell}. Also, one of the mutants, when tested, proved unable to induce transformation. These results suggest that the region including amino acids 430–433 is somehow involved in kinase activity.

Bryant and Parsons [15] have also identified another region important for transformation. By using limiting exonuclease digestion at a *Bgl* II site in *src*, they constructed a mutant, tsCH119, containing a deletion from amino acid 202–255. This mutant is temperature-sensitive for transformation, yet at the restrictive temperature is associated with the plasma membrane and possesses 50% of the normal level of kinase activity as measured in vitro or in vivo. tsCH119 also shows greatly diminished tumorigenicity. These data suggest that tsCH119 has identified a functionally important region of pp60$^{v\text{-}src}$ in the amino-terminal half of the protein which is distinct both from the carboxy-terminal kinase domain and from the myristylation region described above.

QUO VADIS?

The engineered mutants discussed above have provided insights into the workings of pp60$^{v\text{-}src}$ and some direction in terms of where to go next. The tyrosine mutants, with the separation of transformation from tumorigenicity, have provided the tools to examine those cellular parameters which are responsible for the ability of a transformed cell to grow progressively in a host animal. The myristylation mutants potentially provide the ability to identify those specific targets of pp60$^{v\text{-}src}$ whose modification leads to morphological transformation. Mutations at lysine could lead to the identification of properties of pp60$^{v\text{-}src}$ which are independent of kinase activity. Although mutations at serine-17 appear to have no effect on pp60$^{v\text{-}src}$ function, the rationale behind their creation remains valid. Nature rarely does anything without a reason, and the serine mutants, rather than being uninteresting, simply do a better job hiding their imperfections. As more and more is discovered about the structure of pp60$^{v\text{-}src}$, the analysis of tailored mutants will yield increasingly greater insights into the functioning of this protein and others like it. The generation of site-specific mutants is an easy task; the important job is to ask the right questions.

REFERENCES

1. Gillam S, Smith M (1979): Gene 8:81–97.
2. Gillam S, Smith M (1979): Gene 8:99–106.
3. Shortle D, Nathans D (1978): Proc Nat Acad Sci USA 75:2170–2174.
4. Heffron F, So M, McCarthy BJ (1978): Proc Nat Acad Sci USA 75:6012–6016.
5. Bishop JM, Varmus H (1985): In Weiss R, Teich N, Varmus H, Coffin J (eds): "RNA Tumor Viruses" New York: Cold Spring Harbor Laboratory, pp 249–355.
6. Snyder MA, Bishop JM, Colby WW, Levinson AD (1983): Cell 32:891–901.
7. Snyder MA, Bishop JM (1984): Virology 136:375–386.
8. Cross FR, Hanafusa H (1983): Cell 34:597–607.
9. Snyder MA, Bishop JM, McGrath JP, Levinson AD (1985): Mol Cell Biol 5 (in press).
10. Cross FR, Garber EA, Pellman D, Hanafusa H (1984): Mol Cell Biol 4:1834–1842.
11. Pellman D, Garber EA, Cross FR, Hanafusa H (1985): Proc Natl Acad Sci USA 82:1623–1628.
12. Kamps MP, Buss JE, Sefton BM (1985): Proc Nat Acad Sci USA 82 (in press).
13. Bryant DL, Parsons JT (1984): Mol Cell Biol 4:862–866.
14. Bryant D, Parsons JT (1983): J Virol 45:1211–1216.
15. Bryant D, Parsons JT (1982): J Virol 44:683–691.

Protein Engineering, pages 323–336

30

Mapping the Functional Domains in Single-Strand DNA-Binding Proteins Gene 5 and Gene 32

Joseph E. Coleman, Kenneth R. Williams, Garry C. King, Richard V. Prigodich, Yousif Shamoo, and William H. Konigsberg

Department of Molecular Biophysics and Biochemistry, Yale University, New Haven, Connecticut 06510 (J.E.C., K.R.W., G.C.K., Y.S., W.H.K.); Department of Chemistry, Trinity College, Hartford, Connecticut 06106 (R.V.P.)

INTRODUCTION

Much of our current knowledge about structure/function relationships in single-strand DNA-binding proteins has derived from studies on gene 5 protein from bacteriophage M13 (g5P) and gene 32 protein from bacteriophage T4 (g32P).[1] The single-strand DNA binding proteins act in nucleic acid replication and recombination in vivo by forming stoichiometric complexes with little regard for the sequence of the nucleic acid, although preferences for certain sequences are present in some instances [1–5]. Despite this common mode of action, their precise physiological functions vary from protein to protein. Gene 5P acts to shift DNA replication from the double-stranded replicative form to the production of single-strand daughter virions by complexing the emerging single strand and preventing its use as a template for the synthesis of the complementary strand [5,6]. In contrast, g32P is known to stimulate the replication, recombination, and repair of DNA under various conditions. Gene 32P is also capable of autogenous regulation at the translational level by binding to a particular sequence of its own mRNA [3,4]. In common with most members of the class, both proteins bind ssDNA in a highly cooperative manner. More complete descriptions of the functions of g5P and g32P are available in references 2 and 5.

Both proteins, and their interactions with nucleic acids, have been studied with a wide variety of physicochemical techniques. An approach has been developed which combines protein chemistry and NMR methods. Limited proteolysis has determined the minimum DNA-binding domain of g32P, chemical modification has located specific types of amino acid side chain involved in DNA binding, and a selection of one- and two-dimensional ^{1}H-NMR techniques has been used to detect protein–oligonucleotide interactions with both g5P and g32P. These studies have reached a point at which deliberate manipulation of specific

[1]Abbreviations used in this chapter are: G5P, gene 5 protein; G32P, gene 32 protein; G32P*-(A+B) or G32P*, gene 32 protein from which the A (residues 254-301) and B (residues 1–21) have been removed by limited proteolysis, dTyr or dPhe, refers to protein containing perdeuterated tyrosine or phenylalanine residues; TNM, tetranitromethane; NOESY, nuclear Overhauser enhancement spectroscopy.

amino acid sites offers the potential to obtain a better understanding of the ssDNA binding phenomenon. In this respect, site-directed mutagenesis has been used to detect specific residues participating in the protein–DNA interaction and to enable the unequivocal assignment of ^{1}H-NMR resonances. This chapter presents examples of the information obtainable using a combination of protein chemistry and NMR approaches to the single-stranded DNA binding proteins.

CHEMICAL MODIFICATION STUDIES OF GENE 5 PROTEIN

The lesser complexity of g5P (a dimer of Mr 9.7 kd/monomer) has enabled it to serve as a prototype for most of the experiments subsequently applied to g32P [7–10]. In this respect, the refined crystal structure of g5P [15] has been a necessity for much of the detailed interpretation of the solution studies.

A brief statement of the earlier work on the structure of gene 5 protein serves to place the NMR work in context. The amino acid sequence of g5P is shown in Figure 1 [12,13]. Application of the Chou-Fasman rules to this sequence predicted the protein secondary structure to be ~90% β-sheet [7], a prediction subsequently borne out by the crystal structure. Nitration of the protein with TNM modified 3 Tyr residues and prevented DNA binding [7]. Prior binding of DNA prevented the nitration of all three Tyr residues. Peptide mapping showed the nitrated Tyr residues to be residues 26, 41, and 56, underlined in Figure 1. Thus it was initially assumed that there were two buried Tyr residues (34 and 61) not available for nitration and three "free" surface Tyr, potentially available for participation in nucleotide binding [7].

The discrimination between surface and buried residues by TNM is a reactivity often ascribed to the TNM nitration of tyrosyl residues. Subsequent results, however, show that even with such a small protein as gene 5 this distinction is not absolutely achieved. The crystal structure shows three Tyr to be located on a short three-stranded antiparallel β-sheet forming a putative DNA binding groove [14,15], but the three are residues 26, 34, and 41 (circled in Figure 1), rather than 26, 56, and 41 implied by nitration. Residue 56 is mostly buried with one edge exposed to solution, apparently accounting for its nitration. On the other hand, residue 34, exposed to solvent on one face, is relatively occluded at the bottom of the groove, which may account for its failure to nitrate. Acetylation of the Lys residues in g5P with acetylimidazole also prevents DNA binding and acetylation of the g5P–DNA complex dissociates the complex—results strongly implying that at least some of the lysyl side chains participate in DNA binding [7].

CORRELATION OF ^{1}H-NMR STUDIES IN SOLUTION WITH THE CRYSTAL STRUCTURE OF g5P

^{1}H-NMR of g5P–Oligonucleotide Interactions

NMR studies of the interaction of g5P with dpA, $d(pA)_2$, $d(pA)_3$, $d(pA)_4$, and $d(pA)_8$ by one-dimensional ^{1}H-NMR difference methods showed resonances assignable to one Tyr and one Phe to move upfield on dpA or pA binding [8,9,11]. These same resonances moved

```
                 10                      20                      30
M-I-K-V-E-I-K-P-S-Q-A-Q-F-T-T-R-S-G-V-S-R-Q-G-K-P-(Y)-S-L-N-E-Q-L
               40                      50                     60
-C-(Y)-V-D-L-G-N-E-(Y)-P-V-L-V-K-I-T-L-D-E-G-Q-P-A-Y-A-P-G-L-Y-T-V-H
            70                      80                 87
-L-S-S-F-K-V-G-Q-(F)-G-S-L-M-I-D-R-L-R-L-V-P-A-K
```

Fig. 1. *Amino acid sequence of the fd gene 5 protein. The underlined Tyr residues are those nitrated by TNM, while the circled Tyr and Phe residues are those postulated to be involved in DNA binding.*

much further upfield as the binding of $d(pA)_4$ and even further upfield on the binding of $d(pA)_8$. In addition to these resonances, the longer oligonucleotides also shifted signals which apparently derived from two additional Tyr [11]. Representative one-dimensional ^{1}H-NMR difference spectra are shown in Figure 2. From the aromatic ^{1}H-NMR spectrum of the unliganded protein (Figure 2A), the protons of one of the Tyr, in fact the one shifting

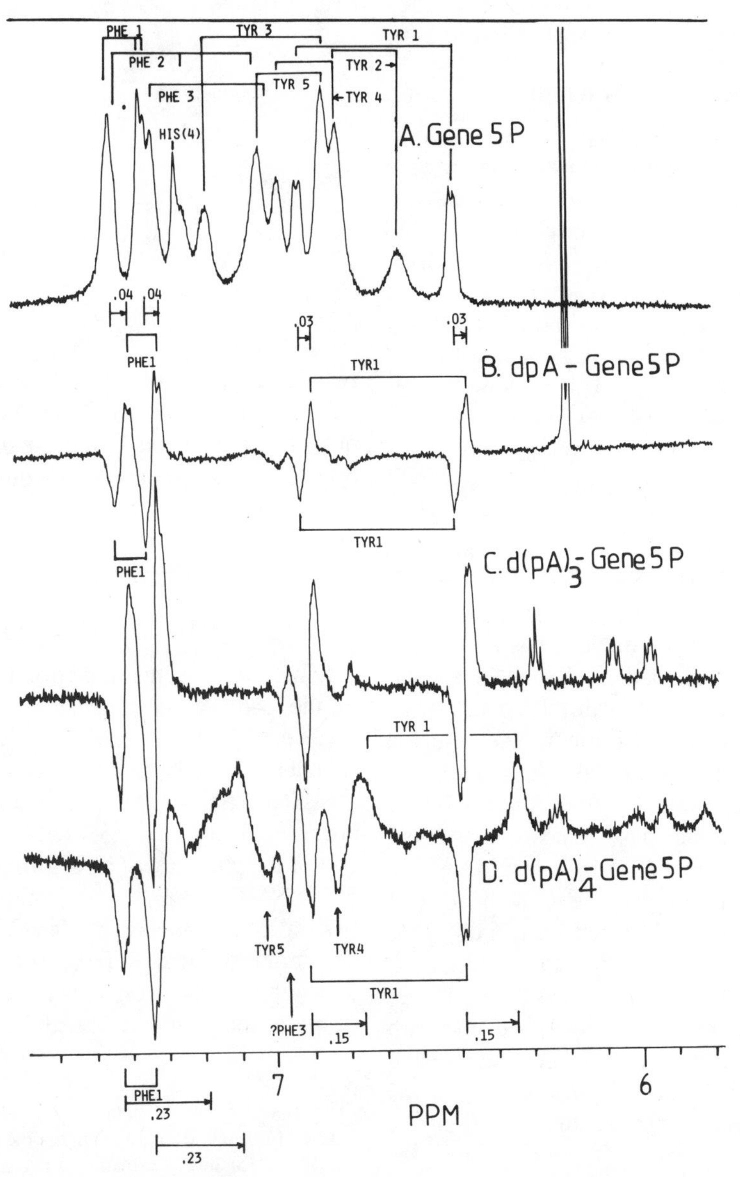

Fig. 2. *^{1}H aromatic proton NMR difference spectra, spectra of oligonucleotide complexes of gene 5 protein minus spectra, of gene 5 protein. The dash is a minus sign, and the nucleotide symbol $d(pA)_n$ refers to the complex with gene 5 protein. (A) Gene 5 protein alone. (B) Spectrum of the pA complex (1:1) minus that of the unliganded protein. (C) Spectrum of the $d(pA)_3$ complex minus that of the unliganded protein. (D) Spectrum of the $d(pA)_4$ complex minus that of the unliganded protein. (From ref. 11.)*

on mononucleotide complex formation (Figure 2B), appear as a doublet of doublets. This must be due to relatively rapid rotation of this Tyr side chain. The original assignment of the additional peaks appearing in the difference spectra of the d(pA)$_4$ complex to other Tyr residues, 4 and 5, is indicated on the difference spectrum of the d(pA)$_4$ complex (Figure 2D).

Crystal structure of Gene 5 protein

The crystal structure of g5P at 2.3 Å resolution shows a tightly interlocked dimer, each monomer consisting of two basic domains, a triple-strand antiparallel β-sheet (residues 36–70) which has been predicted to be the DNA binding surface on the basis of a variety of evidence (Figure 3) (see below) [14,15]. At almost right angles to this is a double-stranded β-sheet (residues 70–87). This "β-hook" holds the two monomers together via interactions between the two hooks. Off one end of the triple-strand sheet is a small, poorly organized domain (residues 15–32) containing Tyr 26, Arg 16, Arg 21, and Lys 24. Thus far it has not been possible to grow satisfactory crystals of a g5P–oligonucleotide complex, and modeling of the DNA interactions has relied on a speculative fitting of the solution data available on the complexes with the crystal structure of the unliganded protein. The resultant conclusions are summarized briefly here.

Of the Tyr residues Tyr 26 on the small, flexible loop is the most exposed to solvent and is therefore likely to be a residue whose 2,6 and 3,5 protons give rise to the doublet of doublets in the NMR spectrum (Figure 2A). Three other significant aromatic residues are found along the groove formed by the triple-stranded β-pleated sheet adjacent to the loop carrying Tyr 26, Phe 73′ from the end of the β-hook of the opposite monomer,[2] Tyr 34 midway along the groove and Tyr 41 at the other end of the groove (Figure 3). Since the one-dimensional NMR and chemical modification data suggested three Tyr and one Phe to be involved in DNA binding, the original model synthesized from the crystallographic and solution data placed 5 nucleotide residues in fully extended conformation along the tri-

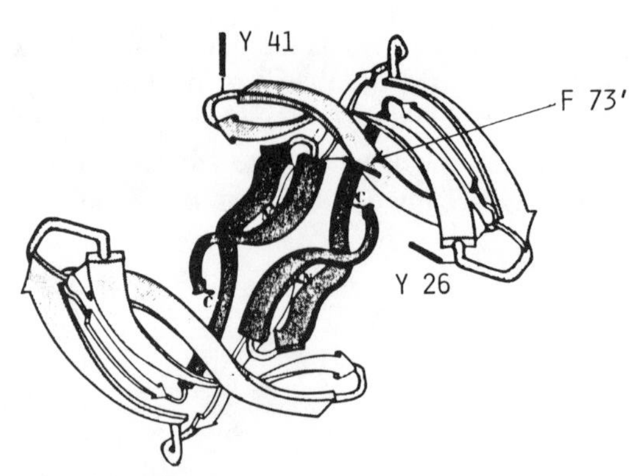

Fig. 3. *Richardson representation of the three-dimensional crystal structure of gene 5 protein. (From ref. 15.)*

ple-stranded β-sheet, one base ring each stacked on Phe 73′, Tyr 34, and Tyr 41 with two of the five bases at the 5′ end stacked on either side of Tyr 26 (Figure 4) [15]. The phosphate backbone is represented as bound by four basic residues, two from the "DNA-binding loop" (residues 15–32) and two from the main body of the protein.

TWO-DIMENSIONAL NOESY SPECTRA OF GENE 5 PROTEIN

In order to obtain a more precise description of the path of an oligonucleotide over the surface of g5P, we have recently applied 2D ^{1}H nuclear Overhauser enhancement spectroscopy (NOESY) to g5P and some oligonucleotide complexes. Despite the relatively large size of the g5P dimer (19K) for the application of 2D NMR methods, an encouraging amount of detailed information has been obtained [16]. The brief summary of the results here will focus only on the aromatic region of the spectrum. A completely assigned 2D ^{1}H NOESY

[2]The superscript prime notation refers to a residue that derives from the second G5P monomer and is used in cases where spatial information is necessary for clarity. Otherwise, residues on each monomer are treated as equivalent. Phenylalanine 73, when discussed as being close to Tyr 26, always refers to Phe 73′, from the opposite monomer. Phe 73 is carried on the end of a β-loop which projects this residue into the proximity of the flexible DNA-binding loop of the other monomer and, hence, close to Tyr 26.

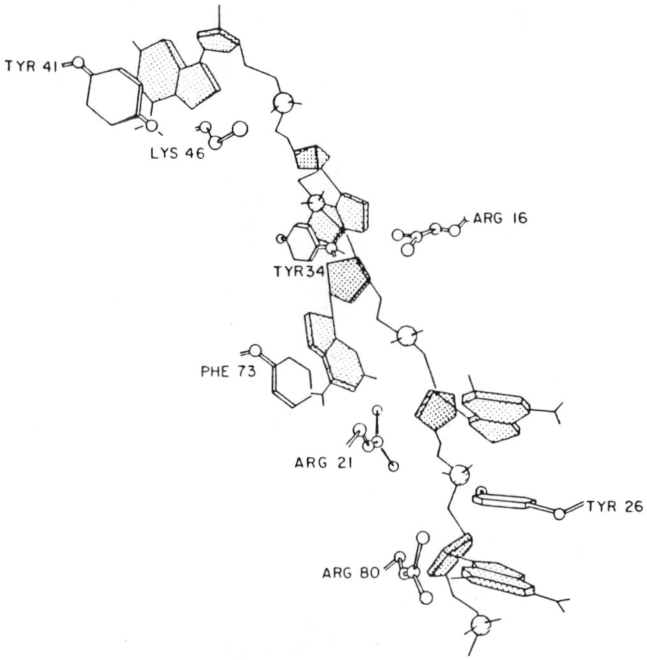

Fig. 4. *Model of the stacking of the bases of a fully extended pentanucleotide on the aromatic residues, Tyr 26, Phe 73, Tyr 34, and Tyr 41, located along the triple-stranded β-sheet shown in Figure 3. (From ref 15.)*

spectrum of unliganded g5P is shown in Figure 5. For the assignment arguments, the reader should consult the original article [16]. A similar NOESY spectrum of the g5P–$d(pA)_8$ complex is shown in Figure 6. One $d(pA)_8$ strand must bind cooperatively to two g5P monomers to form a tetrameric complex containing two bound but oppositely directed $d(pA)_8$ molecules related by a two-fold axis.

The most striking finding from the NOESY spectra is that the protons of only two aromatic residues, Tyr 26 and Phe 73, show significant upfield movement on complexation of the protein with $d(pA)_8$. The five proton resonances of Phe 73 are located at ~7.4 ppm in the unliganded protein and all move upfield ~0.6 ppm. The only other set of proton resonances to shift significantly on complex formation are the 2,6 and 3,5 protons of Tyr 26 which both move upfield ~0.25 ppm and an upfield shift of the methyl signals of Leu 28, which is very close to Tyr 26 (spectrum not shown). The ^{1}H resonances Tyr 34 and Tyr 41 show much smaller shifts on complexation. Hence it seems highly unlikely that the path of the oligonucleotide over the protein surface carries the third and fourth base rings of the oligonucleotide very near to Tyr 34 or Tyr 41 and hence to the distal part of the triple-stranded sheet (if the flexible DNA-binding loop is considered proximal). The reason for the apparent peaks assigned to two additional Tyr residues in the 1-D difference spectrum [11] is due to the movement of the signals (overlapping resonances) of Phe 73 through the center of the spectrum, coupled with very small (0.01–0.06 ppm) movements of the signals of Tyr 34, Phe 13, and Tyr 41. NOEs from the base protons to the aromatic protons of Tyr 26 and Phe 73 show that the first, second, and third bases of $d(pA)_4$ are close to Tyr 26 and Phe 73. In the $d(pA)_4$ complex, intermolecular NOEs are present from the H1′ proton of Ade 1 to the protons of Tyr 26 and from the H2, H8, and H1′ protons of Ade 2 and Ade 3 to Phe 73′. These NOEs are also present in the $d(pA)_8$ complex, (cross-peaks a to e in Figure 6), but assignments to sequential bases are less straightforward for the longer nucleotide.

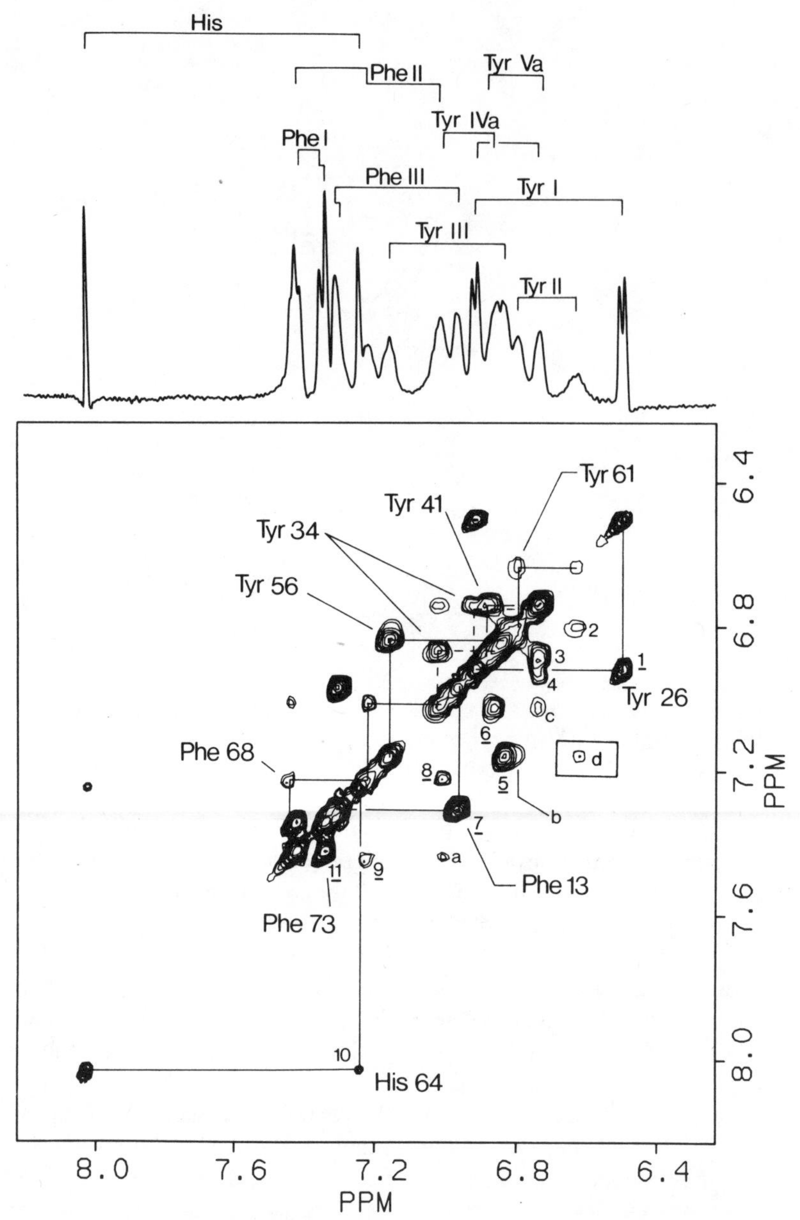

Fig. 5. *Aromatic region from a 200 ms NOESY spectrum of native g5P, with sequence-specific resonance assignments indicated. Nearest-neighbor correlations are numbered 1–11, where the underlined cross-peaks reflect positions at which connectivities can be observed in COSY spectra. Longer-range correlations are designated a–d;, peak d (inset) is contoured at a lower level. The one-dimensional ^1H aromatic spectrum and assignments to residue type are indicated at the top. (From ref. 16.)*

The above findings have led to a flexible clamp model of nucleotide binding in which the nucleotide is "clamped" by intercalation with Tyr 26 in the flexible "DNA-binding loop" and with Phe 73′, more rigidly held to the core of the protein [11]. Two modes of binding are suggested (Figure 7). One appears to be the major mode for longer oligonucleotides with bases 2 and 3 stacked on either side of Phe 73 and base 1 in the same pocket as base 2, but stacked on Tyr 26. These results indicate that Tyr 26, Leu 28, and Phe 73′, in

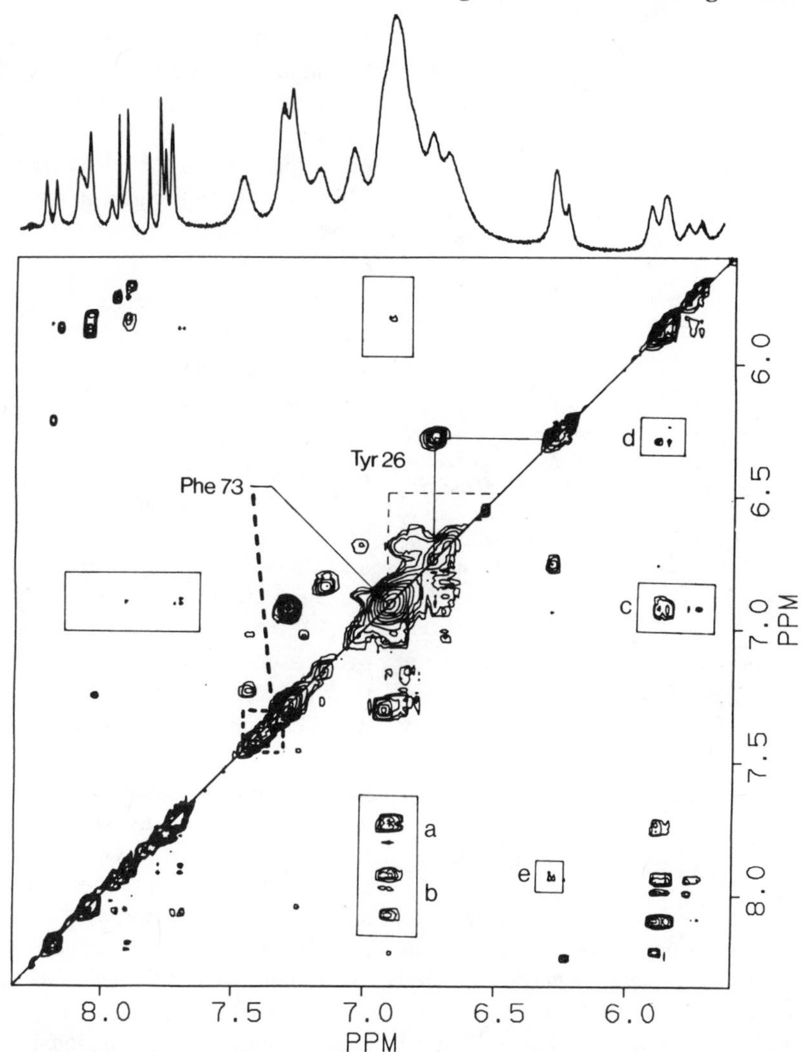

Fig. 6. *Combined 200 ms– and 750 ms– NOESY spectra of a 0.5:1 mixture of d(pA)$_8$ and G5BP. The complexation-induced shift of the Phe 73 resonances is −0.6 ppm, while the upfield shift of the Tyr 26 resonances is −0.27 ppm. The dotted lines indicate the original positions of the resonances from Tyr 26 and Phe 73. The intermolecular NOEs between the base protons (H2, H8, and H1′) of adenine residues 2 and 3 and the protons of Phe 73 are indicated by the cross-peaks a, b, and c. The intermolecular NOEs between an H2 and H1′ of adenine residue 1 and the protons of Tyr 26 are indicated by cross- peaks d and e. (From ref. 16.)*

conjunction with a cluster of the free basic groups from Arg 16, Arg 21, Lys 24, and Lys 46, form the dominant DNA binding domain of g5P.

A second mode, involving placement of one base between Tyr 26 and Phe 73, is suggested by the finding that a mononucleotide carrying a 5′ phosphate, e.g., pA, causes small upfield shifts of the protons of both Tyr 26 and Phe 73′ (Figure 2). Alternate modes of binding of the 5′-phosphate dianion or a monoanionic diester phosphate, e.g., d(A)$_4$, to the positive charges contained in the DNA binding loop may make these alternatives possible. The progressively increasing magnitude of the ring current shifts as the bound nucleotides become longer suggests considerable flexibility in the unstructured domain carrying Tyr 26. By the

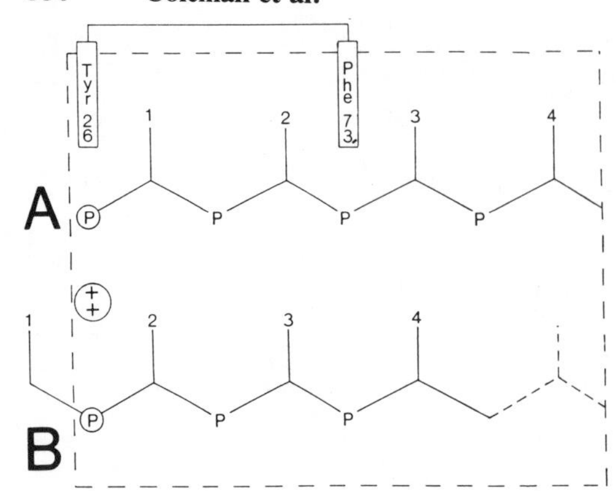

Fig. 7. *Schematic representation of the proposed binding interaction between G5P and oligonucleotides. (A) nucleotides possessing a 5'phosphate. (B) Nucleotides lacking a 5' phosphate. The locations of Tyr 26, Phe 73, and the preferential phosphate site are shown. The minimum "binding" unit is enclosed in dashed lines.*

time full cooperative binding has been achieved in the $d(pA)_8$ complex, the ring current shifts are large enough to suggest nearly complete intercalative clamping.

STRUCTURE AND FUNCTION OF GENE 32 PROTEIN

Based on the amino acid sequence of g32P (Figure 8) and the nucleotide sequence of its gene [4,17,18], the protein contains 301 amino acids with a molecular weight of 33,488. Despite a net charge of approximately -10 at pH 7 and an isoelectric point of 5.0 [19], the charge distribution is quite asymmetric. The NH_2-terminal half has a net charge of $+10$, while that of the COOH-terminal half is -20. Monomeric g32P behaves hydrodynamically as a prolate ellipsoid with an axial ratio of 4:1 and an overall length approaching 120 Å [1]. Since each g32P molecule in a complex with ssDNA may only span a distance of about 33 Å (assuming a stoichiometry of seven nucleotides/protein and an internucleotide spacing of about 4.7 Å (20)), adjacent g32P molecules probably overlap one another to some extent on a ssDNA strand. Although the protein exists as a monomer in dilute (<0.5 μM) solution [21], it undergoes indefinite self-association so that the estimated g32P concentration in vivo (~ 3.0 μM, reference 22) would yield an apparent molecular weight corresponding to a dimer or trimer [21]. It seems likely that the g32P:g32P interactions responsible for indefinite aggregation are related to those involved in cooperative binding to ssDNA [1,20,23].

Limited Proteolysis of g32P: Identification of Functional Domains

Limited proteolysis can be used to generate functionally active fragments of g32P [24–26], facilitating more-detailed structure/function studies. This method takes advantage of the fact that regions spanning residues 9–21 and 253–275 are particularly susceptible to cleavage by a wide variety of proteinases [17,19,26–29]. Cleavage at any point between residues 9 and 21 removes the basic NH_2-terminal "B" region and produces g32P-B, while cleavage between residues 253 and 275 removes the acidic COOH-terminal "A" region, producing g32P-A. Cleavage at both sites results in a 26,000 dalton fragment called g32P-(A+B). The in vitro properties of these cleavage products appear to be identical irrespective of the enzyme used in their preparation. The three fragments have been found to correspond to three "activities" (responsible for g32P:ssDNA, g32P:g32P, and g32P:T4 replication protein interactions).

Gene 32P-A. This fragment binds cooperatively to poly(dT) with an equilibrium constant similar to that of the native protein (10^8–$10^9 M^{-1}$) [30,31]. However, g32P-A is considerably better at destabilizing dsDNA, since it can lower the T_m of T4 dsDNA by 69° under conditions where the native protein has no effect [19,23,32]. Thus, it appears that excision of the A domain removes a kinetic block to helix destabilization [23]; the block may be a (slow) conformational change attendant to cooperative binding [27]. In keeping with this, the A domain is pictured as an arm or flap that partially occludes the ssDNA binding site [31]. While the in vivo significance of the A region vis-à-vis helix destabil-

```
                  10                  20                  30
M-F-K-R-K-S-T-A-E-L-A-A-Q-M-A-K-L-N-G-N-K-G-F-S-S-E-D-K-G-E-W-K
              40                  50                  60
L-K-L-D-N-A-G-N-G-Q-A-V-I-R-F-L-P-S-K-N-D-E-Q-A-P-F-A-I-L-V-N-H
          70                  80                  90
G-F-K-K-N-G-K-W-Y-I-E-T-C-S-S-T-H-G-D-Y-D-S-C-P-V-C-Q-Y-I-S-K-N
       100                 110                 120
-D-L-Y-N-T-D-N-K-E-Y-S-L-V-K-R-K-T-S-Y-W-A-N-I-L-V-V-K-D-P-A-A-P
   130                 140                 150                 160
-E-N-E-G-K-V-F-K-Y-R-F→G-K-K-I-W-D-K-I-N-A-M-I-A-V-D-V-E-M-G-E-T
                   170                 180                 190
-P-V-D-V-T-C-P-W-E-G-A-N-F-V-L-K-V-K-Q-V-S-G-F-S-N-Y-D-E-S-K-F-L
               200                 210                 220
-N-Q-S-A-I-P-N-I-D-D-E-S-F-Q-K-E-L-F-E-Q-M-V-D-L-S-E-M-T-S-K-D-K
           230                 240                 250
-F-K-S-F-E-E-L-N-T-K-F-G-Q-V-M-G-T-A-V-M-G-G-A-A-A-T-A-A-K-K-A-D
       260                 270                 280
-K-V-A-D-D-L-D-A-F-N-V-D-D-F-N-T-K-T-E-D-D-F-M-S-S-S-S-G-S-S-S-S
   290                 300
-A-D-D-T-D-L-D-D-L-L-N-D-L
```

Fig. 8. *Amino acid sequence of the T4 gene 32 protein. The underlined sequence contains six of the eight tyrosyl residues in the protein and a Zn(II) binding site (ligands are boxed) and is believed to constitute a part of the DNA binding domain.*

ization is unclear, it is now apparent that this domain is essential for interactions with other proteins in the T4 replisome [33]. G32P-A does not bind T4 RNA primase or T4 DNA polymerase, unlike g32P [33]. The A domain of g32P may be a prototypical case, in that several other single-strand binding proteins possess acidic COOH terminal regions whose removal causes increased helix destabilization [34–40].

Gene 32P-B. This species does not undergo indefinite self-aggregation [41], and fluorescence quenching studies show that removal of the "B" domain decreases the affinity of the protein for poly(dT) [30] through a reduction of the cooperativity parameter to almost unity [31]. Since the modified protein binds $d(pT)_8$ almost normally [30], it is apparent that the first 15 amino acids are essential for g32P–g32P interactions.

Gene 32P-(A+B). Despite a loss of cooperative protein:protein interactions, g32P-(A+B) binding to ssDNA causes comparable lattice deformation to that observed with g32P [31]. In addition, both species have similar affinities for $d(pT)_8$ [30], suggesting that the ssDNA binding domain resides between residues 22 and 253. Ultraviolet-induced cross-linking experiments [41] have demonstrated that all contact sites lie within the g32P-(A+B) domain. Although an unusual clustering of missense mutations that map between residues 36–125 has led to the suggestion that this region contains the DNA binding domain [42], the ^{1}H-NMR and in vitro mutagenesis experiments detailed below provide the first direct evidence towards identifying individual amino-acid residues at the interface of the g32P: ssDNA complex.

Nitration of Gene 32 Identifies Tyr Residues as Involved in DNA Binding

Nitration of five of the eight Tyr side chains in gene 32P with TNM completely prevents DNA binding as assayed with a circular dichroism binding assay employing single-stranded fd DNA [24]. Binding of fd DNA to g32P prior to treatment with TNM completely prevents nitration of the five Tyr residues [24]. Thus one or more Tyr side chains of the protein appeared to be involved in the nucleotide-binding surface.

DETECTION OF Tyr AND Phe SIDE CHAINS INVOLVED IN g32P-OLIGONUCLEOTIDE INTERACTIONS BY ^{1}H-NMR METHODS

^{1}H-NMR difference methods coupled with selective deuteration of amino acid residues has been successful in identifying amino acid side chains involved in nucleotide binding by g32P [43]. ^{1}H-NMR studies on native g32P are severely limited by the tendency of the protein to oligomerize in solution [44]. Some useful information can be obtained, however, since comparison of the ^{1}H-NMR spectra from g32P and g32P-(A+B) suggests that residues in the A and B domains have considerably more mobility. Resonances corresponding to the AA residues in the A and B domains are narrow in g32P, while all others are broadened beyond detection [5,44]. Removal of the A and B domains by limited proteolysis with trypsin results in the appearance of a typical protein ^{1}H-NMR spectrum expected for a protein of ~26K [44]. While not widely employed, limited proteolysis may be useful in preventing oligomerization in other systems.

Of course, ^{1}H-NMR spectra of g32P* are still very complex, even in the aromatic region. One-dimensional difference spectra show that a large number of aromatic resonances are shifted upfield upon nucleotide binding [44]. While a valuable demonstration of the close approach of nucleotide bases to protein aromatics, the direct difference spectra are too complex for easy interpretation; thus further efforts have been made toward spectral simplification. When the protein of interest is produced in a bacterial system, the use of amino acid auxotrophs allows spectral simplification by pursuance of the maxim "when in doubt, deuterate."

When the ^{1}H-NMR spectrum of dTyr-g32P* is subtracted from that of the native protein, a spectrum of the Tyr residues alone is obtained, as shown in Figure 9A. The most upfield signal in this spectrum is that of a single 3 or 5 proton of a Tyr, and integration of the area under the complete spectrum corresponds closely to that of the 32 protons expected from the eight Tyr residues of the protein. The same technique applied to the protein continuing perdeuterated Phe yields equally good results (Figure 9B).

With the ability to generate difference ^{1}H-NMR spectra of the isolated Tyr or Phe protons, the shifts in these signals on nucleotide binding can then be assessed by superimposing a second set of difference spectra—i.e., those formed by subtracting the ^{1}H spectrum of the oligonucleotide complexes of the perdeuterated protein from the same complexes of the protonated protein. A spectrum of the $d(pA)_6$ complex of dTyr g32P* is pictured in Figure 10A. A hexanucleotide shifts the maximum number of aromatic protons. The difference spectra of the complex is superimposed on the Tyr difference spectra of the unliganded protein to determine which resonances shift in the complex. The darkened resonances are those that shift and resonances attributable to five Tyr are shifted by $d(pA)_6$ (Figure 10A).

Similar spectra for the dPhe g32P* show the dinucleotide to shift no Phe resonances, the tetranucleotide to shift signals corresponding to two Phe residues, and a hexanucleotide to shift the same signal as the tetranucleotide, i.e., those attributable to two Phe residues. The difference spectra for the tetranucleotide complex are shown in Figure 10B. By comparing the sum of the shifted resonances in the Tyr and Phe difference spectra with the total observed in a protonated protein, they are found to account for all the shifts induced by nucleotide binding in the aromatic proton region for the ^{1}H-NMR spectrum, suggesting that the Trp side chains do not closely approach the base rings of a bound nucleotide. This was confirmed by the use of a g32P* containing perdeuterated Trp residues.

With one or two exceptions, the shifts of the Tyr and Phe protons induced by nucleotide binding are upfield and have maximum $\Delta\delta$ values of 0.2–0.3 ppm, which are substantially less than the ~1 ppm upfield shift that might be expected from an intercalation model with ring to ring stacking distances of ~3.4 Å [45]. A more probable model than a "glove-fit" intercalation is one in which the rings of the aromatic amino acids approach the bases in some regular fashion down the nucleotide chain and form part of a series of hydrophobic pockets accepting the base rings. The postulate that these aromatic amino acid side chains participate in the formation of hydrophobic binding pockets for the base rings

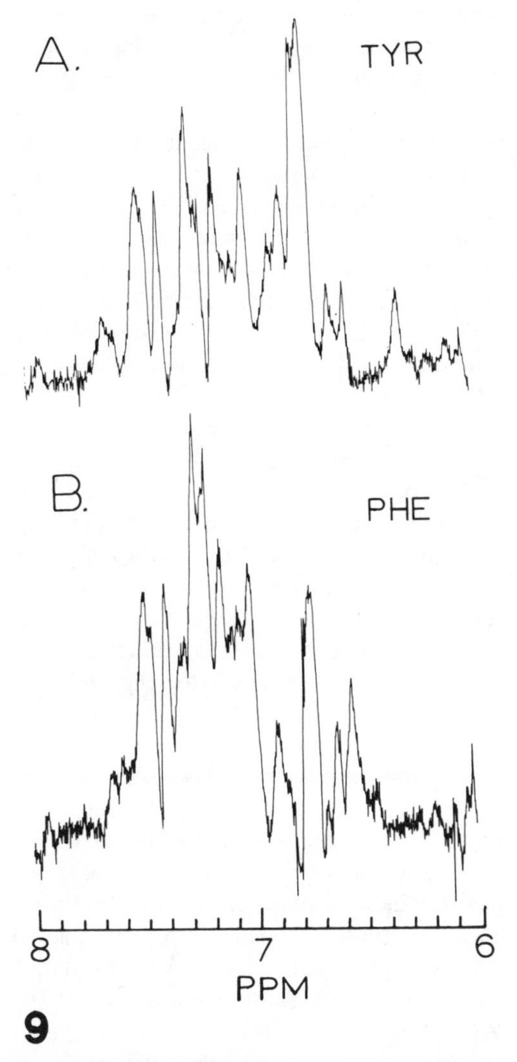

Fig. 9. *Aromatic* [1]*H-NMR (500 MHz) difference spectra. (A) Gene 32P* minus dTyr gene 32P* (containing perdeuterated Tyr). This difference spectrum represents resonances due to tyrosine protons alone. (B) Gene 32P* minus dPhe gene 32P* (containing perdeuterated Phe). This difference spectrum represents resonances due to phenylalanine protons alone. (From ref. 43.)*

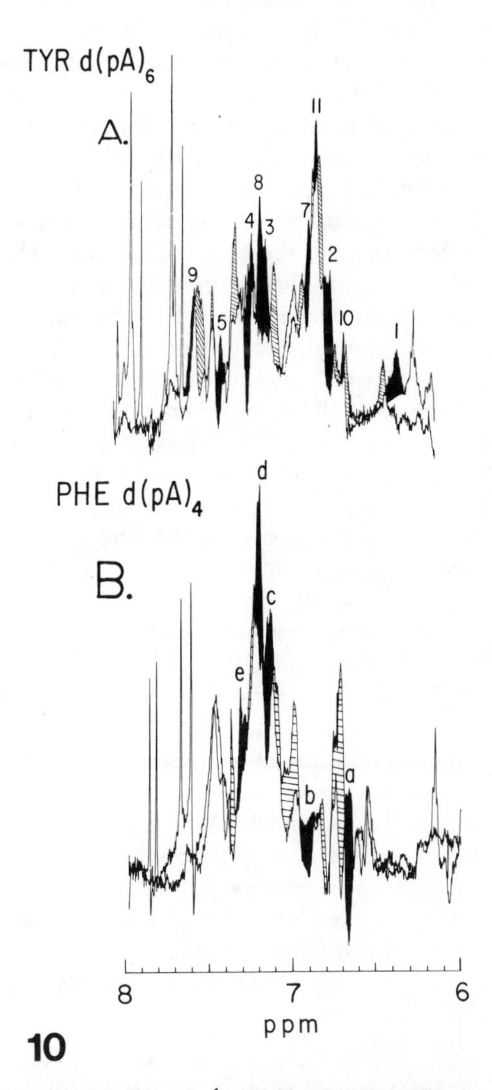

Fig. 10. *(A) Aromatic* [1]*H-NMR (500 MHz) difference spectra of gene 32P*–d(pA)$_6$ complex. Spectrum of Tyr resonance from Figure 9A (envelope of hatched peaks) superimposed on the Tyr resonances of the gene 32P*–d(pA)$_6$ complex generated in the same manner, i.e., gene 32P*–d(pA)$_6$ complex minus gene 32P*–d(pA)$_6$ complex containing perdeuterated Tyr as represented by the envelope over the filled-in peaks. (B) Aromatic* [1]*H-NMR (500 MHz) difference spectra of gene 32P*–d(pA)$_4$ complex. Spectrum of Phe resonances from Figure 9B (envelope of hatched peaks) superimposed on the Phe resonance of the gene 32P*–d(pA)$_4$ complex generated in the same manner—i.e., gene 32P*–d(pA)$_4$ complex minus gene 32P*–d(pA)$_4$ complex containing perdeuterated Phe as represented by the envelope over the filled-in peaks. (From ref. 43.)*

is a more cautious interpretation of the data than the term "intercalation." As base rings are brought near the amino acid rings in a model, the ring current shifts are determined by both distance and angle and can give a wide range of magnitudes as well as shift resonances upfield or downfield. Both phenomena are observed in Fig. 10.

Using a cut-and-weigh method and assuming a standard line shape, the shifted resonances in Figure 10A assignable to tyrosyl protons appear to account for five Tyr residues, while those in Figure 10B account for two Phe residues. Thus the DNA binding surface of g32P* appears to contain five Tyr and two Phe residues which are within 4.5 Å of the base rings of the bound nucleotide. The NMR results, as well as the nitration experiments, focus attention on the amino acid sequence of the molecule extending from Lys^{71} to Trp^{116}, which contains six of the eight Tyr residues in g32P*. This region of the peptide chain must form at least part of the DNA binding domain and is underlined in Figure 8. We have recently discovered that native g32P is a zinc metalloprotein containing 1 gram atom of Zn(II) per mole [46]. Gene 32P* also contains this Zn(II) ion and titration of the protein with an organic mercurial displaces the Zn(II) consequent to reaction with three SH groups. The optical absorption spectrum of the Co(II)-substituted g32P shows a tetrahedral coordination to be present [46]. These observations suggest that Cys^{77}, His^{81}, Cys^{87}, and Cys^{90} in the middle of the putative DNA-binding surface contribute the ligands to the metal ion. Fluorescence quenching used to study the binding of polyd(T) to g32P shows that binding of the polynucleotide is radically altered by removal of the native zinc ion [46].

SITE-DIRECTED MUTAGENESIS AS A MEANS OF ASSIGNING ^{1}H-NMR SIGNALS TO SPECIFIC AMINO ACID RESIDUES INVOLVED IN NUCLEOTIDE BINDING

In cases where a minimal structural perturbation can be introduced, site-directed mutagenesis is a potentially useful method for assigning the ^{1}H resonances from any specific residue. Of course, there are limitations, in that the introduced spectral perturbation must have an observable intensity. The high-quality difference spectra that can be constructed with the newer NMR spectrometers should allow such perturbations to be resolved in (native−mutant) difference spectra. The approach has the best chance of success if directed at relatively infrequently represented residues or at a particular region of the sequence whose resonances can be perturbed by selective perturbations such as ligand binding. Since this is true for the Tyr residues of g32P* that were implicated earlier, plans for the site-directed mutagenesis of the aromatic residues potentially involved in DNA binding were initiated. The first example is the mutation Tyr 115→Ser 115.

The resultant Ser^{115} g325P* still binds to DNA, but with reduced affinity [43]. The (wild-type g32P* minus Ser^{115} g32P*) difference spectrum shows two major peaks (2 and 4 in Figure 11), which are clearly assignable to Tyr^{115} and represent resonances from a Tyr residue that interacts with the nucleotide bases (Figure 3). Repetition of this procedure should allow identification of all the Tyr residues involved in DNA binding. Since there are a relatively large number of Phe residues scattered throughout the sequence, none between 71 and 116, the same approach to identifying the Phe side chains involved in nucleotide binding will be more complex. Several minor peaks in the mutant aromatic ^{1}H NMR difference spectrum (less than one proton) suggest that the $Tyr^{115}\rightarrow Ser^{115}$ mutation has shifted other aromatic resonances slightly (Figure 11). While this does not interfere with interpretation in the present example, such shifts, if large, could complicate the assignment.

CONCLUSIONS

^{1}H NMR has proved to be a powerful probe of those amino acid side chains responsible for oligonucleotide binding to the single-stranded DNA binding proteins. The technique can be applied to proteins as large as 30kD, gene 32 protein, ordinarily considered too large for detailed ^{1}H-NMR studies, since NMR difference spectra can be constructed which isolate individual resonances perturbed by ligand

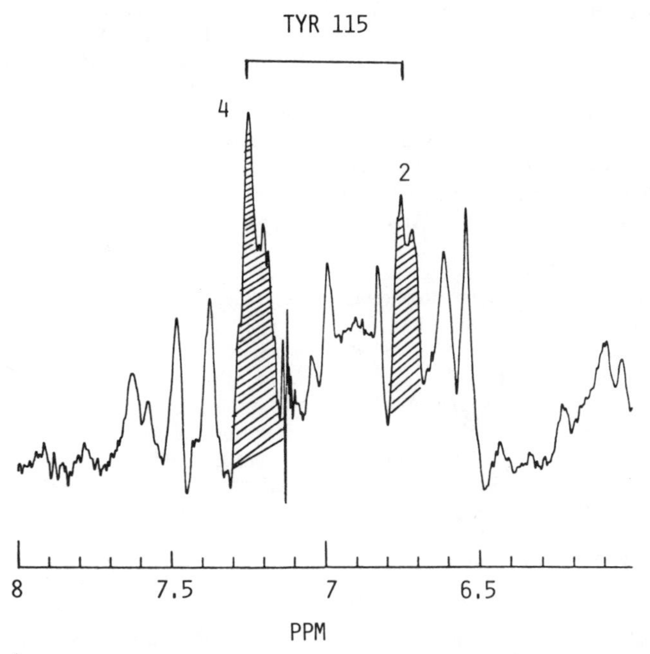

Fig. 11. *Aromatic ^{1}H-NMR (500 MHz) difference spectrum, gene 32P* (wild-type) minus gene 32P* (Ser^{115}mutant). Peaks 2 and 4 in the difference spectrum have the same chemical shifts as the peaks marked 2′ and 4′ in the previous spectra and assigned to Tyr. These peaks shift to positions marked 2 and 4 in the spectra of the nucleotide complex in Figure 10. (Fropm ref. 43.)*

binding. Perdeuteration of one or another set of residues in the protein can also be used to great advantage to simplify the ^{1}H-NMR spectra and helps isolate specific ^{1}H resonances. Despite its size, 19kD, the application of the 2D-NOESY ^{1}H-NMR techniques to the gene 5 protein dimer has allowed the identification of specific residues involved in oligonucleotide binding because of the detection of specific resonance shifts and intermolecular NOEs between H2, H8, and Hl′ proteins of bound adenine nucleotides and the protein made possible by the increased dispersion of cross peaks in the 2D NOESY spectra. These studies illustrate the value of the 2D NMR methods, even if the protein is larger than the cutoff, ~ 10kD, assumed for the application of the more usual conformational analysis of protein structure by the 2D methods.

Chemical modification and protein–nucleotide cross-linking experiments can be useful in focusing on the types of amio acid side chain involved in protein–nucleotide interactions, despite some limitations on specificity. In concert with the more specific ^{1}H-NMR methods, these approaches have proven valuable. ^{1}H-NMR is not useful in probing electrostatic interactions with Lys or Arg residues; and chemical modification has thus far been the only approach applicable.

The correlation of the NMR of the gene 5 protein and its nucleotide complexes with the crystal structure illustrates the complementarity of the NMR and crystallographic techniques in probing protein structure and function.

The picture emerging for the DNA binding domains of single-stranded DNA binding proteins is one consisting of electrostatic interactions between strategically placed Lys and Arg residues and the phosphodiester backbone of the nucleotide; coupled with hydrophobic interactions of the nucleotide bases, with regularly placed aromatic rings from Tyr and Phe residues. In some cases, e.g., gene 5 protein, the aromatic rings may intercalate.

REFERENCES

1. Alberts, BM, Frey, L (1970): Nature 227:1313–18.
2. Alberts BM, Barry J, Bedinger P, Burke RL, Hibner U, Liu CC, Sheridan R (1980): In "Mechanistic Studies on DNA Replication and Genetic Recombinations." Alberts BM, Fox CF (eds): ICN-UCLA Symp Mol Cell Biol, Vol 19. New York: Academic Press, pp 449–71.
3. Krisch HM, Duvoisin RM, Allet B, Epstein RH, (1980): "Mechanistic Studies on DNA Replication and Genetic Recombination." In Alberts, BM, Fox, CF (eds): ICN-UCLA Symp Mol Cel Biol Vol: 19. New York: Academic Press, pp 517–26.
4. Krisch HM, Allet B, (1982): Proc Natl Acad Sci USA 79:4937–41.
5. Coleman JE, Oakley JL (1980): CRC Crit Rev Biochem 7:247–89.
6. Alberts BM, Frey L, Delius H (1972))): J Mol Biol 68:139–152.
7. Anderson RA, Nakashima Y, Coleman JE (1975): Biochemistry 14:907–917.
8. Coleman JE, Anderson RA, Ratcliffe RG, Armitage IM (1976): Biochemistry 15:5419–5430.
9. Coleman JE, Armitage IM (1978) Biochemistry 17:5038–5045.
10. Alma NCM, Harmsen BJM, Hull WE, van der Marel G, van Boom JH, Hilbers CW (1981): Biochemistry 220:4419–4428.
11. O'Connor TP, Coleman JE (1983): Biochemistry 22:3375–3381.
12. Nakashima Y, Dunker AK, Marvin DA, Konigsberg W (1974): FEBS Lett 40:290–292.
13. Nakashima Y, Dunker AK, Marvin DA, Konigsberg W (1974): FEBS Lett, 43:125.
14. McPherson A, Jurnak FA, Wang A, Kolpak F, Rich A, Molineux I, Fitzgerald B (1980): Biophys J 32:155–173.
15. McPherson A, Brayer GD (1985): In A, McPherson Jurnak F (eds): "Biological Macromolecules and Assemblies," Vol 2. John Wiley & Sons New York: pp.323–392.
16. King GC, Coleman JE (1986): Biochemistry (in press).
17. Williams KR, LoPresti M, Setoguchi M, Konigsberg WH (1980): Proc Natl Acad Sci USA 77:4614–17.
18. Williams KR, LoPresti M, Setoguchi M (1981): J Biol Chem 256:1754–62.
19. Hosoda J, Moise H (1978): J Biol Chem 253:7547–55.
20. Delius H, Mantell NJ, Alberts B (1972): J Mol Biol 67:341–50.
21. Carroll RB, Neet K, Goldthwait DA (1975): J Mol Biol 91:275–91.
22. von Hippel PH, Kowalczykowski SC, Lonberg N, Newport JW, Paul LS, Stormo GD, Gold L (1982): J Mol Biol 162:795–818.
23. Jensen DE, Kelly RC, von Hippel PH (1976): Biol Chem 251:7215–28.
24. Anderson RA, Coleman JE (1975): Biochemistry 14:5485–91.
25. Hosoda J, Takacs B, Black C (1974): FEBS Lett 47:338–43.
26. Moise H, Hosoda J (1976): Nature 259:455–58.
27. Williams KR, Konigsberg WH (1978): J Biol Chem 253:2463–70.
28. Tsugita A, Hosoda J (1978): J Mol Biol 122:255–58.
29. Hosoda J, Burke RL, Moise H, Kubota I, Tsugita A, (1980): In:Alberts BM, Fox CG, (eds): "Mechanistic Studies on DNA Replication and Genetic Recombination,"ICN-UCLA Symp Mol Cell Biol, Vol 19. New York: Academic Press, pp.505–513.
30. Spicer EK, Williams KR, Konigsberg WH (1979): J Biol Chem 254:6433–36.
31. Lonberg N, Kowalczykowski SC, Paul LS, von Hippel PH (1981): J Mol Biol 145:123–38.
32. Greve J, Maestre M, Moise H, Hosoda J (1978): Biochemistry 17:893–898.
33. Burke RL, Alberts BM, Hosoda J (1980): J Biol Chem 255:11484–93.
34. Dunn JJ, Studier F (1981): J Mol Biol 148:303–330.
35. Araki H, Ogawa H (1981): Mol Gen Genet 183:66–73.
36. Williams KR, Spicer EK, LoPresti MB, Guggenheimer RA, Chase JW (1983): J Biol Chem 258:3346–55.
37. Chase JW, Merrill BM, Williams KR (1983): Proc Natl Acad Sci USA 80:5480–84.
38. Isackson PJ, Reeck GR (1982): Biochim Biophys Acta 697:378–80.
39. Carballo M, Puigdomenech P, Tancredi T, Palav J, (1984): EMBO J 3:1255–61.
40. Reeck GR, Isackson PJ, Teller DC (1982): Nature 300:76–78.
41. Williams KR, Konigsberg WH (1981): In Chirikjian JG, Papas TS (eds): "Gene, Amplification and Analysis," North Holland, Amsterdam: Elsevier, pp 475–508.
42. Doherty DH, Gauss P, Gold L (1982): Mol Gen Genet 188:77–90.
43. Prigodich RV, Shamoo Y, Williams KR, Chase JW, Konigsberg WH, Coleman JE (1986): Biochemistry 25:3666–3672.
44. Prigodich RV, Casas-Finet J, Williams KR, Konigsberg W, Coleman JE (1984): Biochemistry 23:522–529.
45. Giessner-Prettre C, Pullman B (1976): Biochem Biophys Res Commu 70:578–581.
46. Giedroc DP, Keating KM, Williams KR, Konigsberg WH, Coleman JE (1986): Proc Nat Acad Sci USA (in press).

APPENDIX:
COLOR AND COLOR STEREO REPRESENTATIONS

Protein Engineering, pages 339–354

COLOR AND COLOR STEREO FIGURES FROM TEXT CHAPTERS (PAGES 340–347)

The representations of three dimensional protein structures in two dimensions for publication presents a continuing challenge to authors. Jane and David Richardson, authors of Chapter 12, have pioneered the use of colored ribbons to provide schematic representations of the backbone structures of proteins. More recently the rapid advances of computer graphic techniques have greatly aided the presentation of three-dimensional structures. The computer can be used to generate stereo pair figures of the structures or parts of the structures of proteins to allow the reader to view three-dimensional molecular features.

In the first part of this section we have reproduced color versions of selected figures which have appeared in black and white form in some of the chapters of this book. For some of the figures the use of color or stereo pairs can aid in the presentation of structural elements. The chapter number, title of the article, and authors precede each section of figures to identify where the black and white versions appear in the book.

COLOR STEREO REPRODUCTIONS OF NEWLY DETERMINED STRUCTURES (PAGES 348–354)

The number of structural determinations of proteins and other macromolecular complexes is rapidly expanding. Since the conception of this book the structures of certain viruses, antibody complexes, and additional DNA-binding proteins have appeared. Included in the second portion of the appendix are figures describing four recently determined structures of macromolecules. The authors and references to primary publications describing these structures are included in the text.

STEREO VIEWING

In order to see a stereo pair in three dimensions it is often necessary to use a special stereo viewer which separates the right and the left eye images. It is usually possible to substitute a 10 or 12 inch piece of cardboard for separating the stereo pairs. It is also possible for many readers to train their eyes to see the three-dimensional view directly without the aid of a stereo viewer or piece of cardboard. Technically you must focus your eyes on an imaginary object beyond the plane of the images so that your right and left eyes focus on the separate right and left images, respectively. To do this, first look at the figure in stereo using a viewer. Then, without the viewer, hold the page with the stereo pair at a normal reading distance and relax your eyes so that the stereo pair images appear to move together and become superimposed on one another. Keep staring at the superimposed images until you see the three-dimensional image develop. You must be patient at this point since it usually takes a little time at the beginning. It helps to start with an easy example that you have already seen in stereo.

12
Some Design Principles: Betabellin

Jane S. Richardson and David C. Richardson

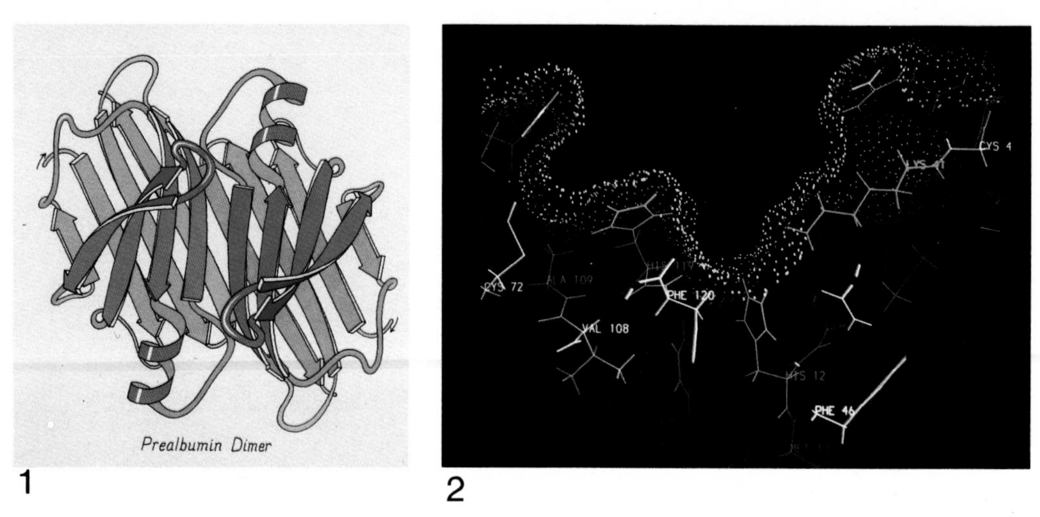

Fig. 1. *Schematic backbone structure of prealbumin, to symbolize the the degree of detail to which one specifies a tertiary structure for the design process.*

Fig. 2. *An all-atom model of ribonuclease A, with dots showing the accessible surface, to symbolize the process of decorating the schematic with side chains, which contribute essentially all the interactions both internally and with solvent and other molecules.*

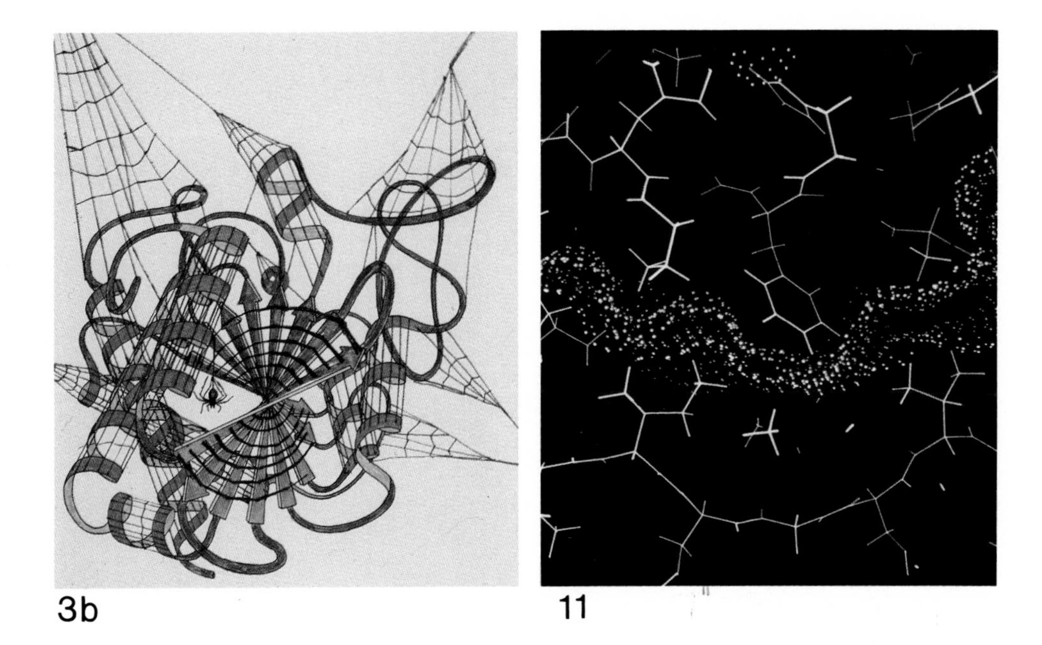

Fig. 3b. *Tertiary structure type that is unsuitable for de novo design and synthesis. Carboxypeptidase A, a parallel alpha/beta protein, for which many structural principles are known but which is much too large and complex. Drawing by Duncan McRee.*

Fig. 11. *Connolly dot surfaces showing the fit of the top beta sheet (green dots) to the bottom β-sheet (purple dots) in Cu, Zn superoxide dismutase.*

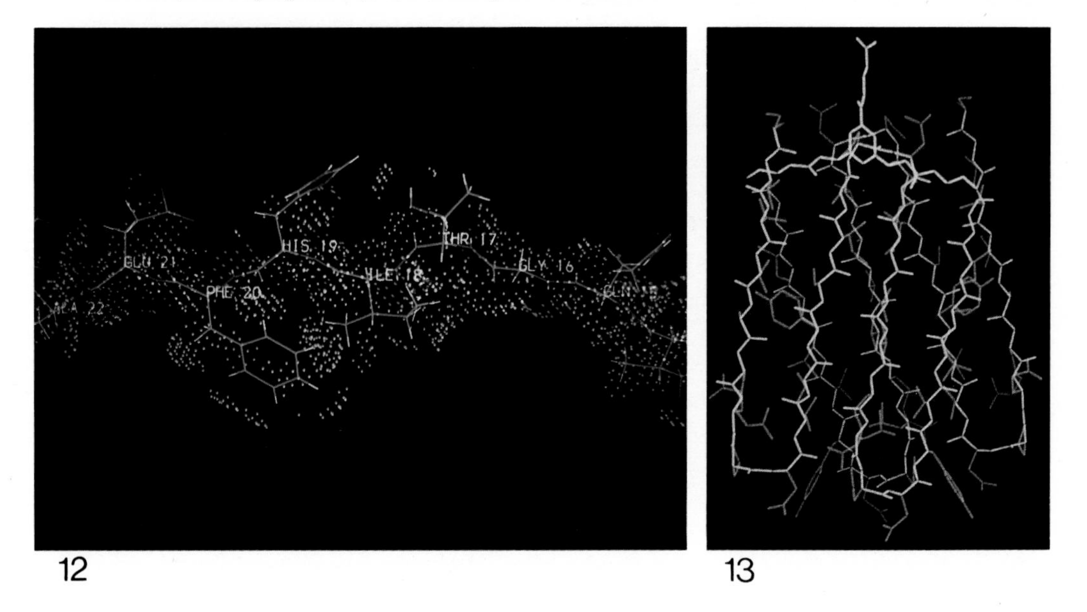

Fig. 12. *Modified dot surfaces that show internal contacts between atoms in a β-strand.*

Fig. 13. *An atomic model of betabellin built on Richard Feldmann's computer graphics system at NIH.*

15

Hydrophobicity and Amphiphilicity in Protein Structure

David Eisenberg, William Wilcox, and Andrew D. McLachlan

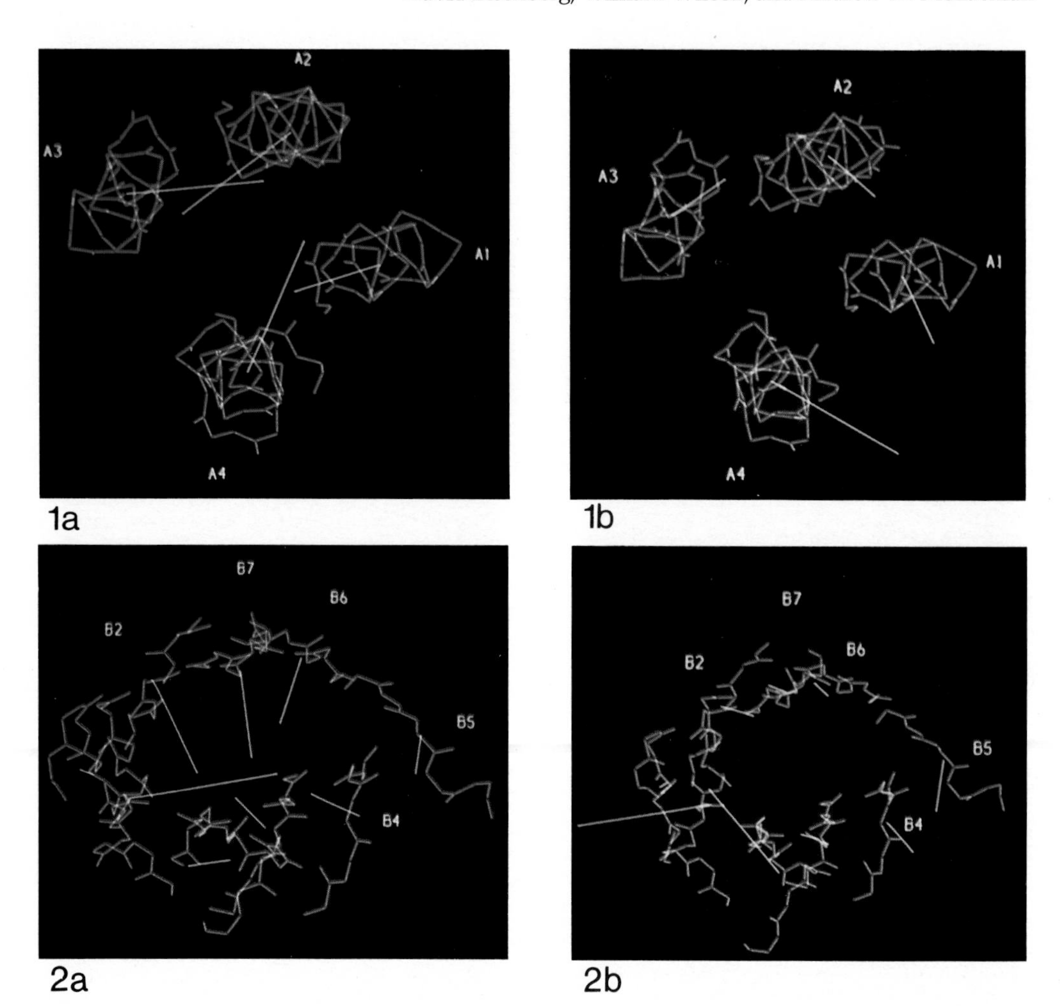

Fig. 1. *Hydrophobic moments of helices in a real and in a misfolded protein. A1 is the first helix in the chain, A2 is the second, and so forth. a) The four α-helical segments from the structure of hemerythrin from* T. dyscritum *[23]. Notice that the hydrophobic moments from the four helices point inwards and oppose each other. Each moment is represented as a line drawn from the center of its segment toward the direction of greater hydrophobicity. b) The incorrectly folded structure of a mouse κ-chain VL domain [25] arranged by Novotný et al. [1] into the structure of hemerythrin. The hydrophobic moments of the four helices are shown by lines emerging from the centers of the helices. These are scaled by a factor of 2 over those of Figure 1a, for better visibility.*

Fig. 2. *Hydrophobic moments of beta sheets in a real and in a misfolded protein. B1 is the first strand of sheet in the polypeptide chain, B2 is the second, and so forth. a) The nine-stranded β-sheet of a mouse κ-chain VL domain [25]. The hydrophobic moment of each strand of the sheet is shown as a line emerging from the center of the strand. The lengths are scaled by a factor of 1.5 over those of Figure 1a. b) The incorrectly folded structure of hemerythrin from* T. dyscritum *[23] as arranged by Novotný et al. [1]. The moments are on a scale 2 times those of Figure 1a.*

18

Theoretical and Experimental Approaches to the Design of Calmodulin-Binding Peptides: A Model System for Studying Peptide/Protein Interactions

Susan Erickson-Viitanen, Karyn T. O'Neil, and William F. DeGrado

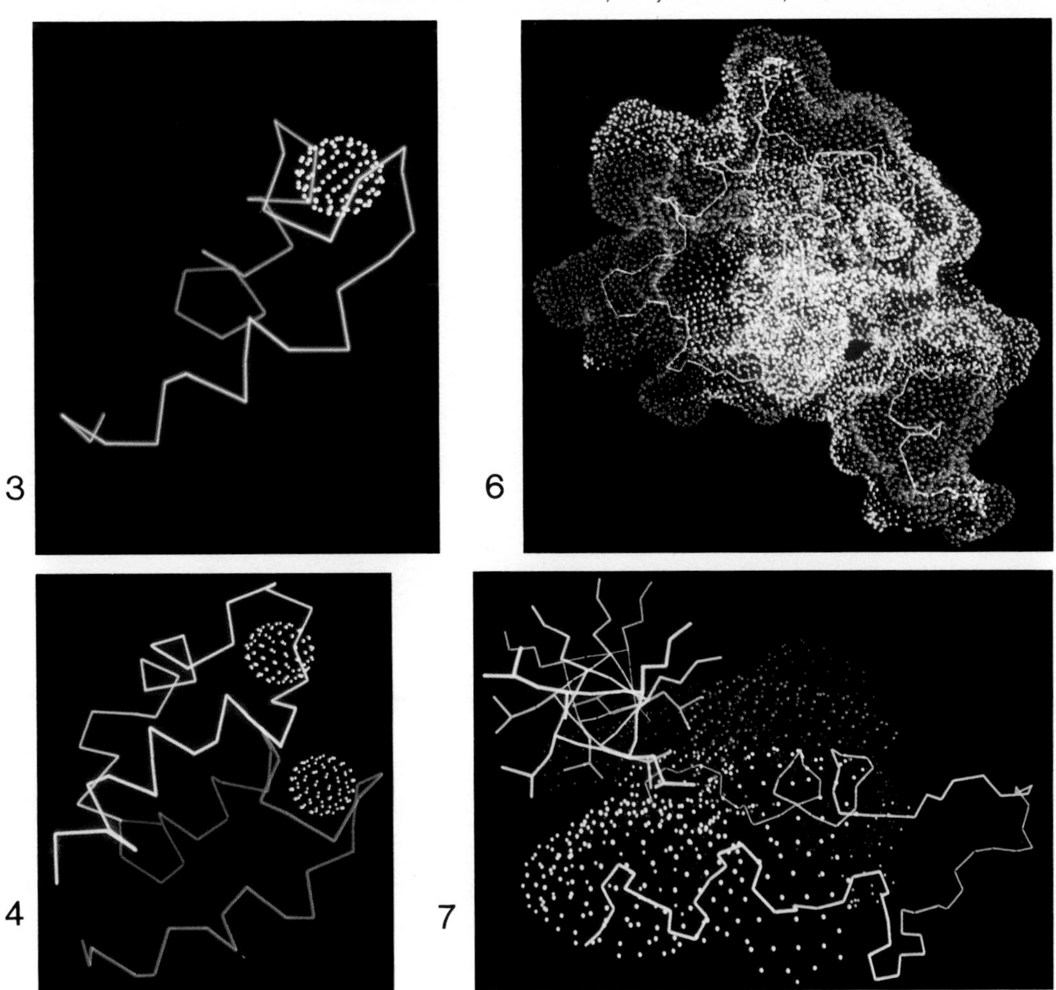

Fig. 3. *The structure of an EF hand. The positions of the C_α atoms are connected by lines. Calcium ions are represented by dots depicting a sphere with a radius 1.8 times that of the van der Waals radius of Ca^{2+}.*

Fig. 4. *Two E–F hands docked together to form an E–F dyad.*

Fig. 6. *Electrostatic potential surface for the second domain of the ICB-based model for calmodulin. Contour levels: red, $V < -10$ kcal/mol; orange, -10 kcal/mol$< V < -3$ kcal/mol; green, -3 kcal/mol $< V < +3$ kcal/mol; blue-green, $+3$ kcal/mol $< V < +10$ kcal/mol; blue, $V > +10$ kcal/mol. The backbone bonds of the structure are shown in white. The second F helix of this domain is in the center with its axis vertically oriented, the second calcium-binding loop is in the upper left, and the second E helix runs along the far left of the diagram. Note the very negative electric potential in the region surrounding the calcium binding loop as well as along the second E helix.*

Fig. 7. *Proposed docking orientation for the basic amphiphilic peptide (Leu–Lys–Lys–Leu–Leu–Lys–Leu)$_2$ to the second domain of the ICB-based model for calmodulin. Hydrophobic residues have been colored green, basic residues blue, and acidic residues red for both the peptide and the surface. The surface shown is a solvent accessible surface generated only for those protein residues in the binding site that are capable of contributing to stabilization of the peptide/protein complex. Note that in this orientation there is the potential for both hydrophobic (green–green) and ionic (blue–red) interactions between the peptide and calmodulin.*

20
Exploring DNA Polymerase I of *E. coli* Using Genetics and X-Ray Crystallography
T.A. Steitz and C.M. Joyce

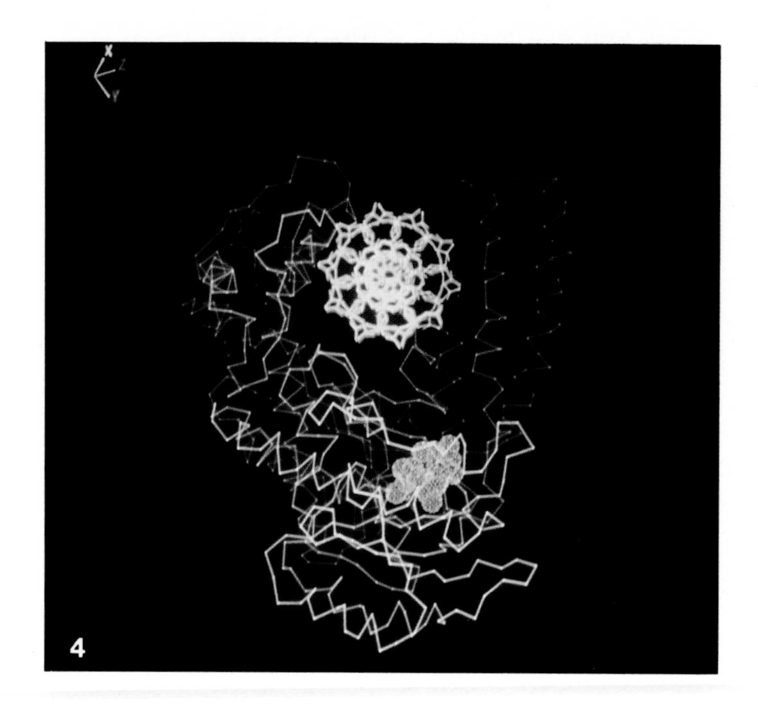

Fig. 4. *Relative orientation of DNA in the cleft and the bound deoxynucleoside monophosphate (in blue).*

Fig. 5. *Color graphic representation of the Klenow fragment bound to a nicked DNA substrate. The protein α-carbon backbone is shown in purple. The green surface represents the side chain of Tyr 776 that is crosslinked to 8-azido-dATP. The length of DNA shown is protected against DNAse I digestion by binding of the Klenow fragment. The template strand is red-orange, the primer strand is white, and the nick lies between the white and yellow strands of DNA. The protected region upstream of the primer terminus appears to extend some distance beyond the boundary of the protein (especially on the template strand). We believe that this represents steric interference between Klenow fragment and DNAse I. Footprinting with the small molecule methidiumpropyl-EDTA indicates that Klenow fragment protects about eight base pairs upstream of the primer terminus, consistent with the model presented here.*

Fig. 6. *Residues that are identical in T7 DNA polymerase and the Klenow fragment shown in orange on a yellow α-carbon backbone. The template strand of the DNA substrate is shown in green, while the primer strand is in blue. There appears to be extensive conservation of side chains pointing into the cleft and in the vicinity of the presumed dNTP binding site labeled by 8-azido-dATP.*

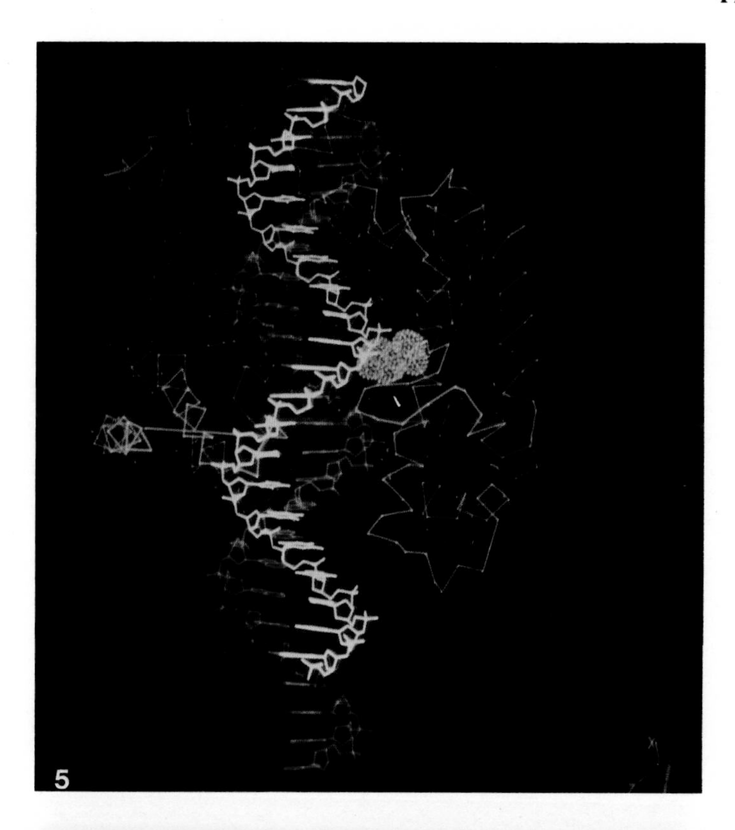
5

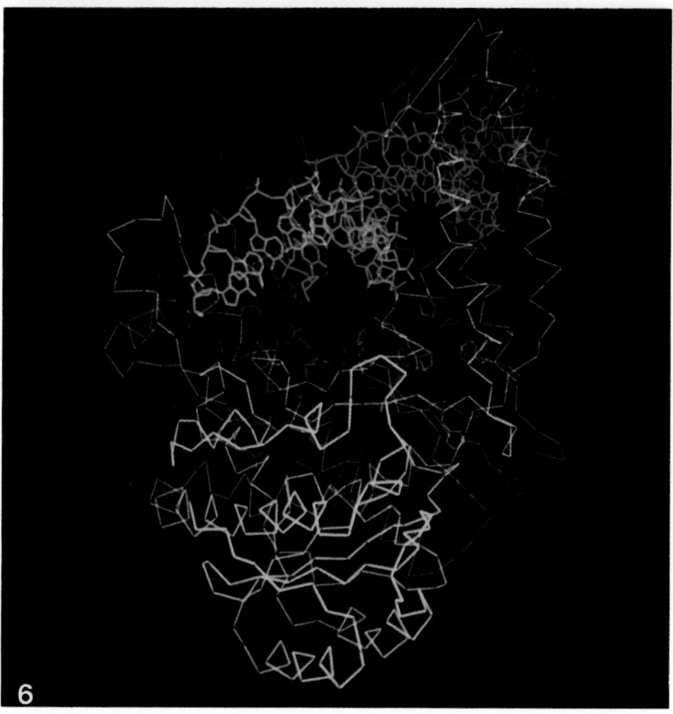
6

21
Development of a Protein Design Strategy for *Eco*RI Endonuclease

John M. Rosenberg, Bi-Cheng Wang, Christin A. Frederick, Norbert Reich, Patricia Greene, John Grable, and Judith McClarin

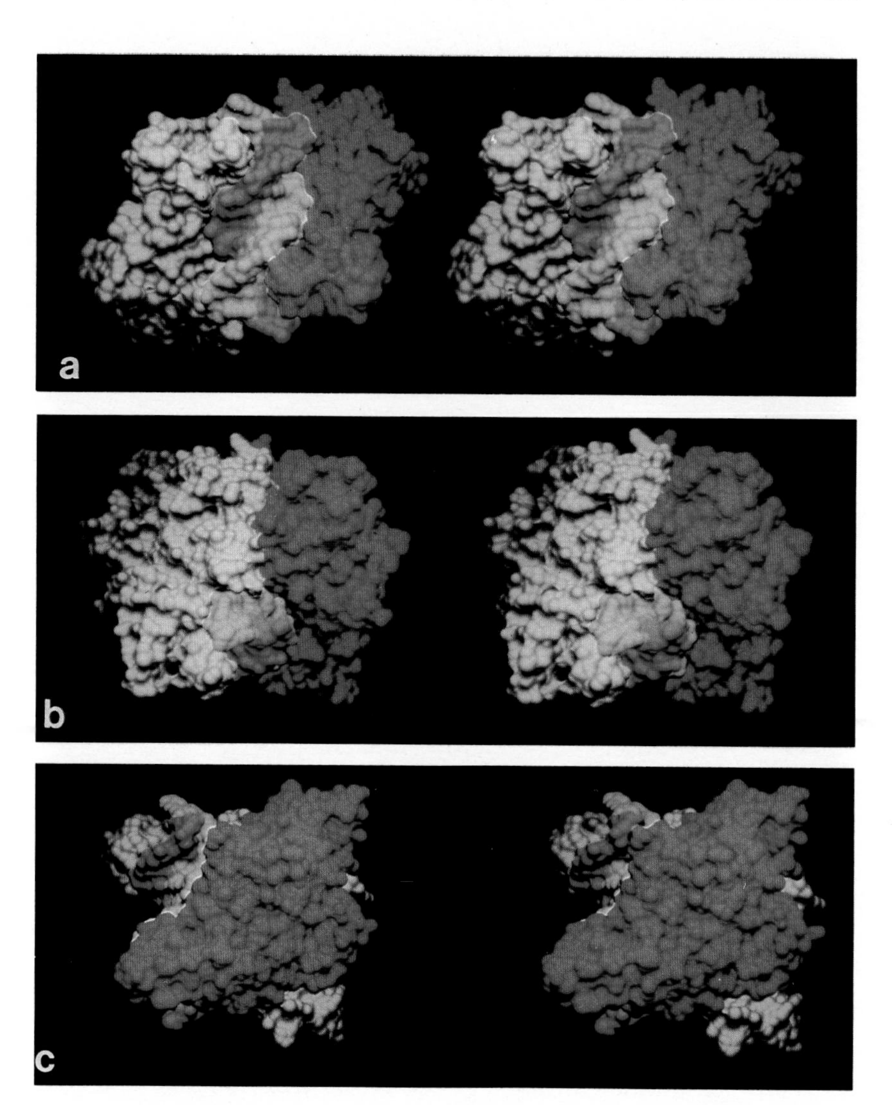

Fig. 2. *Stereo drawings of the solvent accessible surface of the* EcoRI *endonuclease-DNA complex. The subunits of the dimer are colored red and orange and the DNA is colored green and blue. a). The "front" view of the complex, which is a projection down a crystallographic twofold axis. b). The "top" view of the complex, rotated 90° from that in a) so as to view the structure down the c-axis. This results in a view looking approximately down the average DNA helical axis. c). The "side" view of the complex rotated 90° from a) around the c-axis.*

These images were calculated with the programs AMS and RAMS developed by Michael Connolly, modified for use with an Evans and Sutherland PS340 raster graphics system.

23
Redesigning Proteins via Genetic Engineering

William J. Rutter, Stephen J. Gardell, Steven Roczniak, Donald Hilvert,
Stephen Sprang, Robert J. Fletterick, and Charles S. Craik

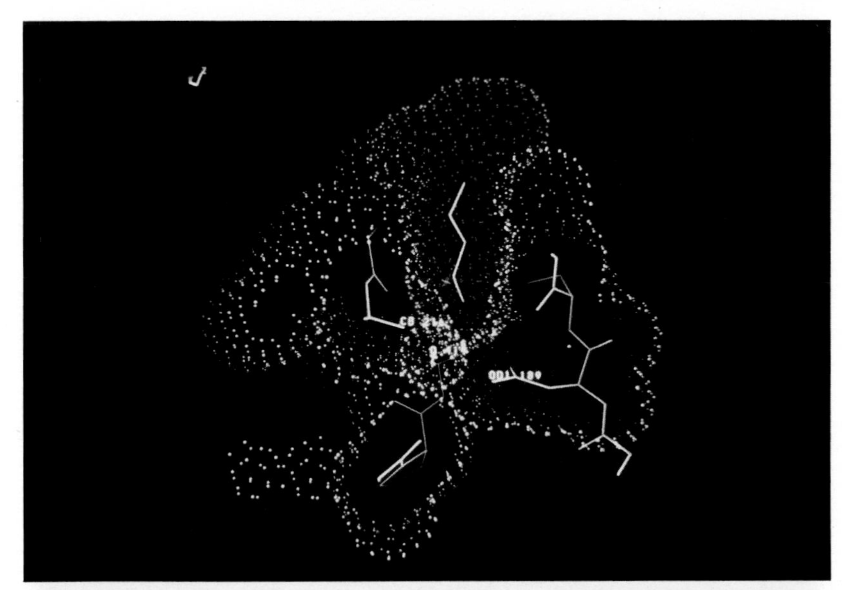

A

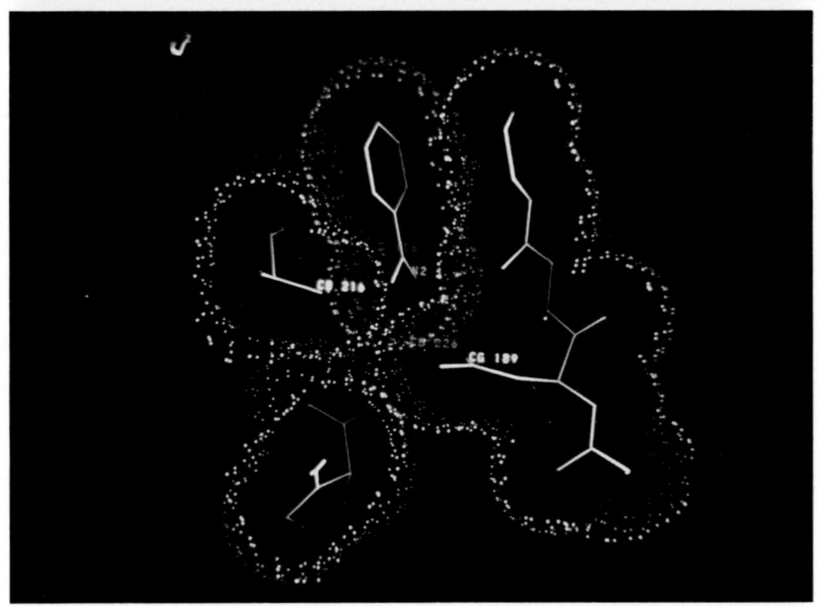

B

Fig. 5. *Space-filling representations of the substrate-binding pocket of mutant trypsins. The polypeptide backbone is shown for the chain segments, including amino acids 189 to 191, 214–217, and 224–227. The atoms are shown with their van der Waals contact surfaces. The enzyme atomic surface is blue and the ligand atomic surface is red. Atomic coordinates for the drawing were obtained from the Brookhaven Protein Data Bank (entry set 3PTB). (A) The Lys15 of pancreatic trypsin inhibitor bound in the trypsin specificity pocket. The van der Waals surface of the beta carbon of Ala216 overlaps that of the water molecule 414 and of Lys15. (B) Benzamidine bound in the trypsin specificity pocket. The van der Waals surface of the beta carbon of Ala226 overlaps that of the guanidinium group of benzamidine. The van der Waals overlap of the gamma carbon of Asp189 with the ligand is appropriate because a hydrogen bond is formed at this contact.*

Understanding Protein Architecture Through Simulated Unfolding

Richard J. Feldmann
Division of Computer Research and Technology
National Institutes of Health
Bethesda, Maryland 20892

Everyone admires Jane Richardson's ribbon representation of protein structure. Using a bold aesthetics, she has simplified protein structure to the point where we can begin to understand the architectural themes in the evolution of proteins on the third planet. Because the ribbon diagrams take considerable time and skill to produce, several people, notably Arthur Lesk, have written programs to automatically make weaker but roughly equivalent representations.

For many years I have wanted to think of proteins as space curves, but I could not find a suitable representation. Several weeks ago, my co-worker Bernard Brooks and I developed a fitting procedure using GEMM (Generate, Emulate, and Manipulate Macromolecules). The atoms of the peptide backbone have been smoothed out to form continuous space curve. It takes about two hours of VAX equivalent time to do the smoothing and about two minutes on the ST-100 array processor. When compared to the carbon alpha representation (which is too simple and harsh) or the normal peptide backbone representation (which is too complex), the smoothed space curve representation has the airy qualities of the original Richardson diagrams.

Using Jane Richardson's monograph as a guide and the Protein Data Bank from Brookhaven National Laboratory, 125 proteins and/or domains have been extracted. Each space curve is colored to show helices and sheets and is oriented to bring the N-terminus and/or the C-terminus of the protein to the front. Using stereo pairs of images, it is possible to begin to think about the late stages of the folding of proteins. In reality, what one does is imagine the unfolding of the protein from the crystal structure.

There are some real surprises. You can really imagine the sequence of motions by means of which the protein folded. The space curve representation makes it easy to think about tugging loops, peptide strands, and domains around in three dimensions. In work to follow, I will systematically pull apart all of these proteins.

The UPATSU will be distributed in these stereo slides, in video tape (¾ inch industrial, VHS, and Beta formats), and as coordinate data sets on magnetic tape.

The two sample stereo pairs are flavodoxin (Fig. 1) and alpha1 antitrypsin (Fig. 2). The flavodoxin stereo pair shows the central beta sheet surrounded by helices. It is easy to imagine unfolding this protein. In the alpha1 antitrypsin stereo pair, the sequence of unfolding would be much more complicated.

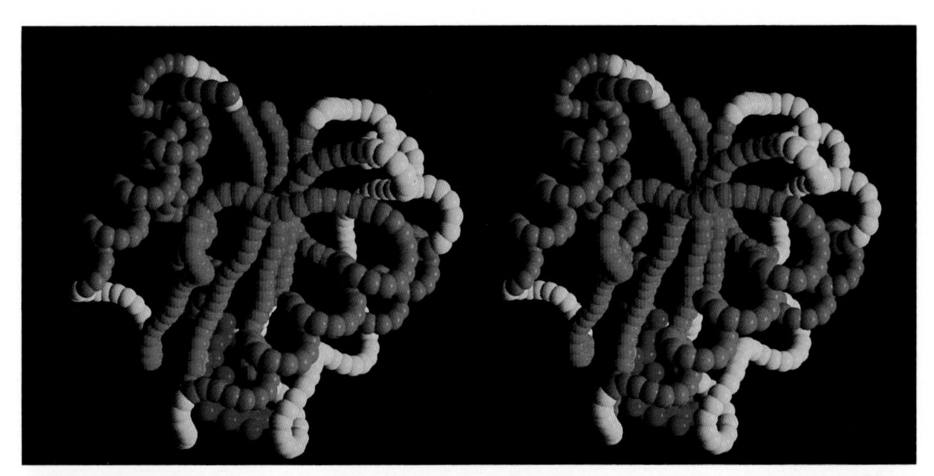

Fig. 1. *Flavodoxin*

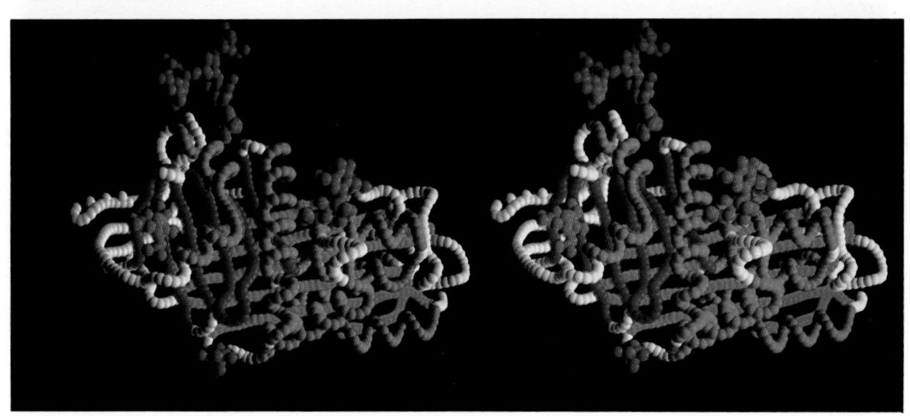

Fig. 2. *Alpha1 antitrypsin*

The Structure of Human Rhinovirus

Michael Rossmann, Edward Arnold, and Gerrit Vriend
Department of Biological Sciences
Purdue University
West Lafayette, Indiana 47097

The structure of human rhinovirus 14 (HRV14), a common cold virus, was solved at a resolution of 3.0 Å using the technique of X-ray diffraction from crystals. The virus is composed of an icosahedrally symmetric protein shell consisting of 60 copies of four distinct polypeptide chains (VP1, VP2, VP3, and VP4, with masses 32,000, 29,000, 26,000, and 7,000, respectively) which surrounds a single-stranded RNA genome ([+] polarity, approximately 7.5 kilobases) (see Rossmann et al. [1985]: Nature [London], 317:145–153).

1

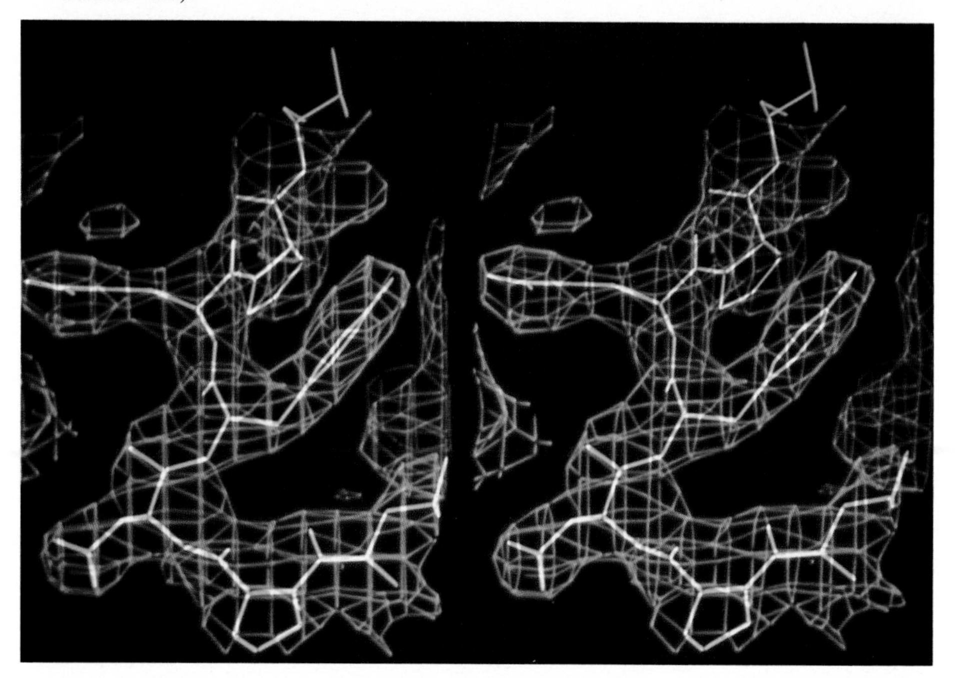

Fig. 1. *A portion of the atomic model (green) built into the 3.0 Å resolution electron density map (orange).*

Fig. 2. *An alpha-carbon backbone representation of the protomeric unit of the virus (containing one copy each of VP1, blue, VP2, green, VP3, red, and VP4, green) decorated with the immunogenic sites which were mapped by sequencing the coat protein region of escape mutants of HRV14 selected for their ability to survive in the presence of neutralizing monoclonal antibodies.*

Fig. 3. *The same representation turned end-on to illustrate the presence of a depression in the viral surface. This cleft or canyon is a likely site for binding to the cellular receptor for rhinoviruses; the neutralizing immunogenic sites reside on ridges of this canyon and can change freely without disrupting the receptor binding.*

Fig. 4. *An alpha-carbon backbone representation of a pentameric morphological unit, which is a central building block in rhinoviral assembly.*

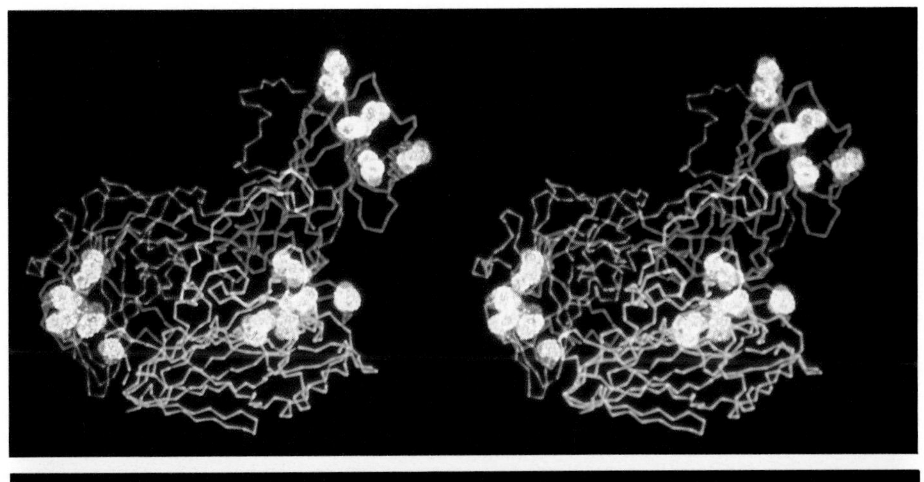

2

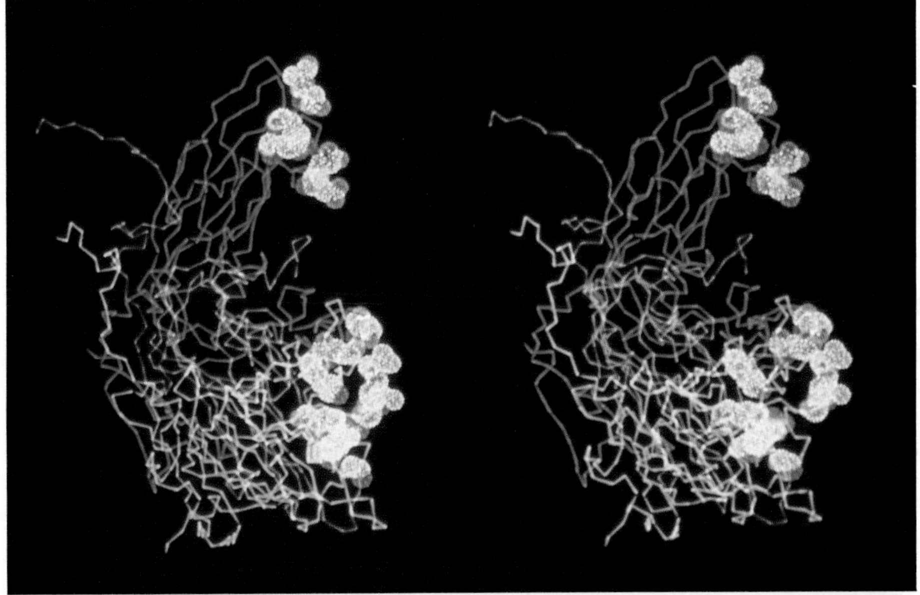

3

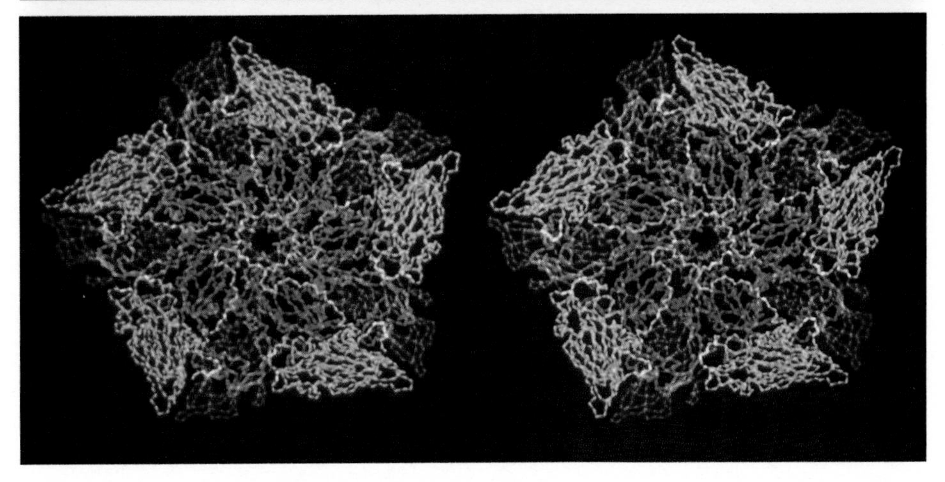

4

Stereoviews of the *trp* Repressor/Operator Complex

Paul Sigler
*Department of Biochemistry
and Molecular Biology
University of Chicago
Chicago, Illinois 60637*

Fig. 1. *(a) Each of the dyad related subunits, consists of 107 amino acids forming six helices linked by short turns. The C_α backbones are represented in blue and red, respectively, for each subunit and reveal the unusual, intertwined molecular fold. The binding of two molecules of L-tryptophan (white) confers upon the repressor the properties which are responsible for tight and specific binding to the operator. The presumed interaction with operator has been modeled by "docking" the repressor with a 20 base-pair regular B-DNA duplex (white) using computer graphics. (b) A full skeletal model of the repressor dimer (blue) with bound tryptophan (red) is shown docked onto the operator (white). The DNA-binding helix-turn-helix motif of each subunit protrudes from the surface of the repressor and fits into the major groove of the operator at two symmetrical positions, separated by one full turn of the DNA duplex.*

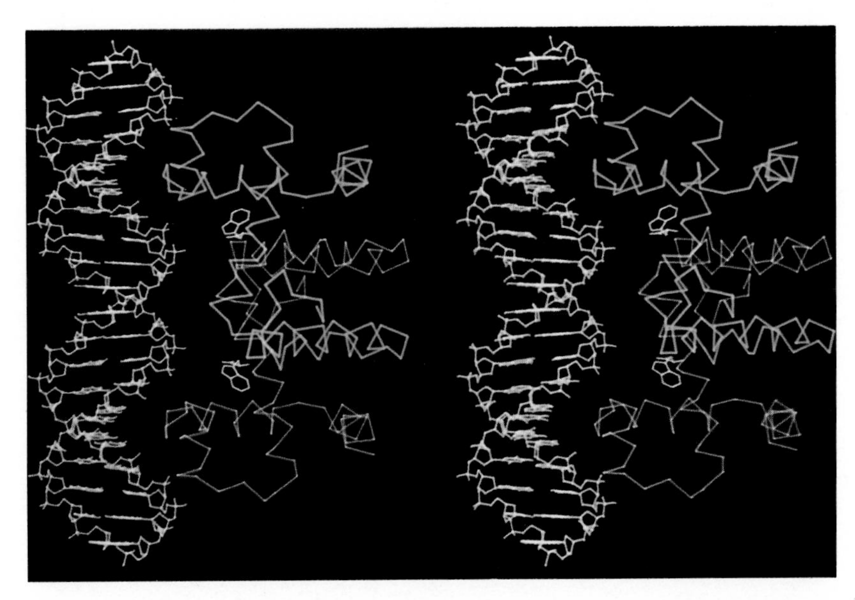

1a

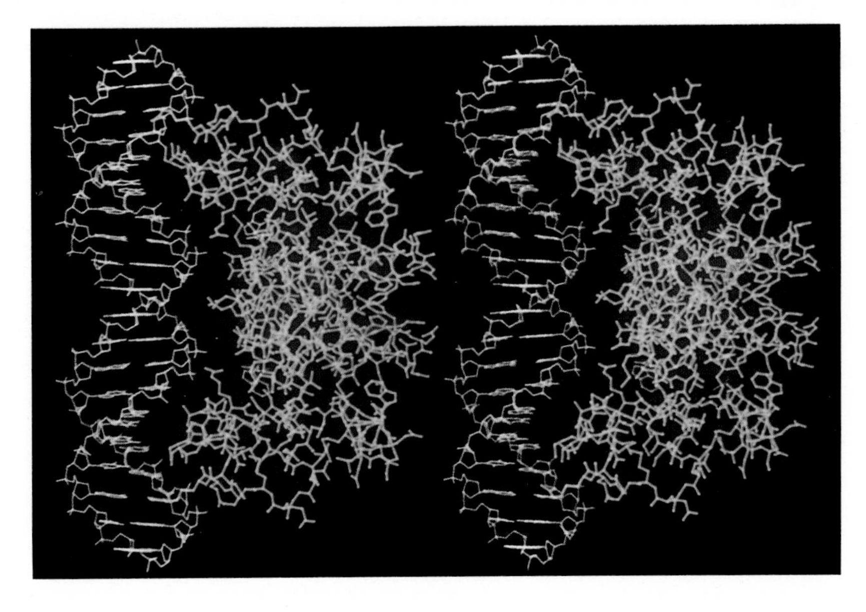

1b

Glutamine Synthetase From *S. typhimurium*

R.J. Almassy, C.A. Janson, and David Eisenberg
Molecular Biology Institute, University of California, Los Angeles, California 90024

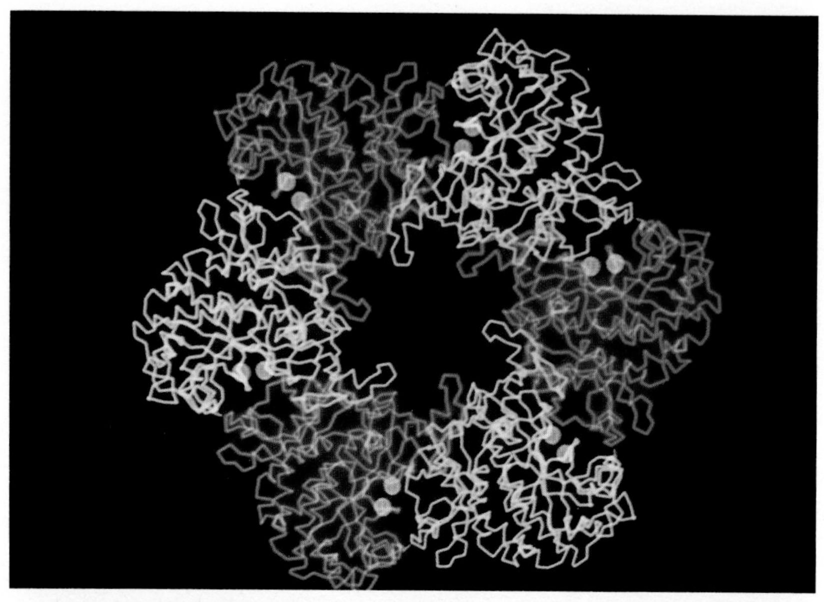

1

2

Fig. 1. *Six of the 12 subunits are shown as line segments between the 468 alpha carbon atoms. Two Mn ions are shown at each active site. For further details please see R.J. Almassy et al. (1986) Nature 323, 304–309.*

Fig. 2. *Six of the 12 subunits are shown as in Figure 1. The small N-terminal folding domain is shown in red and the large C-terminal folding terminal is shown in green.*

Index